ALLE ZEIT WACH

Horst Schubert

Categories

Translated from the German by

Eva Gray

Springer-Verlag Berlin Heidelberg New York 1972

Horst Schubert
Mathematisches Institut der Universität Düsseldorf, Germany

Eva Gray
Urbana, Ill., U.S.A.

Revised and enlarged translation of "Kategorien I" and "Kategorien II" 1970; Heidelberger Taschenbücher, Band 65 und 66

AMS Subject Classification (1970) 18-01

ISBN-13:978-3-642-65366-7 e-ISBN-13:978-3-642-65364-3
DOI: 10.1007/978-3-642-65364-3

Softcover reprint of the hardcover 1st edition 1972

Library of Congress Catalog Card Number 72-83016

Horst Schubert

Categories

Translated from the German by

Eva Gray

Springer-Verlag New York Heidelberg Berlin 1972

Horst Schubert
Mathematisches Institut der Universität Düsseldorf, Germany

Eva Gray
Urbana, Ill., U.S.A.

Revised and enlarged translation of "Kategorien I" and "Kategorien II" 1970; Heidelberger Taschenbücher, Band 65 und 66

AMS Subject Classification (1970) 18-01

ISBN-13:978-3-642-65366-7 e-ISBN-13:978-3-642-65364-3
DOI: 10.1007/978-3-642-65364-3

Library of Congress Catalog Card Number 72-83016

Preface

Categorical methods of speaking and thinking are becoming more and more widespread in mathematics because they achieve a unification of parts of different mathematical fields; frequently they bring simplifications and provide the impetus for new developments. The purpose of this book is to introduce the reader to the central part of category theory and to make the literature accessible to the reader who wishes to go farther. In preparing the English version, I have used the opportunity to revise and enlarge the text of the original German edition.

Only the most elementary concepts from set theory and algebra are assumed as prerequisites. However, the reader is expected to be mathematically sophisticated enough to follow an abstract axiomatic approach. The vastness of the material requires that the presentation be concise, and careful cooperation and some patience is necessary on the part of the reader. Definitions alway precede the examples that illuminate them, and it is assumed that the reader is familiar with some of the algebraic and topological examples (he should not let the other ones confuse him). It is also hoped that he will be able to explain the concepts to himself and that he will recognize the motivation.

At the present time, axiomatic theories play an important role in mathematics. In them one considers sets with a given structure, the "models of the theory" (e. g., real vector spaces, groups, or topological spaces) and structure preserving maps between them (e. g., linear maps, homomorphisms, or continuous maps). The concept of a category is an abstraction of this situation (Chapter 1). Then the concept of a functor is introduced (Chapter 2). It is the abstraction of the concept of a map between one theory and another, which is subject to certain compatibility conditions. In fact, categories are the models of a certain algebraic theory and functors are the corresponding homomorphisms. Then natural transformations are added as a third fundamental concept; they are "compatible" maps between functors. Simple things like the "natural" embedding of a vector space into its bidual space, or the "natural" projection of a group onto its commutator quotient group are examples of this notion.

The above remarks make it plausible that there are categories of categories (with categories as "models" and functors as maps) and categories of functors (with functors as "models" and natural transformations as maps). This raises set theoretical questions. We take the point of view of the "working mathematician" in this dilemma, mentioning briefly (in Chapter 3) and then using an extension of ordinary set theory that serves our purpose. At present, universes seem to be the most useful (in fact, usually only one universe is required, seldom a second one). I trust that the substance of the theory would remain intact even if a revision of the foundations should be necessary.

Additive categories and categories of functors have been included in the presentation from the beginning. The central notion is the concept of a representable functor and its variations: limits and pairs of adjoint functors. This means that special objects are characterized by universal mapping properties. Special cases have long been familiar and Bourbaki has used this concept systematically, but in a different language ("universal problems"). Chapters 4, 7 through 10, 16, 17 and 21 constitute the main framework of the book. I hope that in Chapters 7 and 8 the reader's patience is finally and for the first time rewarded by the insight that "similar" central concepts of different mathematical theories (as, e. g., cartesian, direct, or topological products) can be uniformly grasped and that from the formal concept of the dual category a far reaching duality principle emerges. I have not formulated this principle as a meta theorem — which becomes complicated, if the possibility of partial dualization is to be included. I believe that it can be understood from its applications; in particular, Chapter 8, which is dual to Chapter 7, presents an extended exercise in dualizing.

I hope that the reader will find other sections where familiar material is presented in a unified or simplified manner, or is seen to be a special case of a more general situation. I would hope that the reader would also see (e. g., in 16.6) that the categorical way of looking at things is capable of dispelling psychological barriers created by questions such as, whether in the theory of compact spaces there are (topological) sums, or, whether every equivalence relation generates a quotient in the domain of compact spaces.

The Yoneda Lemma, as a central instrument of category theory, is precented as early as possible (Chapter 4). On the other hand, adjoint functors are not treated until Chapter 16, where this can be done in a coherent form, followed immediately by the important Kan construction (Chapter 17). For a first reading, one can go directly from Chapter 8 to Chapter 16, and then to Chapter 21. In Chapter 18—21, different cases and aspects of adjoint situations and their applications are discussed. These chapters are essentially independent of each other. Note, however, that 18.4 and 19.4.2, as well as the factorization of morphisms in 21.6.2, contain some techniques not presented earlier.

Choosing material for a book always necessitates decisions and, in view of the extensive literature, many more topics might have been included. To incorporate homological algebra, the actual origin of the theory of categories, was out of the question because of the extent of this theory. Therefore, derived categories are also not included. Monoidal and closed categories are certainly beyond the scope of this book. However, the material presented here provides the necessary background for these topics. I hope to have chosen independently of specialized interests and to have captured the central part of the theory — apart from Chapter 20, which has a more specialized character.

In the treatment of everything, a certain amount of completeness was sought, which might make the book useful as a reference work. Whenever feasible, theorems are formulated in such a way that they can be read independently. With respect to terminology and notation, I have taken into account the confused situation in the existing literature by adding appropriate notes in the text as well as in the index.

Since the book is meant to be a textbook, I have not been ashamed to occasionally treat special cases, which later turn up in a more genral context, as, e. g., in the case of 14.2 through 14.4, 16.6.3, 19.4.7, which allow obvious generalization from the point of view of 21.6.2 and even beyond. As regards algebraic structures, I have first supplied an elementary presentation which is convenient for many applications, e. g., the topological ones (Chapter 11). Later, in Chapter 18, Lawvere's theory of equationally defined algebras with finitary operations is treated. Although 18.3 through 18.5 and 18.8.8 in Chapter 21 turn out to be special cases, I consider this historical presentation to be justified, because Lawvere's work has had such a profound influence on much of the further development, and since parts of Chapter 21, as well as other things, have been motivated by it.

I have been helped in many ways in preparing the manuscript. I would like to thank Jens Gamst for suggestions, many discussions, and for looking through parts of the original manuscript. Some of the results in Chapter 9, 19 and 21 are due to Thomas Thode. John Gray read the English version of the manuscript and made many helpful suggestions. And I want particularly to thank Eva Gray who carefully carried out the onerous and unthankful task of translation.

Düsseldorf, March 1972

Horst Schubert

Table of Contents

1. Categories

In the beginning every axiomatic theory is poor in theorems and rich in definitions which must be clarified by examples. Observe however, that every example is an assertion whose verification is in general left to the reader. It is not necessary that the reader be acquainted with all examples.

1.1 Definition of Categories

1.1.1 Definition. A *category* $\mathscr{C}$ consists of the following data:

(i) a class $|\mathscr{C}|$ of *objects* $A, B, C, \ldots,$

(ii) for each ordered pair of objects (A, B) of $\mathscr{C}$ a (possibly empty) set $[A, B]_{\mathscr{C}}$ called the set of *morphisms* from A to B,

(iii) for each ordered triple (A, B, C) of objects of $\mathscr{C}$ a map

$$[B, C]_{\mathscr{C}} \times [A, B]_{\mathscr{C}} \to [A, C]_{\mathscr{C}}$$

called *composition of morphisms*. If $g \in [B, C]_{\mathscr{C}}$, $f \in [A, B]_{\mathscr{C}}$, then the image of the pair (g, f) is designated by $g\,f$ (read g following f); occasionally we also write $g \circ f$.

These data are subject to the following axioms:

(0) The sets $[A, B]_{\mathscr{C}}$ are pairwise disjoint.

(1) *Associativity of composition.* If $h\,g$ and $g\,f$ are both defined,

$$(h\,g)\,f = h(g\,f)$$

holds and hence parantheses are not needed.

(2) *Identities.* For each object B there is an identity morphism $1_B \in [B, B]_{\mathscr{C}}$ for which

$$1_B\,f = f, \qquad g\,1_B = g$$

holds whenever the left side is defined.

Remarks

1.1.2 Later on (in 3.3) we shall elaborate on the use of the words class and set. For now it suffices to say that every set is a class, but not vice versa.

1.1.3 If it is obvious from the context which category is meant, $[A, B]_{\mathscr{C}}$ will be replaced by $[A, B]$. Other notations found in the literature are (A, B), $\mathscr{C}(A, B)$, $\mathrm{Hom}(A, B)$, $\mathrm{hom}(A, B)$, $\mathrm{Mor}(A, B)$, B^A.

1.1.4 $f \in [A, B]$ is usually indicated by $f\colon A \to B$ or $A \xrightarrow{f} B$. A is called the *domain* (*source*) and B the *codomain* (*range, target*) of f.

Axiom (0) means that a morphism uniquely determines its domain and codomain.

1.1.5 According to (iii), two morphisms can be composed if and only if the codomain of the first is the same as the domain of the second, and the composition of

$$A \xrightarrow{f} B \xrightarrow{g} C$$

is denoted by $A \xrightarrow{gf} C$. Writing the composition in this order is almost completely standard, but the reader should be aware that some authors use the opposite convention.

1.1.6 The identity morphism 1_A is uniquely determined by the object A. For, let 1_A and $1'_A$ be identity morphisms for A. Then by (2), $1_A = 1_A 1'_A = 1'_A$. Conversely, A is determined by 1_A because the sets of morphisms are pairwise disjoint.

1.1.7 Using 1.1.6 one can define categories without mentioning objects, replacing them by their identity morphisms. See e.g. [38].

1.1.8 The class of all morphisms of $\mathscr{C}$ is denoted by

$$\operatorname{Mor}\mathscr{C} = \bigcup_{(A,B)\in|\mathscr{C}|\times|\mathscr{C}|} [A, B]_{\mathscr{C}} .$$

1.2 Examples

In the following, when morphisms are maps, their composition is defined in the usual way.

1.2.1 Objects are sets (of a fixed universe, 3.3), and morphisms are maps between them. This category is always denoted by *Ens*.

1.2.2 Objects are abelian groups, and morphisms are homomorphisms between them. This is always denoted by *Ab*.

1.2.3 Objects are left modules over a ring R, and morphisms are the homomorphisms. This is always denoted by ${}_R\mathit{Mod}$, and similarly, Mod_R denotes right modules. 1.2.2 is the special case $R = \mathbf{Z}$, where right and left modules coincide. Another special case is vector spaces over a field. It is generally true for every algebraic structure that its models and the homomorphisms between them form a category. We denote such a category simply by the names of its models, e.g. the category of (multiplicative) groups. For rings we require the existence of an identity and that the homomorphisms preserve it (i.e., 1 is taken into 1). We allow the ring 0 with only one element.

1.2.4 The category *Top* of topological spaces: objects are topological spaces and morphisms are continuous maps.

1.2.5 Objects are non-empty topological spaces with specified basepoints and morphisms are continuous maps which preserve the basepoints. Similarly, there is the category of pointed sets.

1.2.6 Objects are topological spaces and morphisms are homotopy classes of continuous maps. Similarly for pointed spaces, all homotopies are required to preserve basepoints.

1.2.7 The category of sets and relations: objects are sets, morphisms from A to B are the subsets of $A \times B$. Composition is the usual composition of relations; i.e., if $f \in A \times B$, $g \in B \times C$, then

$$g f = \{(a, c) \mid \text{ there is a } b \in B \text{ such that } (a, b) \in f, (b, c) \in g\} .$$

Similarly for groups: multiplicative relations from A to B are subgroups of the direct product $A \times B$.

1.2.8 There are many more examples. We mention topological groups, topological vector spaces over a topological field; in particular, locally convex real or complex vector spaces.

1.2.9 We admit the *empty category* $\emptyset$. It contains no object and hence no morphism.

1.3 Isomorphisms

1.3.1 Definition. A morphism $f: A \to B$ is called an *isomorphism* if there exists a $g: B \to A$ such that $g f = 1_A$, $f g = 1_B$. As $g f = 1_A$, $f g' = 1_B$ implies $g = g'$, g is determined uniquely by f and is called the *inverse* of f; one writes $g = f^{-1}$. A, B are said to be isomorphic if there is an isomorphism $f: A \to B$.

Morphisms in $[A, A]$ are called *endomorphisms* of A. Isomorphic endomorphisms are called *automorphisms*.

1.3.2 Compositions and inverses of isomorphisms are isomorphisms; the automorphisms of an object form a group.

1.4 Further Examples

1.4.1 The endomorphisms of an object, because of their composition, form a *monoid*, i.e., a semigroup with 1. Conversely, every monoid may be considered as a category with only one object (compare 1.1.7). A group can be considered as a category with only one object in which every morphism is an automorphism. Hence monoids and groups are special categories.

1.4.2 A category is called *discrete* if every morphism is an identity morphism. Every class can be thought of as a discrete category.

1.4.3 A category in which each set $[A, B]$ contains at most one element is called a *preordered class*. We write $A \leq B$ for $[A, B] \neq \emptyset$.

If $[A, B] \cup [B, A]$ always contains at most one element, we speak of a *weak ordering*, whereas if $[A, B] \cup [B, A]$ contains exactly one element, the *ordering* is *strong* (strict, linear).

1.5 Additive Categories

1.5.1 A *semi-additive category* is a category in which each set $[A, B]$ is given a commutative, associative addition with 0-element (additive monoid) in such a way that composition of morphisms is distributive on both sides and is compatible with 0-elements; i.e.,

$$(4) \qquad (g_1 + g_2) f = g_1 f + g_2 f; \qquad g(f_1 + f_2) = g f_1 + g f_2,$$

$$(5) \qquad g\,0 = 0; \qquad 0 f = 0.$$

If $[A, B]$ is always an additive group, the category is called *additive* (also preadditive). (5) is then a consequence of (4).

1.5.2 With the usual addition of homomorphisms Ab, ${}_R Mod$, Mod_R are additive categories (1.2.3). In an additive category $[A, A]$ is always a ring and $[A, B]$ and $[B, A]$ are right, resp. left, modules over $[A, A]$. "Right" and "left" are determined by the order of the composition (compare 1.1.5).

1.5.3 A ring (always with 1) is to be regarded as an additive category with only one object.

1.6 Subcategories

1.6.1 A *subcategory* $\mathcal{D}$ of a category $\mathcal{C}$ consists of

(i) a subclass $|\mathcal{D}|$ of the class of objects of $\mathcal{C}$,

(ii) for each ordered pair (A, B) of objects in $|\mathcal{D}|$, a subset $[A, B]_{\mathcal{D}}$ of $[A, B]_{\mathcal{C}}$ such that

(1) for each ordered triple (A, B, C) of objects in $|\mathcal{D}|$, the composition in $\mathcal{C}$ maps $[B, C]_{\mathcal{D}} \times [A, B]_{\mathcal{D}}$ into $[A, C]_{\mathcal{D}}$;

(2) for each $A \in |\mathcal{D}|$, $1_A \in [A, A]_{\mathcal{D}}$.

It is immediate that $\mathcal{D}$ is a category.

The subcategory $\mathcal{D}$ is called *full* if for any two objects A and B in $\mathcal{D}$ all $\mathcal{C}$-morphisms from A to B also belong to $\mathcal{D}$; i.e., if $[A, B]_{\mathcal{D}} = [A, B]_{\mathcal{C}}$ holds. A full subcategory is completely determined by its objects, which can be specified arbitrarily.

Examples

1.6.2 The finite sets determine a full subcategory of *Ens*.

1.6.3 The commutative groups determine a full subcategory of the category of all groups. Similarly, there is the full subcategory of free groups, etc; and, correspondingly, there is the full subcategory of free abelian groups or of torsion groups in *Ab*.

1.6.4 By restricting the objects to spaces with additional properties like Hausdorffness, regularity, complete regularity, compactness etc., one obtains full subcategories of the category *Top* in 1.2.4.

1.6.5 The category *Ens* can be considered as a (non-full) subcategory of the category of sets and relations of 1.2.7 by identifying a map with its graph.

1.6.6 Every category $\mathscr{C}$ contains a discrete (see 1.4.2) subcategory containing all the objects of $\mathscr{C}$.

1.6.7 One obtains subcategories of $\mathscr{C}$ if one takes a single object A of $\mathscr{C}$ and allows as morphisms

a) only 1_A,
b) all automorphisms of A,
c) all endomorphisms of A.

1.6.8 Let $f\colon A \to B$ be a morphism in $\mathscr{C}$ such that $A \neq B$. Then A and B are the objects and 1_A, 1_B, and f are the morphisms of a subcategory of $\mathscr{C}$.

1.7 Problems

1.7.1 Depending on your previous knowledge, discuss the examples in 1.2, 1.4, and 1.5. In particular, verify that 1.4.3 yields the usual definitions. Also 1.4.1 and 1.5.3 can be regarded as definitions.

1.7.2 In a preordered class (as a category) two objects A, B are isomorphic if and only if $A \leq B$ and $B \leq A$.

1.7.3 A strongly ordered set with n elements considered as a category has $\frac{1}{2} n (n + 1)$ morphisms. Find exact upper and lower bounds for the number of morphisms in a weakly ordered and, resp., preordered set of n elements.

1.7.4 A category in which every morphism is an isomorphism is called a *groupoid*. Let $\mathscr{C}$ be a groupoid and for some $X \in |\mathscr{C}|$ let $[A, X] \neq \emptyset$ for all $A \in |\mathscr{C}|$. Prove: If some set of morphisms $[A, B]$ has n elements, then this is so for all of them.

2. Functors

2.1 Covariant Functors

2.1.1 Definition. Let $\mathscr{C}$ and $\mathscr{D}$ be categories. A *functor* $T\colon \mathscr{C} \to \mathscr{D}$, more exactly, a covariant functor, is a map of objects and morphisms.

It assigns to each object $A \in |\mathcal{C}|$ an object $T(A) \in |\mathcal{D}|$ and to each morphism $f: A \to B$ a morphism $T(f): T(A) \to T(B)$ in such a way that

$$(1) \qquad T(1_A) = 1_{T(A)},$$

$$(2) \qquad T(g\,f) = T(g)\,T(f) \quad \text{(provided } g\,f \text{ is defined in } \mathcal{C}).$$

Thus a functor preserves identities and the composition of morphisms. This implies that isomorphisms are also preserved. If $\mathcal{C}$, $\mathcal{D}$ are semi-additive, then $T: \mathcal{C} \to \mathcal{D}$ is called *additive*, provided

$$(3) \qquad T(f_1 + f_2) = T(f_1) + T(f_2) \quad \text{and}$$

$$(4) \qquad T(0) = 0$$

for all 0-morphisms. (4) follows from (3) if $\mathcal{C}$ and $\mathcal{D}$ are additive.

The values of a functor on objects and morphisms are sometimes indicated by "$\mapsto$"; thus $A \mapsto T(A)$, $f \mapsto T(f)$.

Examples

2.1.2 If $\mathcal{C}$, $\mathcal{D}$ are groups (or monoids) regarded as categories with one object, then functors $T: \mathcal{C} \to \mathcal{D}$ are the same as homomorphisms. If $\mathcal{C}$, $\mathcal{D}$ are rings, then additive functors are just (unitary) ring homomorphisms.

2.1.3 The rule "group $\mapsto$ group made abelian, homomorphism $\mapsto$ induced homomorphism" determines a functor from the category of groups into itself (or, resp., into the subcategory of abelian groups).

2.1.4 The rule "topological space $\mapsto$ n-th singular homology group, continuous map $\mapsto$ induced homomorphism" determines a functor, as does "topological space with base point $\mapsto$ n-th homotopy group" (compare 1.2.5). These functors are also defined for the categories in 1.2.6.

2.1.5 Let $\mathcal{C} = {}_R Mod$ and let X be a fixed right R-module. The rule $T(A) = X \otimes_R A$, $T(f) = id_X \otimes_R f$ (id_X is the identity map of X) is a functor.

2.2 Standard Examples

2.2.1 The *identity functor* $\mathrm{Id}_{\mathcal{C}}: \mathcal{C} \to \mathcal{C}$. It maps objects and morphisms of an arbitrary category $\mathcal{C}$ identically onto themselves.

2.2.2 The *inclusion of a subcategory* $\mathcal{D}$ of $\mathcal{C}$ in $\mathcal{C}$; it is denoted by $\subset: \mathcal{D} \to \mathcal{C}$ or $\mathcal{D} \subset \mathcal{C}$.

2.2.3 *Constant functors.* Let $\mathcal{C}$, $\mathcal{D}$ again be arbitrary categories and let $X \in |\mathcal{D}|$. Then the constant functor with value X is given by $T(A) = X$ for all $A \in |\mathcal{C}|$ and $T(f) = 1_X$ for all morphisms f in $\mathcal{C}$. We denote this functor by $X_{\mathcal{C}}$.

2.2.4 *Forgetful functors.* Let the objects of $\mathcal{C}$ be sets with a certain structure (e.g. groups, topological spaces etc.) and let the morphisms be structure-preserving maps (homomorphisms, continuous maps etc.). Then the forgetful functor $U: \mathscr{C} \to \mathscr{E}ns$ assigns to each object its underlying set and to each morphism the corresponding set map. Other forgetful functors forget only a part of the structure, as e.g. $U: {}_R\mathscr{M}od \to \mathscr{A}b$, module $\mapsto$ underlying additive group; or group $\mapsto$ pointed set (with the unit as basepoint).

2.2.5 The *covariant* Hom-*functors* $H^A: \mathscr{C} \to \mathscr{E}ns$. Let A be a fixed object of $\mathscr{C}$. Then $H^A(X) = [A, X]_{\mathscr{C}}$, and if $f: X \to Y$, then $H^A(f)$ is the map $[A, X] \to [A, Y]$ whose value at $u \in [A, X]$ is $f\,u$. We write $[A, f] = H^A(f)$ and $[A, ?] = H^A(?)$.

If $\mathscr{C}$ is additive, then $[A, ?]$ is generally regarded as a functor $\mathscr{C} \to \mathscr{A}b$. This functor is additive.

2.2.6 *Composition of functors.* If $S: \mathscr{C} \to \mathscr{D}$ and $T: \mathscr{D} \to \mathscr{E}$ are functors, one defines the functor TS (T following S) by $A \mapsto T(S(A))$, $f \mapsto T(S(f))$. Besides TS, $T \circ S$ is also used.

2.2.7 Remark. A functor $T: \mathscr{C} \to \mathscr{D}$ defines maps of the sets of morphisms

$$T_{A,B}: [A, B]_{\mathscr{C}} \to [T(A), T(B)]_{\mathscr{D}} \tag{5}$$

by the rule $f \mapsto T(f)$ which produce a mapping of the morphism classes

$$T: \text{Mor}\,\mathscr{C} \to \text{Mor}\,\mathscr{D}\,. \tag{6}$$

A functor may be regarded as a map T of the morphism classes which, as in (1) and (2), satifies the following conditions:

(1′) T maps identity morphisms into identity morphisms.

(2′) If $g\,f$ is defined in $\mathscr{C}$, then so is $T(g)\,T(f)$ in $\mathscr{D}$ and

$$T(g\,f) = T(g)\,T(f)$$

holds.

(1′) determines the action of T upon objects (compare 1.1.6). The composition of functors is then simply the composition of maps.

2.2.8 If $\mathscr{C}$ is the empty category, then there is exactly one functor $\mathscr{C} \to \mathscr{D}$, the "empty functor". Notice that this is analogous to the maps of the empty set in $\mathscr{E}ns$.

2.2.9 Definition. A functor $T: \mathscr{C} \to \mathscr{D}$ is called an *isomorphism* if there is a functor $S: \mathscr{D} \to \mathscr{C}$ such that $S\,T = \text{Id}_{\mathscr{C}}$, $T\,S = \text{Id}_{\mathscr{D}}$. One deduces easily from (1′) and (2′) that T is an isomorphism exactly when (6) is a bijection.

2.3 Contravariant Functors

2.3.1 Definition. A *contravariant functor* $T: \mathscr{C} \to \mathscr{D}$ assigns to each object $A \in |\mathscr{C}|$ an object $T(A) \in |\mathscr{D}|$ and to each morphism $f: A \to B$

a morphism $T(f): T(B) \to T(A)$ in such a way that

(1) $$T(1_A) = 1_{T(A)},$$

(2°) $$T(g\,f) = T(f)\,T(g) \qquad \text{(provided } g\,f \text{ is defined in } \mathscr{C}\text{)}.$$

If $\mathscr{C}$, $\mathscr{D}$ are semi-additive, then T is called *additive* provided 2.1.(3) and (4) are valid.

Thus a contravariant functor reverses the direction of the morphisms while preserving identities and composition and therefore isomorphisms. With proper interchanging 2.2.7 holds; in particular,

(5°) $$T_{A,B}: [A, B]_{\mathscr{C}} \to [T(B), T(A)]_{\mathscr{D}}.$$

Examples

2.3.2 The *contravariant* Hom-*functors* $H_A: \mathscr{C} \to Ens$. Let A be a fixed object of $\mathscr{C}$. Then $H_A(X) = [X, A]_{\mathscr{C}}$, and if $f: X \to Y$, then $H_A(f)$ is the map $[Y, A] \to [X, A]$ whose value at $u \in [Y, A]$ is $u\,f$. We write

$$[f, A] = H_A(f) \quad \text{and} \quad [?, A] = H_A(?)\,.$$

If $\mathscr{C}$ is additive, H_A is generally regarded as a contravariant functor $\mathscr{C} \to Ab$. The classical examples are $\mathscr{C} = {}_R Mod$ or $\mathscr{C} = Mod_R$.

2.3.3 Every constant functor is also a contravariant functor.

2.3.4 Power set $\mathfrak{P}: Ens \to Ens$. This contravariant functor assigns to each set A its power set $\mathfrak{P}(A)$ (the set of subsets of A), and if f: $A \to B$ is a set map, then $\mathfrak{P}(f): \mathfrak{P}(B) \to \mathfrak{P}(A)$ is the map whose value at $X \in \mathfrak{P}(B)$ (i.e. $X \subset B$) is given by $\mathfrak{P}(f)\,(X) = f^{-1}(X)$. The sets $\mathfrak{P}(A)$ have an algebraic structure (Boolean algebra) deriving from intersection, union, and complement of subsets. $\mathfrak{P}(f)$ preserves this structure.

2.3.5 Let K be a field, so that Mod_K is the category of vector spaces over K. The rule "vector space $\mapsto$ dual space, linear map $\mapsto$ transposed map" determines a contravariant functor D: $Mod_K \to Mod_K$. It would be of the form $[?, K]$ except the range category is neither Ens nor Ab. Correspondingly there are contravariant functors D: ${}_R Mod \to Mod_R$ and D: $Mod_R \to {}_R Mod$ for modules over a ring R. They coincide if R is commutative.

2.3.6 The rule "topological space $\mapsto$ n-th singular cohomology group, continuous map $\mapsto$ induced homomorphism" constitutes another classical example. This is where the prefix "co" in category theory originated, see 2.4.6.

2.3.7 As in 2.2.7, the composition of contravariant functors with each other and of co- and contravariant functors is defined as composition of maps. If covariant functors are assigned the variance $+1$, contravariant ones the variance -1, the variance of a composition is the product of the respective variances.

2.4 Dual Categories

2.4.1 For every category $\mathscr{C}$ we define a dual category $\mathscr{C}^0$ (other generally used symbols are $\mathscr{C}^*$, $\mathscr{C}^{\text{op}}$, $\mathscr{C}^{\text{opp}}$) as follows:

The objects of $\mathscr{C}^0$ are those of $\mathscr{C}$ and $[B, A]_{\mathscr{C}^0} = [A, B]_{\mathscr{C}}$. The composition $f\,g$ in $\mathscr{C}^0$ is defined as $g\,f$ in $\mathscr{C}$ (i.e., reverse all arrows and interchange the order of all compositions). Observe that $[A, A]_{\mathscr{C}}$ is also reversed.

Obviously $\mathscr{C}^{00} = \mathscr{C}$. For every category $\mathscr{C}$ the contravariant functor $\text{Op}\colon \mathscr{C} \to \mathscr{C}^0$ maps the objects and morphisms identically onto themselves. (Note that domain and codomain are interchanged.) Op Op = Id (compare 2.2.1).

2.4.2 If $\mathscr{C}$ is a monoid, or a group, or a ring, then $\mathscr{C}^0$ is the opposite monoid, group, or ring. If $\mathscr{C}$ is an abelian group or a commutative ring, then $\mathscr{C}$ and $\mathscr{C}^0$ are the same. Also, every discrete category coincides with its dual.

2.4.3 If $\mathscr{C}$ is (the category associated with) an ordered set, then $\mathscr{C}^0$ is the oppositely ordered set ($\leq$ is replaced by $\geq$). The same holds for preorderings.

2.4.4 Convention: to avoid misunderstandings we write

$$A^0 = \text{Op}(A); \qquad f^0 = \text{Op}(f)\,,$$

i.e., we write A^0 and f^0 if we think of an object A or a morphism f of $\mathscr{C}$ as belonging to $\mathscr{C}^0$. Thus

$$(7) \qquad f\colon A \to B \Leftrightarrow f^0\colon B^0 \to A^0; \qquad (g\,f)^0 = f^0\,g^0\,.$$

Note that $|\mathscr{C}^0| = |\mathscr{C}|$, Mor $\mathscr{C}^0$ = Mor $\mathscr{C}$, Op: Mor $\mathscr{C} \to$ Mor $\mathscr{C}^0$ is the identity map, however the composition of morphisms is different (there are exceptions; compare 2.4.2).

2.4.5 There are two ways in which the introduction of dual categories makes it possible to replace contravariant functors by covariant ones: if $T\colon \mathscr{C} \to \mathscr{D}$ is contravariant, then $T\,\text{Op}\colon \mathscr{C}^0 \to \mathscr{D}$ and $\text{Op}\,T\colon \mathscr{C} \to \mathscr{D}^0$ are covariant, and conversely.

Convention: Replacing a contravariant functor T by a covariant one shall always mean that T is replaced by T Op. We even say that the contravariant functors $\mathscr{C} \to \mathscr{D}$ *are* the covariant functors $\mathscr{C}^0 \to \mathscr{D}$. By "functor" without prefix we mean, as before, a covariant functor.

2.4.6 Dual categories lead to a duality principle which we shall elucidate later with examples. Here we only remark that dual concepts correspond to each other in dual categories, or rather: they are interchanged by Op. Through this interchange each theorem produces its dual theorem ("reversal of all arrows"). Notations like "thing, co-thing" are used for dual concepts. See also 3.6.6.

2.4.7 If $\mathscr{C}$ is a subcategory of $\mathscr{D}$, then $\mathscr{C}^0$ is a subcategory of $\mathscr{D}^0$, and $\mathscr{C}^0$ is full in $\mathscr{D}^0$ if and only if $\mathscr{C}$ is full in $\mathscr{D}$.

2.5 Bifunctors

The examples in 2.2.5 and 2.3.5 suggest the necessity of defining functors of several variables, in particular a functor $[?, ??]_{\mathscr{C}}$ of two variables.

2.5.1 The *product* $\mathscr{C} \times \mathscr{D}$ of the categories $\mathscr{C}$ and $\mathscr{D}$ has as objects the ordered pairs (C, D) of objects $C \in |\mathscr{C}|$ and $D \in |\mathscr{D}|$. The sets of morphisms are defined by

(1) $$[(C, D), (C', D')]_{\mathscr{C} \times \mathscr{D}} = [C, C']_{\mathscr{C}} \times [D, D']_{\mathscr{D}}$$

and composition of morphisms is "componentwise", i.e.,

(2) $$(f', g')\,(f, g) = (f'\,f, g'\,g)\,.$$

One verifies that $1_{(C,D)} = (1_C, 1_D)$, that composition is associative, and that $(\mathscr{C} \times \mathscr{D})^0 = \mathscr{C}^0 \times \mathscr{D}^0$.

2.5.2 A (doubly covariant) *bifunctor* is a functor whose domain is the product of two categories. If $T: \mathscr{C} \times \mathscr{D} \to \mathscr{E}$ is a bifunctor, we denote the image of (C, D), or (f, g), in $\mathscr{E}$ by $T(C, D)$, or $T(f, g)$ respectively. If $f'\,f$ is defined in $\mathscr{C}$ and $g'\,g$ in $\mathscr{D}$, then

(3) $$T(f'\,f, g'\,g) = T(f', g')\,T(f, g)\,.$$

2.5.3 Examples. The tensor product of modules is a bifunctor $Mod_R \times {}_RMod \to Ab$; or ${}_RMod \times {}_RMod \to {}_RMod$, if the ring R is commutative.

If $R: \mathscr{C} \to \mathscr{A}$ and $S: \mathscr{D} \to \mathscr{B}$ are functors, then $(f, g) \mapsto (R(f), S(g))$ determines the bifunctor $R \times S: \mathscr{C} \times \mathscr{D} \to \mathscr{A} \times \mathscr{B}$.

2.5.4 Let $T: \mathscr{C}^0 \times \mathscr{D} \to \mathscr{E}$ be a bifunctor. It assigns to each ordered pair (C, D) of objects $C \in |\mathscr{C}|$, $D \in |\mathscr{D}|$ an object $T(C^0, D)$ in $\mathscr{E}$, and to each ordered pair (f, g) of morphisms $f: C \to C'$ in $\mathscr{C}$ and $g: D \to D'$ in $\mathscr{D}$ a morphism $T(f^0, g): T(C'^0, D) \to T(C^0, D')$. If one defines $S(C, D) = T(C^0, D)$, $S(f, g) = T(f^0, g)$, then S in called a functor of two variables which is contravariant in the first variable and covariant in the second one; briefly, a *contra-co-variant functor*.

Note that in general S is not a functor with domain $\mathscr{C} \times \mathscr{D}$, but another description of the bifunctor T with domain $\mathscr{C}^0 \times \mathscr{D}$. If $f'\,f$ is defined in $\mathscr{C}$ and $g'\,g$ in $\mathscr{D}$, then

(3) $$S(f'\,f, g'\,g) = S(f, g')\,S(f', g)\,.$$

2.5.5 Standard example: given any category $\mathscr{C}$, the Hom-*functor* is a bifunctor $\mathscr{C}^0 \times \mathscr{C} \to \mathit{Ens}$ which, according to 2.5.4, is written as a contra-co-variant functor. As a bifunctor it is defined for objects by $(A^0, B) \mapsto [A, B]_{\mathscr{C}}$ and for morphisms (f^0, g), where $f\colon A \to A'$, $g\colon B \to B'$, by

$$u \mapsto g\,u\,f \quad \text{for} \quad u \in [A', B]_{\mathscr{C}}\,, \tag{4}$$

we write

$$[f, g]_{\mathscr{C}}\colon [A', B]_{\mathscr{C}} \to [A, B']_{\mathscr{C}}\,. \tag{5}$$

Setting $f = 1_A$ (resp. $g = 1_B$) one obtains

$$[1_A, g] = [A, g]\colon [A, B] \to [A, B']\,, \tag{6}$$

$$[f, 1_B] = [f, B]\colon [A', B] \to [A, B]\,, \tag{7}$$

where $[A, g]$ and $[f, B]$ are defined as in 2.2.5 and 2.3.2. One verifies the functorial properties: $[1_A, 1_B]$ is the identity map of $[A, B]$. If $f'\,f$ and $g'\,g$ are defined in $\mathscr{C}$, then (4) implies

$$[f'\,f, g'\,g] = [f, g']\,[f', g]\,. \tag{8}$$

Also,

$$[f, g] = [f, 1_B]\,[1_A, g] = [1_A, g]\,[f, 1_B] \tag{9}$$

holds, so by (6) and (7), if $f\colon A \to A'$ and $g\colon B \to B'$, then the following diagram is commutative

$$\begin{array}{ccc} [A', B] & \xrightarrow{[A', g]} & [A', B'] \\ \downarrow{\scriptstyle [f, B]} & & \downarrow{\scriptstyle [f, B']} \\ [A, B] & \xrightarrow[{[A, g]}]{} & [A, B] \end{array} \tag{10}$$

2.5.6 For a bifunctor $T\colon \mathscr{C} \times \mathscr{D} \to \mathscr{E}$ (with non-empty $\mathscr{C} \times \mathscr{D}$) one generally writes in analogy to (6), (7)

$$T(C, g) = T(1_C, g)\,; \qquad T(f, D) = T(f, 1_D)\,. \tag{11}$$

Then (2) implies immediately that, if in a bifunctor with nonempty domain one variable is fixed at an object, then it becomes a functor of the other variable. To be more precise: every object $C \in |\mathscr{C}|$, resp. $D \in |\mathscr{D}|$, defines a *partial functor*

$$T(C, ?)\colon \mathscr{D} \to \mathscr{E}\,, \quad \text{resp.} \quad T(?, D)\colon \mathscr{C} \to \mathscr{E}\,.$$

Correspondingly one obtains a contravariant functor $S(?, D)$ from a contra-co-variant functor S. The covariant Hom-functors in 2.2.5 and the contravariant Hom-functors in 2.3.2 are therefore partial functors of the contra-co-variant Hom-functors in 2.5.5. The conventions 2.4.5 and 2.5.4 turn out to be consistent and they provide the following notation for the Hom-functor as a bifunctor:

(12) $$[\mathrm{Op}\ ?,\ ??]_{\mathscr{C}} \colon \mathscr{C}^0 \times \mathscr{C} \to \mathit{Ens}\ .$$

2.5.7 First bifunctor criterion. *Let $\mathscr{C}$, $\mathscr{D}$, $\mathscr{E}$ be non-empty categories. For every $A \in |\mathscr{C}|$ let a functor $P_A : \mathscr{D} \to \mathscr{E}$ be given, and for every $X \in |\mathscr{D}|$ a functor $Q_X \colon \mathscr{C} \to \mathscr{E}$. If*

(i) $$P_A(X) = Q_X(A) \quad \text{for all} \quad A \in |\mathscr{C}|, X \in |\mathscr{D}|\,, \quad \text{and if}$$

(ii) $$\begin{array}{ccc} P_A(X) = Q_X(A) & \xrightarrow{Q_X(f)} & Q_X(B) = P_B(X) \\ \downarrow{\scriptstyle P_A(u)} & & \downarrow{\scriptstyle P_B(u)} \\ P_A(Y) = Q_Y(A) & \xrightarrow{Q_Y(f)} & Q_Y(B) = P_B(Y) \end{array}$$

is commutative for every pair (f, u) of morphisms $f \colon A \to B$ in $\mathscr{C}$, $u \colon X \to Y$ in $\mathscr{D}$, then a bifunctor $T \colon \mathscr{C} \times \mathscr{D} \to \mathscr{E}$ is defined by setting

$$T(A, X) = P_A(X)\,, \qquad T(f, u) = Q_Y(f)\, P_A(u)\,.$$

Proof. It is immediate from the assumptions that $T(1_A, 1_X) = 1_{T(A,X)}$. The functor property (3) follows from (ii) if four such rectangles are joined.

2.5.8 It should be clear from 2.5.4 what a co-contra-variant functor is. Bifunctor and co-co-variant functor are synonymous. A contra-contra-variant functor is a contravariant bifunctor (note 2.4.5). It should also be apparent how products of finitely many categories, functors of several variables, partly contravariant, partly covariant, can be defined. None of these will be needed at this point, with the exception of the relation between the contra-co-variant Hom-functors of dual categories; namely,

(13) $$[?,\ ??]_{\mathscr{C}} = [\mathrm{Op}\ ??,\ \mathrm{Op}\ ?]_{\mathscr{C}^0}\,.$$

2.5.9 If $\mathscr{C}$, $\mathscr{D}$, $\mathscr{E}$ are additive, the bifunctor $T \colon \mathscr{C} \times \mathscr{D} \to \mathscr{E}$ is called *biadditive* if all its partial functors are additive. If $\mathscr{C}$ is additive, then so is $\mathscr{C}^0$, and the Hom-functor is generally regarded as a biadditive contra-co-variant functor. $\mathscr{C} = {}_R\mathit{Mod}$ is the classical case.

2.6 Natural Transformations

2.6.1 Definition. Let S, $T \colon \mathscr{C} \to \mathscr{D}$ be functors. A *natural transformation* $\alpha \colon S \to T$ is a map which assigns to every object $A \in |\mathscr{C}|$ a morphism $\alpha_A \colon S(A) \to T(A)$ in $\mathscr{D}$ in such a way that for every morphism $f: A \to B$ the following diagram is commutative

(1)
$$\begin{array}{ccc} S(A) & \xrightarrow{\alpha_A} & T(A) \\ \downarrow{\scriptstyle S(f)} & & \downarrow{\scriptstyle T(f)} \\ S(B) & \xrightarrow{\alpha_B} & T(B) \end{array};$$

i.e., $T(f)\,\alpha_A = \alpha_B\,S(f)$ for every $f: A \to B$ in $\mathcal{C}$. A natural transformation $\alpha: S \to T$ for contravariant functors $S, T: \mathcal{C} \to \mathcal{D}$ is a natural transformation of the covariant functors $S\ \mathrm{Op}$, $T\ \mathrm{Op}: \mathcal{C}^0 \to \mathcal{D}$.

If $\mathcal{C}$ is empty, then there is only the empty functor $\mathcal{C} \to \mathcal{D}$ and it admits only the trivial, empty natural transformation.

Examples

2.6.2 Let $T: \mathcal{C} \times \mathcal{D} \to \mathcal{E}$ be a bifunctor. For $f: A \to B$ in $\mathcal{C}$ and $u: X \to Y$ in $\mathcal{D}$ 2.5(3) implies

(2) $$T(f, u) = T(f, 1_Y)\,T(1_A, u) = T(1_B, u)\,T(f, 1_X)$$

or, using 2.5.7 (ii)

(3)
$$\begin{array}{ccc} T(A, X) & \xrightarrow{T(A, u)} & T(A, Y) \\ \downarrow{\scriptstyle T(f, X)} & \searrow{\scriptstyle T(f, u)} & \downarrow{\scriptstyle T(f, Y)} \\ T(B, X) & \xrightarrow{T(B, u)} & T(B, Y) \end{array}$$

Comparison with (1) shows that $f: A \to B$ in $\mathcal{C}$ and, respectively, $u: X \to Y$ in $\mathcal{D}$ induce natural transformations of the partial functors

$$T(f, ?): T(A, ?) \to T(B, ?); \qquad T(?, u): T(?, X) \to T(?, Y).$$

In $T(f, ?)$, $T(?, u)$ only the operation on all objects $X \in |\mathcal{D}|$, $A \in |\mathcal{C}|$ is considered.

In particular, $f: A \to B$ in $\mathcal{C}$ induces the natural transformations

(4)
$$H^f = [f, ?]: [B, ?] \to [A, ?], \quad \text{i.e.} \quad H^f: H^B \to H^A,$$
$$H_f = [?, f]: [?, A] \to [?, B], \quad \text{i.e.} \quad H_f: H_A \to H_B$$

for the partial Hom-functors (also compare (3) with 2.5.(10)).

2.6.3 For the category of groups there is a natural transformation of the identity functor into the functor "abelianize" (example 2.1.3). It assigns to every group G the "natural" projection $G \to G/G'$ (G' is the commutator subgroup of G).

2.6.4 Example 2.3.4 yields a covariant functor $DD: Mod_K \to Mod_K$ (vector space $\mapsto$ its double dual, linear map $\mapsto$ its double transpose). There is a natural transformation $\alpha: \mathrm{Id} \to D\,D$ which embeds every vector space in its double dual. α_A is an isomorphism for all finite dimensional $A \in |Mod_K|$.

2.6.5 Another classical example is the natural transformation which assigns to every topological space with base point the "natural" mapping of the fundamental group into the first singular homology group.

The desire to formulate precisely what constituted the "naturalness" of the transformations in the last three examples, for instance, was one of the motivating forces in developing the concepts "category", "functor", "natural transformation" by Eilenberg and MacLane (1945).

2.6.6 Remark. The commutativity of diagram (1) is by 2.2(5) equivalent to that of

$$(5)\qquad \begin{array}{ccc} [A, B]_{\mathscr{C}} & \xrightarrow{T_{A,B}} & [T(A), T(B)]_{\mathscr{D}} \\ \big\downarrow{\scriptstyle S_{A,B}} & & \big\downarrow{\scriptstyle [\alpha_A, T(B)]} \\ [S(A), S(B)]_{\mathscr{D}} & \xrightarrow{[S(A), \alpha_B]} & [S(A), T(B)]_{\mathscr{D}} \end{array} \qquad \begin{array}{ccc} f & \mapsto & T(f) \\ \downarrow & & \downarrow \\ S(f) & \mapsto & \alpha_B S(f) = T(f)\,\alpha_A\,. \end{array}$$

2.6.7 For every functor $T\colon \mathscr{C} \to \mathscr{D}$ there is the identity natural transformation $1_T\colon T \to T$ which assigns $1_{T(C)}$ to $C \in |\mathscr{C}|$. If $\alpha\colon S \to T$ and $\beta\colon T \to U$ are natural transformations, then $\beta\alpha$ is the natural transformation defined by the rule $A \mapsto \beta_A \alpha_A$. If α_A is an isomorphism for every $A \in |\mathscr{C}|$, then $\alpha\colon S \to T$ is called a *natural isomorphism* (traditionally this is called a natural equivalence; however, for categories "equivalence" has a different meaning). In this case $A \mapsto \alpha_A^{-1}$ defines a natural transformation $\alpha^{-1}\colon T \to S$, for which $\alpha^{-1}\alpha = 1_S$ and $\alpha\alpha^{-1} = 1_T$ hold. The functors S, T are called isomorphic to each other if there is a natural isomorphism $\alpha\colon S \to T$.

2.6.8 Let S and T be bifunctors $\mathscr{C} \times \mathscr{D} \to \mathscr{E}$. If $\alpha\colon S \to T$ is a natural transformation, then in particular for $f\colon A \to B$ in $\mathscr{C}$ and $u\colon X \to Y$ in $\mathscr{D}$ the diagrams

$$(6)\qquad \begin{array}{ccc} S(A, X) & \xrightarrow{\alpha_{A,X}} & T(A, X) \\ \big\downarrow{\scriptstyle S(f, X)} & & \big\downarrow{\scriptstyle T(f, X)} \\ S(B, X) & \xrightarrow{\alpha_{B,X}} & T(B, X) \end{array} \qquad \begin{array}{ccc} S(B, X) & \xrightarrow{\alpha_{B,X}} & T(B, X) \\ \big\downarrow{\scriptstyle S(B, u)} & & \big\downarrow{\scriptstyle T(B, u)} \\ S(B, Y) & \xrightarrow{\alpha_{B,Y}} & T(B, X) \end{array}$$

are commutative. Conversely, let morphisms $\alpha_{A,X}\colon S(A, X) \to T(A, X)$ in $\mathscr{E}$ be given for every pair $(A, X) \in |\mathscr{C} \times \mathscr{D}|$ in such a way that (6) is always commutative. If the two diagrams (6) are joined vertically, then comparison with (3) shows that $(A, X) \mapsto \alpha_{A,X}$ is a natural transformation $\alpha\colon S \to T$. Therefore, it suffices to verify that $(A, X) \mapsto \alpha_{A,X}$ produces natural transformations for the partial functors.

2.6.9 Second bifunctor criterion. *Let $\mathscr{C}, \mathscr{D}, \mathscr{E}$ be non-empty categories. For every $A \in |\mathscr{C}|$ let a functor $P_A\colon \mathscr{D} \to \mathscr{E}$ be given, and for every f:*

$A \to B$ in $\mathcal{C}$ a natural transformation P_f: $P_A \to P_B$. If we require that

$$(7) \qquad P_{1_C} = 1_{P_C}, \qquad P_{gf} = P_g P_f$$

holds, whenever $g f$ is defined in $\mathcal{C}$, then setting $T(A, X) = P_A(X)$ and $T(f, u) = P_f(Y) P_A(u)$ for f: $A \to B$ in $\mathcal{C}$, u: $X \to Y$ in $\mathcal{D}$, defines a bifunctor T: $\mathcal{C} \times \mathcal{D} \to \mathcal{E}$.

Proof. Assumption (7) means that for every $X \in |\mathcal{D}|$ the rule $A \mapsto P_A(X)$ and $f \mapsto P_f(X)$ defines a functor Q_X: $\mathcal{C} \to \mathcal{E}$. Since P_f is a natural transformation, the assumptions of 2.5.7 are satisfied.

2.7 Problems

2.7.1 Verify the examples 2.1.2, 2.1.3, 2.3.4, 2.3.5, 2.4.2, 2.4.3, 2.6.3, 2.6.4.

2.7.2 (a) Let $\mathcal{C}$ and $\mathcal{D}$ be preordered classes. A map T: $|\mathcal{C}| \to |\mathcal{D}|$ is associated with a functor if and only if T is order preserving; i.e., if $A \leq B$ implies $T(A) \leq T(B)$. T corresponds to a contravariant functor if and only if T is order reversing; i.e., if $A \leq B$ implies $T(A) \geq T(B)$.
(b) Preordered sets and order preserving maps between them form a category; the same is true if one takes both order preserving and order reversing maps.
(c) If $\mathcal{C}$ and $\mathcal{D}$ are preordered or, resp., weakly ordered classes, then $\mathcal{C} \times \mathcal{D}$ is preordered or, resp., weakly ordered. Is a corresponding statement true for strongly ordered classes?
(d) Let $\mathcal{C}$, $\mathcal{D}$ be preordered sets (as categories). Show that the functors $\mathcal{C} \to \mathcal{D}$, together with the natural transformations between them, form a preordered set, which is weakly ordered provided $\mathcal{D}$ is weakly ordered. What can be said if $\mathcal{D}$ is strongly ordered? For functors S, T: $\mathcal{C} \to \mathcal{D}$ give a description of "$S \leq T$ and $S \neq T$" in terms of objects. Consider also the special case where $\mathcal{C}$ is discrete, and apply it to real functions.

2.7.3 If α: $S \to T$ is a natural transformation between functors S, T: $\mathcal{C} \to \mathcal{D}$, and if R: $\mathcal{B} \to \mathcal{C}$ and U: $\mathcal{D} \to \mathcal{E}$ are functors, then the maps $B \mapsto \alpha_{R(B)}$ for $B \in |\mathcal{B}|$ and $C \mapsto U(\alpha_C)$ for $C \in |\mathcal{C}|$ are natural transformations αR: $S R \to T R$ and $U \alpha$: $U S \to U T$.

2.7.4 If a functor T: $\mathcal{C} \to \mathcal{D}$ is composed with the Hom-functor of $\mathcal{D}$, the contra-co-variant functor $[T(?), T(??)]_{\mathcal{D}}$ results. The rule $f \mapsto T(f)$ determines a natural transformation

$$[?, ??]_{\mathcal{C}} \to [T(?), T(??)]_{\mathcal{D}}.$$

2.7.5 (a) Let $T': \mathscr{C}^0 \times \mathscr{D} \to Ens$ be a bifunctor and T the associated contra-co-variant functor. Assuming that $\mathscr{C}$ and $\mathscr{D}$ are disjoint (which is no loss of generality), construct a category $\mathscr{X}$ such that the following holds: $|\mathscr{X}| = |\mathscr{C}| \cup |\mathscr{D}|$, $\mathscr{C}$ and $\mathscr{D}$ are full subcategories of $\mathscr{X}$, and for $C \in |\mathscr{C}|$ and $D \in |\mathscr{D}|$, $[C, D]_{\mathscr{X}}$ is the set of triples (C, D, x) with $x \in$ $\in T(C, D)$, and $[D, C] = \emptyset$.
(b) Consider the special case $T(?, ??) = [F(?), ??]_{\mathscr{D}}$, where $F: \mathscr{C} \to \mathscr{D}$ is a given functor. Let $I: \mathscr{D} \to \mathscr{X}$ be the inclusion. Then there is a functor $R: \mathscr{X} \to \mathscr{D}$ which is uniquely determined by $R\,I = \mathrm{Id}_{\mathscr{D}}$ and $R(C, D, x) = x$ for $(C, D, x) \in [C, D]_{\mathscr{X}}$. And there is a natural transformation $\eta: 1_{\mathscr{X}} \to I\,R$ defined by $\eta_C = (C, F(C), 1_{F(C)})$ and $\eta_D = 1_D$.
(c) What can be said if $T(?, ??) = [?, G(??)]_{\mathscr{C}}$ for a given functor $G: \mathscr{D} \to \mathscr{C}$?

2.7.6 Show that for functors $S, T: \mathscr{C} \to \mathscr{D}$ there is a canonical bijection between natural transformations $\alpha: S \to T$ and maps $\bar{\alpha}$: Mor $\mathscr{C} \to$ Mor $\mathscr{D}$ which satisfy

$$\bar{\alpha}(g\,f) = \bar{\alpha}(g)\,S(f) = T(g)\,\bar{\alpha}(f)\,.$$

3. Categories of Categories and Categories of Functors

3.1 Preliminary Remarks

The composition of functors in 2.2.6 suggests the study of categories whose objects are categories and whose morphisms are functors. 2.2.7 leads to categories whose objects are functors $\mathscr{C} \to \mathscr{D}$ and whose morphisms are natural transformations. However, familiar antinomies like "the set of all sets" or "the set of all sets not containing themselves as an element" show that precise definitions are necessary. Three possibilities exist, namely:

3.1.1 One uses the set theory of von Neumann-Bernays-Gödel as a basis. Their fundamental concept is that of a "class". Sets are those classes which are elements of classes. There are also classes, "non-sets", which are not an element of a class. There exists the universal class which contains all sets as elements. More details can be found in the appendix of J. L. Kelley: General Topology or in J. Schmidt: Mengenlehre I.

3.1.2 One tries to base mathematics not on an axiomatic set theory but instead on an axiomatic theory of the "category of categories" which encompasses set theory as the theory of discrete categories. Unfortunately, the axioms in [52] are not adequate for the results claimed there.

3.1.3 One expands the (usual) set theory of Zermelo-Fraenkel by introducing universes as suggested by Grothendieck; i. e., one admits unaccessible cardinals (Tarski). In [6] Brinkmann-Puppe give an account of this approach. We shall do no more than point out a few facts which (hopefully) will suffice for the understanding of what follows.

In 3.1.1 there are, apart from the ordinary categories in the sense of our definition in 1.1.1, "large" categories where $[A, B]_{\mathscr{C}}$ is only required to be a class. In 3.1.2 there exist not only the categories that are elements of the universal category, but also "large" categories that are not elements, but instead subcategories, of the universal category. In 3.1.3 a reduction of large categories to ordinary ones is possible. There are only sets here, but some of them are special sets, the universes, whereby a universe is the universal class of a model of the set theory in 3.1.1. Large categories of a universe are ordinary ones of a higher universe. To be able to understand the following, one has to be aware that the set theory used here has no "primitive (ur-) elements"; elements of sets, or classes resp., are always themselves sets.

3.2 Universes

3.2.1 A *universe* $\mathfrak{U}$ is a set (of sets) subject to the following conditions:

(1) $A \in \mathfrak{U} \Rightarrow A \subset \mathfrak{U}$.

(2) $A \in \mathfrak{U}$ and $B \in \mathfrak{U} \Rightarrow \{A, B\} \in \mathfrak{U}$ (set with the elements A, B).

(3) $A \in \mathfrak{U} \Rightarrow \mathfrak{P}(A) \in \mathfrak{U}$ (power set).

(4) If $J \in \mathfrak{U}$ and if $f: J \to \mathfrak{U}$ is a map, then $\bigcup_{j \in J} f(j) \in \mathfrak{U}$.

That is: if for a family of sets that are elements of $\mathfrak{U}$ the index set is also an element of $\mathfrak{U}$, then the union of the family is also an element of $\mathfrak{U}$.

3.2.2 From these conditions one can deduce: If $A \in \mathfrak{U}$, then every subset of A is also an element of $\mathfrak{U}$. For any two sets A and B that are elements of $\mathfrak{U}$, $A \times B$ and B^A (the set of all maps of A into B) are also elements of $\mathfrak{U}$; and the product set $\prod_{j \in J} A_j$ is an element of $\mathfrak{U}$ if J and all A_j are elements of $\mathfrak{U}$. In short: The usual constructions of set theory, carried out with elements of $\mathfrak{U}$, lead to elements of $\mathfrak{U}$.

3.2.3 We require as an axiom: every set is an element of a universe. Thus, in particular, every universe is an element of a higher universe.

3.3 Conventions

We make use of universes, but we choose a language which allows us to a large extent to use the language in 3.1.1. Let a fixed universe $\mathfrak{U}$, which contains the set $\boldsymbol{N}$ of natural numbers (and therefore $\boldsymbol{Z}$, $\boldsymbol{Q}$, $\boldsymbol{R}$, $\boldsymbol{C}$) be chosen from now on. If a change of universe is indicated, we shall say so explicitly.

3.3.1 *Sets* (more exactly: $\mathfrak{U}$-sets) are the elements of $\mathfrak{U}$.

3.3.2 *Classes* (more exactly: $\mathfrak{U}$-classes) are the subsets of $\mathfrak{U}$. Note that sets are classes but not vice versa.

3.3.3 Groups, rings, modules, topological spaces etc. (more exactly: $\mathfrak{U}$-groups, ... etc.) always have $\mathfrak{U}$-sets as their underlying sets.

3.3.4 Using the conventions 3.3.1, 3.3.2, 3.3.3, all the preceding examples have to be stated more precisely. The categories "of" sets, groups, modules, ..., etc. are always to be based on $\mathfrak{U}$. Here group homomorphisms may not be regarded simply as maps of the underlying sets, but these set maps must be indexed in such a way that the same set maps are regarded as different morphisms if the group structures are different (compare the proof in 3.5.1 below). The same applies to modules, ..., etc.

3.3.5 To adjust the *definition of categories* (more precisely, of $\mathfrak{U}$-categories) one has to go beyond 3.3.2 and 3.3.4, which amounts to an elimination of objects. Henceforth, a category $\mathcal{C}$ consists of a $\mathfrak{U}$-class Mor $\mathcal{C}$ (its class of morphisms) and a composition rule (which is a partially defined associative binary operation with left and right identities for each element). In particular, this composition determines the class of identity morphisms of $\mathcal{C}$, denoted by 1-Mor $\mathcal{C}$, and the partitioning of Mor $\mathcal{C}$ into the classes $[A, B]_{\mathcal{C}}$, which are required to be $\mathfrak{U}$-sets. The objects merely function as indices. If the class of objects $|\mathcal{C}|$ is replaced by an isomorphic one (i.e. an equipotent one), nothing in the category changes. Thus it is unnecessary that $|\mathcal{C}|$ be a $\mathfrak{U}$-class; however, $|\mathcal{C}|$ is isomorphic to a $\mathfrak{U}$-class, namely 1-Mor $\mathcal{C}$. It therefore is superfluous to retain objects together with their 1-morphisms. However, we keep them to avoid unwieldy formulations. A set, group, ..., etc. is different from its identity morphism.

3.3.6 A *category* $\mathcal{C}$ is called *small*, more precisely: $\mathfrak{U}$-small, if 1-Mor $\mathcal{C}$ is a $\mathfrak{U}$-set. This is equivalent to Mor $\mathcal{C}$ being a $\mathfrak{U}$-set. We may then assume, without any loss in generality, that the objects form a $\mathfrak{U}$-set. We will make use of this without further mentioning it. The same applies to classes of objects of arbitrary $\mathfrak{U}$-categories.

3.3.7 Chapters 1 and 2 will be regarded from now on as subject to the preceding conventions. However, it still has to be verified that the product of two categories $\mathscr{C}$ and $\mathscr{D}$ is a category. In fact, 2.5(1) does describe $\mathfrak{U}$-sets. These are pairwise disjoint because $\{[C, C']_{\mathscr{C}}\}$ and $\{[D, D']_{\mathscr{D}}\}$ are classes of pairwise disjoint sets. Since the sets in 2.5(1) are all elements of $\mathfrak{U}$, they form a $\mathfrak{U}$-class. Their union Mor $(\mathscr{C} \times \mathscr{D})$ is then also a $\mathfrak{U}$-class (compare 3.2(1)). 1-Mor $(\mathscr{C} \times \mathscr{D})$ is a subclass.

3.4 Functor Categories

3.4.1 Lemma. *If $\mathscr{C}$ is a small category, then* Mor $\mathscr{C} = \bigcup [A, B]_{\mathscr{C}}$ *and* 1-Mor $\mathscr{C} = \{1_A \mid A \in |\mathscr{C}|\}$ *and also* $\prod [A, B]_{\mathscr{C}}$ *are sets. Unions and products have to range over all pairs* $(A, B) \in |\mathscr{C}| \times |\mathscr{C}|$.

Since $|\mathscr{C}|$ can be replaced by 1-Mor $\mathscr{C}$, this follows immediately from 3.2.2. This, with 3.3.7 also implies the following:

3.4.2 Proposition. *If $\mathscr{C}$ and $\mathscr{D}$ are small categories, then $\mathscr{C} \times \mathscr{D}$ is also a small category.*

3.4.3 Proposition. *Let $\mathscr{C}$ be a small category and $\mathscr{D}$ an arbitrary category. Then the functors $\mathscr{C} \to \mathscr{D}$ are the objects and their natural transformations are the morphisms of a category. The composition of the morphisms is that of the natural transformations. This category is denoted by $[\mathscr{C}, \mathscr{D}]$. If $\mathscr{C}$ and $\mathscr{D}$ are small, then so is $[\mathscr{C}, \mathscr{D}]$.*

Proof. If $\mathscr{C}$ is empty, then $[\mathscr{C}, \mathscr{D}]$ has exactly one element and its identity morphism. If $\mathscr{D}$ is empty but not $\mathscr{C}$, then $[\mathscr{C}, \mathscr{D}]$ is empty. Now let $\mathscr{C}$ and $\mathscr{D}$ be non-empty and let $S, T\colon \mathscr{C} \to \mathscr{D}$ be functors. Since $\mathscr{C}$ is small, $M_S = \bigcup [S(A), S(B)]_{\mathscr{D}}$, $M_T = \bigcup [T(A), T(B)]_{\mathscr{D}}$, and $N = \prod [S(A), T(A)]_{\mathscr{D}}$ are sets. S defines a map $\Phi_S\colon \text{Mor}\, \mathscr{C} \to M_S$, namely $f \mapsto S(f)$, and S is completely determined by $\Phi_S \in [\text{Mor}\, \mathscr{C}, M_S]_{Ens}$; Φ_T is defined accordingly. A natural transformation $\alpha\colon S \to T$ is determined by a suitable element of N. The natural transformations $S \to T$ could therefore be considered as subsets of N. However, $\alpha\colon S \to T$ can also be a natural transformation of another pair of functors. We therefore consider natural transformations as triples $(\Phi_S, \Phi_T, \alpha) \in [\text{Mor}\, \mathscr{C}, M_S] \times [\text{Mor}\, \mathscr{C}, M_T] \times N$. By this, $[\mathscr{C}, \mathscr{D}]$ satisfies condition 1.1.1 (0). 1.1.1 (iii) follows immediately from 2.6.7. If $\mathscr{D}$ is also small, then by 2.2.7 and 3.2.2 the functors $\mathscr{C} \to \mathscr{D}$ form a subset of $[\text{Mor}\, \mathscr{C}, \text{Mor}\, \mathscr{D}]_{Ens}$, which proves the final assertion.

3.4.4 Proposition. *Let $\mathscr{C}$ and $\mathscr{D}$ be small categories and let $\mathscr{E}$ be an arbitrary category. Then there is a canonical isomorphism*

(1) $$\Phi\colon [\mathscr{C}, [\mathscr{D}, \mathscr{E}]] \xrightarrow{\approx} [\mathscr{C} \times \mathscr{D}, \mathscr{E}]\,.$$

Proof. We leave the discussion of the trivial cases, where $\mathscr{C}$, $\mathscr{D}$, or $\mathscr{E}$ are empty, to the reader. So let $\mathscr{C}$, $\mathscr{D}$, and $\mathscr{E}$ be non-empty. By

3.4.2 and 3.4.3, the indicated categories exist. The second bifunctor criterion 2.6.9 assigns to every functor $S: \mathcal{C} \to [\mathcal{D}, \mathcal{E}]$ a bifunctor $R: \mathcal{C} \times \mathcal{D} \to \mathcal{E}$ in such a way that, for $A \in |\mathcal{C}|$ and $f \in \operatorname{Mor} \mathcal{C}$, $S(A) = R(A, ?)$ and $S(f) = R(f, ?)$. This defines Φ on objects, and Φ is a bijection on object classes by 2.6.9 and 2.6.2. Now let $S': \mathcal{C} \to [\mathcal{D}, \mathcal{E}]$ be a functor, let $R' = \Phi(S')$ and let $\alpha: S \to S'$ be a natural transformation of functors. For every $A \in |\mathcal{C}|, \alpha_A: S(A) \to S'(A)$ is a natural transformation of functors $\mathcal{D} \to \mathcal{E}$, namely $\alpha_A: R(A, ?) \to R'(A, ?)$. For $X \in |\mathcal{D}|$, this defines $\alpha_{A,X}: R(A, X) \to R'(A, X)$. We claim now that this defines a natural transformation of bifunctors. For $u: X \to Y$ in $\mathcal{D}$, the diagram

(2)
$$\begin{array}{ccc} R(A, X) & \xrightarrow{\alpha_{A,X}} & R'(A, X) \\ \Big\downarrow{\scriptstyle R(A,u)} & & \Big\downarrow{\scriptstyle R'(A,u)} \\ R(A, Y) & \xrightarrow{\alpha_{A,Y}} & R'(A, Y) \end{array}$$

is commutative because α_A is a natural transformation. For $f: A \to B$ in $\mathcal{C}$, the diagram

(3)
$$\begin{array}{ccc} S(A) & \xrightarrow{\alpha_A} & S'(A) \\ \Big\downarrow{\scriptstyle S(f)} & & \Big\downarrow{\scriptstyle S'(f)} \\ S(B) & \xrightarrow{\alpha_B} & S'(B) \end{array} \quad = \quad \begin{array}{ccc} R(A, ?) & \xrightarrow{\alpha_A} & R'(A, ?) \\ \Big\downarrow{\scriptstyle R(f,?)} & & \Big\downarrow{\scriptstyle R'(f,?)} \\ R(B, ?) & \xrightarrow{\alpha_B} & R'(B, ?) \end{array}$$

is commutative, in particular for $? = X$. By 2.6.8, $(A, X) \mapsto \alpha_{A,X}$ is a natural transformation $\Phi(\alpha)$ of bifunctors. Conversely, if a natural transformation $R \to R'$ is given by $(A, X) \mapsto \alpha_{A,X}$, then (2) yields a natural transformation $\alpha_A: S(A) \to S'(A)$ and from (3) one obtains $\alpha: S \to S'$. Thus Φ gives a bijection $\Phi_{S,S'}: [S, S'] \to [\Phi(S), \Phi(S')]$. This, together with what was proved above, shows Φ to be a bijection of the morphism classes. The definition of $\Phi(\alpha)$ implies immediately that Φ is a functor.

3.4.5 It is obvious that $(A, X) \mapsto (X, A)$, $(f, u) \mapsto (u, f)$ defines an isomorphism

(4) $$\tau: \mathcal{C} \times \mathcal{D} \xrightarrow{\cong} \mathcal{D} \times \mathcal{C}$$

for arbitrary categories. The rule $R \mapsto R\tau$ for $R: \mathcal{D} \times \mathcal{C} \to \mathcal{E}$ produces an isomorphism

(5) $$[\tau, \mathcal{E}]: [\mathcal{D} \times \mathcal{C}, \mathcal{E}] \xrightarrow{\cong} [\mathcal{C} \times \mathcal{D}, \mathcal{E}]$$

for small categories $\mathcal{C}$ and $\mathcal{D}$. Thus 3.4.4 finally yields

(6) $$[\mathcal{C}, [\mathcal{D}, \mathcal{E}]] \cong [\mathcal{D}, [\mathcal{C}, \mathcal{E}]] .$$

3.5 The Category of Small Categories

3.5.1 Proposition. *There is a category whose objects are small categories, and whose morphisms are functors between such categories. The composition of the morphisms is that of the functors. This category is called the category of small categories and is denoted by cat.*

Proof. A small category is completely described by its set of morphisms Mor $\mathscr{C}$ and their composition. The graph of this composition, i. e. the set of triples (u, v, w), where $w = v\,u$, is a subset of Mor $\mathscr{C} \times$ $\times$ Mor $\mathscr{C} \times$ Mor $\mathscr{C}$; i.e., it is an element $\gamma_{\mathscr{C}}$ of $\mathfrak{P}_{\mathscr{C}} = \mathfrak{P}$ (Mor $\mathscr{C} \times$ Mor $\mathscr{C} \times$ $\times$ Mor $\mathscr{C}$). If $\mathscr{C}$ and $\mathscr{D}$ are small, then we regard a functor $T: \mathscr{C} \to \mathscr{D}$ as a triple $(\gamma_{\mathscr{C}}, \gamma_{\mathscr{D}}, T) \in \mathfrak{P}_{\mathscr{C}} \times \mathfrak{P}_{\mathscr{D}} \times [\text{Mor}\,\mathscr{C}, \text{Mor}\,\mathscr{D}]_{Ens}$. The proof is then completed as in 3.4.3.

3.5.2 If $\mathscr{C}$, $\mathscr{D}$ are small categories, then $[\mathscr{C}, \mathscr{D}]_{cat}$ is precisely the set of objects of the functor category $[\mathscr{C}, \mathscr{D}]$, hence the notation. The definition of isomorphisms in *cat* agrees with 2.2.9. We also use $1_{\mathscr{C}}$ for $\text{Id}_{\mathscr{C}}$.

3.6 Large Categories

3.6.1 Convention. A comparison of 2.6.9 with 3.4.4 shows that the restriction to small categories in 3.4 and 3.5 is unsatisfactory. (Even so it is useful in some contexts). We agree therefore to a permanent choice of a universe $\mathfrak{V}$ containing $\mathfrak{U}$ as an element. Thus every $\mathfrak{U}$-category is a small $\mathfrak{V}$-category, and the results about small $\mathfrak{V}$-categories imply results for arbitrary $\mathfrak{U}$-categories.

3.6.2 We use *ENS* for the category of $\mathfrak{V}$-sets and their maps, while, as before, *Ens* denotes the category of $\mathfrak{U}$-sets. *Ens* is a full subcategory of *ENS*; by enlarging the universe one obtains no new maps between the previously given sets. Corresponding remarks are true for other categories: for every mathematical structure whose models are sets with this structure (groups, modules, topological spaces etc.) the category consisting of the $\mathfrak{U}$-models of the structure and the structure preserving maps between them is a full subcategory of the corresponding $\mathfrak{V}$-category. In particular this is true for the $\mathfrak{U}$-category *Ab* of abelian groups and their homomorphisms. We denote the corresponding $\mathfrak{V}$-category by *AB*.

3.6.3 By 3.4.3 one obtains from a pair of $\mathfrak{U}$-categories $\mathscr{C}$, $\mathscr{D}$ a small $\mathfrak{V}$-category $[\mathscr{C}, \mathscr{D}]$ and thus by 3.4.4 one deduces the

Bifunctor theorem. *If $\mathscr{C}$, $\mathscr{D}$, $\mathscr{E}$ are categories, then there exists a canonical isomorphism*

$$(1) \qquad \Phi: [\mathscr{C}, [\mathscr{D}, \mathscr{E}]] \xrightarrow{\cong} [\mathscr{C} \times \mathscr{D}, \mathscr{E}]$$

as described in 3.4.4. Similarly, 3.4.5 is valid.

3.6.4 In (1) we have not explicitly mentioned $\mathfrak{U}$ or $\mathfrak{V}$. This is, in fact, unnecessary, due to a fact from the theory of sets with universes, namely: given any finite set of categories (in any universes), there is always a universe containing these categories.

The notation $\mathscr{D}^{\mathscr{C}}$ is also used for the functor category $[\mathscr{C}, \mathscr{D}]$. (1) then takes on the form of an exponential law.

3.6.5 Proposition. *There is a category Cat whose objects are $\mathfrak{U}$-categories and whose morphisms are functors between these $\mathfrak{U}$-categories. The composition of morphisms is that of functors.*

By 3.6.1 and 3.6.2, *Cat* is a full subcategory of the category *CAT* of small $\mathfrak{V}$-categories.

3.6.6 *Cat* possesses a duality functor J: $Cat \to Cat$ defined by $\mathscr{C} \mapsto \mathscr{C}^0$, $T \mapsto \mathrm{Op}\, T\, \mathrm{Op}$. Note that if $T: \mathscr{C} \to \mathscr{D}$ is covariant, then so is $\mathrm{Op}\, T\, \mathrm{Op}$: $\mathscr{C}^0 \to \mathscr{D}^0$. J must not be confused with Op: $Cat \to Cat^0$. Obviously $JJ = \mathrm{Id}_{Cat}$ is true. J leads to an expansion of the duality principle mentioned in 2.4.6. Concepts and theorems are called dual (to each other) if they generate each other through application of J ("dualization of all categories involved"). Examples will come up later.

3.6.7 Convention. In talking about categories we shall continue to mean $\mathfrak{U}$-categories (*legitimate categories*), except where categories of the form $[\mathscr{C}, \mathscr{D}]$, $[\mathscr{C}, [\mathscr{D}, \mathscr{E}]]$, etc., or *Cat, ENS, AB* are concerned which are characterized by the notation; or if we explicitly state otherwise.

Remarks

3.6.8 If $S: \mathscr{C} \to \mathscr{C}'$, $T: \mathscr{D} \to \mathscr{D}'$ are functors, then the rule $(\mathscr{C}, \mathscr{D}) \mapsto$ $\mapsto \mathscr{C} \times \mathscr{D}$, $(S, T) \mapsto S \times T$ defines a functor X: $cat \times cat \to cat$, $Cat \times$ $\times\, Cat \to Cat$ resp.. Using $(\mathscr{C}, \mathscr{D}) \mapsto \mathscr{D} \times \mathscr{C}$, $(S, T) \mapsto T \times S$ yields a corresponding result. Thus τ in 3.4.5 becomes an isomorphism of bifunctors.

3.6.9 The Hom-functors of *cat* and *Cat* can be changed to bifunctors $cat^0 \times cat \to cat$ and $Cat^0 \times Cat \to CAT$ by the rule $(\mathscr{C}, \mathscr{D}) \mapsto [\mathscr{C}, \mathscr{D}]$. Here *CAT* denotes the category of small $\mathfrak{V}$-categories. Similarly, by $(\mathscr{C}, \mathscr{D}, \mathscr{E}) \mapsto [\mathscr{C}, [\mathscr{D}, \mathscr{E}]]$ and $(\mathscr{C}, \mathscr{D}, \mathscr{E}) \mapsto [\mathscr{C} \times \mathscr{D}, \mathscr{E}]$ one gets trifunctors $Cat^0 \times Cat^0 \times Cat \to CAT$. Φ as given in 3.6.3 then becomes an isomorphism of trifunctors (compare later with 16.1.3).

3.7 The Evaluation Functor

3.7.1 There is a bifunctor

$$E: [\mathscr{C}, \mathscr{D}] \times \mathscr{C} \to \mathscr{D},$$

called the *evaluation functor*, associated with the functor category $[\mathscr{C}, \mathscr{D}]$. It is defined on objects by $(T, A) \mapsto T(A)$ and on morphisms as the diagonal of the commutative diagram

(1)

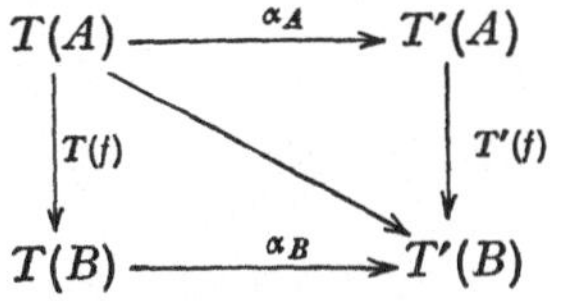

In particular,

(2) $$E(\alpha, A) = \alpha_A; \quad E(T, f) = T(f) .$$

It follows immediately from 2.6.9 that E is a bifunctor. E is an object of $[[\mathscr{C}, \mathscr{D}] \times \mathscr{C}, \mathscr{D}]$.

3.7.2 Proposition

$$E = \Phi\,(1_{[\mathscr{C}, \mathscr{D}]}) ,$$

where Φ is the isomorphism in 3.6.3.

Proof. According to 3.6.3 and 3.4.4, the bifunctor $R = \Phi(1_{[\mathscr{C}, \mathscr{D}]})$ satisfies the following

$$R(T, ?) = 1_{[\mathscr{C}, \mathscr{D}]}(T) = T\colon \mathscr{C} \to \mathscr{D};$$

$$R(\alpha, ?) = 1_{[\mathscr{C}, \mathscr{D}]}(\alpha) = \alpha .$$

A comparison with (2) and (1) then shows that $R = E$.

3.8 The Additive Case

The preceding remarks apply to additive categories and functors.

3.8.1 If $\mathscr{C}$ and $\mathscr{D}$ are additive categories, then we can consider the category *Add* $(\mathscr{C}, \mathscr{D})$ of additive functors $\mathscr{C} \to \mathscr{D}$. It is a full subcategory of $[\mathscr{C}, \mathscr{D}]$. Both categories can again be considered as additive categories; addition for natural transformations $\alpha, \beta\colon S \to T$ being defined by $(\alpha + \beta)_A = \alpha_A + \beta_A$.

If $\mathscr{C}, \mathscr{D}, \mathscr{E}$ are additive, then the proof of 3.4.4 and 3.6.3 produces an isomorphism of $Add(\mathscr{C}, Add(\mathscr{D}, \mathscr{E}))$ with the category of biadditive functors $\mathscr{C} \times \mathscr{D} \to \mathscr{E}$. This is done by an appropriate "restriction" of Φ, as can be seen from 2.5.7, 2.6.9 and 3.4.4 where (using different formulations) a bifunctor is constructed from its partial functors.

3.8.2 Analogous to 3.5, one can consider the category of small additive categories and additive functors as well as the analogue of 3.6.5. This does not yield subcategories of *cat* or *Cat*. Different additive categories can collapse into the same category if one disregards the additive structure of the sets of morphisms. Therefore, in carrying over the proof of 3.5.1 the sets $\mathfrak{P}_{\mathscr{C}}$ have to be replaced by others.

3.8.3 If $\mathscr{C}$ and $\mathscr{D}$ are additive, then according to 3.7 one obtains a biadditive evaluation functor

$$E\colon Add\,(\mathscr{C}, \mathscr{D}) \times \mathscr{C} \to \mathscr{D}\,.$$

Here the analogue of 3.7.2 is valid.

3.9 Problems

3.9.1 Write explicit axioms for a category as described in 3.3.5.

3.9.2 Let $T\colon \mathscr{C} \to \mathscr{C}$ be a functor. T is called an endomorphism or, resp., an automorphism of $\mathscr{C}$ if T is an endomorphism or, resp., an automorphism in Cat (compare 1.3.1). T is called an *inner equivalence* if there is an isomorphism $\alpha\colon 1_{\mathscr{C}} \to T$ in $[\mathscr{C}, \mathscr{C}]$, and T is called an *auto-equivalence* if there is a functor $S\colon \mathscr{C} \to \mathscr{C}$ such that ST and TS are inner equivalences. The natural transformations $1_{\mathscr{C}} \to 1_{\mathscr{C}}$ (i. e., the endomorphisms of $1_{\mathscr{C}}$) constitute what we shall call the *center* of $\mathscr{C}$.

(a) Let $\mathscr{C}$ be small. Show that the auto-equivalences or, resp., the inner equivalences, the automorphisms, the inner automorphisms (i. e., the automorphisms that are inner equivalences) form monoids, where multiplication is composition of functors. How are these monoids related to each other? Show by examples that, in general, these monoids are different. What can be said if $\mathscr{C}$ is not small?

(b) Discuss the above notions if $\mathscr{C}$ is a group. Here the center is also a group.

(c) Discuss the analogous additive case and the special case where $\mathscr{C}$ is a ring.

3.9.3 Let $U\colon \mathscr{B} \to \mathscr{C}$ be a functor. The functor $[U, \mathscr{D}]\colon [\mathscr{C}, \mathscr{D}] \to [\mathscr{B}, \mathscr{D}]$ is constructed by means of the rule $T \mapsto TU$, $\alpha \mapsto \alpha\, U$ (see 2.7.3). Similarly construct the functor $[\mathscr{A}, U] : [\mathscr{A}, \mathscr{B}] \to [\mathscr{A}, \mathscr{C}]$.

3.9.4 Verify 3.6.8 and 3.6.9.

3.9.5 Check 3.8.1 through 3.8.3.

4. Representable Functors

4.1 Embeddings

4.1.1 Definition. A *functor* $T\colon \mathscr{C} \to \mathscr{D}$ is called *faithful* if the induced maps (see 2.2.7)

(1) $$T_{A,B}\colon [A, B]_{\mathscr{C}} \to [T(A), T(B)]_{\mathscr{D}}$$

are injective for every pair $(A, B) \in |\mathscr{C}| \times |\mathscr{C}|$.

T is called *full* if (1) is always surjective, and T is called *fully faithful* if (1) is always bijective.

4.1.2 Notice that a faithful or a fully faithful functor may map two different objects into the same one. The maps $|\mathcal{C}| \to |\mathcal{D}|$ and Mor $\mathcal{C} \to$ Mor $\mathcal{D}$ which are defined by $A \mapsto T(A)$ and $f \mapsto T(f)$ need not be injective. For a faithful functor, $|\mathcal{C}| \to |\mathcal{D}|$ is injective if and only if Mor $\mathcal{C} \to$ Mor $\mathcal{D}$ is injective.

4.1.3 A functor $T: \mathcal{C} \to \mathcal{D}$ is called an *embedding* if

$$T: \text{Mor}\, \mathcal{C} \to \text{Mor}\, \mathcal{D}$$

is injective. For an embedding $T: \mathcal{C} \to \mathcal{D}$ the objects $T(A)$ and the morphisms $T(f)$ form a subcategory of $\mathcal{D}$, which is full when T is full.

4.1.4 In general the objects $T(A)$ and morphisms $T(f)$ do not constitute a subcategory, as is shown by a counterexample:

$$\mathcal{C}: \quad A \xrightarrow{f} B\,, \quad C \xrightarrow{g} D;$$

$$\mathcal{D}: \quad X \xrightarrow{u} Y\,, \quad Y \xrightarrow{v} Z\,, \quad X \xrightarrow{w} Z \quad \text{where} \quad w = v\,u\,.$$

$\mathcal{C}$ has four identity morphisms and two others, f and g. $\mathcal{D}$ has three identity morphisms and three others, u, v, w with $w = v\,u$. $T: \mathcal{C} \to \mathcal{D}$ is defined by $T(f) = u$, $T(g) = v$. $T(g)\,T(f)$ is not in the image of T. T is faithful, but it is not an embedding.

4.1.5 Proposition. *Let $T: \mathcal{C} \to \mathcal{D}$ be a fully faithful functor. If f: $A \to B$ in $\mathcal{C}$, then $T(f)$ is an isomorphism if and only if f is an isomorphism.*

Proof. Let $T(f): T(A) \to T(B)$ be an isomorphism whose inverse is u. Since T is fully faithful, there is exactly one morphism $g: B \to A$ such that $T(g) = u$. It then follows that $T(g\,f) = u\,T(f) = 1_{T(A)}$, which implies $g\,f = 1_A$, again because T is fully faithful. Similarly one finds that $f\,g = 1_B$, so that f is an isomorphism whose inverse is g. The converse is true for arbitrary functors (2.1.1).

4.2 Yoneda Lemma

Let $\mathcal{C}$ be a non-empty category, and let $A \in |\mathcal{C}|$. We consider the functor $H^A = [A, ?]_{\mathcal{C}}$ and another functor $T: \mathcal{C} \to Ens$. Let $\alpha: H^A \to T$ be a natural transformation. We look at it at the "point" A; i. e., $\alpha_A: [A, A] \to T(A)$. $\alpha_A(1_A)$ is a well defined element in $T(A)$.

4.2.1 Lemma. *The "Yoneda map" $Y: [H^A, T]_{[\mathcal{C}, Ens]} \to T(A)$, which is defined by $\alpha \mapsto \alpha_A(1_A)$, is bijective.*

Proof. First let $\alpha: H^A \to T$ be given and set $\alpha_A(1_A) = x \in T(A)$. For arbitrary B and $f: A \to B$ the diagram

$$(1) \qquad \begin{array}{ccc} [A, A] & \xrightarrow{\alpha_A} & T(A) \\ \downarrow{\scriptstyle [A, f]} & & \downarrow{\scriptstyle T(f)} \\ [A, B] & \xrightarrow{\alpha_B} & T(B) \end{array} \qquad\qquad \begin{array}{ccc} 1_A & \mapsto & x \\ \downarrow & & \downarrow \\ f & \mapsto & T(f)\,(x) \end{array}$$

is commutative, so that

$$\alpha_B(f) = T(f)\,(x) = T(f)\,(\alpha_A(1_A))\,. \tag{2}$$

α_B is therefore defined by $f \mapsto T(f)\,(x)$ and α is completely determined by $x = \alpha_A(1_A)$. This shows that Y is injective.

Now let $x \in T(A)$ be given and define α_B by (2) for all $B \in |\mathscr{C}|$. It has to be shown that for an arbitrary $g\colon B \to C$ the diagram

$$\begin{array}{ccc} [A, B] & \xrightarrow{\alpha_B} & T(B) \\ \Big\downarrow{\scriptstyle [A,g]} & & \Big\downarrow{\scriptstyle T(g)} \\ [A, C] & \xrightarrow{\alpha_C} & T(C) \end{array} \tag{3}$$

is commutative. But for $f \in [A, B]$

$$\alpha_C[A, g]\,(f) = \alpha_C(g\,f) = T(g\,f)\,(x) = T(g)\,T(f)\,(x) = T(g)\,\alpha_B(f)\,,$$

so the proof is complete.

4.2.2 Proposition. *For $A \in |\mathscr{C}|$, $f \in \operatorname{Mor} \mathscr{C}$, the rule $A \mapsto H^A$, $f \mapsto H^f$ defines a full embedding $H^*\colon \mathscr{C}^0 \to [\mathscr{C}, Ens]$ called the Yoneda embedding.*

Proof. We may assume $\mathscr{C} \neq \emptyset$. If A and B are different objects, then the functors H^A and H^B are different, because $[A, A] \cap [B, A] = \emptyset$. If $f \in [C, A]$, then $H^f = [f, ?]$ is a natural transformation $H^A \to H^C$ (compare 2.6.(4)). By 4.2.1, $Y(H^f) = H^f_A(1_A) = [f, A]\,(1_A)$. But $[f, A]\colon [A, A] \to [C, A]$ is described by $u \mapsto u\,f$, so using $u = 1_A$, one concludes that

$$Y(H^f) = f\,, \tag{4}$$

which, together with 4.2.1, shows that the functor H^* is fully faithful.

If $\mathscr{C}$ is replaced by $\mathscr{C}^0$ and $\mathscr{C}^{00} = \mathscr{C}$ is taken into account, then one obtains the Yoneda embedding $H_*\colon \mathscr{C} \to [\mathscr{C}^0, Ens]$ defined by $A \mapsto H_A$, $f \mapsto H_f$.

Every (small) category $\mathscr{C}$ may be regarded as a full subcategory of the functor category $[\mathscr{C}^0, Ens]$.

4.2.3 The Hom-functor of the category $[\mathscr{C}, Ens]$ is a bifunctor $[\mathscr{C}, Ens]^0 \times [\mathscr{C}, Ens] \to ENS$. Using the embedding $\operatorname{Op} H^* \operatorname{Op}\colon \mathscr{C} \to$ $\to [\mathscr{C}, Ens]^0$, from 4.2.2 one obtains a bifunctor $[H^?, ??]_{[\mathscr{C}, Ens]}\colon \mathscr{C} \times$ $\times [\mathscr{C}, Ens] \to ENS$. We also consider the evaluation functor E with the transposition τ (3.7 and 3.4.5): $E\tau\colon \mathscr{C} \times [\mathscr{C}, Ens] \to Ens$.

4.2.4 Theorem. *The Yoneda map $Y(\alpha) = \alpha_A(1_A)$, where $\alpha\colon H^A \to T$, is a bifunctor isomorphism*

$$Y\colon [H^?, ??]_{[\mathscr{C}, Ens]} \xrightarrow{\cong} E\tau(?, ??)\,. \tag{5}$$

This uses $Ens \subset ENS$.

Proof. If $\mathscr{C} = \emptyset$, both sides in (5) are the empty functor. So let $\mathscr{C} \neq \emptyset$. According to 4.2.1, Y is a bijection at every point $(A, T) \in$

$\in |\mathscr{C} \times [\mathscr{C}, Ens]|$. It remains to be shown that Y is a natural transformation; and by 2.6.8 it suffices to show this to be true for the partial functors. If $\beta\colon T \to R$ is a natural transformation, then the diagram

$$(6)\quad \begin{array}{ccccccc} [H^A, T] & \xrightarrow{[H^A,\beta]} & [H^A, R] & \qquad & \alpha & \mapsto & \beta\alpha \\ \downarrow Y & & \downarrow Y & & \downarrow & & \downarrow \\ T(A) & \xrightarrow{\beta_A} & R(A) & & \alpha_A(1_A) & \mapsto & \beta_A(\alpha_A(1_A)) = (\beta\alpha)_A(1_A) \end{array}$$

is commutative. If $f\colon A \to B$, then the diagram

$$\begin{array}{ccccccc} [H^A, T] & \xrightarrow{[H^f, T]} & [H^B, T] & \qquad & \alpha & \mapsto & \alpha H^f \\ \downarrow Y & & \downarrow Y & & \downarrow & & \downarrow \\ T(A) & \xrightarrow{T(f)} & T(B) & & \alpha_A(1_A) & \mapsto & T(f)\,(\alpha_A(1_A)) \end{array}$$

is commutative: $Y(\alpha H^f) = (\alpha H^f)_B(1_B) = \alpha_B(H^f_B(1_B)) = \alpha_B(Y(H^f)) = \alpha_B(f)$ by (4) and the definition of Y. From this and from (2) it follows that $Y(\alpha H^f) = T(f)(\alpha_A(1_A))$, and the theorem is proved.

4.3 The Additive Case

If $\mathscr{C}$ is an additive category, then H^A can be considered as an additive functor $\mathscr{C} \to Ab$. If $T\colon \mathscr{C} \to Ab$ is additive, then 4.2.1 is valid for $Add(\mathscr{C}, Ab)$ instead of for $[\mathscr{C}, Ens]$. For, the additivity of T and 4.2(2) imply $\alpha_B(f_1 + f_2) = \alpha_B(f_1) + \alpha_B(f_2)$. Furthermore, the sum $\alpha + \beta$ of two natural transformations $\alpha, \beta\colon H^A \to T$ is given by the rule $(\alpha + \beta)_B = \alpha_B + \beta_B$ for all $B \in |\mathscr{C}|$, whereby $\alpha_B + \beta_B$ is the sum of two homomorphisms between additive groups. Bearing in mind that in 4.2(2) $T(f)$ is also a homomorphism, one obtains:

4.3.1 Lemma. *Let $\mathscr{C}$ be an additive category and $T\colon \mathscr{C} \to Ab$ an additive functor. Then the Yoneda map $Y\colon [H^A, T]_{Add(\mathscr{C}, Ab)} \to T(A)$ defined by the rule $\alpha \mapsto \alpha_A(1_A)$ is an isomorphism of additive groups.*

Employing the forgetful functor $U\colon Ab \to Ens$ one obtains the following: if T is additive, then every natural transformation $UH_A \to UT$ is of the form $\{U(\alpha_A)\}$ for a suitable $\alpha = \{\alpha_A\}$.

This, however, does not hold for arbitrary pairs of additive Ab-valued functors.

4.3.2 Proposition. *If $\mathscr{C}$ is additive, then $H^*\colon \mathscr{C}^0 \to Add(\mathscr{C}, Ab)$ is a full additive embedding and so is $H_*\colon \mathscr{C} \to Add(\mathscr{C}^0, Ab)$.*

4.3.3 Proposition. *If $\mathscr{C}$ is additive, then for $\alpha\colon H^A \to T$ the Yoneda map $Y(\alpha) = \alpha_A(1_A)$ determines an isomorphism*

$$Y\colon [H^?, ??]_{Add(\mathscr{C}, Ab)} \cong\!\!\Rightarrow E\tau(?, ??)$$

of the biadditive functors $\mathscr{C} \times Add(\mathscr{C}, Ab) \to AB$ with $Ab \subset AB$.

The proofs of 4.3.2 and 4.3.3 follow immediately from those of 4.2.2 and 4.2.4.

4.4 Representable Functors

4.4.1 Definition. A functor $T: \mathcal{C} \to Ens$ is called *representable* if, for some $A \in |\mathcal{C}|$, T is isomorphic to H^A. A is then called a representing object for T. A *representation* of T is an isomorphism $\varrho: H^A \to T$.

It follows from 4.2.1 that a representation of T is completely determined by A and $\varrho_A(1_A) \in T(A)$. Therefore, representations are described by giving the pair $(A, \varrho_A(1_A))$; $\varrho_A(1_A)$ is called the *universal element* of the representation. We say: T is represented by $(A, \varrho_A(1_A))$. The representation is provided by 4.2 (2).

4.4.2 It follows from 4.2(2) that, if A and $x \in T(A)$ are given, then the natural transformation $H^A \to T$ determined by (A, x) is an isomorphism if and only if for every $y \in T(B)$, with arbitrary B, there is exactly one $f: A \to B$ such that $T(f)(x) = y$.

4.4.3 Examples. Let $\mathcal{C}$ be the category Top of topological spaces and let $U: \mathcal{C} \to Ens$ be the forgetful functor. Every space consisting of a single point is a representing object for U. If $\mathcal{C}$ is the category of (multiplicative) groups and U again the forgetful functor, then every free cyclic group (i. e. isomorphic to the additive group of the integers) is a representing object. Note that a representable functor T can take non-isomorphic objects into the same set. Also, the choice of a representing object A may not be replaced by a choice of $T(A)$: a countable set can be given various group structures.

4.4.4 Proposition. (*a*) *If S, T: $\mathcal{C} \to Ens$ are isomorphic (to each other), then S is representable if and only if T is representable. More exactly: an isomorphism $\xi: T \to S$ yields a bijection of representations by means of $\varrho \mapsto \xi\,\varrho$.*

(b) *H^A and H^B are isomorphic if and only if A and B are isomorphic. More exactly: $u \mapsto H^u$ yields a bijection between the isomorphisms $A \to B$ and $H^B \to H^A$.*

(c) *If T is represented by $(A, \varrho_A(1_A))$ and $(B, \sigma_B(1_B))$, then there is exactly one morphism $u: A \to B$ with $T(u)\,(\varrho_A(1_A)) = \sigma_B(1_B)$, and u is an isomorphism.*

Proof. (a) and (b) follow immediately from 4.2.2. In (c) the uniqueness of u is guaranteed by 4.2 (2) as follows: one has $u = \varrho_B^{-1}\,\sigma_B(1_B)$; $\varrho^{-1}\,\sigma: H^B \to H^A$ is an isomorphism with $Y(\varrho^{-1}\,\sigma) = u$, and because of 4.2.2, u is an isomorphism. $u^{-1} = \sigma_A^{-1}\,\varrho_A(1_A)$ also holds.

4.4.5 Proposition. *Let $T: \mathcal{C} \to Ens$ be a representable functor and A a representing object for T. Let S: $\mathcal{C} \to Ens$ be an arbitrary functor.*

Then there exists a bijection between the set of natural transformations $T \to S$ and the set $S(A)$. If T is represented by $(A, \varrho_A(1_A))$, then $\alpha \mapsto \alpha_A(\varrho_A(1_A))$ provides such a bijection.

This is just another formulation of 4.2.1 which we shall also refer to as the Yoneda lemma. It contains in particular an assertion about the natural transformations of a representable functor into itself.

4.4.6 The contravariant case results if $\mathscr{C}$ is replaced by its dual category $\mathscr{C}^0$. 2.5(13) implies that a contravariant functor $T: \mathscr{C} \to Ens$ is representable if it is isomorphic to a functor H_A. A representation is again determined completely by a pair $(A, \varrho_A(1_A))$. 4.4.4 and 4.4.5 are transferred accordingly, whereby in 4.4.5 "functor" is to be replaced by "contravariant functor". 4.2(2) is valid for $f: B \to A$, and so is 4.4.2.

4.4.7 Example. The contravariant functor "power set" $\mathfrak{P}$ in 2.3.4 is representable. A set of two elements is a representing object, each of the two one element subsets is a universal element of a representation.

4.4.8 The additive case. If $\mathscr{C}$ is additive, then $H^A: \mathscr{C} \to Ab$ and $H_A: \mathscr{C}^0 \to Ab$ are always additive. The preceding remarks carry over smoothly if Ens is replaced by Ab. All such representable functors are always additive.

One also considers functors $T: \mathscr{C} \to Ens$. Such a functor is called representable if it is isomorphic to a functor UH^A, where U is the fortgetful functor. The contravariant case is analogous.

4.4.9 Example. Let $\mathscr{C} = Ab$ and let M be a given set. For $A \in |Ab|$, let $T(A)$ be the set of mappings of M into $U(A)$; $T(A) = [M, U(A)]_{Ens}$. For $f \in \operatorname{Mor} Ab$ let $T(f) = [M, U(f)]_{Ens}$. Then $T = H^M U$. This functor is represented by the free additive group F with basis M and the inclusion $M \subset U(F)$. Similar results hold for ${}_RMod$ and Mod_R. This example can be regarded as a definition of "free" over M. It can be carried over to other, even non-additive categories, e. g. the category of groups.

4.4.10 Theorem. *Let $\mathscr{C}$ be an additive category. The additive functor $T: \mathscr{C} \to Ab$ is representable if and only if $UT: \mathscr{C} \to Ens$ is representable ($U: Ab \to Ens$ is the forgetful functor).*

Proof. Let $(A, \varrho_A(1_A))$ be a representation of UT. $\varrho_A(1_A)$ is an element of the group $T(A)$ and by 4.3.1 it defines a natural transformation $\bar{\varrho}: H^A \to T$, whereby $U(\bar{\varrho}_B) = \varrho_B$ for all $B \in |\mathscr{C}|$. $\bar{\varrho}_B$ is thus always a bijective homomorphism in Ab. Therefore $\bar{\varrho}$ is an isomorphism. The converse is evident.

4.5 Partially Representable Bifunctors

4.5.1 Proposition. *Let $R: \mathscr{C} \times \mathscr{D} \to Ens$ be a bifunctor. Assume that for every $A \in |\mathscr{C}|$ there is a representation $\varrho_A: H^{G(A)} \to R_A$ of the partial functor $R_A(?) = R(A, ?): \mathscr{D} \to Ens$. The rule $A \mapsto G(A)$ can be extended to a contravariant functor $G: \mathscr{C} \to \mathscr{D}$ in such a way that $(A, X) \mapsto \varrho_{A,X}$ is an isomorphism $\varrho: [G(?), ??]_{\mathscr{D}} \to R(?, ??)$ of bifunctors. This determines G uniquely.*

Proof. For $f: A \to B$ in $\mathscr{C}$, $\varrho_B^{-1} R_f \varrho_A: H^{G(A)} \to H^{G(B)}$ is a natural transformation by 2.6.2, where we have written R_f instead of $R(f, ?)$. By 4.2.2 there is exactly one morphism $u: G(B) \to G(A)$ in $\mathscr{D}$ such that $H^u = \varrho_B^{-1} R_f \varrho_A$. Let $G(f) = u$. Then it follows from 4.2.2 again that G is a contravariant functor. By 2.6.8 one confirms that $\varrho = \{\varrho_{A,X}\}$ is a bifunctor transformation. Because ϱ_A is an isomorphism, so is every $\varrho_{A,X}: [G(A), X] \to R(A, X)$ and therefore also ϱ. From 2.6.8 it follows that $\varrho_B H^{G(f)} = R_f \varrho_A$ must hold in any case, hence G is determined uniquely.

4.5.2 If in the preceding paragraph R is a contra-co-variant functor, then G is covariant. Thus 4.4.9 produces a functor $F: Ens \to Ab$ which assigns to every set M the free abelian group with basis M. Furthermore, there is an isomorphism of contra-co-variant functors

$$\varphi: [F(?), ??]_{Ab} \xrightarrow{\cong} [?, U(??)]_{Ens}$$

where the Hom-functor of Ab has Ens as its codomain.

Corresponding situations arise for ${}_R Mod$ and the category of groups. In 16.4 we shall discuss such "adjoint situations" where for the functors $T: \mathscr{C} \to \mathscr{D}$, $S: \mathscr{D} \to \mathscr{C}$ there exists an isomorphism $\varphi: [S(?), ??]_{\mathscr{C}} \xrightarrow{\cong} [?, T(??)]_{\mathscr{D}}$ of contra-co-variant functors.

4.5.3 Let $\mathscr{C}$ and $\mathscr{D}$ be arbitrary non-empty categories. As suggested by 4.5.1, one can assign to every functor $T: \mathscr{C} \to \mathscr{D}$ the contra-co-variant functor $\Phi(T) = [T(?), ??]_{\mathscr{D}}$. Then, if $\alpha: T \to S$ is a natural transformation, the rule $(A, X) \mapsto [\alpha_A, X]$ defines a natural transformation $\Phi(\alpha): \Phi(S) \to \Phi(T)$, as is easily confirmed using 2.6.8. Thus there exists a contravariant functor

$$\Phi: [\mathscr{C}, \mathscr{D}] \to [\mathscr{C}^0 \times \mathscr{D}, Ens],$$

where on the right we are using convention 2.5.4. Making use of convention 2.4.5, there is the following result:

4.5.4 Proposition. *Φ Op is a full embedding $[\mathscr{C}, \mathscr{D}]^0 \to [\mathscr{C}^0 \times \mathscr{D}, Ens]$.*

Proof. (a) Φ is injective on objects. If $S, T: \mathscr{C} \to \mathscr{D}$ are different, then there is an $f: A \to B$ in $\mathscr{C}$ such that $S(f) \neq T(f)$. If $S(B) \neq T(B)$, then $\Phi(S) \neq \Phi(T)$ because $[S(B), T(B)] \cap [T(B), T(B)] = \emptyset$. If

$S(B) = T(B)$, then $\Phi(S) \neq \Phi(T)$ follows from $[S(f), T(B)]\,(1_{T(B)}) = S(f) \neq [T(f), T(B)]\,(1_{T(B)})$.

(b) Φ is faithful. If α, β are different natural transformations $T \to S$, then there is an $A \in |\mathscr{C}|$ such that $\alpha_A \neq \beta_A$ and

$$[\alpha_A, S(A)] \neq [\beta_A, S(A)] \,.$$

(c) Φ is full. A natural transformation $\xi = \{\xi_{A,X}\}: \Phi(S) \to \Phi(T)$ yields for every $A \in |\mathscr{C}|$ a natural transformation

$$\xi_A = \{\xi_{A,X}\}_{X \in |\mathscr{D}|}: [S(A), ??] \to [T(A), ??] \,.$$

By 4.2.2 there is exactly one morphism α_A: $T(A) \to S(A)$ with $\xi_A = [\alpha_A, ??]$. Assuming $\alpha = \{\alpha_A\}$ to be a natural transformation $T \to S$, the assertion follows from $\Phi(\alpha)_{A,X} = [\alpha, X] = \xi_{A,X}$. Now, for every $X \in |\mathscr{D}|$, $\{[\alpha_A, X]\} = \{\xi_{A,X}\}$ is a natural transformation $[S(?), X] \to [T(?), X]$, so that the remaining assertion is proved by the next lemma.

4.5.5 Lemma. *Let S, T: $\mathscr{C} \to \mathscr{D}$ be functors. For every $A \in |\mathscr{C}|$ let there be given a morphism α_A: $T(A) \to S(A)$. If $\{[\alpha_A, X]\}$ is a natural transformation for every $X \in |\mathscr{D}|$, then $\{\alpha_A\}$ is a natural transformation.*

Proof. Let $f: A \to B$ be an arbitrary morphism in $\mathscr{C}$. For $X = S(B)$ the diagram

$$\begin{array}{ccc} [S(B), S(B)] & \xrightarrow{[\alpha_B, S(B)]} & [T(B), S(B)] \\ \Big\downarrow{\scriptstyle [S(f), S(B)]} & & \Big\downarrow{\scriptstyle [T(f), S(B)]} \\ [S(A), S(B)] & \xrightarrow{[\alpha_A, S(B)]} & [T(A), S(B)] \end{array}$$

is then commutative. Using $1_{S(B)} \in [S(B), S(B)]$ one obtains $\alpha_B\, T(f) = S(f)\, \alpha_A$ and thus the desired result.

4.5.6 Remark. $[\mathscr{C}, \mathscr{D}]^0$ is isomorphic to $[\mathscr{C}^0, \mathscr{D}^0]$ by means of the rule $T^0 \mapsto \mathrm{Op}\; T\; \mathrm{Op}$ for functors and $\{\alpha_A\}^0 \mapsto \{\alpha_A{}^0\}$ for natural transformations $\alpha = \{\alpha_A\}$ as morphisms of $[\mathscr{C}, \mathscr{D}]$. $[\mathscr{C}^0, \mathscr{D}^0]$ can thus be viewed as the dual category of $[\mathscr{C}, \mathscr{D}]$. Using 4.5.3, 4.5.4 one obtains this way a full embedding

$$[\mathscr{C}, \mathscr{D}] \to [\mathscr{C} \times \mathscr{D}^0, \mathit{Ens}] \,,$$

if in addition $\mathscr{C}^0$ and $\mathscr{D}^0$ are replaced by their duals. The interchange τ in 3.4.5 then produces the full embedding

$$[\mathscr{C}, \mathscr{D}] \to [\mathscr{D}^0 \times \mathscr{C}, \mathit{Ens}] \,.$$

It is identical with $T \mapsto [??, T(?)]_{\mathscr{D}}$ and $\alpha \mapsto \{[X, \alpha_A]\}$.

4.5.7 The additive case. If $\mathscr{C}$ and $\mathscr{D}$ are additive categories and if R: $\mathscr{C} \times \mathscr{D} \to \mathit{Ab}$ is a biadditive functor, then there is an additive contravariant functor G: $\mathscr{C} \to \mathscr{D}$ according to 4.5.1. This follows from

$R_{f+g} = R_f + R_g$ (compare 3.8.1) by 4.3.2. According to 4.5.4 one obtains a full embedding $Add(\mathscr{C}, \mathscr{D})^0 \to Biadd(\mathscr{C}^0 \times \mathscr{D}, Ab)$ and according to 4.5.6 the full embedding $Add(\mathscr{C}, \mathscr{D}) \to Biadd(\mathscr{D}^0 \times \mathscr{C}, Ab)$, where *Biadd* are categories of biadditive functors.

4.6 Problems

4.6.1 Let $T: \mathscr{C} \to \mathscr{D}$ be a functor.

(a) Describe the smallest subcategory $\mathscr{D}'$ of $\mathscr{D}$ which contains the objects $T(A)$ and the morphisms $T(f)$ for all $A \in |\mathscr{C}|$ and for $f \in \operatorname{Mor} \mathscr{C}$.

(b) What are the facts, if T is full?

(c) What are the facts, if T is an injective map for the classes of objects?

4.6.2 Determine and describe the endomorphisms of the representable functors in 4.4.3, 4.4.7, and 4.4.9.

4.6.3 Fill in the details in 4.5.1 and 4.5.3.

4.6.4 Prove the statement at the end of 4.5.6.

4.6.5 Carry out 4.5.7.

4.6.6 Consider the ring $\mathbf{Z}$ as an additive category. Then $Add(\mathbf{Z}, Ab) \cong$ $\cong Ab$ in a canonical way. Considering the additive groups $\mathbf{Z} \oplus \mathbf{Z}$ (direct sum) and $\mathbf{Z}$ as objects of $Add(\mathbf{Z}, Ab)$, give an example of a natural transformation $U(\mathbf{Z} \oplus \mathbf{Z}) \to U(\mathbf{Z})$ which has not the form $U\alpha$. ($U: Ab \to Ens$ is the forgetful functor.)

5. Some Special Objects and Morphisms

5.1 Monomorphisms

5.1.1 A morphism m in the category $\mathscr{C}$ is called a *monomorphism* if for all pairs (f, g) of $\mathscr{C}$

$$(1) \qquad m f = m g \Rightarrow f = g .$$

Of course, $m f = m g$ can only hold if $m f$ and $m g$ are both defined and if f and g gave the same domain. A monomorphism is a "left cancellable" morphism.

5.1.2 For $\mathscr{C} = Ens$ monomorphic is the same as injective. For $\mathscr{C} = Ab$, ${}_R Mod$, for the category of groups and the category of topological spaces being monomorphic is equivalent to the underlying set map being injective. However, it is not always true that a forgetful functor (if it exists) transforms monomorphisms into injective maps. Trivial counterexamples are built as follows:

Let $f\colon A \to B$ be any morphism in $\mathscr{C}$ with $A \neq B$. If the subcategory $\mathscr{D}$ of $\mathscr{C}$ consists of the objects A, B and the morphisms 1_A, 1_B, f, then f is a monomorphism in $\mathscr{D}$.

5.1.3 If $m\colon A \to B$ is a monomorphism in $\mathscr{C}$, then m is also a monomorphism in every subcategory of $\mathscr{C}$ containing m.

5.1.4 The following are equivalent:
(a) $m\colon A \to B$ is monomorphic in $\mathscr{C}$.
(b) For *all* $X \in |\mathscr{C}|$ $[X, m]\colon [X, A] \to [X, B]$ is injective.

5.1.5 (a) If f is an isomorphism, then f is a monomorphism.

(b) If f and g are monomorphisms and $g\,f$ is defined, then $g\,f$ is a monomorphism.

(c) If $g\,f$ is a monomorphism, then f is a monomorphism.

5.1.3, 5.1.4, 5.1.5 follow immediately from the definition. In 5.1.5 (c) one may not conclude that g is a monomorphism, since if $f\colon A \to B$ is an inclusion of non-empty sets, then there is always a g such that $g\,f = 1_A$.

5.1° Epimorphisms

5.1.1° A morphism h in $\mathscr{C}$ is called an *epimorphism* if h° is a monomorphism in $\mathscr{C}^\circ$; i. e., if

$$f\,h = g\,h \Rightarrow f = g \qquad (1^\circ)$$

holds in $\mathscr{C}$. Epimorphisms are "right cancellable".

5.1.2° For $\mathscr{C} = \mathit{Ens}$ epimorphic is equivalent to surjective, the same holds for $\mathscr{C} = \mathit{Ab}$, ${}_R\mathit{Mod}$. It is also true, but not evident, in the category of groups (see 7.9.2). In the category of Hausdorff spaces it is sufficient to require $f(A)$ to be dense in B for $f\colon A \to B$ to be an epimorphism.

5.1.3° If $h\colon A \to B$ is an epimorphism in $\mathscr{C}$, then h is also an epimorphism in every subcategory of $\mathscr{C}$ containing h.

5.1.4° The following are equivalent:
(a) $h\colon A \to B$ is an epimorphism in $\mathscr{C}$.
(b) For *all* $X \in |\mathscr{C}|$ $[h, X]\colon [B, X] \to [A, X]$ is *injective*.

5.1.5° (a) If f is an isomorphism, then f is an epimorphism.

(b) If f, g are epimorphisms and $g\,f$ is defined, then $g\,f$ is an epimorphism.

(c) If $g\,f$ is an epimorphism, then g is an epimorphism.

5.2 Retractions and Coretractions

5.2.1 $r: A \to B$ in $\mathscr{C}$ is called a *retraction* if there is an $s: B \to A$ such that $r\,s = 1_B$. $s: B \to A$ is called a *coretraction* (or *section*) if there is an $r: A \to B$ such that $r\,s = 1_B$.

Thus $r\,s = 1_B$ implies both that r is a retraction and s a coretraction. Note, however, that for a given retraction r there are in general various s with $r\,s = 1_B$, and analogously for coretractions, as is seen easily in *Ens*.

5.2.2 Every retraction is an epimorphism and every coretraction is a monomorphism. The converse is in general not true.

The proof is furnished by 5.1.5⁰ and 5.1.5. *Ab* provides examples of epimorphisms that are not retractions, and in *Ens* $\emptyset \subset A$ for $A \neq \emptyset$ is a monomorphism but not a coretraction.

5.2.3 Every functor preserves retractions and coretractions.

5.2.4 The following are equivalent:
(a) $r: A \to B$ is a retraction in $\mathscr{C}$.
(b) For all $X \in |\mathscr{C}|$ $[X, r] : [X, A] \to [X, B]$ is surjective.
(c) $[B, r]: [B, A] \to [B, B]$ is surjective.

5.2.5 (a) If f is an isomorphism, then f is a retraction.
(b) If f and g are retractions and $g\,f$ is defined, then $g\,f$ is a retraction.
(c) If $g\,f$ is a retraction, then g is a retraction.

5.2.6 Monomorphism — epimorphism, retraction — coretraction are first examples of dual pairs of concepts.

5.2.7 In *Ens* every epimorphism is a retraction and every monomorphism with non-empty domain is a coretraction.

5.3 Bimorphisms

5.3.1 A morphism f in $\mathscr{C}$ is called a *bimorphism* (dimorphism) if f is both a monomorphism and an epimorphism.

5.3.2 Every isomorphism is a bimorphism. The converse need not be true as is seen in *Top* where every bijective continuous map is a bimorphism. In a preordered class (as a category) every morphism is a bimorphism. Compare also 5.1.2.

5.3.3 A category is called *balanced* if every bimorphism is an isomorphism. *Ens*, *Ab*, ${}_R\mathit{Mod}$ are balanced, and so is the category of groups.

5.3.4 Every monomorphic retraction is an isomorphism, and so is every epimorphic coretraction.

Proof. From $r s = 1_B$ for r: $A \to B$ it follows that $r s r = r$. If r is a monomorphism, then $s r = 1_A$ follows.

5.4 Terminal and Initial Objects

5.4.1 An object P of the category $\mathscr{C}$ is called *terminal* (point, null) if for every object $A \in |\mathscr{C}|$ there is exactly one morphism $A \to P$.

5.4.2 Examples. In *Ens* every set consisting of one element is terminal. In *Top* it is any space of one point, in the category of groups it is every group of one element, and correspondingly in *Ab* and ${}_R Mod$. There are also terminal objects in *cat* and *Cat*. A category need not have a terminal object. If an ordered set is considered as a category, then a terminal object, if it exists, is the greatest element.

5.4.3 A terminal object is a representing object for a constant contravariant functor $\mathscr{C} \to Ens$ which assigns to all objects the same one-element set. For any two terminal objects P, P' there is exactly one morphism $P \to P'$ and it is an isomorphism.

5.4.4 A morphism whose domain is a terminal object is a coretraction.

5.4.1⁰ An object Q of the category $\mathscr{C}$ is called *initial* (copoint, conull) if it is terminal in $\mathscr{C}^0$; i. e., if in $\mathscr{C}$ there is exactly one morphism $Q \to A$ for every object A.

5.4.2⁰ Examples. In *Ens* the empty set is initial, in *Top* the empty space, in the category of groups, as well as in *Ab* und ${}_R Mod$, it is the group of one element. In *Cat* and in *cat* the empty category is initial. For an ordered set, considered as a category, an initial object is the least element. In the category of rings (with unit) $\mathbf{Z}$ is initial.

5.4.3⁰ An initial object of $\mathscr{C}$ is a representing object of the constant (covariant) functor which assigns to all objects of $\mathscr{C}$ the same set of one element.

5.4.4⁰ A morphism whose codomain is an initial object is a retraction.

5.5 Zero Objects

5.5.1 An object of a category $\mathscr{C}$ is called a *zero object* if it is both terminal and initial.

5.5.2 There is a unique isomorphism between each pair of zero objects of a category. If zero objects exist, we shall fix one and denote it by 0.

5.5.3 Examples. In the category of groups every group of one element is a zero object, and similarly in *Ab* und ${}_R Mod$. The notation is taken from *Ab*.

5.5.4 We denote the category of sets, and topological spaces respectively, with a distinguished element (base point) by Ens_*, and Top_* respectively (compare 1.2.5). Both categories have zero objects.

5.5.5 Let $\mathscr{C}$ be a category with a zero object 0. If A and B are any objects, then there is exactly one morphism $A \to B$ which factors through 0; i. e., it can be represented in the form $A \to 0 \to B$. We call it the 0-morphism and it is usual to denote it by 0 (better $0_{A,B}$). $0: A \to B$ does not depend on the choice of a zero object 0 in $\mathscr{C}$. If $0'$ is another zero object, consider $A \to 0 \to 0' \to B$.

5.5.6 If $\mathscr{C}$ is a semi-additive category, then in every set of morphisms $[A, B]$ there is a neutral element with respect to addition which is also referred to as a 0-morphism. This is consistent with 5.5.5 since both concepts coincide if $\mathscr{C}$ has a zero object.

5.5.7 In a category with a zero object the range of the contra-covariant Hom-functor and its partial functors is often understood to be Ens_*.

5.5.8 "Isomorphism", "bimorphism", and "zero object" are first examples of selfdual concepts.

5.6 Problems

5.6.1 Verify 5.1.2.

5.6.2 In the category of rings $\boldsymbol{Z} \subset \boldsymbol{Q}$ (field of the rational numbers) is a bimorphism. (Hint: For a ring consider the group of elements with (twosided) inverses). What are the facts in the category of small additive categories and additive functors, provided rings are regarded as small additive categories?

5.6.3 Carry out 5.2.4 und 5.2.5.

5.6.4 A functor $T: \mathscr{C} \to Ens$ is representable if and only if the following category has an initial object: objects are pairs (A, a) with $A \in |\mathscr{C}|$ and $a \in T(A)$, morphisms from (A, a) to (B, b) are triples (a, b, f) with f: $A \to B$ in $\mathscr{C}$ and $T(f)\,(a) = b$.

5.6.5 Let the category $\mathscr{C}$ have a zero object 0.
(a) $A \in |\mathscr{C}|$ is a zero object if and only if 1_A is a zero morphism.
(b) If $m: A \to 0$ is a monomorphism, then A is a zero object.

5.6.6 If the category $\mathscr{D}$ has a terminal object (or, resp., an initial or a zero object), then so does the category $[\mathscr{C}, \mathscr{D}]$.

6. Diagrams

We have already had occasion to consider commutative diagrams. They were all rectangles. We shall now provide the means for discussing more general diagrams. To grasp the "form" of a diagram we use the concept of a diagram scheme.

6.1 Diagram Schemes and Diagrams

6.1.1 Definition. A *diagram scheme* Σ consists of two sets Ve and Ar and two maps $o, e: Ar \to Ve$. The elements of Ve are called *vertices* and those of Ar *arrows*; for $a \in Ar$ $o(a)$ is called the *origin* and $e(a)$ the *end* of a. We say that a is an arrow from $o(a)$ to $e(a)$. Σ is finite if Ve and Ar are finite.

A diagram scheme is simply an oriented graph.

6.1.2 Examples. If $\mathscr{C}$ is a small category, then one obtains the "*underlying diagram scheme* of $\mathscr{C}$" as follows: let $Ve = |\mathscr{C}|$ and Ar be Mor $\mathscr{C}$ and for $f: A \to B$ one sets $o(f) = A$ and $e(f) = B$. Thus one disregards the composition of morphisms.

Finite diagram schemes are often represented by drawings, whereby vertices are points and arrows are just that, e. g.

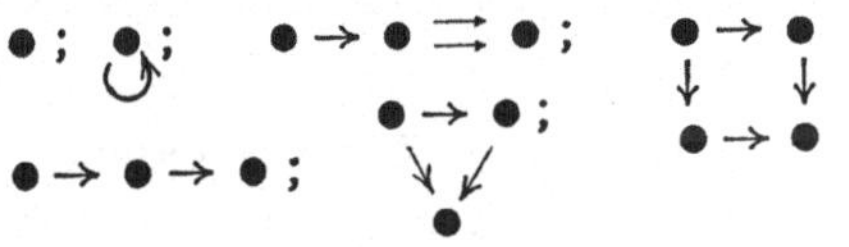

(1)

6.1.3 Definition. Let Σ be a diagram scheme and $\mathscr{C}$ a category (not necessarily small). A *diagram* D in $\mathscr{C}$ of *type* Σ is a map of Σ into $\mathscr{C}$ defined as follows: if i is a vertex of Σ, then $D(i)$ is an object of $\mathscr{C}$, and if a is an arrow of Σ with origin i_1 and end i_2, then $D(a)$ is a morphism in $\mathscr{C}$ with domain $D(i_1)$ and codomain $D(i_2)$. We write $D: \Sigma \to \mathscr{C}$.

A *natural transformation for diagrams* of type Σ in $\mathscr{C}$ is defined by transferring the definition for natural transformations of functors in the the obvious way. One obtains a category $[\Sigma, \mathscr{C}]$ which is analogous to a functor category.

If $\mathscr{B}$ is a small category and Σ the underlying diagram scheme, $[\mathscr{B}, \mathscr{C}]$ and $[\Sigma, \mathscr{C}]$ are in general different. $[\mathscr{B}, \mathscr{C}]$ can be considered as a (full) subcategory of $[\Sigma, \mathscr{C}]$.

6.1.4 A *diagram* is *finite* if it belongs to a finite diagram scheme. In this case the diagram is most often given by a drawing, as we have done for diagrams of rectangular type. Unless the given case requires it. objects and morphisms need not all or always be given specific names,

In particular, drawings as in 6.1.2 may be read as interrelations between not specifically designated objects and morphisms of a category $\mathscr{C}$.

6.1.5 It is clear how diagrams between diagram schemes are defined. This yields a category whose objects are diagram schemes and whose morphisms are diagrams.

6.2 Diagrams with Commutativity Conditions

6.2.1 A *path* w in a diagram scheme Σ is a finite sequence of arrows $a_1, a_2, \ldots, a_n$ such that $e(a_i) = o(a_{i+1})$ for $i = 1, 2, \ldots, n-1$. n (≥ 1) is called the *length* of w. For such a path we write $w = a_n a_{n-1} \ldots a_2 a_1$ and define $o(w) = o(a_1)$ as *origin* and $e(w) = e(a_n)$ as the *end* of w. A path is *closed* if its origin and end coincide.

6.2.2 There is an obvious *composition of paths*. If $w = a_n a_{n-1} \ldots a_1$ and $v = a'_m a'_{m-1} \ldots a'_1$ are two paths with $e(w) = o(v)$, then $a'_m a'_{m-1} \ldots a'_1 a_n a_{n-1} \ldots a_1$ is again a path which we denote by $v\,w$ (v following w). Obviously this composition of paths is associative, more exactly: if u, v, w are paths, and if $u\,v$ und $v\,w$ are paths, then so are $u(vw)$ and $(u\,v)\,w$, and $u(v\,w) = (u\,v)\,w$, so that parantheses are unnecessary. Every path is composed of paths of length 1 in a unique way.

If a path w is of the form $u_2 u_1$ or $u_3 u_2 u_1$ where u_1, u_2, u_3 are paths, then u_1, u_2 or u_1, u_2, u_3 resp. are *subpaths* of w; w is to be considered as a subpath of itself.

6.2.3 Every diagram $D\colon \Sigma \to \mathscr{C}$ can be extended in an obvious way to paths in Σ: if $w = a_n a_{n-1} \ldots a_1$ is a path in Σ, then one sets $D(w) = D(a_n)\,D(a_{n-1}) \ldots D(a_1)$ using the composition of morphisms in $\mathscr{C}$. The composition of paths in Σ is thus carried over to the composition of morphisms in $\mathscr{C}$.

The extension described above is compatible with natural transformations between diagrams, more precisely: If $D_1, D_2\colon \Sigma \to \mathscr{C}$ are diagrams, and if $\alpha\colon D_1 \to D_2$ is a natural transformation, and w a path in Σ, then the rectangle

$$\begin{array}{ccc} D_1(o(w)) & \xrightarrow{\alpha_{o(w)}} & D_2(o(w)) \\ {\scriptstyle D_1(w)}\downarrow & & \downarrow{\scriptstyle D_2(w)} \\ D_1(e(w)) & \xrightarrow{\alpha_{e(w)}} & D_2(e(w)) \end{array}$$

is commutative, i. e. $\alpha_{e(w)}\,D_1(w) = D_2(w)\,\alpha_{o(w)}$, as is shown by using induction with respect to the length of w.

For a diagram scheme Σ the vertices and the paths of Σ form almost a category, the just described extension of a diagram $D\colon \Sigma \to \mathscr{C}$ is almost a functor; only the identity morphisms are missing.

6.2.4 Conventions. If Σ is a diagram scheme, we construct its *trivial extension* Σ_0 by adding to every vertex i of Σ an identity arrow 1_i whose origin and end are both i itself. The set $\{1_i\}$ and the set Ar of arrows of Σ are disjoint.

The trivial extension D_0 of a diagram $D: \Sigma \to \mathscr{C}$ is constructed as follows: one defines $D_0(1_i) = 1_{D(i)}$, $D_0 \mid \Sigma = D$ and then extends D_0 to the paths in Σ_0 as in 6.2.3.

Introduction of a suitable equivalence relation for the paths in Σ_0 will now make a category out of Σ_0 with 1_i as identity morphism for all $i \in Ve$. And a diagram $D: \Sigma \to \mathscr{C}$ will thus become a functor. This is a special case of a more general fact to be discussed in 6.3.

6.2.5 Definitions. A *commutativity condition* for the diagram scheme Σ is a pair of paths (v, w) in the trivial extension Σ_0 of Σ, where v and w have the same origin and the same end, i. e. $o(v) = o(w)$, $e(v) = = e(w)$. A diagram $D: \Sigma \to \mathscr{C}$ satisfies the commutativity condition (v, w) if for the trivial extension D_0 of D, $D_0(v) = D_0(w)$ holds.

For every path v in Σ_0 there are the *trivial commutativity conditions* $(v\, 1_{o(v)}, v)$ and $(1_{e(v)}\, v, v)$. Every diagram $D: \Sigma \to \mathscr{C}$ satisfies all trivial commutativity conditions.

A diagram $D: \Sigma \to \mathscr{C}$ is called *commutative* if D satisfies all possible commutativity conditions; i. e., if for every pair (v, w) of paths in Σ_0 with $o(v) = o(w)$ and $e(v) = e(w)$, $D_0(v) = D_0(w)$.

6.2.6 Remarks. (a) For a diagram $D: \Sigma \to \mathscr{C}$ to be commutative it is sufficient to require that

(i) $D(v) = D(w)$ für every pair (v, w) of paths in Σ with $o(v) = o(w)$ and $e(v) = e(w)$,

(ii) $D(u) = 1_{D(o(u))}$ for every closed path u in Σ.

(b) The commutative rectangles considered up to now are in fact commutative diagrams because of (a).

(c) The following proposition is a typical example showing how by means of monomorphisms or epimorphisms known commutativity conditions imply other ones.

6.2.7 Proposition. *In the prismatic diagram*

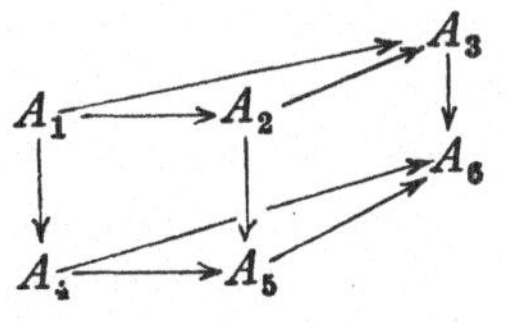

the top face and the three sides are assumed to be commutative. If $A_1 \to A_4$ is an epimorphism, then the bottom is also commutative. If the bottom and the sides are commutative, and if $A_3 \to A_6$ is a monomorphism, then the top is commutative.

Proof. Let f_j^i be the morphism $A_i \to A_j$ of the diagram. Then $f_6^5 f_5^4 f_4^1 = f_6^5 f_5^2 f_2^1 = f_6^3 f_3^2 f_2^1 = f_6^3 f_3^1 = f_6^4 f_4^1$ holds for the first claim, and hence $f_6^5 f_5^4 = f_6^4$ if f_4^1 is an epimorphism. The second claim is dual to the first one after a suitable change of indices.

6.3 Diagrams as Presentations of Functors

6.3.1 Let Σ be a diagram scheme and K a (possible empty) set of commutativity conditions for Σ. A *diagram* is said to be of *type* Σ/K if it is of type Σ and satisfies all commutativity conditions of K.

If $\mathscr{C}$ is a category, then the diagrams of type Σ/K in $\mathscr{C}$ together with their natural transformations form a category, which we denote by $[\Sigma/K, \mathscr{C}]$. It is a full subcategory of $[\Sigma, \mathscr{C}]$.

6.3.2 Proposition. *Let Σ be a diagram scheme and let K be a set of commutativity conditions for Σ. There exists a (small) category $\mathscr{P}(\Sigma/K)$, the path category belonging to Σ and K, and a diagram $\Delta\colon \Sigma \to \mathscr{P}(\Sigma/K)$ with the following universal property:*

If $\mathscr{C}$ is any category, then

(i) *If $D\colon \Sigma \to \mathscr{C}$ is a diagram of type Σ/K, then there is exactly one functor $T_D\colon \mathscr{P}(\Sigma/K) \to \mathscr{C}$ with $D = T_D \Delta$.*

(ii) *There is an isomorphism*

$$[\Sigma/K, \mathscr{C}] \xrightarrow{\approx} [\mathscr{P}(\Sigma/K), \mathscr{C}] ,$$

where the map for objects is given by the rule $D \mapsto T_D$ in (i).

Proof. The objects of $\mathscr{P}(\Sigma/K)$ are the vertices of Σ. If u_1 and u_2 are two paths in the trivial extension Σ_0 of Σ, then u_1 and u_2 are to be called equivalent (with respect to K) if there exists a finite sequence of paths $u_1 = w_0, w_1, \ldots, w_n = u_2$ in Σ_0 such that w_i is constructed from w_{i-1} (for $i = 1, 2, \ldots, n$) as follows: some subpath v_1 of w_{i-1} that belongs to a commutativity condition (v_1, v_2) or (v_2, v_1), which is in K or is a trivial one, is replaced by v_2. One verifies easily that this really defines an equivalence relation for the paths in Σ_0 and that equivalent paths have the same origin and end. Let $[i_1, i_2]_{\mathscr{P}(\Sigma/K)}$ be the set of equivalence classes of paths in Σ_0 with origin i_1 and end i_2. The composition of paths in Σ_0 induces a composition of the equivalence classes; this follows immediately from the definition. Thus one has $\mathscr{P}(\Sigma/K)$ as a category: the trivial commutativity conditions imply that the equivalence class of the identity arrow 1_i of the vertex i is the identity morphism of the object i of $\mathscr{P}(\Sigma/K)$.

$\Delta\colon \Sigma \to \mathscr{P}(\Sigma/K)$ is the identity map of the set Ve of vertices of Σ, and Δ maps every arrow a of Σ into the equivalence class of the path a (of length 1). By construction, Δ is a diagram of type Σ/K in $\mathscr{P}(\Sigma/K)$.

If $D: \Sigma \to \mathcal{C}$ is a diagram satisfying all commutativity conditions of K, then equivalent paths in Σ_0 have the same image in the trivial extension D_0 of D. (i) follows from this immediately.

If, conversely, $T: \mathcal{P}(\Sigma/K) \to \mathcal{C}$ is a functor, then $D = T\Delta$ is a diagram of type Σ/K. By 6.2.3, every natural transformation $\alpha = \{\alpha_i\}: D_1 \to D_2$ between diagrams D_1, D_2 of type Σ/K in $\mathcal{C}$ is also a natural transformation $T_{D_1} \to T_{D_2}$, and conversely, which implies (ii).

Examples and Remarks

6.3.3 The construction of $\mathcal{P}(\Sigma/K)$ contains the construction of a group from generators and relations as a special case.

The equivalence relation for paths introduced in the proof of 6.3.2 is the smallest one for which the two paths of every commutativity condition of K, as well as those of every trivial commutativity condition, are equivalent and which is also compatible with the composition of paths (congruence relation). Different sets of commutativity conditions can yield the same path category; this is so in particular for K and the *saturation* of K, which consists of all commutativity conditions satisfied by Δ.

6.3.4 Instead of $\mathcal{P}(\Sigma/\emptyset)$ we write $\mathcal{P}(\Sigma)$ and call it the *path category* of Σ. Here 6.3.2 supplies the special case mentioned at the end of 6.2.4. $\mathcal{P}(\Sigma)$ is also called the *free category* over Σ.

6.3.5 If K consists of all commutativity conditions which are possible in Σ, then there is at most one morphism $i_1 \to i_2$ for any pair (i_1, i_2) of objects of $\mathcal{P}(\Sigma/K)$. $\mathcal{P}(\Sigma/K)$ is then a preordered set.

6.3.6 Let $\mathcal{C}$ be a small category and Σ the underlying diagram scheme (6.1.2). If for K the set of all pairs $(u\,v, w)$, for which $u\,v = w$ in $\mathcal{C}$, is chosen (here $u\,v$, or, resp., w are paths of length 2, resp. 1, in Σ), then one obtains an evident isomorphism $\mathcal{P}(\Sigma/K) \xrightarrow{\approx} \mathcal{C}$. This remains true even if $\mathcal{C}$ is not small; however, then the construction of $\mathcal{P}(\Sigma/K)$ in general necessitates a change of universe.

6.3.7 If $\mathcal{C}$ is small, then a functor $\mathcal{C} \to \mathcal{D}$ can be considered as a diagram with commutativity conditions. Conversely, every diagram of type Σ/K may be considered as an abbreviated description of a functor by 6.3.2. This is how "presentation of functors" is to be understood.

If with every small category $\mathcal{C}$ the set of diagrams of type Σ/K in $\mathcal{C}$ is associated (for fixed Σ/K), then every functor $T: \mathcal{C} \to \mathcal{D}$ induces a map of these sets. In this way one obtains a functor *cat* $\to$ *Ens*. 6.3.2 says in particular, that this functor is representable. A corresponding statement is valid for *Cat* $\to$ *ENS* for arbitrary $\mathfrak{U}$-categories.

6.4 Quotients of Categories

6.4.1 Let $\mathcal{C}$ be a category. For every set of morphisms $[A, B]_{\mathcal{C}}$ let there be given an equivalence relation $\sim$ such that: if $g\,f$ is defined in $\mathcal{C}$, and if $g \sim g'$, $f \sim f'$, then $g\,f \sim g'\,f'$. Then there is a category $\mathcal{Q}$, which has the same objects as $\mathcal{C}$, and for which $[A, B]_{\mathcal{Q}}$ is the set of equivalence classes of $[A, B]_{\mathcal{C}}$. There is then a functor $P\colon \mathcal{C} \to \mathcal{Q}$ which takes every morphism of $\mathcal{C}$ into its equivalence class. One calls $\mathcal{Q}$ a *quotient* of $\mathcal{C}$ and P the corresponding *projection*.

6.4.2 A classical example of this is the transition from *Top* to the homotopy category, where the morphisms are homotopy classes of continuous maps. There is the analogous case with a base-point. One can say in general: If a functor $T\colon \mathcal{C} \to \mathcal{D}$ is injective on the classes of objects, set $f \sim f'$ if $T(f) = T(f')$. In this way one obtains a quotient of $\mathcal{C}$ through which T can be factored.

6.4.3 Proposition. *Let Σ be a diagram scheme, and let K_1 and K_2 be sets of commutativity conditions for Σ. Then $\mathcal{P}(\Sigma/K_1 \cup K_2)$ is a quotient of $\mathcal{P}(\Sigma/K_1)$.*

This follows immediately from 6.3.2.

6.4.4 Let $\mathcal{C}$ be a small category. We ussume that an equivalence relation is given on the class of objects of $\mathcal{C}$, and further a set of pairs of morphisms (f, f') is given where the domains and codomains of f and f' are equivalent. If in the diagram scheme underlying $\mathcal{C}$ the vertices in every equivalence class are identified with one vertex, a diagram scheme Σ is obtained. As the set K of commutativity conditions one takes the ones supplied by $\mathcal{C}$ as in 6.3.6, together with all pairs of identity morphisms for pairs of equivalent objects, and given pairs (f, f'). One obtains a functor $P\colon \mathcal{C} \to \mathcal{P}(\Sigma/K)$ with the following universal property:

If $T\colon \mathcal{C} \to \mathcal{D}$ is a functor, for which equivalent objects of $\mathcal{C}$ and the morphisms of every pair (f, f') have the same image in $\mathcal{D}$, then T is of the form $T = S\,P$, where S is determined uniquely by T.

We say here, too, that $\mathcal{P}(\Sigma/K)$ is a *quotient* of $\mathcal{C}$ and we call P a *projection*.

If $\mathcal{C}$ is an arbitrary category, then there is a corresponding construction by going to a higher universe $\mathfrak{V}$. The quotient which is constructed need not be a $\mathfrak{U}$-category. One example of this is to identify all objects in *Ens* with one of them. However, 6.4.1 is a special case of this construction up to an evident isomorphism.

6.4.5 Remark. Let $T\colon \mathcal{C} \to \mathcal{D}$ be a functor between arbitrary categories $\mathcal{C}$, $\mathcal{D}$. If objects, resp. morphisms, are said to be equivalent, if they have the same image under T, then T is of the form $T = S\,P$, where P and S are defined as in 6.4.4. S need not be an embedding as is shown, for instance, by the non-constant functor $\mathbf{2} \to \mathbf{Z}_2$ ($\mathbf{2}$ is defined below in 6.5.1 and $\mathbf{Z}_2$ is the group with two elements as a category).

6.5 Classes of Mono-, resp., Epimorphisms

6.5.1 The set consisting of the integers 0 and 1 in their natural order forms the category $\mathbf{2}$. It is isomorphic to $\mathscr{P}(\Sigma)$, if Σ is the diagram scheme $\bullet \to \bullet$.

If $\mathscr{C}$ is any category, then there is a bijection between the objects of $[\mathbf{2}, \mathscr{C}]$ and the morphisms of $\mathscr{C}$, and between the morphisms of $[\mathbf{2}, \mathscr{C}]$ and commutative rectangles of the form

$$\begin{array}{ccc} A_0 & \xrightarrow{f} & A_1 \\ {\scriptstyle t_0}\downarrow & & \downarrow{\scriptstyle t_1} \\ B_0 & \xrightarrow{g} & B_1 \end{array}$$

in $\mathscr{C}$, where (t_0, t_1) is a natural transformation of functors $\mathbf{2} \to \mathscr{C}$.

6.5.2 There is the functor "*domain*" Δ^0: $[\mathbf{2}, \mathscr{C}] \to \mathscr{C}$, which assigns to every object f of $[\mathbf{2}, \mathscr{C}]$ its domain and to the morphism (t_0, t_1) of $[\mathbf{2}, \mathscr{C}]$ the morphism t_0 of $\mathscr{C}$. Analogously there is the functor "*codomain*" Δ^1: $[\mathbf{2}, \mathscr{C}] \to \mathscr{C}$, where $\Delta^1(f)$ is the codomain of f.

6.5.3 Let X be a fixed object of $\mathscr{C}$. The functors $\mathbf{2} \to \mathscr{C}$, which take the number 1 into X, and the natural transformations, which take 1 into 1_X, form a subcategory of $[\mathbf{2}, \mathscr{C}]$. Thus, up to an evident isomorphism of categories, objects in this category are $\mathscr{C}$-morphisms with codomain X and morphisms are possible completions of

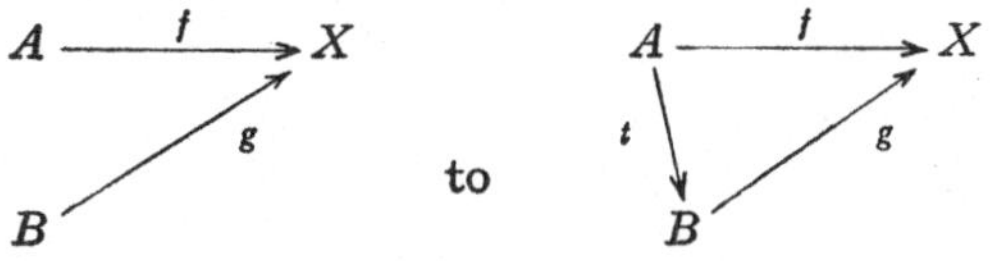

in $\mathscr{C}$, such that the triangle is commutative, i.e., $f = g\,t$.

One calls this category the category $\mathscr{C}/X$ of *objects over* X. The restriction of Δ^0: $[\mathbf{2}, \mathscr{C}] \to \mathscr{C}$ to a functor $\mathscr{C}/X \to \mathscr{C}$ is also denoted by Δ^0.

6.5.4 We specialize further by admitting as objects only monomorphisms with codomain X. If $f = g\,t$ is a monomorphism, then 5.1.5 implies that t is also a monomorphism. Further, for given f, g there is at most one t with $f = g\,t$, since g is a monomorphism. One concludes (compare 1.4.3):

The monomorphisms with codomain X form a preordered class.

By considering *Ens*, *Ab*, *Top* it is clear, that in general one can not replace monomorphisms by their domains here. There can be different monomorphisms $A \to X$.

6.5.5 There is always an equivalence relation for a preordering "$\leq$" in a class K; namely: for $a, b \in K$ one defines $a \sim b$, if $a \leq b$ and $b \leq a$. The corresponding equivalence classes need not be sets and within the same universe it then makes no sense to talk of the class of these equivalence classes. However, one can do so in a higher universe. If K is a set, then the preordering induces an order in the set of equivalence classes and hence also for every system of their representatives.

6.5.6 In the case of 6.5.4 $f \leq g$ and $g \leq f$ means, that the uniquely determined morphism t in $\mathscr{C}$ with $f = g\,t$ is an isomorphism in $\mathscr{C}$: there is an s with $g = f\,s$, and since $f = f\,s\,t$, $g = g\,t\,s$, with f and g monomorphisms, $s\,t$ and $t\,s$ are identity morphisms in $\mathscr{C}$.

6.5.7 For some special categories $\mathscr{C}$ there is a canonical choice of representatives of the classes of equivalent monomorphisms with codomain X; namely: For $\mathscr{C} = Ens$ it is the inclusion of subsets of X, for $\mathscr{C} = Ab$ it is the inclusion of subgroups, for $\mathscr{C} = Top$ one takes those morphisms whose underlying set map is an inclusion.

6.5.8 The word *subobjects* is used for the classes of equivalent monomorphisms with codomain X or also for a complete system of their representatives. It can be misunderstood in two ways: the sub"objects" are not objects in $\mathscr{C}$ (compare 6.5.4) and, as shown by the example of *Top*, not only the monomorphisms corresponding to the inclusions of subspaces are obtained. There are various suggestions for a definition of "proper" subobjects.

6.5.2⁰—6.5.7⁰ Dualizing ($\mathscr{C}$ replaced by $\mathscr{C}^0$) yields statements about morphisms with a fixed domain X (6.5.3⁰), in particular epimorphisms (6.5.4⁰). These also form a preordered class. The corresponding equivalence classes are often called *quotients* of X.

6.6 Problems

6.6.1 Determine $\mathscr{P}(\Sigma)$ for the diagram schemes in 6.1.2 and also $\mathscr{P}(\Sigma/K)$ if K is the set of all commutativity conditions.

6.6.2 Determine $\mathscr{P}(\Sigma/K)$ for all possible sets K of commutativity conditions for the following two diagram schemes

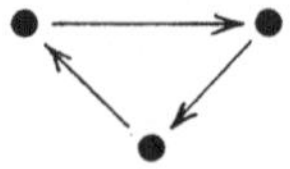

6.6.3 Let $\mathscr{C}$ be a strongly ordered set of n elements (as a category). If all objects are identified as described in 6.4.4 (and only identity morphisms), then one obtains as the quotient category a free monoid with $n-1$ generators.

6.6.4 Consider *Ens*, or *Ab*, as a small $\mathfrak{V}$-category, and let Σ be the underlying diagram scheme in $\mathfrak{V}$. $\mathscr{P}(\Sigma)$ is then not isomorphic to any $\mathfrak{U}$-category.

6.6.5 (a) A functor $T: \mathscr{C} \to \mathscr{D}$ is a monomorphism in *Cat* if and only if T is an embedding

(b) If $P: \mathscr{C} \to \mathscr{D}$ is the projection of $\mathscr{C}$ onto a quotient $\mathscr{D}$ in *Cat*, then P is an epimorphism in *Cat*.

(c) A functor $T: \mathscr{C} \to \mathscr{D}$, which is not a surjection on the classes of objects, is not an epimorphism in *Cat*. (Hint: Let $X \in |\mathscr{D}|$. By adding an object X', $\mathscr{D}$ can be extended to a category $\mathscr{D}'$ in such way that $\mathscr{D}$ is a full subcategory of $\mathscr{D}'$ and so that X' is isomorphic to X).

(d) There is a category $\mathscr{D}$ and a functor $T: \mathbf{2} \to \mathscr{D}$ such that in *Cat* T is a bimorphism, but not an isomorphism.

7. Limits

7.1 Definition of Limits

7.1.1 Let $\mathscr{C}$ be a category and Σ a diagram scheme. For $A \in |\mathscr{C}|$, let A_Σ be the constant diagram which takes all the vertices of Σ into A and all the arrows into 1_A. A morphism $f: A \to B$ in $\mathscr{C}$ induces a natural transformation $f_\Sigma: A_\Sigma \to B_\Sigma$ which assigns the morphism f to every vertex of Σ. If $T: \Sigma \to \mathscr{C}$ is an arbitrary diagram, then a natural transformation $\xi: A_\Sigma \to T$ consists of morphisms $\xi_i: A \to T(i)$ for all vertices i of Σ, such that

(1) $\xi_{e(a)} = T(a)\, \xi_{o(a)}$

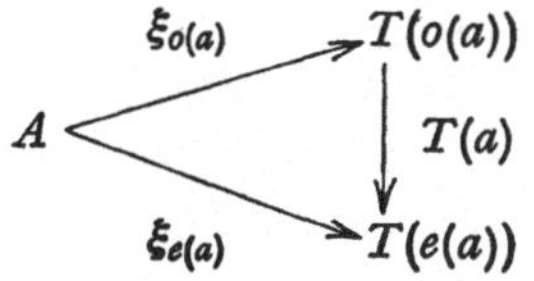

holds for all arrows a of Σ. ($o(a)$ and $e(a)$ are origin and end of a, respectively). A natural transformation $A_\Sigma \to T$ may be considered as an element $\xi = \{\xi_i\}_{i \in Ve}$ of the product set

(2) $$\prod_{i \in Ve} [A, T(i)] \quad (Ve \text{ is the set of vertices of } \Sigma)$$

that satisfies conditions (1). (1) may also be formulated as

(1′) $$[A, T(a)]\,(\xi_{o(a)}) = \xi_{e(a)}\,.$$

7.1.2 Definition. A *limit* (*projective limit, inverse limit,* left root, infimum) (L, λ) of the diagram $T: \Sigma \to \mathscr{C}$ consists of an object L of $\mathscr{C}$ and a natural transformation $\lambda: L_\Sigma \to T$ with the following property: given any natural transformation $\xi: A_\Sigma \to T$, there is exactly one morphism $f: A \to L$ such that

(3) $\qquad \xi = \lambda f_\Sigma$

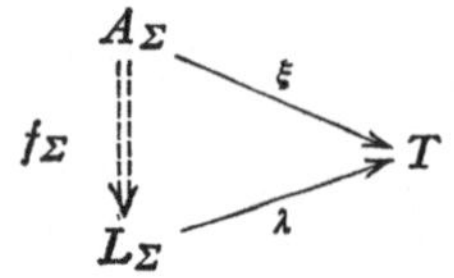

Σ may be empty. If it is, every A_Σ, as well as every T, is the empty diagram, and f_Σ, λ, ξ are the trivial identity natural transformation. A limit, if it exists, consists of a terminal object of $\mathscr{C}$ with the empty natural transformation. (1) and (1′) are vacuous and the set product in (2) is to be understood as the set $\{\emptyset\}$ consisting of the single element $\emptyset$.

7.1.3 Proposition. *A limit* (L, λ) *of* $T: \Sigma \to \mathscr{C}$ *(if it exists) represents the contravariant functor* $N_T(?) = [(?)_\Sigma, T]_{[\Sigma, \mathscr{C}]}$; *i.e., for* $A \in |\mathscr{C}|$, N_T *is the set of natural transformations* $A_\Sigma \to T$, *and for* $f: A \to B$, $N_T(f)$ *is the map* $N_T(B) \to N_T(A)$ *which is described by* $\eta \mapsto \eta f_\Sigma$. *Conversely, every representation of* N_T *produces a limit of* T.

Proof. By 4.4.2 and 4.4.6, asserting that $\varrho: [?, L] \to N_T(?)$ is an isomorphism of contravariant functors with $\varrho_L(1_L) = \lambda \in N_T(L)$ is equivalent to saying that given $\xi \in N_T(A)$, there is exactly one morphism $f: A \to L$ such that $\xi = N_T(f)(\lambda)$. Since $N_T(f)(\lambda) = \lambda f_\Sigma$, this is just what 7.1.2 says, even for Σ empty.

7.1.4 Remarks. $N_T(A)$ is that subset of the product (2) which is characterized by conditions (1). $N_T(f)$ is given by the restriction of the map

(4) $$\Pi\,[f, T(i)]: \Pi\,[B, T(i)] \to \Pi\,[A, T(i)]$$

which is described by $\{\beta_i\} \mapsto \{\beta_i f\}$.

Let T be a diagram of type Σ which satisfies a (possibly empty) set K of commutativity conditions. (In the case of A_Σ these are obviously satisfied). Then $A \mapsto A_\Sigma$, $f \mapsto f_\Sigma$ define a functor $S: \mathscr{C} \to [\Sigma/K, \mathscr{C}]$ and

(5) $$N_T(?) = [S(?), T]_{[\Sigma/K, \mathscr{C}]}$$

holds. A limit (L, λ) of T is a representation of N_T; i.e., it is the isomorphism

(6) $$\varrho: [?, L]_{\mathscr{C}} \xrightarrow{\cong} [S(?), T]_{[\Sigma/K, \mathscr{C}]}$$

which is characterized by $\varrho_L(1_L) = \lambda$. Note that N_T Op is a partial functor of the bifunctor

(7) $$[S\,\mathrm{Op}(?), ??]: \mathscr{C}^0 \times [\Sigma/K, \mathscr{C}] \to \mathit{Ens}\,.$$

7.1.5 Proposition. *If $T: \Sigma \to \mathcal{C}$ has a limit (L, λ), then L is determined by T up to an isomorphism. Given two limits (L, λ) and (M, μ) of T, there is exactly one morphism $u\ L \to M$ such that $\lambda = \mu\, u_\Sigma$. u is an isomorphism. In particular, λ is determined by T and L up to an automorphism of L.*

This follows immediately from 7.1.3 and 4.4.4.

7.1.6 If in the above discussion Σ is a small category, then T is a functor. If Σ is an arbitrary category, then a functor $T: \Sigma \to \mathcal{C}$ may perfectly well have a limit and definition 7.1.2 makes sense as stated. In this case we speak of a *large limit.* In (1), (2), 7.1.3, 7.1.4 the universe may have to be changed, but, in any case, 7.1.5 can be proved directly from 7.1.2.

7.1.7 Proposition. *Let Z be a terminal object of $\mathcal{C}$. If Σ is an arbitrary category or a diagram scheme, then the limit of Z_Σ is $(Z, \{1_Z\})$.*

In fact, Z_Σ is terminal in $[\Sigma, \mathcal{C}]$. If Z is not terminal, then no similar claim can be made in general, as will become clear in 7.3.

7.1.8 Proposition. *Let Σ be a category with an initial object Q. The functor $T: \Sigma \to \mathcal{C}$ has the limit $(T(Q), \{T(q)\})$, where q ranges over all morphisms with domain Q in Σ.*

This follows directly from the definitions.

7.1.9 Lemma. *Let $\omega: S \to T$ be a natural transformation of diagrams $S, T: \Sigma \to \mathcal{C}$ with commutativity conditions K. If $\xi: A_\Sigma \to S$ is a natural transformation and (L, λ) a limit of T, then there is exactly one morphism $f: A \to L$ such that $\omega\, \xi = \lambda\, f_\Sigma$ holds. If ω is a monomorphism in $[\Sigma/K, \mathcal{C}]$ and (A, ξ) a limit of S, then f is also a monomorphism.*

Proof. The first claim follows directly from definition 7.1.2. Then let $f\,u = f\,v$ for $u, v: B \to A$. This implies $\lambda\, f_\Sigma\, u_\Sigma = \lambda\, f_\Sigma\, v_\Sigma$, and therefore, $\omega\, \xi\, u_\Sigma = \omega\, \xi\, v_\Sigma$. If ω is a monomorphism, $\xi\, u_\Sigma = \xi\, v_\Sigma$ follows and, if (A, ξ) is a limit of S, also $u = v$.

7.1.10 Observation. In 7.1.9 ω is a monomorphism provided ω_i is a monomorphism for every vertex i of Σ. The converse need not be true.

7.1.11 Remark. For $T: \Sigma \to \mathcal{C}$, $\lambda: L_\Sigma \to T$ is called a *weak limit* if, for any $\xi: A_\Sigma \to T$, there is at least one morphism $f: A \to L$ for which $\xi = \lambda\, f_\Sigma$. 7.1.3 through 7.1.9 are not valid for weak limits. Later weak equalizers (7.2), products (7.3), pullbacks (7.8) are to be understood in this way.

7.2 Equalizers

We consider the special diagram scheme $\bullet \rightrightarrows \bullet$. A corresponding diagram T is of the form $A \overset{f}{\underset{g}{\rightrightarrows}} B$; and a natural transformation ξ:

$C_\Sigma \to T$ is determined completely by a morphism $h\colon C \to A$ with $f\,h = g\,h$. Thus 7.1.2 leads to the following:

7.2.1. Definition. Let $f, g\colon A \to B$ be two morphisms with the same domain A and the same codomain B. An *equalizer* (also *difference kernel* or even kernel) (K, k) of the pair (f, g) is a morphism $k\colon K \to A$ such that

(i) $f\,k = g\,k$,
(ii) for every morphism $v\colon Y \to A$ with $f\,v = g\,v$ there is exactly one morphism $w\colon Y \to K$ such that $v = k\,w$.

7.2.2 Proposition. *Every equalizer is a monomorphism. Every epimorphic equalizer is an isomorphism. Every coretraction is an equalizer.*

Proof. Let $k\colon K \to A$ be an equalizer of $f, g\colon A \to B$. If $k\,w_1 = k\,w_2$, then, setting $v = k\,w_1$, the first claim follows from 7.2.1 (ii). If k is an epimorphism, $f = g$ follows and 1_A is also an equalizer of f and g. According to 7.1.5, k is an isomorphism. Finally, let $s\colon K \to A$ be a coretraction and t a corresponding retraction. One confirms easily that s is an equalizer of 1_A and $s\,t$.

7.2.3 Definition. The category $\mathscr{C}$ is said *to have equalizers* if every pair of morphisms with the same domain and codomain has an equalizer.

7.2.4 By 7.1.5, equalizers are only determined up to being preceded by an isomorphism. Often, however, a canonical choice presents itself, as is the case in the following examples of categories with equalizers. We indicate the chosen equalizers.

An equalizer for $A \underset{g}{\overset{f}{\rightrightarrows}} B$ in *Ens* is the coincidence set $\{a \mid f(a) = g(a)\}$ with its inclusion in A. Every subset appears as an equalizer. In *Ab* one obtains as equalizers all subgroups, in ${}_R\mathit{Mod}$ all submodules, and in the category of groups all subgroups with their inclusions (see 7.9.2). In *Top* the same holds for subspaces and in the category of Hausdorff spaces for closed subspaces.

There are equalizers in *cat* and in *Cat*. For functors $F, G\colon \mathscr{A} \to \mathscr{B}$ one obtains as an equalizer the inclusion of the subcategory of $\mathscr{A}$ in which F and G coincide. It may be empty.

The example *Top* shows that the converse of 7.2.2 is not true (compare 5.1.2). Equalizers represent a possibility, compared with 6.5.8, of defining "better subobjects".

7.2.5 Definition. We assume that the category $\mathscr{C}$ has o-morphisms (compare 5.5.5, 5.5.6). A *kernel* of $f\colon A \to B$ is an equalizer of f and $o\colon A \to B$. $\mathscr{C}$ has kernels if every morphism has a kernel.

7.2.6 Proposition. *In an additive category, $k: K \to A$ is an equalizer of $f, g: A \to B$ if and only if k is a kernel of $f - g$. An additive category has equalizers if and only if it has kernels.*

This follows immediately from the fact that $f v = g v$ is equivalent to $(f - g) v = 0$.

7.2.7 A counterexample. In the category of groups the (canonically chosen) kernels are inclusions of normal subgroups, whereas all inclusions of subgroups are equalizers. Thus, the existence of a zero object does not imply the coincidence of equalizers and kernels.

7.3 Products

Products are limits of diagrams whose diagram scheme is discrete, i.e. without arrows. Such a diagram is just a family of objects whose index set is the diagram scheme.

7.3.1 Definition. Let $\{A_i\}_{i \in I}$ be a family of objects of the category $\mathscr{C}$. A *product* of this family is an object X with morphisms $pr_i: X \to A_i$ such that for any family $\{f_i: Y \to A_i\}_{i \in I}$ there is exactly one morphism $f: Y \to X$ with $pr_i f = f_i$ for all i.

Most often the object X is denoted by $\prod_{i \in I} A_i$ (or $\prod A_i$ if the context excludes any confusion), and pr_i is called the i-th *projection* of the product. A *finite product* is a product whose index set I is finite. I may be empty; we then have a terminal object in $\mathscr{C}$. The category $\mathscr{C}$ has products, or, resp., finite products, if every family of objects in $\mathscr{C}$ whose indices form a set (resp. a finite set) has a product.

7.3.2 Examples. For *Ens* the usual products are also products according to 7.3.1. Note, however, that the projections must be included as part of the characterizing data. Also, there is here a canonical choice amongst the products (compare 7.1.5 and 7.2.4) as is the case in the following examples, where the projections are evident. For *Top* products are topological products. For *Ab*, ${}_R\mathit{Mod}$, and the category of groups products are the usual "direct products". There are corresponding facts for models of other algebraic structures for which the algebraic operations are defined "coordinatewise" on the product set, e.g. rings. *cat* and *Cat* have products: Let $\{\mathscr{C}_i\}_{i \in I}$ be a family of categories. Objects of $\prod \mathscr{C}_i$ are families $\{A_i \mid A_i \in |\mathscr{C}_i|\}_{i \in I}$ of objects. The set of morphisms from $\{A_i\}$ to $\{B_i\}$ is the product set $\prod [A_i, B_i]_{\mathscr{C}_i}$. If all $\mathscr{C}_i$ are small, then $\prod \operatorname{Mor} \mathscr{C}_i$ is a set and $\prod \operatorname{Mor} \mathscr{C}_i = \operatorname{Mor} \prod \mathscr{C}_i$ holds.

7.3.3 Proposition. *Let $\{pr_i: X \to A_i\}$ and $\{q_i: Y \to B_i\}$ be products in $\mathscr{C}$ belonging to the same index set I. If for every $i \in I$ a mor-*

phism $f_i\colon A_i \to B_i$ *is given, then there is exactly one morphism* $f\colon X \to Y$ *with* $q_i f = f_i pr_i$ *for all* i. *One writes* $f = \prod f_i$. *If all the* f_i *are monomorphisms, then so is* f.

This is an often used special case of 7.1.9.

7.3.4 Proposition. *Let* $\{pr_i\colon X \to A_i\}_{i \in I}$ *be a product in* $\mathscr{C}$, *and let* $k \in I$. *If for every* $i \in I$ *there is a morphism* $f_i\colon A_k \to A_i$, *then* $pr_k\colon X \to A_k$ *is a retraction.*

Proof. Define $g\colon A_k \to X$ by $pr_i g = f_i$ for $i \neq k$ and by $pr_k g = 1_{A_k}$.

The assumption of this proposition is satisfied in particular if $\mathscr{C}$ possesses 0-morphisms, but also in *Ens*, *Top*, and *cat* provided no A_i is empty.

7.3.5 We assume that the category $\mathscr{C}$ has a terminal object Z. Then, for $i \neq k$ let $A_i = Z$ and let A_k be a given object A. If one chooses for pr_i the only morphism from A to Z and for pr_k 1_A, then A is a product of the family. If the product of the family is determined differently (compare 7.3.2), then pr_k is, in any case, an isomorphism.

7.3.6 Comparison of 7.3.1 and 7.1.4 shows that there is an isomorphism

$$\theta\colon\ [?, \prod A_i]_{\mathscr{C}} \xrightarrow{\cong} \prod\, [?, A_i]_{\mathscr{C}}\,,$$

where on the right there are products of sets, or set maps respectively, as in 7.1.4 (4).

7.4 Complete Categories

7.4.1 Definition. A category $\mathscr{C}$ is called *complete* (also left complete), or *finitely complete*, if every diagram, or every finite diagram resp., has a limit in $\mathscr{C}$.

By *finite limits* we mean limits of finite diagrams, including the empty one.

Hence a finitely complete category has a terminal object and is therefore not empty.

7.4.2 Theorem. *A category* $\mathscr{C}$ *is complete, or finitely complete, if and only if it possesses equalizers and products, or, resp., finite products.*

Proof. Equalizers and products are special cases of limits. Now let us assume that they exist in $\mathscr{C}$ and that $T\colon \Sigma \to \mathscr{C}$ is any diagram. Let Ve and Ar be the sets of vertices and of arrows of Σ. By assumption, there are products

(1) $$X = \prod_{i \in Ve} T(i) \quad \text{with projections } pr_i\colon\ X \to T(i)\,,$$

(2) $$Y = \prod_{a \in Ar} T(e(a)) \quad \text{with projections } q_a\colon\ Y \to T(e(a))\,,$$

($o(a)$ and $e(a)$ are the origin and end of a). We consider the two morphisms $v, w: X \to Y$ defined by

(3) $$q_a v = pr_{e(a)}: X \to T(e(a)) ,$$

(4) $$q_a w = T(a)\, pr_{o(a)}: X \to T(o(a)) \to T(e(a)) .$$

A natural transformation $\{\xi_i\}: A_\Sigma \to T$ defines a morphism $\xi: A \to X$ with $pr_i \xi = \xi_i$ which, because of (1) and (2), satisfies

(5) $$v\,\xi = w\,\xi: A \to X \to Y .$$

For,

(6) $$pr_{e(a)}\,\xi = T(a)\, pr_{o(a)}\,\xi , \quad \text{i.e.,}$$
$$\xi_{e(a)} = T(a)\,\xi_{o(a)}: A \to T(e(a)) \quad \text{for all} \quad a \in Ar ,$$

says precisely, that $\{\xi_i\}: A_\Sigma \to T$ is a natural transformation. (Compare this with 7.1 (1), (2) and 7.3.6). Thus it follows that if $k: L \to X$ is an equalizer of v and w, then (L, λ) with $\lambda = \{pr_i\, k\}$ is a limit of T. If Σ is finite, then X and Y are finite products. This concludes the proof of the theorem.

This proof together with 7.2.4 and 7.3.2 yields the following

7.4.3 Corollary. *The following categories are complete and there is for them a canonical choice of limits*: *Ens*, *Ens*$_*$, *Top*, *Top*$_*$, *Ab*, $_R$*Mod*, *Mod*$_R$, *cat*, *Cat*, *the category of groups.*

The same is true for the category of rings. One easily finds examples of complete subcategories of *Top* and *Ab*. Examples of finitely complete categories are the category of finite sets, of finite groups, of finitely generated abelian groups, of finite dimensional vector spaces over a field, of Lie groups, or of finite dimensional Lie algebras.

Note that in *Cat* only diagram schemes of the universe $\mathfrak{U}$ are admissible. Instead of complete one should here say $\mathfrak{U}$-complete, to be more precise. Other notions of completeness may be introduced by restricting the cardinality of the vertex and arrow sets of the diagrams, e.g. countably complete etc.

7.4.4 In the examples mentioned above, the proof in 7.4.2 also furnishes a description of the limits, e.g. in *Ens* as a subset L of a product (together with the projections restricted to L): $k: L \to \Pi\, T(i)$ is here an inclusion and L consists of those elements $\{x_i\}$ of $\Pi\, T(i)$ for which by (6)

(7) $$x_{e(a)} = T(a)\,(x_{o(a)}) \quad \text{for all} \quad a \in Ar$$

is valid. For *Top*, *Ab*, $_R$*Mod* etc. a corresponding description is valid; similarly for *cat* and *Cat*, if $T(i)$ is understood to be the set resp. the class of all morphisms of the category $T(i)$ and if $T(a)$ is the corresponding map of the morphism classes as in 2.2.7.

For categories with a canonical choice of limits, such a limit (L, λ) is usually designated just by the object L, assuming that there is no doubt about the proper λ.

7.4.5 The proof of 7.4.2 shows:

If $\mathcal{C}$ is a complete (finitely complete) category and $S: \mathcal{C} \to \mathcal{D}$ a functor which preserves products (finite products) and equalizers, then S preserves all (all finite) limits, i.e., carries (finite) limits into limits. If in $\mathcal{C}$ and $\mathcal{D}$ there is a canonical choice of limits, then a corresponding statement is true.

The latter applies, e.g., to the forgetful functors ${}_R Mod \to Ab$, $Ab \to Ens_*$, $Ens_* \to Ens$, $Top \to Ens$ etc. Also for $\mathfrak{U}$-diagrams to the inclusions $Ab \to AB$, $Ens \to ENS$, $cat \to Cat$ etc..

7.5 Limits in Functor Categories

7.5.1 Remark. As the proof of 6.3.2 shows, every diagram $T: \Sigma \to \mathcal{C}$ of the type Σ/K can be considered as a functor $\mathcal{P}(\Sigma/K) \to \mathcal{C}$. Then the definition of limits 7.1.2 and 6.3.2 imply immediately that this does not result in any change in the limits (compare also 7.1.4). However, the construction of limits in the proof of 7.4.2 is in general simpler for diagrams than for the associated functors. In particular, $\mathcal{P}(\Sigma/K)$ need not be finite if Σ is finite.

If the use of functors allows simpler formulations, we replace diagrams by their associated functors. By a finite limit one then has to mean a limit whose diagram is finite.

7.5.2 Theorem. *Let $\mathcal{D}$ be a category and $\mathcal{C}$ a complete (finitely complete) category. The functor category $[\mathcal{D}, \mathcal{C}]$ is then complete (finitely complete). If there is a canonical choice of limits in $\mathcal{C}$, then there is a canonical choice for $[\mathcal{D}, \mathcal{C}]$.*

Completeness is to be understood with regard to the fixed universe $\mathfrak{U}$, even if $\mathcal{D}$ is not small.

Proof. We may assume that $\mathcal{D} \neq \emptyset$. If Z is terminal in $\mathcal{C}$, then $Z_{\mathcal{D}}$ is terminal in $[\mathcal{D}, \mathcal{C}]$ (compare 7.1.7). Now let $T_0: \Sigma \to [\mathcal{D}, \mathcal{C}]$ be a nonempty diagram (finite diagram). By 7.5.1, instead of T_0 we may consider a functor $T': \mathcal{Y} \to [\mathcal{D}, \mathcal{C}]$, where $\mathcal{Y} = \mathcal{P}(\Sigma/K)$ for some set K of commutativity conditions.

If X is an object and $p: X \to X'$ a morphism of $\mathcal{Y}$, then $T'(X)$ is a functor $\mathcal{D} \to \mathcal{C}$ and $T'(p): T'(X) \to T'(X')$ is a natural transformation. By 3.4.4 or 3.6.3, there is a bifunctor $T: \mathcal{Y} \times \mathcal{D} \to \mathcal{C}$ corresponding to T', and corresponding to $A \in |\mathcal{D}|$ there is the partial functor $T_A: \mathcal{Y} \to \mathcal{C}$. One has

$$T_A(X) = T'(X)(A) \quad \text{and} \quad T_A(p) = (T'(p))_A;$$

i.e., one considers T' "at the point $A \in |\mathcal{D}|$". Furthermore, $f: A \to B$ in $\mathcal{D}$ produces a natural transformation $T_f: T_A \to T_B$. If $F': \mathcal{D} \to \mathcal{C}$ is any functor, then a bifunctor $F: \mathcal{Y} \times \mathcal{D} \to \mathcal{C}$ is associated with the constant functor $F'_{\mathcal{Y}}: \mathcal{Y} \to [\mathcal{D}, \mathcal{C}]$, for which $F_A = F(?, A) = (F'(A))_{\mathcal{Y}}$; i.e., its partial functors with respect to the objects of $\mathcal{D}$ are constant and $F(?, f) = (F'(f))_{\mathcal{Y}}$.

A natural transformation $\alpha': F'_{\mathcal{Y}} \to T'$ produces a natural transformation $\alpha: F \to T$ of bifunctors. At the point A, α induces the natural transformation $\alpha_A: F'(A)_{\mathcal{Y}} \to T_A$ of functors $\mathcal{Y} \to \mathcal{C}$ and, by 3.4.4 (3),

(1) $$T_f \alpha_A = \alpha_B F'(f)_{\mathcal{Y}}: F'(A)_{\mathcal{Y}} \to T_B .$$

Now, for every $A \in |\mathcal{D}|$, let a limit (L_A, λ_A) of T_A be chosen, (if possible in $\mathcal{C}$, make a canonical choice). Then for every A there is a uniquely determined morphism $u_A: F'(A) \to L_A$ with

(2) $$\alpha_A = \lambda_A (u_A)_{\mathcal{Y}} .$$

Corresponding to $T_f: T_A \to T_B$ there is, by 7.1.9, a uniquely determined morphism $L_f: L_A \to L_B$ such that

(3) $$T_f \lambda_A = \lambda_B (L_f)_{\mathcal{Y}} .$$

From this it follows that the rule $A \mapsto L_A$, $f \mapsto L_f$ defines a functor $L: \mathcal{D} \to \mathcal{C}$ and that there is a natural transformation $\lambda: L_{\mathcal{Y}} \to T'$ which at the point A is determined by λ_A. The theorem now follows from (2), if we show additionally that $\{u_A: F'(A) \to L_A\}$ is a natural transformation $u: F' \to L$. But we have

$$\lambda_B (L_f)_{\mathcal{Y}} (u_A)_{\mathcal{Y}} \overset{(3)}{=} T_f \lambda_A (u_A)_{\mathcal{Y}} \overset{(2)}{=} T_f \alpha_A$$
$$\overset{(1)}{=} \alpha_B F'(f)_{\mathcal{Y}} \overset{(2)}{=} \lambda_B (u_B)_{\mathcal{Y}} F'(f)_{\mathcal{Y}} .$$

Since (L_B, λ_B) is a limit of T_B, it follows that $L_f u_A = u_B F'(f): F'(A) \to$ $\to L_B$, which completes the proof.

7.5.3 Remarks. Taking 7.5.1 into account, what we have actually proved is the following:

If diagrams in $\mathcal{C}$ of a given type Σ/K always have a limit, then the diagrams in $[\mathcal{D}, \mathcal{C}]$ of this type also always have limits and they are constructed "pointwise".

This is also the case for an empty Σ. For, let $L: \mathcal{D} \to \mathcal{C}$ be a functor which assigns to every object of $\mathcal{D}$ a terminal object in $\mathcal{C}$, then L is isomorphic to $Z_{\mathcal{D}}$, where Z is a terminal object of $\mathcal{C}$.

If there is no canonical choice of limits in $\mathcal{C}$, then 7.5.2 is only valid provided that the axiom of choice is assumed for the universe $\mathfrak{U}$ as a $\mathfrak{V}$-set. The limits for $T_A: \mathcal{Y} \to \mathcal{C}$ do not in general form $\mathfrak{U}$-sets.

If for every diagram $\Sigma \to \mathcal{C}$, that satisfies the commutativity conditions K, a limit is chosen, then by 7.1.9 a functor $\mathrm{Lim}\colon [\Sigma/K, \mathcal{C}] \to \mathcal{C}$ is produced and with the functor S of 7.1.4 there is an isomorphism

$$(4) \qquad \varrho\colon [?, \mathrm{Lim}\,(??)]_{\mathcal{C}} \overset{\cong}{\Rightarrow} [S(?), ??]_{[\Sigma/K, \mathcal{C}]}$$

of contra-co-variant functors. The natural transformation $\mathrm{Lim}(T)_\Sigma \to \to T$ belonging to the limit object $\mathrm{Lim}\,(T)$ is $\varrho\,(1_{\mathrm{Lim}\,(T)})$.

With the isomorphism 3.4 (6) one obtains a functor

$$[\Sigma/K, [\mathcal{D}, \mathcal{C}]] \overset{\cong}{\Rightarrow} [\mathcal{D}, [\Sigma/K, \mathcal{C}]] \xrightarrow{[\mathcal{D}, \mathrm{Lim}]} [\mathcal{D}, \mathcal{C}]\,.$$

The proof of 7.5.2 shows that T_0 goes into its limit object L. Another formulation of the "pointwise" construction in 7.5.2 is the following:

7.5.4 If $\mathcal{C}$ is (finitely) complete and $\mathcal{D}$ not empty, then, for all $A \in |\mathcal{D}|$, the evalution functor $E_A\colon [\mathcal{D}, \mathcal{C}] \mapsto \mathcal{C}$, which is described by the rule $F \mapsto F(A)$, $\alpha \mapsto \alpha_A$, preserves (finite) limits.

7.5.5. 7.5.2 through 7.5.4 are correspondingly valid for $Add(\mathcal{D}, \mathcal{C})$ if $\mathcal{C}$ and $\mathcal{D}$ are additive categories. The proof of 7.5.2 carries over if one additionally shows that L is an additive functor. But this follows if one looks at 7.5.2 (3) for an arbitrary object X of $\mathcal{Y}$. For $f, g\colon A \to B$ one gets

$$(3\,\mathrm{a}) \qquad T_{f+g}(X)\,\lambda_{A,X} = \lambda_{B,X}\,L_{f+g}\,.$$

The additivity of L thus follows from that of $T'(X)\colon \mathcal{D} \to \mathcal{C}$. The proof of 7.5.2 even shows much more: the embedding $Add(\mathcal{D}, \mathcal{C}) \to [\mathcal{D}, \mathcal{C}]$ preserves (finite) limits if $\mathcal{C}$ is (finitely) complete.

Limits of additive functors are additive.

7.6 Double Limits

If in 7.6.2 $\mathcal{C}$ is complete and $\mathcal{D}$ a small category, then the functor $L\colon \mathcal{D} \to \mathcal{C}$, given there as a limit object of $T_0\colon \Sigma \to [\mathcal{D}, \mathcal{C}]$, has itself a limit (M, μ). 7.5.3 suggests that this gives a limit of the bifunctor $T\colon \mathcal{P}(\Sigma/K) \times \mathcal{D} \to \mathcal{C}$ which is associated with $T_0\colon \Sigma \to [\mathcal{D}, \mathcal{C}]$. This is in fact true. We start with an auxiliary observation.

7.6.1 Lemma. *Let Σ' be a subscheme of the diagram scheme Σ (i.e., vertices and arrows of Σ' are also vertices and arrows of Σ). A diagram $T\colon \Sigma \to \mathcal{C}$ determines a subdiagram $T' = T|\Sigma'$. If (L', λ') is a limit of T' and $\xi\colon A_\Sigma \to T$ is a natural transformation, then there is exactly one morphism $f\colon A \to L'$ such that $\xi_i = \lambda'_i\, f$ for all vertices i of Σ'.*

This follows immediately from 7.1.2 because by restriction ξ becomes a natural transformation $\xi'\colon A_{\Sigma'} \to T'$.

7.6.2 Proposition. *Let $\mathcal{X}$ and $\mathcal{Y}$ be small non-empty categories and $T\colon \mathcal{X} \times \mathcal{Y} \to \mathcal{C}$ a functor. For every $U \in |\mathcal{X}|$ let $(L(U), \lambda(U))$ be a limit of the partial functor $T(U, ?)\colon \mathcal{Y} \to \mathcal{C}$. For $w\colon U \to V$ in $\mathcal{X}$ there is* *exactly one morphism* $L(w)\colon L(U) \to L(V)$ *in* $\mathscr{C}$ *with* $T(w, ?)\,\lambda(U) = \lambda(V)\,\big(L(w)\big)_{\mathscr{Y}}$. *The rule* $U \mapsto L(U)$, $w \mapsto L(w)$ *defines a functor* $L\colon \mathscr{X} \to \mathscr{C}$. *T has a limit if and only if L has a limit. If* (M, μ) *is a limit of L, then*

$$(M, \{\big(\lambda(U)\big)_Z\,\mu_U\}_{(U,Z)\in|\mathscr{X}|\times|\mathscr{Y}|})$$

is a limit of T, and every limit of T can be represented in this way.

Proof. The proof that L is a functor is as in 7.5.2. It also follows that $\{\big(\lambda(U)\big)_Z\,\mu_U\}$ is a natural transformation $M_{\mathscr{X}\times\mathscr{Y}} \to T$, provided $\mu\colon M_{\mathscr{X}} \to L$ is a natural transformation. If, conversely, $\xi\colon A_{\mathscr{X}\times\mathscr{Y}} \to T$ is a natural transformation, then, by 7.6.1, for every $U \in |\mathscr{X}|$ there is exactly one morphism $f_U\colon A \to L(U)$ with $\xi_{(U,Z)} = \big(\lambda(U)\big)_Z\,f_U$, and $\{f_U\}\colon A_{\mathscr{X}} \to L$ is a natural transformation (compare 7.1.9). Then, by the definition of limits, 7.1.2, the proof is complete.

7.6.3 Applying 7.6.2 twice an interchange of the limits of diagrams of fixed types $\mathscr{X}$ and $\mathscr{Y}$ is obtained, provided every diagram of type $\mathscr{X}$ and of type $\mathscr{Y}$ has a limit in $\mathscr{C}$. Because of 7.1.2 and 7.1.7 this is also true if $\mathscr{X}$ or $\mathscr{Y}$ is empty.

Limits commute with limits.

We shall make use of this often.

7.6.4 Proposition. *Let* $\mathscr{B}$ *be an arbitrary and* $\mathscr{C}$ *a complete category. If* $\mathscr{L}[\mathscr{B}, \mathscr{C}]$ *is the full subcategory of* $[\mathscr{B}, \mathscr{C}]$ *whose objects are those functors which preserve the limits that exist in* $\mathscr{B}$, *then* $\mathscr{L}[\mathscr{B}, \mathscr{C}]$ *is complete. Limits are formed as in* $[\mathscr{B}, \mathscr{C}]$; *in particular they are preserved by the inclusion* $\mathscr{L}[\mathscr{B}, \mathscr{C}] \subset [\mathscr{B}, \mathscr{C}]$.

With respect to limits $\mathscr{L}[\mathscr{B}, \mathscr{C}]$ *is closed in* $[\mathscr{B}, \mathscr{C}]$.

Proof. For $\mathscr{B} = \emptyset$ the claim is trivial, so we assume $\mathscr{B} \neq \emptyset$. If $R\colon \mathscr{X} \to \mathscr{L}[\mathscr{B}, \mathscr{C}]$ is a diagram, then R must have a limit (M, μ) in $[\mathscr{B}, \mathscr{C}]$. It has to be shown that $M\colon \mathscr{B} \to \mathscr{C}$ preserves limits. So let $S\colon \mathscr{Y} \to \mathscr{B}$ be a diagram with a limit (N, ν) in $\mathscr{B}$.

7.5.1 allows us to assume that $\mathscr{X}$ and $\mathscr{Y}$ are small categories. Then there is a bifunctor $T\colon \mathscr{X} \times \mathscr{Y} \to \mathscr{C}$, obtained from $R \times S\colon \mathscr{X} \times \mathscr{Y} \to [\mathscr{B}, \mathscr{C}] \times \mathscr{B}$ by applying the evaluation functor of 3.7. We first assume that $\mathscr{X}$ and $\mathscr{Y}$ are not empty. For $X \in |\mathscr{X}|$, $T(X, ?) = R(X)$ preserves limits and M is constructed "pointwise". By 7.6.2, one gets $M(N)$ and an associated natural transformation as a limit of T by first constructing the limits of the partial functors $T(X, ?)$ with respect to S and then the limit with respect to R. One obtains the same by first constructing the limits with respect to R, thus getting $M\,S$, and then the limit with respect to S. By 7.6.2, $(M(N), M\,\nu)$ with $M\,\nu = \{M(\nu_Y)\}$ and $Y \in |\mathscr{Y}|$ is a limit of $M\,S$.

The cases where $\mathscr{X}$ or $\mathscr{Y}$ are empty follow similarly, making use of 7.1.7. One has to take into account that isomorphic functors have isomorphic limits.

7.6.5 Additional remarks. 7.6.4 can be generalized. Let $\mathfrak{K}$ be a class of diagram schemes with commutativity conditions and $\mathfrak{K}[\mathscr{B}, \mathscr{C}]$ the full subcategory of $[\mathscr{B}, \mathscr{C}]$ whose objects are the functors $\mathscr{B} \to \mathscr{C}$ which preserve limits (those existing in $\mathscr{B}$) for all diagrams whose type is in $\mathfrak{K}$. We mention in particular the full subcategories $l[\mathscr{B}, \mathscr{C}]$, ${}_\pi[\mathscr{B}, \mathscr{C}]$, ${}_\Pi[\mathscr{B}, \mathscr{C}]$ whose objects are the functors which preserve finite limits existing in $\mathscr{B}$, or, resp., finite products, resp., products. Every such category $\mathfrak{K}[\mathscr{B}, \mathscr{C}]$ is complete and closed with respect to limits in $[\mathscr{B}, \mathscr{C}]$, provided $\mathscr{C}$ is complete.

A further generalization results if $\mathscr{C}$ is only presumed to have limits for diagrams belonging to some class $\mathfrak{K}'$ of diagram schemes with commutativity conditions. We mention in particular: if $\mathscr{C}$ is finitely complete, then $l[\mathscr{B}, \mathscr{C}]$ is finitely complete and closed with respect to finite limits.

The proof of 7.6.4 works in these cases too, if only $\mathscr{X}$ or, resp., $\mathscr{X}$ and $\mathscr{Y}$ are chosen appropriately.

Since limits of additive functors are additive (7.5.5), corresponding assertions can be made for the corresponding full subcategories of $Add(\mathscr{B}, \mathscr{C})$.

7.7 Criteria for Limits

Let again $\mathscr{C}$ be any category and $T\colon \Sigma \to \mathscr{C}$ a diagram. For $A \in |\mathscr{C}|$, $H^A T$ is a diagram in *Ens*. It has a limit, since *Ens* is complete. This limit can be described as in 7.4.4, where $T(i)$ and $T(a)$ have to be replaced by $[A, T(i)]_\mathscr{C}$ and $[A, T(a)]_\mathscr{C}$. Comparison with 7.1.3, 7.1.4, and 7.1(1') yields:

7.7.1 Proposition. *The limit of $H^A T$ is the set $N_T(A)$ of natural transformations $A_\Sigma \to T$ with the evident maps q_i: $N_T(A) \to [A, T(i)]$ given by the rule $\xi \mapsto \xi_i$ (assuming Σ to be non-empty).*

7.7.2 H^A takes a natural transformation $\omega = \{\omega_i\}\colon D_\Sigma \to T$ into a natural transformation $H^A\omega = \{[A, \omega_i]\}$. For $g \in [A, D]$, $[A, \omega_i]\,(g) = \omega_i\, g$. Therefore, $H^A\omega$ produces a map

(1) $$\omega^A\colon [A, D] \to N_T(A) \quad \text{with} \quad \omega^A(g) = \omega\, g_\Sigma\,,$$

(2) $$q_i\, \omega^A = [A, \omega_i] = H^A(\omega_i)\,.$$

If Σ is empty, then (1) is a map into a set of one element and (2) is void.

A comparison of (1), (2) with 7.7.1 and 7.1.2 shows that ω^A yields precisely the unique factorization of $H^A\,\omega$ through the limit $N_T(A)$ of $H^A\,T$. For $f\colon A \to B$ and ω^B of (1), the diagram

$$(3)\qquad \begin{array}{ccc} [B,D] & \xrightarrow{\;\omega^B\;} & N_T(B) \\ \Big\downarrow{\scriptstyle [f,D]} & & \Big\downarrow{\scriptstyle N_T(f)} \\ [A,D] & \xrightarrow{\;\omega^A\;} & N_T(A) \end{array} \qquad\qquad \begin{array}{ccc} h & \mapsto & \omega\, h_\Sigma \\ \downarrow & & \downarrow \\ h\,f & \mapsto & \omega(h\,f)_\Sigma \end{array}$$

is commutative. We get:

For $? \in |\mathscr{C}|$, $\omega^?$ is a natural transformation $H_D \to N_T$; namely, the one for which $\omega^D(1_D) = \omega \in N_T(D)$.

7.7.3 Theorem. *Let $T\colon \Sigma \to \mathscr{C}$ be a diagram, $\mathscr{C} \neq \emptyset$, and $\lambda\colon L_\Sigma \to T$ a natural transformation. (L, λ) is a limit of T if and only if for all $A \in |\mathscr{C}|$ $(H^A(L), H^A\,\lambda)$ is a (not necessarily canonically chosen) limit of $H^A T$.*

Proof. If in 7.7.2 (D, ω) is replaced by (L, λ) it follows from (1) and (2) that λ^A is an isomorphism if and only if $(H^A(L), H^A\,\lambda)$ is a limit of $H^A T$. A comparison of 7.7.2 and 7.1.3 then completes the proof.

7.7.4 Corollary. *Every representable (covariant) functor $F\colon \mathscr{C} \to \mathit{Ens}$ takes limits (those existing in $\mathscr{C}$) into limits (not necessarily canonically chosen ones).*

Proof. Let $\varrho\colon H^A \to F$ be an isomorphism and (L, λ) a limit of $T\colon \Sigma \to \mathscr{C}$. Then $(H^A(L), \{H^A(\lambda_i)\})$ is a limit of $H^A T$. Since ϱ is a natural transformation and an isomorphism at every point $B \in |\mathscr{C}|$, there is an isomorphism $\varrho\, T\colon\ H^A T \to F\,T$ of diagrams and there are isomorphisms for the limits, so that $(F(L), \{F(\lambda_i)\})$ is obtained as a limit of $F\,T$.

7.7.5 Corollary. *A functor $S\colon \mathscr{C} \to \mathscr{D}$ preserves limits if and only if $H^X S$ preserves limits for all $X \in |\mathscr{D}|$.*

Proof. If S preserves limits, then so does $H^X S$ by 7.7.4. For the converse, let (L, λ) be a limit of $T\colon \Sigma \to \mathscr{C}$. Then $(H^X S(L),\ H^X S\,\lambda)$ is a limit of $H^X S\ T$ by assumption, and $(S(L),\ S\,\lambda)$ is a limit of $S\ T$ by 7.7.3.

7.7.6 Theorem. *Let $S\colon \mathscr{C} \to \mathscr{D}$ be a fully faithful functor and $T\colon \Sigma \to \mathscr{C}$ a diagram. Then let $\lambda\colon L_\Sigma \to T$ be a natural transformation. If $S\,\lambda\colon S(L)_\Sigma \to S\ T$ is a limit, then $(L\ \lambda)$ is a limit of T.*

"Fully faithful functors reflect limits".

Proof. A natural transformation $\xi\colon A_\Sigma \to T$ is taken into $S\ \xi\colon S(A)_\Sigma \to \to S\ T$ by S. If $(S(L), S\,\lambda)$ is a limit of $S\ T$, then there is exactly one $u\colon S(A) \to S(L)$ such that $S\,\xi = (S\,\lambda)\, u_\Sigma$. Since S is fully faithful, there is exactly one $f\colon A \to L$ with $S(f) = u$; and $\xi = \lambda\, f_\Sigma$.

7.7.7 Remarks. 7.7.3 allows us to refer relations between limits in $\mathscr{C}$ back to those in *Ens*, provided the limits exist in $\mathscr{C}$. As an example, the fact, that the formation of products (finite products) is associative and commutative up to isomorphisms, carries over from *Ens* to arbitrary complete (finitely complete) categories. It would also have been sufficient to prove 7.6.2 for *Ens*.

7.7.3 through 7.7.6 are also valid for large limits (as far as these exist; compare 7.1.7, 7.1.8), where Σ is a diagram of a higher universe or an arbitrary category. This is seen by changing the universe and taking into account that $(H^A(L), H^A\lambda)$ is still in *Ens*.

7.7.8 The additive case. If $\mathscr{C}$ is an additive category, then 7.7.1 is valid in *Ab* if $N_T(A)$ is provided with the group structure originating from $\Pi[A, T(i)]$. ω^A in (1) is then a homomorphism and therefore a morphism in *Ab*. $N_T(?)$ can be regarded as a functor with codomain *Ab*; (3) is valid in *Ab* and 7.7.3, 7.7.4 are valid for representable functors $\mathscr{C} \to Ab$. 7.7.5 is valid for additive S with values of H^X in *Ab*. This implies (compare also 7.7.7):

7.7.9 Proposition. *The forgetful functor $Ab \to Ens$ preserves and reflects all existing limits, including large limits.*

Here "reflects" is to be understood as in 7.7.6, namely an existing natural transformation is recognized as a limit.

7.8 Pullbacks

Pullbacks are an important special case of finite limits. Here the diagrams have the form

(1) $$A \xrightarrow{f} C \xleftarrow{g} B .$$

A natural transformation from a corresponding constant diagramm D_Σ to (1) is determined completely by two morphisms $u\colon D \to A$, $v\colon D \to B$ with $f u = g v$.

7.8.1 Definition. Let $f\colon A \to C$, $g\colon B \to C$ be two morphisms with the same codomain. A *pullback* (also *cartesian square*, or *fibre product*) for the pair (f, g) is a commutative rectangle

(2) $$\begin{array}{ccc} P & \xrightarrow{r} & B \\ {\scriptstyle s}\downarrow & & \downarrow{\scriptstyle g} \\ A & \xrightarrow{f} & C \end{array} \qquad g r = f s$$

with the following property: if $u\colon D \to A$, $v\colon D \to B$ are morphisms with $f u = g v$, then there is exactly one morphism $w\colon D \to P$ with $u = s w$ and $v = r w$.

A category is said to have pullbacks if every pair of morphisms in it with the same codomain has a pullback.

7.8.2 Proposition. *Let* (2) *be a pullback. If f is a monomorphism, then r is a monomorphism. If f is a retraction, then r is a retraction.*

Proof. First, let f be a monomorphism and let $w_1, w_2 : D \to P$ be given such that $r\,w_1 = r\,w_2$. Then $f\,s\,w_1 = f\,s\,w_2$ follows because of $g\,r = f\,s$, and further we get $s\,w_1 = s\,w_2$ since f is a monomorphism. So $w_1 = w_2$ follows because (2) is a pullback. Hence r is a monomorphism. Now let f be a retraction with the corresponding coretraction t. By means of $t\,g\colon B \to A$ and 1_B and since $f\,t\,g = g = g\,1_B$, one obtains a coretraction for r.

7.8.3 Remarks. Given $f\colon A \to C$, pullbacks allow us to define an operation of "induced" morphism taking morphisms with codomain C to morphisms with codomain A, whereby 7.8.2 (for g instead of f) guarantees that monomorphisms are taken into monomorphisms. Inverse images of "subobjects" are a special case. Induced fibrations are another classical special case. The pullback (2) can also be described as follows: (r, f) is a natural transformation from $s\colon P \to A$ to $g\colon B \to C$ (compare 6.5.1) such that every natural transformation (v, f) from $u\colon D \to A$ to $g\colon B \to C$ with *the same* f can be factored uniquely through (r, f).

The transfer back of morphisms with codomain C given by pullbacks is associative, up to an isomorphism, as the following result shows.

7.8.4 Proposition. *In the diagram*

$$(3)\qquad \begin{array}{ccccc} X & \xrightarrow{x} & Y & \xrightarrow{y} & Z \\ {\scriptstyle r}\downarrow & \mathrm{I} & {\scriptstyle s}\downarrow & \mathrm{II} & \downarrow{\scriptstyle t} \\ A & \xrightarrow{a} & B & \xrightarrow{b} & C \end{array}$$

the right rectangle II *is assumed to be a pullback. Then the whole rectangle is a pullback if and only if the left rectangle* I *is a pullback.*

Proof. Applying the definition twice one finds the whole rectangle to be a pullback, if I and II are pullbacks. Now let the whole rectangle be a pullback. If $u\colon M \to A$, $v\colon M \to Y$ are morphisms with $a\,u = s\,v$, then $b\,a\,u = t\,y\,v$, and there is exactly one $w\colon M \to X$ with $y\,x\,w = y\,v$ and $r\,w = u$. $y\,x\,w = y\,v$ and $s\,x\,w = a\,r\,w = s\,v$ imply $x\,w = v$, because II is a pullback.

Remark. If in (3) the whole rectangle and I are pullbacks, II need not be a pullback. There is a counterexample in *Ens* in which X, Y, Z, B are sets with two elements and A and C are sets of one element.

7.8.5 In a finitely complete category the pullback (2) can be constructed as follows: Let ($pr_1 : X \to A$, $pr_2 : X \to B$) be a product of A and B and let $k: P \to X$ be an equalizer of ($f\,pr_1$, $g\,pr_2$). Then f, g, $s = pr_1\,k$, and $r = pr_2\,k$ constitute a pullback. This, e.g., is the explicit construction of induced fibrations. We agree to construct pullbacks in this way (and not according to 7.4.2) for all finitely complete categories with a canonical choice of products and equalizers.

If the category $\mathscr{C}$ has a terminal object Z, then products of two objects are special pullbacks: instead of f and g in (2) one takes the only morphism $A \to Z$ and, resp., $B \to Z$. We note also: if in $\mathscr{C}$ any two objects have a product, then this is so for any finite number (not zero) of objects. This follows from 7.7.7.

7.8.6 If in (2) f and g are monomorphisms, then by 7.8.2 and 5.1.5 $f\,s = g\,r$ is a monomorphism, which is called the *intersection* of f and g. If any two monomorphisms with the same codomain in $\mathscr{C}$ have an intersection, $\mathscr{C}$ is said to have *finite intersections*. For, 7.7.7 implies that every diagram consisting of finitely many (not zero) monomorphisms with the same codomain has a limit, provided that this is true for any two such monomorphisms, and the intersection of the empty family of monomorphisms with codomain A is 1_A. (In *Ens*, intersections are, up to isomorphisms, intersections of subsets with the corresponding inclusions). Similarly one says that $\mathscr{C}$ has arbitrary intersections if every family of monomorphisms with the same codomain (as a diagram) has a limit. As in 7.8.2 one shows that everything is a monomorphism.

7.8.7 Let f, g: $A \to B$ be two morphisms and ($pr_1 : X \to A$, $pr_2 : X \to B$) a product of A and B. Corresponding to $(1_A, f)$ and $(1_A, g)$ there are two uniquely determined morphisms f', g': $A \to X$. The following holds:

$k: K \to A$ is an equalizer of f and g if and only if

$$(4)\qquad \begin{array}{ccc} K & \xrightarrow{k} & A \\ {\scriptstyle k}\downarrow & & \downarrow{\scriptstyle g'} \\ A & \xrightarrow{f'} & X \end{array} \qquad\qquad \begin{array}{ccc} K & \xrightarrow{k} & A \\ {\scriptstyle k}\downarrow & & \downarrow{\scriptstyle \binom{1_A}{g}} \\ A & \xrightarrow{\binom{1_A}{f}} & A \sqcap B \end{array}$$

is a pullback (the notation used on the right will be explained in 12.2.1).

Proof. If u, v: $D \to A$ are morphisms, then $f'\,u = g'\,u$ is equivalent to $u = v$ and $f\,u = g\,v$, as is shown by pr_1, pr_2. This implies that (4) can not remain commutative if the two morphisms originating from K are replaced by different ones. Then the claim follows from the definition.

In (4), f' and g' are monomorphisms, as pr_1 shows. Hence the pullback is an intersection. Altogether we now have:

7.8.8 Proposition. *For a category $\mathcal{C}$ the following are equivalent:*
(a) *$\mathcal{C}$ is finitely complete.*
(b) *$\mathcal{C}$ has finite products and equalizers.*
(c) *$\mathcal{C}$ has finite products and intersections.*
(d) *$\mathcal{C}$ has a terminal object and pullbacks.*

The equivalence of (a) and (b) is part of 7.4.2.

7.8.9 The diagram

$$\begin{array}{ccc} A & \xrightarrow{1_A} & A \\ {\scriptstyle 1_A}\downarrow & & \downarrow{\scriptstyle u} \\ A & \xrightarrow{u} & B \end{array}$$

is a pullback if and only if $u\colon A \to B$ is a monomorphism.

Both just say that, for every pair of morphisms $f, g\colon X \to A$, $u f = u g$ can only be valid if $f = g$. From this follows the

Proposition. *If the functor $T\colon \mathcal{C} \to \mathcal{D}$ preserves pullbacks (as far as they exist in $\mathcal{C}$), then it preserves monomorphisms.*

7.9 Problems

7.9.1 Interpret limits as terminal objects of suitable categories and compare with 7.1.3 and 5.6.4.

7.9.2 Let G be a group and $i\colon H \rightarrowtail G$ the inclusion of a subgroup. Let the set R consist of all left cosets $g H = \{g h \mid h \in H\}$ of G and an additional element ∞. Let K denote the group of permutations of R. Let $\tau\colon R \to R$ be the permutation which interchanges ∞ with the coset $H = 1 \cdot H$ and leaves all other elements fixed.
(a) For every $g' \in G$, the map $g H \mapsto (g' g) H$, $\infty \mapsto \infty$ is a permutation $\sigma(g')$ of R.
(b) The map $g' \mapsto \sigma(g')$ is a homomorphism $\sigma\colon G \to K$.
(c) The map $g' \mapsto \tau\, \sigma(g')\, \tau$ is a homomorphism $\varrho\colon G \to K$.
(d) i is an equalizer of σ and ϱ.
(e) In the category of groups every monomorphism is an equalizer and every epimorphism is a surjective map. (Use 7.2.2.)

7.9.3 Verify the examples in 7.2.4 and 7.3.2.

7.9.4 What are equalizers, products, limits in a preordered class (as a category)?

7.9.5 Carry out the details in 7.5.5 and 7.6.2.

7.9.6 Let $\mathcal{C} = \prod \mathcal{C}_\nu$ be a product of non-empty categories with projections $pr_\nu\colon \mathcal{C} \to \mathcal{C}_\nu$. A diagram $T\colon \Sigma \to \mathcal{C}$ has a limit (L, λ) if and only if every diagram $pr_\nu T$ has a limit $(L_\nu, \lambda(\nu))$. One then has $L_\nu = pr_\nu(L)$ and $\lambda(\nu) = pr_\nu \lambda$ up to isomorphisms.

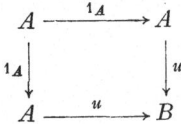

7.9.7 Give a counterexample as suggested by the remark in 7.8.4.

7.9.8 Fill in the details in 7.8.5 and in 7.8.6

7.9.9 Let

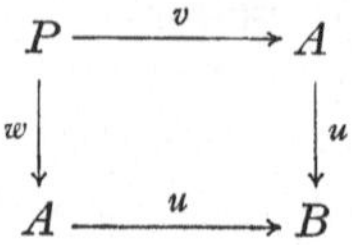

be a pullback.

(a) If $v = w$, then u is a monomorphism and v is an isomorphism.

(b) v and w are retractions with a uniquely determined common coretraction.

(c) If v is a monomorphism, then v is an isomorphism, u a monomorphism and $v = w$.

Remark. For any u, the pair (v, w) is called a *kernel pair* of u. We shall discuss kernel pairs in 18.4.

7.9.10 In *Ens*, products of epimorphisms are epimorphisms, because they are retractions. However, an epimorphic natural transformation $\alpha: D_1 \to D_2$ between diagrams $D_1, D_2: \Sigma \to Ens$ need not induce an epimorphism between the limits. Give an example.

8. Colimits

By dualizing chapter 7 one obtains colimits with corresponding properties. Diagram schemes and categories are replaced by their duals ("reverse the arrows"). However, there is one exception: whenever Hom-functors are involved, *Ens* may not be dualized. Instead, the partial covariant Hom-functors have to be replaced by contravariant ones and vice versa. We do not go into all the details and leave the job of completing the results to the reader. He should note, though, that dualizing the propositions involves different interpretations of existing results. Proofs, however, need not be repeated.

8.1 Definition of Colimits

8.1.1 We now consider a natural transformation $\xi: T \to A_\Sigma$ for a diagram $T: \Sigma \to \mathscr{C}$. 7.1 (1), (2), (1') become

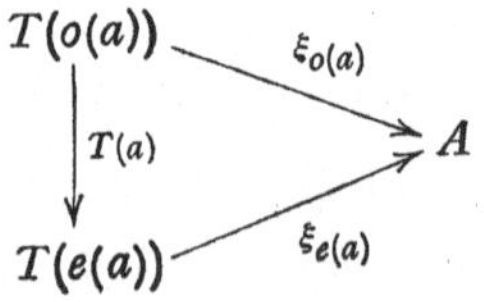

(1) $\xi_{o(a)} = \xi_{e(a)}\, T(a)$

(2) $\xi \in \prod_{i \in Ve} [T(i), A]$

(1') $[T(a), A]\, (\xi_{e(a)}) = \xi_{o(a)}\,.$

Note that T is again covariant, since Σ and $\mathscr{C}$ have taken the place of their duals.

8.1.2 Definition. A *colimit* (*inductive limit, direct limit,* right root, supremum) for the diagram $T\colon \Sigma \to \mathscr{C}$ consists of an object L and a natural transformation $\lambda\colon T \to L_\Sigma$ with the following property: for any natural transformation $\xi\colon T \to A_\Sigma$ there is exactly one $f\colon L \to A$ in $\mathscr{C}$ such that

(3) $$\xi = f_\Sigma \lambda$$

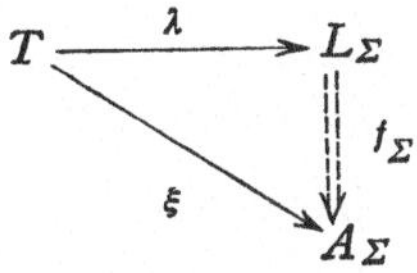

If Σ is empty, a colimit of T, if it exists, consists of an initial object of $\mathscr{C}$ and the trivial natural transformation of the empty diagram.

8.1.3 Proposition. (*i*) (L, λ) *is a colimit of* $T\colon \Sigma \to \mathscr{C}$ *if and only if* (Op (L), Op λ) *is a limit of* Op T Op: $\Sigma^0 \to \mathscr{C}^0$.

(ii) *If it exists, a colimit* (L, λ) *of* $T\colon \Sigma \to \mathscr{C}$ *represents the following functor* $N^T\colon \mathscr{C} \to Ens$: *for* $A \in |\mathscr{C}|$ $N^T(A)$ *is the set of natural transformations* $T \to A_\Sigma$ *and for* $f\colon A \to B$ $N^T(f)$ *is the map* $N^T(A) \to N^T(B)$ *which is given by the rule* $\xi \mapsto f_\Sigma \xi$. *Conversely, every representation of* N^T *gives a colimit of* T.

The above remarks should make this immediately clear. In (ii) *Ens* is not to be replaced by the dual category, but now N^T is covariant. Note in (i) that through the change from Σ to Σ^0 the origin and end of every arrow are interchanged, so that the change from T to Op T Op really does make (1) correspond to 7.1.1 (1).

8.1.4 $N^T(A)$ is a subset of the product (2) (again a product, since *Ens* is not dualized), namely that subset which is characterized by the conditions (1). $N^T(f)$ is produced by restricting the map

(4) $$\Pi\,[T(i), f]\colon \Pi\,[T(i), A] \to \Pi\,[T(i), B]\,,$$

given by $\{\beta_i\} \mapsto \{f\beta_i\}$.

If, as in 7.1.4, $S\colon \mathscr{C} \to [\Sigma/K, \mathscr{C}]$ is the functor defined by $A \mapsto A_\Sigma$, $f \mapsto f_\Sigma$ and if T satisfies the commutativity conditions K, then a colimit (L, λ) of T is a representation of

(5) $$N_T(?) = [T, S(?)]_{[\Sigma/K, \mathscr{C}]}\,,$$

namely the isomorphism

(6) $$\varrho\colon [L, ?]_{\mathscr{C}} \xrightarrow{\approx} [T, S(?)]_{[\Sigma/K, \mathscr{C}]}$$

which is characterized by $\varrho_L(1_L) = \lambda$.

8.2 Coequalizers

8.2.1 Definition. Let $f, g: A \to B$ be two morphisms with the same codomain B and the same domain A. A *coequalizer* (*difference cokernel*, even cokernel) (C, c) is a morphism $c: B \to C$ such that

(i) $c f = c g$,

(ii) for every morphism $v: B \to Y$ with $v f = v g$ there is exactly one morphism $w: C \to Y$ such that $v = w c$.

8.2.2 Every coequalizer is an epimorphism. Every monomorphic coequalizer is an isomorphism. Every retraction is a coequalizer.

8.2.3 A category has coequalizers if the dual category has equalizers.

8.2.4 Coequalizers are only determined up to being followed by an isomorphism. In the following examples, however, there is a canonical choice.

Given $A \overset{f}{\underset{g}{\rightrightarrows}} B$ in *Ens* let Q be the smallest equivalence relation on the set B with respect to which $f(a)$ and $g(a)$ are equivalent for every $a \in A$. The projection $c: B \to B/Q$ is a coequalizer of f and g (the same in Ens_*). In *Top* the same construction works if B/Q is provided with the identification topology with respect to c. In *Ab* Q is the image of A by $f - g$ and $c: B \to B/Q$ is the projection onto the factor group. Analogous assertions can be made for ${}_RMod$ and Mod_R. In the category of rings, Q is the (two sided) ideal generated by the elements $f(a) - g(a)$; in the category of groups, Q is the normal subgroup generated by the elements $f(a)\, g^{-1}(a)$.

There is also a canonical choice of coequalizers in *cat*. Let $F, G: \mathcal{A} \to \mathcal{B}$ be functors between small categories. Then consider the smallest equivalence relation provided by F and G on $|\mathcal{B}|$ and Mor $\mathcal{B}$, that is compatible with the composition of morphisms in $\mathcal{B}$ and with dentities (compare 6.4.4). With this, a coequalizer of F and G is provided by the construction in 6.4.4. For instance, let $\mathcal{B}$ be the category **2** of 6.5.1, let $\mathcal{A}$ be a terminal category, i. e., it has only one morphism (which is therefore the identity morphism), and let F and G be the two embeddings of $\mathcal{A}$ in $\mathcal{B}$. The coequalizer is a category with one object, and the set of morphisms is isomorphic to the additive monoid of the non-negative integers. In this way one obtains an example of an epimorphism in *cat* which, although an epimorphism on the objects, is not an epimorphism on any Hom-set.

8.2.5 If the category $\mathcal{C}$ has zero morphisms, then a *cokernel* of $f: A \to B$ is a coequalizer of f and $0: A \to B$.

8.2.6 In an additive category, $c: B \to C$ is a coequalizer of $f, g: A \to B$ if and only if c is a cokernel of $f - g$.

8.3 Coproducts

8.3.1 Definition. Let $\{A_i\}_{i \in I}$ be a family of objects of the category $\mathscr{C}$. A *coproduct* (also *sum, direct sum*) of this family is an object Y with morphisms $in_i: A_i \to Y$ such that for any family $\{f_i: A_i \to Z\}_{i \in I}$ there is exactly one morphism $f: Y \to Z$ with $f\, in_i = f_i$. We use the notation $\coprod_{i \in I} A_i$ or, for short $\coprod A_i$ (in some categories $\bigoplus A_i$ or ΣA_i is also used), and in_i is called the i-th *injection* of the coproduct. If I is empty, the coproduct is an initial object. The category $\mathscr{C}$ has coproducts (finite coproducts) if the dual category $\mathscr{C}^0$ has products (finite coproducts).

8.3.2 Examples. In *Ens* canonically chosen coproducts are the "disjoint unions". They are constructed as follows: let Y be the subset of $I \times A_i$ consisting of the elements (i, a) with $a \in A_i$ and let $in_i : A_i \to Y$ be the injective map $a \mapsto (i, a)$. For any two sets A and B (in this order) one gets $A \sqcup B$ as a coproduct by taking $I = \{1, 2\}$ and setting $A_1 = A$, $A_2 = B$. The corresponding procedure yields the topological sum in *Top*. In Ens_* and Top_* the basepoints of the subsets $in_i(A_i)$ have to be identified with each other (this is called a *bouquet of sets* or, resp., *spaces*). In *Ab* and ${}_R Mod$ coproducts are the usual "direct sums". Their objects are subgroups or, resp., submodules of the corresponding direct products. If here I is a finite set, then the objects of the product and coproduct coincide; we shall discuss this more thoroughly later. In the category of groups coproducts are the free products. (This is why the name "direct sum" for coproducts can be confusing). In the category of commutative rings (with 1) coproducts are the tensor products over $\mathbf{Z}$ for rings as $\mathbf{Z}$-algebras. The category of rings also has coproducts. *cat* has coproducts. If $\{\mathscr{C}_i\}$ is a family in *cat*, then $|\coprod \mathscr{C}_i|$ (resp., Mor $\coprod \mathscr{C}_i$) is the set coproduct $\coprod |\mathscr{C}_i|$ (resp., $\coprod$ Mor $\mathscr{C}_i$) with the obvious composition of morphisms.

8.3.3 Let $\{in_i : A_i \to Y\}$ and $\{j_i : B_i \to Z\}$ be coproducts in $\mathscr{C}$ with the same index set I. If for every $i \in I$ there is given a morphism $f_i : A_i \to B_i$, then there is exactly one morphism $f: Y \to Z$ with $j_i f_i = f\, in_i$ for all i. We write $f = \coprod f_i$. If all f_i are epimorphisms, then f is an epimorphism.

8.3.4 There is an isomorphism

$$\theta: [\coprod A_i, ?]_{\mathscr{C}} \overset{\cong}{\Rightarrow} \prod [A_i, ?]_{\mathscr{C}} .$$

Note that *Ens* has not been dualized here.

8.4 Cocomplete Categories

8.4.1 A category $\mathscr{C}$ is *cocomplete* (also right complete) or, resp., *finitely cocomplete*, if the dual category is complete or, resp., finitely complete.

8.4.2 A category is cocomplete (finitely cocomplete) if and only if it has coequalizers and coproducts (finite coproducts).

8.4.3 Proposition. *The following categories are cocomplete with a canonical choice of colimits: Ens, Ens_*, Top, Top_*, Ab, $_R Mod$, Mod_R, cat, the category of groups.*

These are examples of finitely cocomplete categories: the category of finite sets, of countable sets, of finite abelian groups, of finitely generated groups in *Ab* or, resp., modules in $_R Mod$.

8.4.4 7.4.2 yields, by 8.4.1, a description of canonically chosen colimits in *Ens*. For $T\colon \Sigma \to Ens$ a quotient L of the coproduct $\amalg\, T(i)$ is a colimit. The quotient is taken with respect to the smallest equivalence relation on $\amalg\, T(i)$ for which

$$(o(a),\, x) \sim \big(e(a),\, T(a)\,(x)\big) \quad \text{with} \quad x \in T\big(o(a)\big)$$

for all arrows a of Σ. The corresponding maps $\lambda_i\colon T(i) \to L$ are described by passing from $x \in T(i)$ to the class of (i, x). The same description is valid for *Top* and, analogously, for Ens_*, Top_*, *cat*. For *Ab* and $_R Mod$ 8.2.4 and 8.3.2 have to be used.

8.4.5 A functor $T\colon \mathscr{C} \to \mathscr{D}$ preserves colimits if Op T Op: $\mathscr{C}^0 \to \mathscr{D}^0$ preserves limits.

The forgetful functor $_R Mod \to Ab$ preserves colimits, since it preserves coproducts and cokernels; 8.2.5 is valid. The forgetful functor $Top \to Ens$ preserves colimits. The forgetful functor $Ab \to Ens$ does not preserve colimits, in particular it does not preserve coproducts. The inclusions $Ens \to ENS$, $Ab \to AB$ preserve $\mathfrak{U}$-colimits.

˙8.5 Colimits in Functor Categories

8.5.1 Let $\mathscr{C}$ be a cocomplete (finitely cocomplete) category. The functor category $[\mathscr{D}, \mathscr{C}]$ is then also cocomplete (finitely cocomplete). The construction of colimits is done "pointwise", i.e. at every "point" $A \in |\mathscr{D}|$.

This follows from 7.5.2 if Σ, $\mathscr{Y}$, $\mathscr{D}$, $\mathscr{C}$ are all replaced by their duals. For diagrams of type Σ/K in $[\mathscr{D}, \mathscr{C}]$ it is again sufficient just to assume the existence of the corresponding colimits in $\mathscr{C}$.

8.5.2 A choice of colimits of type Σ/K in $\mathscr{C}$ gives an isomorphism

$$(1) \qquad \varrho\colon [\mathrm{Colim}\,(\,?\,?),\ ?]_{\mathscr{C}} \xrightarrow{\approx} [\,?\,?,\ S(\,?)]_{[\Sigma/K,\,\mathscr{C}]}$$

which has to be considered as an isomorphism of bifunctors $[\Sigma/K, \mathscr{C}]^0 \times \times\, \mathscr{C} \to Ens$.

8.5.3 If $\mathscr{D}$ and $\mathscr{C}$ are additive, 8.5.1 is valid as well for $Add\,(\mathscr{D}, \mathscr{C})$, since for every additive category the dual category is also additive and Op is here an additive contravariant functor.

Colimits of additive functors are additive.

8.6 Double Colimits

8.6.1 Lemma. *Let Σ' be a subscheme of the diagram scheme Σ and T: $\Sigma \to \mathscr{C}$ a diagram. If (L', λ') is a colimit of the subdiagram $T' = T|\Sigma'$ and if ξ: $T \to A_\Sigma$ is a natural transformation, then there is exactly one morphism f: $L' \to A$ with $\xi_i = f\lambda'_i$ for all vertices i of Σ'.*

8.6.2 Proposition 7.6.2 and its consequence 7.6.3 can be dualized immediately. *Colimits* (of fixed type $\mathscr{X}$) *comuute with colimits* (of fixed type $\mathscr{Y}$; existence assumed). 7.6.4 and 7.6.5 can also be dualized.

8.6.3 In general, limits do not commute with colimits. In *Ens* this is especially not true for finite products and coproducts (however, compare later with 9.5).

8.7 Criteria for Colimits

8.7.1 Preliminary remarks. In the following we also have to consider contravariant diagrams T: $\Sigma \to \mathscr{C}$. They are (compare 2.4.5) ordinary diagrams T Op: $\Sigma^0 \to \mathscr{C}$, whereby Op is extended in the obvious way to diagram schemes, which is compatible with 6.3.2. It is then clear what natural transformations, limits and colimits of contravariant diagrams are. In 7.1 (1), (1') and 8.1 (1), (1') $o(a)$ and $e(a)$ have to be interchanged. A_Σ and A_{Σ^0} coincide, as do f_Σ and f_{Σ^0}.

Op: $\mathscr{C} \to \mathscr{C}^0$ interchanges limits and colimits. For the contravariant Hom-functor $H_A = [?, A]_\mathscr{C}$ one has H_A Op $= [A^0, ??]_{\mathscr{C}^0} = H^{A^0}$ (with $A^0 = \mathrm{Op}(A)$). Dualizing 7.7 will therefore again produce limits in *Ens*.

8.7.2 Proposition. *Let T: $\Sigma \to \mathscr{C}$ be a diagram and let $A \in |\mathscr{C}|$. The limit of $H_A T$ is the set $N^T(A)$ of natural transformations $T \to A_\Sigma$ with the maps $N^T(A) \to [T(i), A]$ given by the rule $\xi \mapsto \xi_i$ (unless Σ is empty).*

8.7.3 Theorem. *Let T: $\Sigma \to \mathscr{C}$ be a diagram and λ: $T \to L_\Sigma$ a natural transformation. (L, λ) is a colimit of T if and only if for all $A \in |\mathscr{C}|$ $(H_A(L), H_A\lambda)$ is a limit of $H_A T$.*

8.7.4 Corollary. *Every representable contravariant functor F: $\mathscr{C} \to Ens$ takes colimits (as far as they exist in $\mathscr{C}$) into limits.*

8.7.5 Corollary. *A functor S: $\mathscr{C} \to \mathscr{D}$ preserves colimits if and only if $H_X S$ takes colimits into limits for every $X \in |\mathscr{D}|$.*

8.7.6 Theorem. *Fully faithful (covariant) functors reflect colimits.*

Here "reflect" is to be understood as in 7.7.6 and 7.7.9; namely, that an existing natural transformation is recognized as a limit or, resp., colimit. For T: $\Sigma \to \mathscr{C}$ and a fully faithful functor F: $\mathscr{C} \to \mathscr{D}$, $F\,T$ may possess a limit or colimit even if T does not. On the other hand, a fully faithful functor need not even preserve finite limits or colimits (compare 10.2.6 later). We shall continue to use "reflect" in the above sense.

8.7.7 If $\mathscr{C}$ is additive, 8.7.2 through 8.7.4 are valid as well with *Ab* instead of *Ens*, also, 8.7.5 holds for additive *S*.

8.8 Pushouts

8.8.1 Definition. Let $f\colon A \to B$, $g\colon A \to C$ be two morphisms with the same domain. A *pushout* (also *cocartesian square, amagalmated sum,* or even *fibre sum*) for the pair (f, g) is a commutative square

$$(1) \qquad \begin{array}{ccc} A & \xrightarrow{f} & B \\ {\scriptstyle g}\downarrow & & \downarrow{\scriptstyle s} \\ C & \xrightarrow{r} & Q \end{array} \qquad s\,f = r\,g$$

with the following property: if $u\colon B \to X$, $v\colon C \to X$ are morphisms with $u\,f = v\,g$, then there is exactly one morphism $w\colon Q \to X$ with $w\,s = u$, $w\,r = v$.

8.8.2 If in the pushout (1) f is an epimorphism, then r is an epimorphism. If f is a coretraction, then r is a coretraction.

8.8.3 Sections 7.8.3 through 7.8.5 can easily be dualized; in particular also 7.8.5: in finitely cocomplete categories the pushout (1) is constructed by first forming the coproduct $B \sqcup C$ with the injections in_1, in_2 and then the coequalizer of $in_1 f$ and $in_2 g$.

8.8.4 If f and g in (1) are epimorphisms, then the epimorphism $s\,f = r\,g$ is called the *cointersection* of f and g.

8.8.5 The duals of 7.8.7 through 7.8.9 are evident. In the future we shall mention dualizations only in special cases.

8.9 Problems

8.9.1 Check the examples 8.2.4 and 8.3.2.

8.9.2 The category of finite groups has coequalizers and is not finitely cocomplete, e. g., there is no coproduct of $\mathbf{Z}_2$ with itself ($\mathbf{Z}_2$ as multiplicative group with two elements. Hint: Consider a group with two generators a, b satisfying $a^2 = b^2 = (a\,b)^n = 1$, i. e., the group of motions of a regular n-gon in the plane = the n-th dihedral group).

8.9.3 Find examples to demonstrate that in *Ens* finite products do not commute with finite coproducts. Furthermore show, that in *Ens* coproducts of pullbacks are again pullbacks. What are the corresponding facts in *cat* and in *Top*?

8.9.4 7.9.6 is valid accordingly for colimits in $\prod \mathscr{C}_\nu$. Why is this claim dual to 7.9.6? What happens in $\mathscr{C}_1 \sqcup \mathscr{C}_2$?

8.9.5 In *Cat* every projection of a category onto a quotient in the sense of 6.4.1 is a coequalizer.

9. Filtered Colimits

9.1 Connected Categories

9.1.1 Definition. A *category* is called *connected* if for any two objects A, B there are finitely many objects $A = A_0, A_1, \ldots, A_{2n} = B$ such that there are morphisms $A_{2j-2} \to A_{2j-1}$ and $A_{2j} \to A_{2j-1}$ for $j = 1, 2, \ldots, n$.

If $\mathscr{C}$ is any category, consider the smallest equivalence relation on objects, with respect to which any two objects connected by a morphism are equivalent. Every such equivalence class determines a full subcategory of $\mathscr{C}$. These subcategories are called *connected components* of $\mathscr{C}$.

9.1.2 The following can be verified: every connected component is connected and as a subcategory of $\mathscr{C}$ it is maximal with respect to this property. $\mathscr{C}$ is a coproduct of its connected components. If $\mathscr{C}$ has an initial or a terminal object, then $\mathscr{C}$ is connected.

Warning. An infinite product of connected categories need not be connected.

9.1.3 Proposition. *Let $\mathscr{X}$ be a connected non-empty category and $\mathscr{C}$ an arbitrary category. For $A \in |\mathscr{C}|$ the constant functor $A_{\mathscr{X}}$ has $(A, \{1_A\})$ as a limit and as a colimit.*

9.1.4 Proposition. *Let $\mathscr{X}$ be a small category with connected components $\mathscr{X}_j$. If $\mathscr{C}$ is a cocomplete category and $T\colon \mathscr{X} \to \mathscr{C}$ a functor, then the colimit of T is the coproduct of the colimits of the functors $T|\mathscr{X}_j$.*

Both propositions are easily deduced from the definitions and 8.6.1.

9.1.5 Remark. 9.1.1 through 9.1.4 carry over easily to diagram schemes. A diagram scheme is called connected if the category $\mathscr{P}(\Sigma)$ is connected (see 6.3).

9.2 On the Calculation of Limits and Colimits

We give preference to colimits in the following because they occur more frequently in applications. We first consider colimits for arbitrary functors.

9.2.1 Let $D\colon \Sigma \to \mathscr{X}$ be a functor. For $Y \in |\mathscr{X}|$, the following category is denoted by Y/D: objects are pairs (i, u) with $i \in |\Sigma|$ and $u\colon Y \to D(i)$ in $\mathscr{X}$, morphisms from (i, u) to (i', u') are triples (u, u', a) with $a\colon i \to i'$ in Σ and $D(a)\, u = u'$

(1)

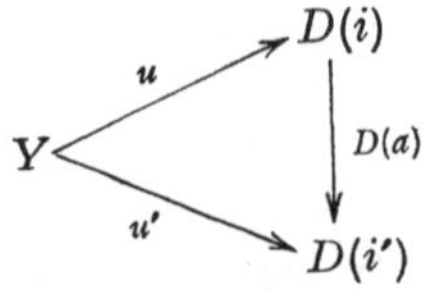

with the obvious composition

$$(u', u'', a')\,(u, u', a) = (u, u'', a'\,a)\,.$$

The notion (Y, D) is often used instead of Y/D.

9.2.2. Definition. The *functor* $D\colon \Sigma \to \mathscr{X}$ is called *final* if Y/D is non-empty and connected for every $Y \in |\mathscr{X}|$. D is called *initial* if $\mathrm{Op}\, D\, \mathrm{Op}\colon \Sigma^0 \to \mathscr{X}^0$ is final. A diagram $D\colon \Sigma \to \mathscr{X}$ is final if its extension to the functor $\mathscr{P}(\Sigma) \to \mathscr{X}$ is final.

9.2.3 Theorem. *Let the functor $D\colon \Sigma \to \mathscr{X}$ be final. If $T\colon \mathscr{X} \to \mathscr{C}$ is any functor with domain $\mathscr{X}$, then for every natural transformation $\alpha\colon TD \to A_\Sigma$ there is exactly one $\beta\colon T \to A_{\mathscr{X}}$ with*

(2) $$\beta_{D(i)} = \alpha_i$$

for all $i \in |\Sigma|$. Here (A, β) is a colimit of T if and only if (A, α) is a colimit of TD.

Proof. We refer back to 8.1.1 (1), (2). $\alpha = \{\alpha_i\colon TD(i) \to A\}$ with

(3) $$\alpha_{o(a)} = \alpha_{e(a)}\, TD(a)$$

for every morphism $a\colon o(a) \to e(a)$ in Σ. Now let $Y \in |\mathscr{X}|$. By assumption, there is a $u\colon Y \to D(i)$ for a suitable $i \in |\Sigma|$. If we set

(4) $$\beta_Y = \alpha_i\, T(u)\,,$$

then β_Y does not depend on the choice of (i, u) in Y/D. For, if (1) is a morphism in Y/D, then by (3)

$$\alpha_{i'}\, T(u') = \alpha_{i'}\, TD(a)\, T(u) = \alpha_i\, T(u)\,.$$

By definition 9.1.1, this implies that every object of Y/D produces the same morphism β_Y by means of (4). Now let $w\colon Z \to Y$ be an $\mathscr{X}$-morphism and again $u\colon Y \to D(i)$. Because of what we just proved and (4)

(5) $$\beta_Z = \alpha_i\, T(u\, w) = \alpha_i\, T(u)\, T(w) = \beta_Y\, T(w)\,,$$

so that $\beta = \{\beta_Y\}\colon T \to A_{\mathscr{X}}$ is a natural transformation. (2) is valid on account of (4) for $Y = D(i)$ and $u = 1_Y$. By α and (2), β is uniquely determined, since (4) has to be valid in any case. Conversely, β:

$T \to A_{\mathfrak{X}}$ determines α: $TD \to A_{\Sigma}$ uniquely by (2). So there is a bijection Φ_A: $N^{TD}(A) \to N^T(A)$ (possibly in ENS). Since $f_{\Sigma}\alpha = \{f\alpha_i\}$ and $f_{\mathfrak{X}}\beta = \{f\beta_Y\}$ for f: $A \to B$ in $\mathscr{C}$, (2) and (4) imply that $\{\Phi_A\}$ is an isomorphism $N^{TD} \to N^T$. Hence the last claim of the theorem follows from 8.1.3 (if necessary after a change of universe).

9.2.4 Remarks. It can be shown with some effort, that the converse of 9.2.3 is true, so that definition 9.2.2 and the assertion of theorem 9.2.3 are equivalent.

In applications of 9.2.3, D is frequently the inclusion of a subcategory Σ of $\mathfrak{X}$. One then talks about a *final subcategory*. If Σ is a full subcategory or, resp., a full small subcategory, one only gives $|\Sigma|$, calling it a *final class* or, resp., *set* in $\mathfrak{X}$. The name cofinal is also used for this, however, we avoid it in order to reserve the prefix "co" for dualizations. If it exists, a terminal object of $\mathfrak{X}$ is a final set. Therefore, 7.1.8 is a special case of the dual of 9.2.3.

9.3 Filtered Categories

Classical examples of colimits are, apart from coproducts and cokernels, those where T: $\Sigma \to \mathscr{C}$ is a functor and Σ an ordered, directed (to the right) set (i. e., for i_1, i_2 in $|\Sigma|$ there is always an i_3 with $i_1 \leq i_3$ and $i_2 \leq i_3$). They were introduced as direct limits and were later also named inductive limits. Meanwhile it has proved useful to replace ordered sets by a generalization. We start with some auxiliary notions.

9.3.1 Definition. A *pencil* with domain A is a non-empty family of morphisms $\{s_j$: $A \to B_j\}$ with the same domain A, where the indices form a $\mathfrak{U}$-set. A *copencil* with codomain B is defined dually. A pencil or, resp., copencil is called finite if its index set is finite and non-empty.

9.3.2 A *generalized pullback* for the copencil $\{s_j$: $B_j \to C)\}$ consists of this copencil and of a pencil $\{\lambda_j$: $L \to B_j\}$ such that $s_j\lambda_j$: $L \to C$ is independent of j, and so that for every pencil $\{\xi_j$: $A \to B_j\}$ with this property there is exactly one morphism f: $A \to L$ with $\xi_j = \lambda_j f$ for all j. I. e., $(L, \{\lambda_j\})$ is a limit for the copencil. Generalized pushouts are dual.

A category $\mathscr{C}$ has generalized pullbacks if for every copencil there is a generalized pullback. If $\mathscr{C}$ has a terminal object, then $\mathscr{C}$ is complete if and only if $\mathscr{C}$ has generalized pullbacks. This follows as in 7.8.8.

9.3.3 Proposition. *Let $\{\lambda_j$: $L \to B_j$; s_j: $B \to C\}$ be a generalized pullback. If s_j is a monomorphism for all $j \neq k$, then λ_k is a monomorphism.*

Proof. If there are $w_1, w_2\colon A \to L$ with $\lambda_k w_1 = \lambda_k w_2$, then, for all j, $s_j \lambda_j w_1 = s_j \lambda_j w_2$. This implies $\lambda_j w_1 = \lambda_j w_2$ for all j by assumption. The definition of generalized pullbacks then yields $w_1 = w_2$.

9.3.4 Definition. A *category* $\mathcal{X}$ is called *pseudofiltered* (quasifiltered) if it is non-empty and if

(i) Every diagram of the form 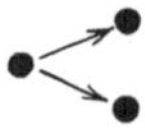 has a commutative completion of the form

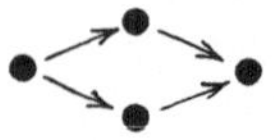

(ii) For every diagram $Y_1 \overset{u}{\underset{v}{\rightrightarrows}} Y_2$ there is a morphism $w\colon Y_2 \to Y_3$ with $w\,u = w\,v$.

$\mathcal{X}$ is called *strongly pseudofiltered* if $\mathcal{X}$ is non-empty and if (ii) is valid together with a strengthened version of (i):

($\mathrm{i_s}$) For every pencil $\{s_j\colon Y \to Z_j\}$ there is a commutative completion, i. e., a copencil $\{t_j\colon Z_j \to Z\}$ such that $t_j\, s_j$ is independent of j.

$\mathcal{X}$ is called *filtered* or, resp., *strongly filtered*, if again $\mathcal{X}$ is non-empty and if besides (i) and (ii) or, resp., ($\mathrm{i_s}$), (ii),

(iii) For any two objects Y_1, Y_2 there is an object Z with morphisms $Y_1 \to Z$, $Y_2 \to Z$

is valid.

9.3.5 Remarks. It can be deduced from (i) by induction, that every finite pencil has a commutative completion. Furthermore, (i) follows from (ii) and (iii).

If $\mathcal{X} \neq \emptyset$ satisfies (i) and if it has finite weak coproducts, then $\mathcal{X}$ is filtered. For, there is a strengthened form of (iii) from which (ii) can be deduced by means of (i) following the pattern of the dual of 7.8.7.

9.3.6 Examples. Discrete non-empty categories are strongly pseudofiltered. Cocomplete categories and categories with a terminal object are strongly filtered. The same is true for every full subcategory of *Ens* with only one object (endomorphisms of a set as morphisms). Finitely cocomplete categories are filtered, also preordered, directed (to the right) non-empty sets, in particular strongly ordered ones.

9.3.7 The connected components of a (strongly) pseudofiltered category $\mathcal{X}$ are (strongly) filtered.

Proof. Let $Y_1 \sim Y_2$ if there is an object Z with morphisms $Y_1 \to Z$, $Y_2 \to Z$. This relation is obviously reflexive and symmetrical. Transitivity follows from (i). For, if $Y_1 \sim Y_2$ and $Y_2 \sim Y_3$, then there is a commutative diagram

(1)

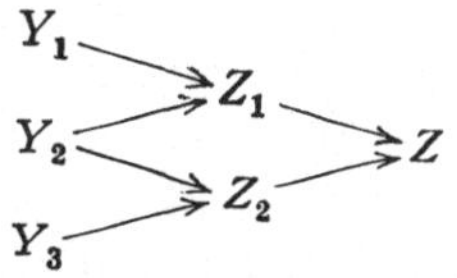

By 9.1.1, the full subcategories corresponding to the equivalence classes are just the connected components of $\mathcal{X}$.

9.3.8 In strongly pseudofiltered categories the following sharper version of (ii) is valid: if $\{u_j : Y_1 \to Y_2\}$ is a family of morphisms with the same codomain and domain, then there is a w: $Y_2 \to Z$ such that $w\, u_j$ is independent of j.

Proof. Choose a fixed index k. For every j there is a w_j: $Y_2 \to Z_j$ with $w_j\, u_k = w_j\, u_j$. The pencil $\{w_j\}$ has a commutative completion $\{t_j: Z_j \to Z\}$ and $w = t_j\, w_j$ has the desired property.

9.4 Filtered Colimits

9.4.1 Definition. We shall call the colimit of a functor T: $\mathcal{X} \to \mathcal{C}$, where $\mathcal{X}$ is a small filtered category, a *filtered colimit*. *Pseudofiltered* or *strongly filtered* or *strongly pseudofiltered colimits* are defined similarly.

9.4.2 8.4.4 yields a direct description of pseudofiltered colimits in *Ens*. If $\mathcal{X}$ is a small, pseudofiltered category and T: $\mathcal{X} \to Ens$ a functor, one forms the coproduct

(2) $$\coprod T(Y) = \{(Y, a) \mid Y \in |\mathcal{X}|,\ a \in T(Y)\}$$

and lets $(Y_1, a_1) \sim (Y_2, a_2)$ if there are morphisms u_1: $Y_1 \to Z$, u_2: $Y_2 \to Z$ in $\mathcal{X}$ such that

(3) $$T(u_1)\,(a_1) = T(u_2)\,(a_2)\,.$$

9.3.7 (1) then implies that this is an equivalence relation. It is the smallest equivalence relation under which for u: $Y \to Z$ the pairs (Y, a) and $(Z, T(u)\,(a))$ are always equivalent, since this implies (3). Let $[Y, a]$ be the equivalence class of (Y, a). These classes are the elements of the colimit object L. The corresponding natural transformation $T \to L_{\mathcal{X}}$ consists of the maps $T(Y) \to L$ given by the rule $a \mapsto [Y, a]$.

9.4.3 There is a similar description in *Top* if the colimit object L is supplied with the identification topology with respect to the corresponding map $\coprod T(Y) \to L$.

9.4.4 For 9.4.2 and 9.4.3, only 9.3.4 (i) need be assumed, (ii) is not used. The equivalence relation (3) induces an equivalence relation on every set $T(Y)$. If 9.3.4 (ii) is satisfied, then for any two equivalent elements a_1, a_2 of $T(Y)$ there is a morphism $u\colon Y \to Z$ in $\mathcal{X}$ such that $T(u)(a_1) = T(u)(a_2)$. If 9.3.4 (ii) is not satisfied, then this need not be the case. Let $\mathcal{X}$ be the cyclic multiplicative group of order 2 (as a category). Let $T\colon \mathcal{X} \to Ab$ take the only object of $\mathcal{X}$ into the group $\mathbf{Z}_4$, and the two morphisms of $\mathcal{X}$ into the two automorphisms of $\mathbf{Z}_4$. $\mathcal{X}$ then satisfies 9.3.4 (i) but not (ii). Applying the forgetful functor $Ab \to Ens$ gives the desired counterexample.

9.4.5 Lemma. *Let $\mathcal{X}$ be a (not necessarily small) strongly pseudofiltered category and $T\colon \mathcal{X} \to Ens$ a functor with a colimit (L, λ). Further, let $\{[Y, a_j]\}$ be a family of elements in L with representing families $\{(Y, a_j)\}$ and $\{(Y, a_j')\}$ in the set $T(Y)$. Then there is a morphism $u\colon Y \to Z$ in $\mathcal{X}$ such that $T(u)(a_j) = T(u)(a_j')$ for all j.*

If $\mathcal{X}$ is pseudofiltered, then the corresponding statement is true for finite families.

Proof. For every j there is a Z_j with morphisms $u_j\colon Y \to Z_j$, $u_j'\colon Y \to Z_j$ so that $T(u_j)(a_j) = T(u_j')(a_j')$. 9.3.4 (ii) allows us to choose Z_j such that $u_j = u_j'$. Now $\{u_j\colon Y \to Z_j\}$ is a pencil with a commutative completion $\{v_j\colon Z_j \to Z\}$. $u = v_j u_j$ then has the desired property.

9.4.6 Lemma. *Let $\mathcal{X}$ be a strongly filtered category and $T\colon \mathcal{X} \to Ens$ a functor with a colimit (L, λ). Further, let $\{[Y_j, a_j]\}$ be a family of elements in L. Then there is a $Z \in |\mathcal{X}|$ such that all members of the family have representatives in $T(Z)$.*

If all $[Y_j, a_j]$ coincide as elements of L, and if (Y_j, a_j) is a representative of $[Y_j, a_j]$, then Z with morphisms $u_j\colon Y_j \to Z$ can be chosen such that $T(u_j)(a_j)$ is independent of j.

If X is filtered, corresponding statements hold for finite families.

Proof. Let (Y_j, a_j) be a representative of $[Y_j, a_j]$ and k a fixed index. For every j there is a Z_j with morphisms $v_j\colon Y_k \to Z_j$, $w_j\colon Y_j \to Z_j$. For the pencil $\{v_j\}$ there is a commutative completion $\{t_j\colon Z_j \to Z\}$, which proves the first claim. If the elements $[Y_j, a_j]$ of L coincide, then because of (3), Z_j, v_j, w_j can be chosen such that, for every j, $T(v_j)(a_k) = T(w_j)(a_j)$. Since $T(t_j v_j)(a_k)$ is independent of j, the second claim follows with $u_j = t_j w_j$.

Remark. The second claim is also valid if $\mathcal{X}$ is strongly pseudofiltered, because then all Y_j belong to the same connected component of $\mathcal{X}$.

9.4.7 Filtered colimits can be described in Ab as in Ens. 9.3.7 and 9.1.4 reduce the pseudofiltered case to the filtered one in any case. So let $\mathcal{X}$ be a small filtered category and $T\colon \mathcal{X} \to Ab$ a functor. We deviate from 8.4.4 and replace 9.4.2 (2) not by the direct sum, but we first construct the colimit for UT by 9.4.2, where $U\colon Ab \to Ens$ is the forgetful functor. Addition for the colimit object L is defined as follows:

Let α, β be elements of L. By 9.4.6 they have representatives a, b in a set $T(Y)$. We set

$$\alpha + \beta = [Y, a] + [Y, b] = [Y, a + b] . \tag{4}$$

It has to be shown that this defines an addition on L. If a', b' are also representatives on $T(Y)$, then $(Y, a + b)$ and $(Y, a' + b')$ are equivalent by 9.4.5 and (3). If further a', b' are representatives of α and β on $T(Y')$, then there are morphisms $u\colon Y \to Z$, $u'\colon Y' \to Z$, and $T(u)\,(a + b)$ and $T(u')\,(a' + b')$ are equivalent by what we just proved above. 9.4.6 makes the addition defined in (4) associative and thus L is an additive group. The maps given by the rule $a \mapsto [Y, a]$ for $a \in T(Y)$ are now homomorphisms, and it follows easily from (3) that we now have a colimit in Ab.

9.4.8 The constructions of filtered colimits in Ab can be carried out for other algebraic structures, where the algebraic operations on the colimit object L are defined as in (4). This applies in particular to ${}_R Mod$, the category of groups, rings, R-algebras over a commutative ring R. This also implies that forgetful functors between such categories preserve and reflect filtered colimits. In particular:

Proposition. *The forgetful functor $Ab \to Ens$ preserves and reflects filtered colimits.*

9.4.9 Remark. 9.2.3 contains as a special case a well known theorem for direct limits in the classical sense. In a pseudofiltered category $\mathscr{X}$ a subcategory $\mathscr{Y}$ is final if

(i) For $X \in |\mathscr{X}|$ there is always a morphism $X \to Y$ with $Y \in |\mathscr{Y}|$.

(ii) For any two morphisms $u_1\colon X \to Y_1$, $u_2\colon X \to Y_2$ in $\mathscr{X}$ with $Y_1, Y_2 \in |\mathscr{Y}|$ there are morphisms $p_1\colon Y_1 \to Y$, $p_2\colon Y_2 \to Y$ in $\mathscr{Y}$ with $p_1 u_1 = p_2 u_2$.

For a full subcategory $\mathscr{Y}$ of $\mathscr{X}$ (ii) follows from (i). This is true in particular, if $\mathscr{X}$ is a preordered set (directed to the right) and if the subset $\mathscr{Y}$ is considered with the induced preordering.

9.4.10 In dualizing filtered colimits it is customary not to replace the domain of the functor $T\colon \mathscr{X} \to \mathscr{C}$ explicitly by its dual but instead to talk about a contravariant functor. A particular reason for this is, that in applications $\mathscr{X}$ is frequently the ordered set of natural numbers. By a *pseudofiltered* or, resp., *filtered limit* we therefore mean a limit of a contravariant functor $T\colon \mathscr{X} \to \mathscr{C}$, where $\mathscr{X}$ is again a pseudofiltered or, resp., a filtered small category. Inverse or projective limits in the classical sense are special cases ($\mathscr{X}$ is an ordered set, directed to the right).

9.5 Commutativity Theorems

9.5.1 Proposition. *In Ens strongly pseudofiltered colimits commute with generalized pullbacks, and pseudofiltered colimits commute with pullbacks.*

Proof. Let $\mathscr{X}$ be a small, strongly pseudofiltered category and

(1) $$\{\alpha_j: L \to R_j,\ \beta_j: R_j \to S\}$$

a generalized pullback in $[\mathscr{X}, Ens]$. Construct colimits for the functors $L, R_j, S: \mathscr{X} \to Ens$ according to 9.4.2. If the colimit objects and the maps induced between them by α_j, β_j are denoted by bars, then

(2) $$\{\bar{\alpha}_j: \bar{L} \to \bar{R}_j,\ \bar{\beta}_j: \bar{R}_j \to \bar{S}\}$$

is certainly commutative, i. e., $\bar{\beta}_j \bar{\alpha}_j$ is independent of j. It has to be shown that (2) is even a generalized pullback.

Because of the "pointwise" construction of limits in $[\mathscr{X}, Ens]$, (1) is at every point $Y \in |\mathscr{X}|$ a generalized pullback in *Ens*, and by 9.3.2 and 7.4.4, $L(Y)$ can be described by

(3) $$L(Y) = \{\{r_j\} \mid r_j \in R_j(Y),\ \beta_{j,Y}(r_j) \text{ is independent of } j\},$$

where, for every index k, $\alpha_{k,Y}(\{r_j\}) = r_k$. The proof is now completed by showing that the elements of $\bar{L}$ can be described in a unique way as

(4) $$\{\bar{r}_j\} \mid \bar{r}_j \in \bar{R}_j,\ \bar{\beta}_j(\bar{r}_j) \quad \text{independent of } j$$

with $\bar{\alpha}_k(\{\bar{r}_j\}) = \bar{r}_k$.

As $\bar{\alpha}_j, \bar{\beta}_j$ are induced by α_j, β_j, every element $\{r_j\}$ of $L(Y)$ determines a family (4) with $\bar{r}_j = [Y, r_j]$. If $(Y, \{r_j\})$ and $(Y', \{r_j'\})$ are representatives of the same element of $\bar{L}$, then appropriately chosen morphisms $u: Y \to Z$, $u': Y' \to Z$ in $\mathscr{X}$ yield $L(u)(\{r_j\}) = L(u')(\{r_j'\}) \in L(Z)$. Since α_j, β_j are natural transformations of functors, $L(u)(\{r_j\}) = \{R_j(u)(r_j)\} = \{R_j(u')(r_j')\}$, and it follows that every element of $\bar{L}$ determines a family (4) in a unique way.

Now, conversely, assume that the family (4) is given and let (Y_j, a_j) be a representative of $\bar{r}_j$ for every j. Since $\bar{\beta}_j$ is induced by β_j, $(Y_j, \beta_{j,Y_j}(a_j))$ is a representative of $\bar{z} = \bar{\beta}_j(\bar{r}_j)$. By 9.4.6, there is a $Z \in |\mathscr{X}|$ with morphisms $u_j: Y_j \to Z$ such that

(5) $$z = S(u_j)\, \beta_{j,Y_j}(a_j) \in S(Z)$$

is independent of j. Let $r_j = R_j(u_j)(a_j) \in R_j(Z)$. Because all β_j are natural transformations of functors, $\beta_{j,Z}(r_j) = z$. And since $[Z, r_j] = [Y_j, a_j] \in \bar{R}_j$, (3) implies that $\{r_j\}$ is an element of $L(Z)$ with $[Z, r_j] = \bar{r}_j$ for all j. So $[Z, \{r_j\}]$ is an element of $\bar{L}$ associated with the given family (4). It remains to be shown that there is only one such element in $\bar{L}$.

So let $[Z', \{r'_j\}]$ also be an element of $\overline{L}$ with representative $\{r'_j\} \in L(Z')$ and $[Z', r'_j] = \overline{r}_j = [Z, r_j]$ for all j. Here

$$[Z, \beta_{j,Z}(r_j)] = \overline{\beta}_j(\overline{r}_j) = [Z', \beta_{j,Z'}(r'_j)]$$

is valid. Now there are $u: Z \to Y$, $u': Z' \to Y$ in $\mathcal{X}$. As α_j and β_j are natural transformations, we can assume $Z = Z' = Y$ to simplify the notation; we then have $[Z, r_j] = [Z, r'_j]$ for all j. For every j there is a $W_j \in |\mathcal{X}|$ with morphisms $u_j: Z \to W_j$, $v_j: Z \to W_j$, so that $R_j(u_j)(r_j) = R_j(v_j)(r'_j)$, and by 9.3.4 (ii), W_j can be chosen such that $u_j = v_j$. The pencil $\{u_j: Z \to W_j\}$ has a commutative completion $\{t_j: W_j \to W\}$. One has

$$R_j(t_j u_j)(r_j) = R_j(t_j u_j)(r'_j) \in R_j(W) .$$

Since $t_j u_j$ does not depend on j, $[Z, \{r_j\}]$, $[Z, \{r'_j\}]$ and $[W, \{R_j(t_j u_j)(r_j)\}]$ represent the same element of $\overline{L}$, which proves the first part of the proposition. The second part follows similarly.

9.5.2 Theorem. *In Ens filtered colimits (of fixed type) commute with finite limits (of fixed type) and strongly filtered colimits with limits.*

In ENS colimits of functors $T: \mathcal{X} \to ENS$, where $\mathcal{X}$ is a strongly filtered $\mathfrak{U}$-category, commute with limits of diagrams $D: \Sigma \to ENS$, where Σ is small with respect to $\mathfrak{U}$.

Remark. Note that by definition filtered categories are not empty.

Proof. Let Z be terminal in *Ens*, so that it is a set of one element. If $\mathcal{X}$ is connected and non-empty, $(Z, \{1_Z\})$ is by 9.1.3 a colimit of the constant functor $Z_{\mathcal{X}}$. From this and from 9.5.1 it follows that strongly filtered colimits commute with products in *Ens*. The commutativity with arbitrary limits follows thus from 9.5.1 by 7.8.8. The first part of the theorem follows also from 7.8.8 and 9.5.1. The theorem for *ENS*, too, follows from the proof of 9.5.1.

9.5.3 Proposition. *In Ab pseudofiltered colimits commute with finite limits and strongly filtered colimits commute with limits.*

Proof. The second claim, as well as the first claim for filtered colimits, follows by means of the forgetful functor $U: Ab \to Ens$ from 7.7.9, 9.4.8, and 9.5.2. By 9.3.7 and 9.1.4, the first claim follows in general, because in *Ab* coproducts, i.e. direct sums, commute with finite products and kernels (compare later with 14.5.5).

9.5.4 Remarks. 9.5.3 carries over to ${}_R Mod$. If pseudofiltered is replaced by filtered, then by 7.4.4, 7.4.5, and 9.4.8, corresponding statements are valid for categories given by other algebraic structures as, e.g., groups or rings.

In *Ens* and *Ab* pullbacks do not commute with coequalizers. Let $p: \mathbf{Z}_4 \twoheadrightarrow \mathbf{Z}_2$ be the only epimorphism and let e, a be the two automorphisms of $\boldsymbol{Z}_4$, where $a^2 = e$ is the identity morphism. There are two natural transformations of the pullback square corresponding to $\boldsymbol{Z}_4 \xrightarrow{p} \boldsymbol{Z}_2 \xleftarrow{p} \boldsymbol{Z}_4$ into itself given by applying e or a to both copies of $\boldsymbol{Z}_4$. The coequalizer of these natural transformations does not yield a pullback square. If this is preceded by applying the forgetful functor $Ab \to Ens$, then a counterexample in Ens is obtained.

9.5.5 Universal colimits. Let $\mathscr{C}$ be a finitely complete and Σ a small category. For a functor $T\colon \Sigma \to \mathscr{C}$, let there be given a natural transformation $\alpha\colon T \to A_\Sigma$ and a $\mathscr{C}$-morphism $u\colon B \to A$. Then there is a pullback in $[\Sigma, \mathscr{C}]$

(6)
$$\begin{array}{ccc} P & \xrightarrow{\xi} & B_\Sigma \\ {\scriptstyle\varrho}\downarrow & & \downarrow{\scriptstyle u_\Sigma} \\ T & \xrightarrow{\alpha} & A_\Sigma \end{array}$$

We assume that T has the colimit (L, λ). α induces a uniquely determined morphism $f\colon L \to A$ with

(7)
$$f_\Sigma \lambda = \alpha\,.$$

There are pullbacks for f and u:

(8)
$$\begin{array}{ccc} R & \xrightarrow{g} & B \\ {\scriptstyle h}\downarrow & & \downarrow{\scriptstyle u} \\ L & \xrightarrow{f} & A \end{array} \qquad\qquad \begin{array}{ccc} R_\Sigma & \xrightarrow{g_\Sigma} & B_\Sigma \\ {\scriptstyle h_\Sigma}\downarrow & & \downarrow{\scriptstyle u_\Sigma} \\ L_\Sigma & \xrightarrow{f_\Sigma} & A_\Sigma \end{array}$$

(7) produces a natural transformation $(\mu, \lambda, (1_B)_\Sigma, (1_A)_\Sigma)$ of the pullback (6) into the right one of (8), just because the latter is a pullback.

Definition. The *colimit* (L, λ) *of* T *is called universal,* if for every choice of A, B, $u\colon A \to B$ and $\alpha\colon T \to A_\Sigma$ the procedure described above results in a colimit (R, μ) of P.

There is a different way to describe this. Consider the morphism $u\colon B \to A$ as a fibration with base A and the morphism (here h) opposite u in a pullback of the form of the left side of (8) as the fibration induced by the change of base with respect to f. If one observes that (6) is a pullback at every point $X \in |\Sigma|$, then the definition reads as follows. :

Universal colimits are those which commute with induced fibrations.

By 9.1.3 and 9.5.3 filtered colimits in Ab are universal. In Ens even more holds:

9.5.6 Theorem. *Colimits in Ens are universal.*

Proof. Because of 7.5.3, the pullback (6) is given at every point $X \in |\Sigma|$ by

(9) $$P(X) = \{(t, b) \mid \alpha_X(t) = u(b)\}$$

with the projections $(t, b) \mapsto t$ and $(t, b) \mapsto b$. For $q: X \to Y$ in Σ, $P(q)$ is described by

(10) $$(t, b) \mapsto (T(q)(t), b) .$$

The colimit of P exists and, by 8.4.4, it is given by equivalence classes of triples (X, t, b), where $(t, b) \in P(X)$, with respect to the equivalence relation generated by

(11) $$(X, t, b) \sim (Y, T(q)(t), b) \quad \text{for all} \quad q: X \to Y \text{ in } \Sigma .$$

L consists of equivalence classes of pairs (X, t), where $t \in T(X)$, with respect to the equivalence relation generated by

(12) $$(X, t) \sim (Y, T(q)(t)) \quad \text{foı all} \quad q: X \to Y \text{ in } \Sigma .$$

Here

(13) $$f[X, t] = \alpha_X(t) .$$

A comparison of (11) and (12) shows that, because of (9) and (13), the following holds:

If the left pullback in (8) is described as in 7.8.5, then R is the colimit object of P, and h and g are the evident projections. This proves the theorem for non-empty Σ. If in (8) $L = \emptyset$, then $R = \emptyset$, which completes the proof in case Σ is empty.

9.5.7 Remarks. 9.5.6 is not valid in Ab, finite coproducts in Ab are not universal.

In pullbacks as in (8) on the left, one frequently writes $R = L \sqcap_A B$ assuming that there is no doubt about the morphisms. 9.5.6 can then be expressed as follows

$$(\text{Colim } T) \sqcap_A B = \text{Colim } (T(X) \sqcap_A B)$$

An analogous notation $Q = B \sqcup_A C$ is used for pushouts.

9.6 Problems

9.6.1 For $n \in \boldsymbol{N}$ let the category $\mathscr{C}_n$ have exactly $2n + 1$ objects $A_0, A_1, \ldots, A_{2n}$ and $4n + 1$ morphisms, where apart from the identity morphisms, there is one morphism $A_{2\nu-1} \to A_{2\nu-2}$ and $A_{2\nu-1} \to A_{2\nu}$ for $\nu = 1, 2, \ldots, n$. Prove that $\prod_{n=1}^{\infty} \mathscr{C}_n$ is not connected.

9.6.2 Prove the following

Proposition. *Let $\mathscr{X}$ be a non-empty category and $\{Y_j\}_{j\in J}$ a set of objects of $\mathscr{X}$ such that*

(i) *For every $X \in |\mathscr{X}|$ there is at least one morphism $X \to Y_j$ for a suitable Y_j.*

(ii) *For every pair $(j, k) \in J \times J$ there exists in $\mathscr{X}$ a weak product (compare 7.1.11) with projections*

$$pr_{jk,1}\colon Y_j \sqcap Y_k \to Y_j \quad \text{and} \quad pr_{jk,2}\colon Y_j \sqcap Y_k \to Y_k .$$

If $\mathscr{C}$ is a cocomplete category and $T\colon \mathscr{X} \to \mathscr{C}$ a functor, then T has a colimit which is constructed as follows: let

$$p, q\colon \amalg\, T(Y_j \sqcap Y_k) \to \amalg\, T(Y_j)$$

be the two morphisms defined by

$$p\, in_{jk} = in_j\, T(pr_{jk,1}) \quad \text{and} \quad q\, in_{jk} = in_k\, T(pr_{jk,2}) ,$$

and let $c\colon \amalg\, T(Y_j) \to L$ be a coequalizer of p and q. The set of morphisms $\{\xi_j = c\, in_j\}$ can be extended in exactly one way to a natural transformation $\xi\colon T \to L_{\mathscr{X}}$, and (L, ξ) is a colimit of T.

Hints for the proof: (a) Construct a category Σ and a functor $D\colon \Sigma \to \mathscr{X}$ such that the morphisms of the form $D(a)$ are: the identity morphisms of all Y_j, of all $Y_j \sqcap Y_k$ and all projections $pr_{jk,1}$, $pr_{jk,2}$.

(b) Show that D is final.

(c) Show that TD has the colimit it is said to have.

Question: Why is it essential that all weak products $Y_j \sqcap Y_j$ exist?

9.6.3 Prove the statements in 9.3.5 and 9.3.6

9.6.4 Prove that 9.4.7 actually does provide a colimit in *Ab*.

9.6.5 Prove 9.4.9.

9.6.6 Verify the counterexample in 9.5.4.

9.6.7 Find an example to document the first remark in 9.5.7.

9.6.8 Prove: In *Ens* equalizers commute with coproducts. What are the facts in *Cat* and *Top*?

9.6.9 A *cardinal* α is called *regular* if the following holds: If $\{M_j\}_{j\in J}$ is a non-empty family of sets, such that the cardinality of J and of each M_j is smaller than α, then the cardinality of $\bigcup M_j$ is smaller than α. A *category* is called α-filtered if 9.3.4 (ii), (iii) are valid as well as (i_s) for all pencils whose index set has a cardinality less than α. If (L, λ) is the limit of a diagram $D\colon \Sigma \to \mathscr{C}$, then (L, λ) is called an α-limit if the set of vertices and arrows of Σ is of cardinality less than α. Let α be an infinite, regular cardinal. Show:

In Ens α-limits commute with α-filtered colimits; i. e., with colimits of functors $T\colon \mathscr{X} \to Ens$, where $\mathscr{X}$ is a small α-filtered category.

Consider the special cases where α is the cardinality of $\boldsymbol{N}$ or, resp., the smallest non-countable cardinal.

10. Setvalued Functors

10.1 Properties Inherited from the Codomain Category

10.1.1 As a mnemonic device and heuristic principle it may be stated that "nice" properties of a category $\mathscr{C}$ are inherited by the functor categories $[\mathscr{D}, \mathscr{C}]$. If, e. g., $\mathscr{C}$ has an initial, terminal or zero object, then the same is true for $[\mathscr{D}, \mathscr{C}]$ as is shown by the appropriate constant functors. If $\mathscr{C}$ is additive, then $[\mathscr{D}, \mathscr{C}]$ is also additive, if addition of natural transformations $\alpha, \beta \colon S \to T$ between functors $S, T \colon \mathscr{D} \to \mathscr{C}$ is defined "pointwise" with respect to $\mathscr{D}$, i. e., if $(\alpha + \beta)_A = \alpha_A + \beta_A$ for every $A \in |\mathscr{D}|$. By 7.5.2, 7.5.3, and 8.5.1, the existence of limits and colimits (of a fixed type or in general) is inherited by $[\mathscr{D}, \mathscr{C}]$ from $\mathscr{C}$. The "pointwise" construction employed for this implies other inherited properties.

10.1.2 Proposition. *Let $\mathscr{C}$ be finitely complete and have filtered colimits. If in $\mathscr{C}$ finite limits commute with filtered colimits, then they also commute in $[\mathscr{D}, \mathscr{C}]$.*

Proof. Because of 7.8.8 and 9.1.3, it suffices to show that pullbacks commute with filtered colimits. Let

$$(1)\qquad \begin{array}{ccc} P & \xrightarrow{\alpha} & R \\ {\scriptstyle\gamma}\downarrow & & \downarrow{\scriptstyle\delta} \\ T & \xrightarrow{\beta} & S \end{array}$$

be a pullback in $[\mathscr{X}, [\mathscr{D}, \mathscr{C}]]$, where $\mathscr{X}$ is small and filtered. Let $(\overline{P}, \pi)$ be a colimit of P, and correspondingly $(\overline{T}, \tau)$, $(\overline{R}, \varrho)$, $(\overline{S}, \sigma)$ colimits of T, R, S. Then by the definition of colimits, there is a commutative square in $[\mathscr{D}, \mathscr{C}]$:

$$(2)\qquad \begin{array}{ccc} \overline{P} & \xrightarrow{\bar{\alpha}} & \overline{R} \\ {\scriptstyle\bar{\gamma}}\downarrow & & \downarrow{\scriptstyle\bar{\delta}} \\ \overline{T} & \xrightarrow{\bar{\beta}} & \overline{S} \end{array}$$

where $\bar{\alpha}, \bar{\beta}, \bar{\gamma}, \bar{\delta}$ are determined by the corresponding natural transformations for functors $\mathscr{X} \to [\mathscr{D}, \mathscr{C}]$ in (1). Here $(\pi, \tau, \varrho, \sigma)$ is a natural transformation of (1) into the square which is obtained from (2) by adding the index $\mathscr{X}$ everywhere. Furthermore, there is a pullback in $[\mathscr{D}, \mathscr{C}]$, namely

(3)
$$\begin{array}{ccc} M & \xrightarrow{x} & \overline{R} \\ {\scriptstyle m}\downarrow & & \downarrow{\scriptstyle \overline{\delta}} \\ \overline{T} & \xrightarrow{\overline{\beta}} & \overline{S} \end{array},$$

and there is a uniquely determined morphism $j\colon \overline{P} \to M$ with $x\,j = \overline{\alpha}$ and $m\,j = \overline{\gamma}$. Now, j is a natural transformation of functors $\overline{P}$, M: $\mathscr{D} \to \mathscr{C}$, and (2), (3) can be evaluated at any point $A \in |\mathscr{D}|$. By the assumption for $\mathscr{C}$, $j_A\colon \overline{P}(A) \to M(A)$ is an isomorphism. Therefore j is an isomorphism and hence (2) a pullback.

Remark. Obviously, there is an analogous proposition for limits and strongly filtered colimits.

10.1.3 Proposition. *Let $\mathscr{C}$ be cocomplete and finitely complete. If colimits in $\mathscr{C}$ are universal, then they are universal in $[\mathscr{D}, \mathscr{C}]$.*

The proof of 10.1.2 can be applied with suitable modifications.

10.1.4 Proposition. *Let $\mathscr{C}$ be finitely complete. A natural transformation $\beta\colon S \to T$ of functors $\mathscr{D} \to \mathscr{C}$ is a monomorphism in $[\mathscr{D}, \mathscr{C}]$ if and only if $\beta_A\colon S(A) \to T(A)$ is a monomorphism for every $A \in |\mathscr{D}|$. If $\mathscr{C}$ is finitely cocomplete, then β is an epimorphism if and only if every β_A is an epimorphism.*

Proof. $\mathscr{D} \neq \emptyset$ may be assumed. For the first statement we consider the pullback

$$\begin{array}{ccc} R & \xrightarrow{\sigma} & S \\ {\scriptstyle \varrho}\downarrow & & \downarrow{\scriptstyle \beta} \\ S & \xrightarrow{\beta} & T \end{array}$$

in $[\mathscr{D}, \mathscr{C}]$. There is exactly one natural transformation $\tau\colon S \to R$ of functors $R, S\colon \mathscr{D} \to \mathscr{C}$ with $\sigma\tau = \varrho\tau = 1_S$. Here τ is an isomorphism if and only if this is true for τ_A for all $A \in |\mathscr{D}|$ (see 2.6.7 and 3.4.3). Thus the first statement follows from 7.8.9; the second is its dual.

10.1.5. Corollary. *If $\mathscr{C}$ is finitely complete, finitely cocomplete and balanced (i.e., every bimorphism is an isomorphism), then the same is the case for $[\mathscr{D}, \mathscr{C}]$.*

10.1.6 The assumptions in 10.1.2 through 10.1.5 are satisfied in particular by $\mathscr{C} = Ens$. In Ens there is a canonical factorization for every morphism $f\colon M \to N$, namely $M \overset{f'}{\twoheadrightarrow} f(M) \overset{i}{\rightarrowtail} N$, where $f = i\,f'$

and i is the inclusion of the image. f' is an epimorphism. If

$$\begin{array}{ccc} M & \xrightarrow{f} & N \\ {\scriptstyle u}\downarrow & & \downarrow{\scriptstyle v} \\ P & \xrightarrow{g} & Q \end{array}$$

is commutative in *Ens*, by restricting v one gets a map $v': f(M) \to g(P)$ with $g' u = v' f'$ and $j v' = v i$, where $g = j g'$ is the canonical factorization of g. Now there is another consequence.

Proposition. *Every morphism* $\alpha: S \to T$ *in* $[\mathscr{C}, Ens]$ *admits a canonical factorization* $\alpha = \iota \pi$, *where* π *is an epimorphism and* ι *a monomorphism.* (*Compare later with* 12.4.10).

10.1.7 Proposition. *In* $[\mathscr{C}, Ens]$ *every epimorphism* $\pi: S \to H^X$ *is a retraction.*

Proof. By 10.1.4, there is an $a \in S(X)$ with $\pi_X(a) = 1_X$, and by 4.2.1, there exists an $\alpha: H^X \to S$ with $\alpha_X(1_X) = a$, and $\pi \alpha = 1_{H^X}$.

10.1.8 Proposition. *Let* $\mathscr{C}$ *be a* $\mathfrak{U}$*-category. The full embedding* $Ens \to ENS$ *induces a full embedding* $i: [\mathscr{C}, Ens] \to [\mathscr{C}, ENS]$. *This embedding preserves and reflects limits and colimits and therefore also mono- and epimorphisms. If* T *is an object of* $[\mathscr{C}, Ens]$ *and* $\alpha: S \to i(T)$ *a monomorphism in* $[\mathscr{C}, ENS]$, *then there is a monomorphism* $\mu: R \to T$ *in* $[\mathscr{C}, Ens]$ *and an isomorphism* $\varrho: S \to i(R)$ *with* $\alpha = i(\mu)\,\varrho$. *Analogously every epimorphism* $\alpha: i(T) \to S$ *is of the form* $\alpha = \varrho\, i(\mu)$, *where* ϱ *is an isomorphism and* μ *an epimorphism. A functor* $T: \mathscr{C} \to Ens$ *is representable if and only if* $i(T): \mathscr{C} \to ENS$ *is representable.*

Proof. That i is a full embedding follows immediately from the fact that *Ens* is a full subcategory of *ENS*. Limits and colimits are preserved because of the "pointwise" construction. They are reflected because i is fully faithful. The next two statements claim that there are no new "subobjects" and "quotients". The first follows from 10.1.6 with images, and the latter correspondingly with quotient sets, since for every set in *Ens* all quotient sets belong to *Ens*. The last statement is again implied by the fact that i is a full embedding.

10.1.9 The additive case. If $\mathscr{C}$ and $\mathscr{D}$ are additive categories, then 10.1.1 through 10.1.5 are valid for $Add(\mathscr{D}, \mathscr{C})$, since limits and colimits of additive functors are additive. For arbitrary $\mathscr{C}$, 10.1.6 carries over to $[\mathscr{C}, Ab]$, and for additive $\mathscr{C}$ to $Add\,(\mathscr{C}, Ab)$. 10.1.7 also transfers to $Add\,(\mathscr{C}, Ab)$. Similarly, 10.1.8 can be transferred to embeddings associated with the inclusion $Ab \to AB$.

The meaning of 10.1.8 is, roughly, that arbitrariness in the choice of a universe does not affect the results.

10.2 The Yoneda Embedding $H_*: \mathscr{C} \to [\mathscr{C}^0, Ens]$

10.2.1 Proposition. *Let $\mathscr{C}$ be a small category. Every functor T: $\mathscr{C} \to Ens$ in $[\mathscr{C}, Ens]$ is a colimit of representable functors; more precisely, it is the colimit object of a functor $F: \Sigma \to [\mathscr{C}, Ens]$, where Σ is a small category and where every object of Σ goes to a covariant* Hom-*functor (of the form H^A with $A \in |\mathscr{C}|$). If $\mathscr{C}$ is not small, then the corresponding statement holds for a category Σ.*

Proof. If T is the constant functor $\Phi_{\mathscr{C}}$, then T is initial in $[\mathscr{C}, Ens]$ and colimit of the trivial functor F, where Σ is the empty category. Now let $T \neq \Phi_{\mathscr{C}}$.

We take the following category for Σ: Objects are natural transformations $\alpha: H^A \to T$ for $A \in |\mathscr{C}|$, morphisms from $\beta: H^B \to T$ to $\alpha: H^A \to T$ are triples (β, α, f) with $f: A \to B$ in $\mathscr{C}$ and $\beta = \alpha H^f$.

(1) 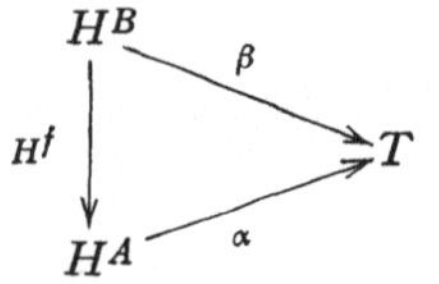

$f: A \to B$

$F: \Sigma \to [\mathscr{C}, Ens]$ is the functor which takes $\alpha: H^A \to T$ into H^A and (β, α, f) into H^f. (1) gives a natural transformation $\lambda: F \to T_\Sigma$ with $\lambda_\alpha = \alpha$. Now let $R: \mathscr{C} \to Ens$ be any functor and $\psi: F \to R_\Sigma$ a natural transformation of functors $\Sigma \to [\mathscr{C}, Ens]$. ψ_α is a morphism in $[\mathscr{C}, Ens]$, so $F(\alpha) \to R$ is a natural transformation $\alpha': H^A \to R$. If β' has the corresponding meaning for β, then

(2) 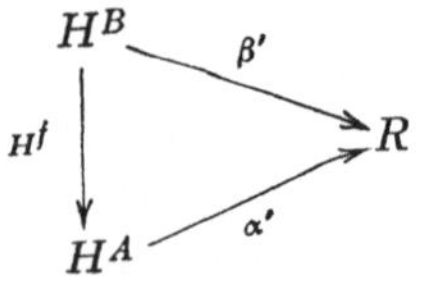

is commutative (compare 8.1.1 (1)). The Yoneda map Y of 4.2.1 now determines a map

$$\sigma_A: T(A) \to R(A)$$

by means of the rule

(3) $$a \mapsto (Y^{-1}(a) = \alpha) \mapsto \alpha' \mapsto Y(\alpha').$$

By theorem 4.2.4, there is an isomorphism Y: $[H^?, T] \to T(?)$, and (1) means that $T(f)\big(Y(\alpha)\big) = Y(\beta)$ (compare 4.2.4 (7)). From (2) $R(f)\big(Y(\alpha')\big) = Y(\beta')$ follows similarly. (3) then implies that $\{\sigma_A\}$ is a natural transformation $\sigma: T \to R$. For any $\alpha: H^A \to T$, $\sigma\alpha = \alpha'$ holds because of (3) and 4.2 (2). Conversely, $\sigma\alpha = \alpha'$ implies (3), so σ is determined uniquely and (T, λ) is a colimit of F.

Now let $\mathscr{C}$ be small. Then $|\Sigma|$ is the union of the disjoint sets $[H^A, T]$ for $A \in |\mathscr{C}|$ and is thus a $\mathfrak{U}$-set. It now follows that the morphisms from β to α form a $\mathfrak{U}$-set, and these sets are pairwise disjoint. If $\mathscr{C}$ is not small, then Σ is isomorphic to a $\mathfrak{U}$-category via the Yoneda map Y. Namely, Y determines bijections

(4) $\{\alpha \mid \alpha\colon H^A \to T,\ A \in |\mathscr{C}|\} \rightarrowtail\!\!\twoheadrightarrow \{(A, a) \mid a \in T(A),\ A \in |\mathscr{C}|\}$,

(5) $\{(\beta, \alpha, f) \mid \beta = \alpha\, H^f\} \rightarrowtail\!\!\twoheadrightarrow \{(f, T(f))\colon (A, a) \to (B, b) \mid b = T(f)\, a\}$.

Actually, one recognizes that the proposition only expresses the fact that T is determined by the elements of all $T(A)$ and the way the maps $T(f)$ work. The statement made by it is not this triviality, but the relation to the concept of colimits.

10.2.2 Corollary. *Let $\mathscr{C}$ be small. If $R, T\colon \mathscr{C} \to Ens$ are functors, then $[T, R]_{[\mathscr{C}, Ens]}$ is a limit object of the contravariant functor $G\colon \Sigma \to Ens$ for which $G(\alpha) = R(A)$ and $G(\beta, \alpha, f) = R(f)$. Here Σ, α and β have the same meaning as in* (1). *If $\mathscr{C}$ is not small, one obtains $[T, R]$ as a limit in ENS.*

Proof. If T is initial and thus Σ empty, then $[T, R]$ has only one element and is therefore terminal in *Ens*. Now let T not be initial. By 10.2.1 and 8.7.3, $[T, R]$ is a limit object of the contravariant functor $[F(?), R]\colon \Sigma \to Ens$, and by the Yoneda map 4.2.4 $[F(?), R]$ is isomorphic to $R(F(?)) = G$.

10.2.3 Remark. The category Σ used in 10.2.1 and 10.2.2 can also be described as follows: Consider in $[\mathscr{C}, Ens]$ the category of "objects over T" as in 6.5.3 and in it the subcategory generated by the Yoneda embedding $H^*\colon \mathscr{C}^0 \to [\mathscr{C}, Ens]$ of 4.2.2. From now on we write $\mathscr{C}^0/T$ or H^*/T for this category.

10.2.4 The contravariant functors $\mathscr{C} \to Ens$ "are" the covariant functors $\mathscr{C}^0 \to Ens$. If in 10.2.1 and 10.2.2 $\mathscr{C}$ is replaced by $\mathscr{C}^0$, then by the remarks in 8.7.1, one gets corresponding statements for contravariant functors $\mathscr{C} \to Ens$, where H^A, H^f have to be replaced by H_A, H_f, as well as $\beta = \alpha\, H^f$ by $\alpha = \beta\, H_f$. Note that one has again a colimit in 10.2.1 and a limit in 10.2.2.

10.2.5 Theorem. *The Yoneda embedding $H_*\colon \mathscr{C} \to [\mathscr{C}^0, Ens]$ preserves limits.*

Proof. First let Z be terminal in $\mathscr{C}$. Then for every $A \in |\mathscr{C}|$, $H_Z(A) = [A, Z]$ has only one element. This implies that H_Z is terminal in $[\mathscr{C}^0, Ens]$. Now let $T\colon \Sigma \to \mathscr{C}$ be a non-empty diagram. Since $[\mathscr{C}^0, Ens]$ is complete, $H_*\, T\colon \Sigma \to [\mathscr{C}^0, Ens]$ has a limit, which is constructed pointwise. One has $H_*\, T = [?, T(??)]_{\mathscr{C}}$ with ? in $\mathscr{C}$ and ?? in Σ. At $A \in |\mathscr{C}|$ one gets $H^A\, T$, and because of 7.7.1, $N_T(A)$ is obtaind as the limit with evident projections. It follows that $N_T(?)$ is a limit object of $H_*\, T$. According to 7.1.4, a limit (L, λ) of T in $\mathscr{C}$ is given by the isomorphism $\varrho: H_L \to N_T$ with $\varrho_L(1_L) = \lambda$, which completes the proof.

10.2.6 Note. H_* reflects limits and colimits because it is fully faithful. However, in general, H_* does not preserve colimits. This does not follow from 10.2.1 through 10.2.3, because either $\mathscr{C}$ is small and hence in general not cocomplete, or colimits are formed with respect to a category that is not small.

Let $\mathscr{C} = Ab$, $A = B = \mathbf{Z}$, $C = \mathbf{Z}_2$, and let $m: A \to B$ be multiplication by 2 and $c: B \to C$ the cokernel of m. The embedding H_*: $Ab \to [Ab^0, Ens]$ produces the following diagram:

$$[?, \mathbf{Z}]_{Ab} \underset{[?,0]}{\overset{[?,m]}{\rightrightarrows}} [?, \mathbf{Z}]_{Ab}\,.$$

The colimit at the place $\mathbf{Z}_2$ is the set of one element $[\mathbf{Z}_2, \mathbf{Z}]$ (with its identity morphism). On the other hand $[\mathbf{Z}_2, \mathbf{Z}_2]$ has two elements. So $[?, \mathbf{Z}_2]$ is not a colimit of the diagram.

Later (in 17.3.2) it will turn out that, in a certain sense, H_* forgets the existing colimits in $\mathscr{C}$.

10.2.7 The additive case. If $\mathscr{C}$ is an additive category, then 10.2.5 is valid analogously for $Add(\mathscr{C}, Ab)$ and $[\mathscr{C}, Ab]$. The proof uses the additive version of 7.7.1 given in 7.7.8. 10.2.1 and 10.2.2 are not valid for $[\mathscr{C}, Ab]$, since colimits of additive functors are always additive (8.5.3). This makes itself felt in the proof through the fact that the Yoneda map is not available for $[\mathscr{C}, Ab]$. For $Add(\mathscr{C}, Ab)$, 10.2.1 and 10.2.2 are in general valid only if $\mathscr{C}$ possesses finite products. Then one can conclude that σ_α in (3) is a homomorphism (compare later 12.2.6 and 17.2.10). Otherwise this need not be the case. If $\mathscr{C}$ has as its only object the additive group $\mathbf{Z}$ of the integers and as morphisms the endomorphisms of $\mathbf{Z}$, then $Add(\mathscr{C}, Ab)$ is isomorphic to Ab, and, e.g., 10.2.1 is not true for the functor which assigns to the only object of $\mathscr{C}$ the group $\mathbf{Z} \oplus \mathbf{Z}$ in Ab. Instead, one obtains as a colimit, according to 8.2.4 and 8.3.2, a direct sum (coproduct) of countably many summands $\mathbf{Z}$.

10.3 The General Representation Theorem

10.3.1 Definition. A *functor* T: $\mathscr{C} \to Ens$ is called *proper* if in $\mathscr{C}$ there is a set $\mathfrak{D}$ of objects with the following property:

If X is any object of $\mathscr{C}$ and if $x \in T(X)$, then for a suitable object D in $\mathfrak{D}$ there is a $d \in T(D)$ and a morphism $f: D \to X$ with $x = T(f)\,(d)$.

$\mathfrak{D}$ is then called a *dominating set* for T. The corresponding definition is made for additive categories and additive functors with values in Ab.

10.3.2 The definition admits a "functorial" formulation. For a functor T: $\mathscr{C} \to Ens$ and an arbitrary set $\mathfrak{D}$ of objects of $\mathscr{C}$, let $\mathfrak{D}(T)$ be the set of all pairs (D, δ) consisting of a $D \in \mathfrak{D}$ and a natural transformation $\delta\colon H^D \to T$. Further, let $a\colon \mathfrak{D}(T) \to [\mathscr{C}, Ens]$ be the map $(D, \delta) \mapsto$ $\mapsto H^D$; i.e., $a(D, \delta) = H^D$. There is a canonical morphism

$$\psi(\mathfrak{D}, T)\colon \coprod_{\mathfrak{D}(T)} a(D, \delta) \to T$$

with $\psi(\mathfrak{D}, T)\, in_{(D,\delta)} = \delta\colon H^D \to T$. (If $\mathfrak{D}(T)$ is empty, the initial functor has to be taken as the coproduct, naturally). Now one has the

Proposition. *$T\colon \mathscr{C} \to Ens$ is proper with dominating set $\mathfrak{D}$ if and only if $\psi(\mathfrak{D}, T)$ is an epimorphism.*

Proof. By 10.1.4 (with $[\mathscr{C}, Ens]$ instead of $[\mathscr{D}, \mathscr{C}]$), the morphism $\psi(\mathfrak{D}, T)$ is an epimorphism if and only if it is an epimorphism at every place $X \in |\mathscr{C}|$, i.e., if for every $x \in T(X)$ there is a $(D, \delta) \in \mathfrak{D}(T)$ and an $f \in H^D(X) = [D, X]_{\mathscr{C}}$ such that $\delta_X(f) = x$. By the Yoneda lemma 4.2.1 (2), this is exactly the condition of the definition.

10.3.3 Examples. If T is representable with representing object A, then T is proper with dominating set $\{A\}$.

If $\mathscr{C}$ is small, then every functor $\mathscr{C} \to Ens$ is proper.

10.3.4 Let $\mathscr{C}$ be a non-empty category. For $A \in |\mathscr{C}|$, $H^A\colon \mathscr{C} \to Ens$ preserves all existing limits in $\mathscr{C}$ by 7.7.4. The Yoneda embedding $H^*\colon$ $\mathscr{C}^0 \to [\mathscr{C}, Ens]$ induces therefore a full embedding $H^*_{\mathfrak{K}}\colon \mathscr{C}^0 \to \mathfrak{K}[\mathscr{C}, Ens]$, where $\mathfrak{K}[\mathscr{C}, Ens]$ has the same meaning as in 7.6.5 and H^* is the composite of $H^*_{\mathfrak{K}}$ and the inclusion

$$\mathfrak{K}[\mathscr{C}, Ens] \subset [\mathscr{C}, Ens]\,.$$

From 7.6.5 and 10.2.5 it follows immediately that $H^*_{\mathfrak{K}}$ preserves the existing limits in $\mathscr{C}^0$.

10.3.5 Proposition. *With the notations from above the following holds: Colimits in $\mathscr{C}^0$ of diagrams whose type is dual to one of the class $\mathfrak{K}$ are preserved by $H^*_{\mathfrak{K}}\colon \mathscr{C}^0 \to \mathfrak{K}[\mathscr{C}, Ens]$; in particular, colimits are preserved by $H^*_{\mathscr{I}}\colon \mathscr{C}^0 \to \mathscr{L}[\mathscr{C}, Ens]$, finite colimits by $H^*_l\colon \mathscr{C}^0 \to l[\mathscr{C}, Ens]$, finite coproducts by $H^*_\pi\colon \mathscr{C}^0 \to \pi[\mathscr{C}, Ens]$.*

Proof. Let Σ be a diagram scheme in $\mathfrak{K}$ and $D\colon \Sigma \to \mathscr{C}$ a diagram with a limit (L, λ) in $\mathscr{C}$. This is equivalent to saying that $(\mathrm{Op}(L), \mathrm{Op}\,\lambda)$ is a colimit of $\mathrm{Op}\, D\, \mathrm{Op}\colon \Sigma^0 \to \mathscr{C}^0$. For an arbitrary functor $T\colon \mathscr{C} \to Ens$ one has $[H^*\,\mathrm{Op}\, D, T] \cong T\,D\colon \Sigma \to Ens$ by Theorem 4.2.4. For $T \in |\mathfrak{K}[\mathscr{C}, Ens]|$, $(T(L), T\lambda)$ is a limit of $T\,D$ and thus $([H^L, T],$ $[H^\lambda, T])$ is a limit of $[H^*\,\mathrm{Op}\, D, T]$, again by 4.2.4. $H^*\,\mathrm{Op}\, D$ may be considered as a contravariant diagram $H^*_{\mathfrak{K}}\,\mathrm{Op}\, D\colon \Sigma \to \mathfrak{K}[\mathscr{C}, Ens]$. By 8.7.5, (H^L, H^λ) is then a colimit of the diagram $H^*_{\mathfrak{K}}\,\mathrm{Op}\, D\,\mathrm{Op}$ in $\mathfrak{K}[\mathscr{C}, Ens]$, which completes the proof.

10.3.6 Remark. By 7.6.4, the inclusion $\mathscr{L}[\mathscr{C}, Ens] \subset [\mathscr{C}, Ens]$ preserves limits, but in general not colimits (compare 10.2.6). The preceding proof together with 8.7.5 yields the fact that $\mathscr{L}[\mathscr{C}, Ens]$ is the largest full subcategory of $[\mathscr{C}, Ens]$ for which H^* induces an embedding that preserves colimits.

10.3.7 Corollary. *If $\mathscr{C}$ is finitely complete and if $T: \mathscr{C} \to Ens$ preserves finite limits, then $\mathscr{C}^0/T$ in 10. 2. 3. is filtered.*

Proof. $\mathscr{C}^0/T$ can here be formed in $l[\mathscr{C}, Ens]$. Because of 10.3.5, the conditions (i), (ii), (iii) of 9.4.3 follow therefore in succession from the existence of pullbacks, equalizers, and finite products in $\mathscr{C}$. Furthermore $T \neq \Phi_{\mathscr{C}}$, since T preserves terminal objects and therefore $\mathscr{C}^0/T \neq$ $\neq \Phi$.

10.3.8 Proposition. *Let $\mathscr{C}$ be finitely complete, and let $T: \mathscr{C} \to Ens$ preserve finite limits. Then T is proper if and only if T is a colimit object of a (small) diagram of representable functors in $[\mathscr{C}, Ens]$.*

Proof. Let T be proper with a dominating set $\mathfrak{D}$, and let $\mathfrak{D}/T$ be the full subcategory of $\mathscr{C}^0/T$ whose objects are the natural transformations $\delta: H^D \to T$ for $D \in \mathfrak{D}$. By 9.4.9, 10.3.7 and the Yoneda lemma 4.2.1, the condition of definition 10.3.1 says that $\mathfrak{D}/T$ is final in $\mathscr{C}^0/T$, so that 9.2.3 together with 10.2.1 yields one part of the proof.

If, conversely, T is a colimit object of a diagram of representable functors, then the construction in 8.4.2 shows T to be the codomain of an epimorphism from a coproduct of a set of representable functors. By 10.3.2, this implies that T is proper.

10.3.9 Theorem. *Let $\mathscr{C}$ be a complete category. A functor $T: \mathscr{C} \to Ens$ is representable if and only if*

(i) *T is proper,*

(ii) *T preserves limits.*

Proof. If A is a representing object for T, then $\{A\}$ is a dominating set for T, and by 7.7.4, T preserves limits.

If, conversely, (i) and (ii) are assumed for T, then T is not the initial functor $\Phi_{\mathscr{C}}$, because $\mathscr{C}$ has a terminal object and T preserves limits. By 10.3.8, T is a colimit object of a diagram $D: \Sigma \to [\mathscr{C}, Ens]$, so that every $D(i)$ for $i \in |\Sigma|$ is of the form H^A for a suitable $A \in |\mathscr{C}|$. Therefore, there is an associated contravariant diagram $D': \Sigma \to \mathscr{C}$, which has a limit (L, λ) on account of the completeness of $\mathscr{C}$. By 10.3.5, $H^*_{\mathscr{I}}$ Op D' then has the colimit $(H^L, \{H^{\lambda_i}\})$. Now, since T is in $\mathscr{L}[\mathscr{C}, Ens]$, $H^*_{\mathscr{I}}$ Op D' is taken into D by $\mathscr{L}[\mathscr{C}, Ens] \subset [\mathscr{C}, Ens]$, and since this full embedding reflects colimits (8.7.6), H^L and T are isomorphic.

10.3.10 The analogues of the above statements, and in particular of 10.3.9, for additive categories $\mathscr{C}$ and additive functors $\mathscr{C} \to Ab$ are valid. Here 10.3.7 enables one to get the additive version of 10.2.1 needed for

10.3.8 from 9.4.8 and the "pointwise" construction of colimits in functor categories. The additive version of 10.3.9 follows immediately from 4.4.10 and 7.7.9.

10.3.11 The condition that T be proper is essential in 10.3.9. Let $\mathscr{C}$ be the ordered class of cardinals in $\mathfrak{U}$. $\mathscr{C}^0$ is complete (compare 14.1). A terminal functor $\mathscr{C}^0 \to Ens$ preserves limits and is not representable, since $\mathscr{C}$ does not have a terminal object. See, however, later in 10.6.5.

Let $\mathscr{C}$ be an arbitrary category. The functor F in 10.2.1 has the form $H^* F'$ with $F' \colon \Sigma \to \mathscr{C}^0$. Now one has: T is representable if and only if Op F' Op has a limit and if T preserves this limit. This follows by generalizing 7.6.5 and 10.3.5.

10.4 Projective and Injective Objects

10.4.1 Definition. An *object* P of a category $\mathscr{C}$ is called *projective* if H^P preserves epimorphisms. This is equivalent to: for every diagram

(1)
$$\begin{array}{ccc} & & P \\ & & \downarrow{\scriptstyle g} \\ A & \xrightarrow{f} & B \end{array}$$

where f is an epimorphism, there is at least one morphism $h \colon P \to A$ with $f\,h = g$.

10.4.2 Examples. In Ens and Ens_* every object is projective, in Top and Top_* every discret space, in Ab every free additive group, in ${}_RMod$ every free module. In the category of groups every free group is projective. Every initial object is trivially projective.

For $A \in |\mathscr{C}|$, H^A is projective in $[\mathscr{C}, Ens]$. Even the following holds (compare the dual of 7.8.9):

10.4.3 Proposition. *If $F \colon \mathscr{C} \to Ens$ is representable, then $H^F = [F, ?]_{[\mathscr{C}, Ens]} \colon [\mathscr{C}, Ens] \to ENS$ preserves colimits with respect to $\mathfrak{U}$.*

Proof. If $D \colon \Sigma \to [\mathscr{C}, Ens]$ is a diagram and $A \in |\mathscr{C}|$, then by 4.2.4 there is the Yoneda isomorphism

$$H^{H^A}(D) = [H^A, D(?)]_{[\mathscr{C}, Ens]} \cong D(?)(A)\,.$$

This means that D is considered at the „point" A, and then the pointwise construction of colimits in functor categories implies the statement for $F = H^A$ and thus in general.

Note that for a representable F, H^F is isomorphic to a functor $[\mathscr{C}, Ens] \to Ens$.

This proposition is valid correspondingly for $Add(\mathscr{C}, Ab)$ if $\mathscr{C}$ is additive.

10.4.4 Proposition. *Every coproduct of projective objects is projective. Every retract of a projective object is projective.*

Proof. The first statement follows immediately from the definition of coproducts. If $r: U \to P$ is a retraction, then there is an $i: P \to U$ with $r\,i = 1_P$. If we have case (1) and if U is projective, then there is a $k: U \to A$ with $f\,k = g\,r$ and one has $f\,k\,i = g\,r\,i = g$, so that P is projective.

10.4.5 Remark. Here and in the following we understand by coproducts and products the corresponding colimit or, resp., limit objects. As before, an initial or, resp., terminal object (if it exists) is to be regarded as a coproduct or, resp., product, with an empty index set.

10.4.6 Proposition. *If $j: A \to P$ is an epimorphism and P projective, then j is a retraction. If the category $\mathscr{C}$ has a zero object, then a coproduct $P = \coprod P_i$ is projective only if every P_i is projective.*

Proof. The first statement follows from (1) with $f = j$ and $g = 1_P$, and by 10.4.4, the second one is implied by the fact that the injections for a coproduct are here coretractions (dual of 7.3.4), so that the P_i are retracts of P.

10.4.7 Definition. A *category $\mathscr{C}$ has* (enough) *projectives* if every object is a quotient of a projective one; i.e., if for every $A \in |\mathscr{C}|$ there is always an epimorphism $P \twoheadrightarrow A$ with P projective.

The categories Ab, ${}_RMod$ and the category of groups have projectives (so do, trivially, Ens and Top). In Ab the projectives are exactly the free additive groups, in the category of groups exactly the free groups. In both cases every group is the quotient of a free one, so by 10.4.6 a projective one is isomorphic to a subgroup of a free one and hence itself free. If R is not a principal ideal ring, then projective objects in ${}_RMod$ are not necessarily free modules. In the category of rings the polynomial ring $\mathbf{Z}[x]$ is free, but not projective, as is shown by the epimorphism $\mathbf{Z} \subset \mathbf{Q}$.

10.4.1° Definition. An *object Q* of the category $\mathscr{C}$ is called *injective* if it is projective in $\mathscr{C}^0$; i.e., if $[?, Q]_{\mathscr{C}}$ carries monomorphisms into epimorphisms or, equivalently, if for every diagram

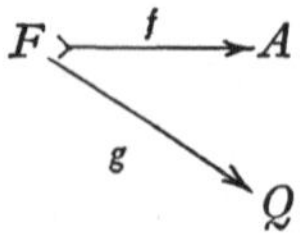

where f is a monomorphism, there is an $h: A \to Q$ with $h\,f = g$.

10.4.2° In Ens and in Ens_* every non-empty object is injective, and so is every non-empty space with the coarsest topology in Top.

In the full subcategory of *Top*, whose objects are the compact spaces, the unit interval is injective (by the Tietze-Urysohn extension theorem).

In *Ab* the additive group $\boldsymbol{Q}$ of rational numbers and the factor group $\boldsymbol{Q}/\boldsymbol{Z}$ are injective, also $\boldsymbol{R}$ (real numbers) and $\boldsymbol{R}/\boldsymbol{Z}$. *Ab* has (enough) injectives; i.e., every additive group can be embedded in an injective one. ${}_R Mod$ has injectives for any R (see 15.3.5). 10.4.4 and 10.4.6 can be dualized.

10.5 Generators and Cogenerators

10.5.1 Definition. A set $\mathfrak{G}$ of objects of a category $\mathscr{C}$ is called a *generating set* if for every pair of different morphisms $f, g\colon A \to B$ with the same domain and codomain there is a morphism $h\colon G \to A$ with $f\,h \neq g\,h$ and with $G \in \mathfrak{G}$. An object G is called a *generator* if $\{G\}$ is a generating set. This is equivalent to H^G being faithful; i.e., to H^G: $\mathscr{C} \to Ens$, or, resp., $H^G\colon \mathscr{C} \to Ab$ for additive $\mathscr{C}$, being an embedding.

Remark. This definition is void if $\mathscr{C}$ is a preordered class. Then every set of objects, even the empty one, is a generating set.

Instead of "generating" the word "coseparating" might be better. First, it would account for the fact that different elements of $[A, B]$ can still be recognized as different, and, second, it would constitute a systematic use of the prefix "co" for objects appearing as contravariant arguments of the Hom-functor. Grothendieck has a different definition of generator, however, in most applications the two definitions coincide (see 20.6.1).

10.5.2 Examples. The following are examples of generators: *Ens*, every non-empty set; *Top*, every discrete non-empty space; *Ab*, $\boldsymbol{Z}$; ${}_R Mod$, ${}_R R$; groups, every free cyclic group; rings (with unit), the polynomial ring $\boldsymbol{Z}[x]$. Apart from $\boldsymbol{Z}[x]$, the just mentioned generators are also projective. For a small category $\mathscr{C}$, $|\mathscr{C}|$ is a generating set and the set of all H^A is a generating set for $[\mathscr{C}, Ens]$ or, resp., for $Add(\mathscr{C}, Ab)$ in the additive case. For, let $\beta, \gamma\colon T \to R$ be natural tranformations such that for every $\alpha\colon H^A \to T$ one always has $\beta\alpha = \gamma\alpha$; then Theorem 4.2.4 implies that $\beta = \gamma$. If $\mathscr{C}$ is not small, then the universe has to be changed for $[\mathscr{C}, Ens]$ anyhow.

10.5.3 Proposition. *A coproduct of generators with a non-empty index set is a generator. If $\mathfrak{G}$ is a generating set and if for every non-initial $A \in |\mathscr{C}|$ there is a morphism $G \to A$ for every object $G \in \mathfrak{G}$, then the coproduct of all objects of $\mathfrak{G}$ (if it exists) is a generator.*

This follows immediately from the definition.

10.5.4 Proposition. *Let $\mathscr{C}$ have coproducts (and thus an initial object). A set $\mathfrak{G}$ of objects is a generating set if and only if for every $A \in |\mathscr{C}|$ the following holds: For*

$$(1) \qquad G_A = \coprod_{e \in \bigcup_{G \in \mathfrak{G}} [G, A]} G_e, \qquad G_e \text{ is the domain of } e,$$

the morphism $\pi_A: G_A \to A$ defined by $\pi_A\, in_e = e$ is an epimorphism.

This follows immediately from the definition of coproducts and from 10.5.1. Note that $\bigcup [G, A]$ may be empty.

10.5.5 Corollary. *A category with coproducts and a non-empty generating set of projective objects has projectives.*

10.5.1° Definition. A set $\mathfrak{G}$ of objects of the category $\mathscr{C}$ is called a *cogenerating set* if $\mathfrak{G}$ is a generating set for $\mathscr{C}^0$; i.e., if for every pair of different morphisms $f, g: A \to B$ there is a morphism $h: B \to G$ with $G \in \mathfrak{G}$ and $h f \neq h g$. An object G is called a *cogenerator* if $\{G\}$ is a cogenerating set. This is equivalent to the contravariant functor H_G being faithful; i.e., to H_G Op: $\mathscr{C}^0 \to Ens$ (resp. $\to Ab$) being an embedding.

10.5.2° Examples. In *Ens* every set of at least two elements is a cogenerator, and so is every space with the coarsest topology and containing at least two points in *Top*. In the full subcategory of *Top*, whose objects are the completely regular spaces, the unit interval is a cogenerator. In *Ab* $\mathbf{Q}/\mathbf{Z}$ is an injective cogenerator. ${}_R Mod$ always has an injective cogenerator, namely $[R_R, \mathbf{Q}/\mathbf{Z}]_{Ab}$, where first only the additive group of R is considered and then $[R, \mathbf{Q}/\mathbf{Z}]$ is made into a left R-module via the right operation of R on itself (see 15.3.5).

10.5.3 through 10.5.5 can be dualized; for 10.5.4 $\prod G_e$ is to be taken with $e \in \bigcup [A, G]$.

10.6 Well-powered Categories

10.6.1 Definition. A *category* $\mathscr{C}$ is called *well-powered* if for every object A the equivalence classes of monomorphisms with codomain A (see 6.5.4 through 6.5.8) have a set as a complete set of representatives. "For every object the subobjects form a set".

Instead of "well-powered" some authors use "locally small". However, other authors use "locally small" in a different sense.

10.6.2 Every small category is well-powered. *Ens, Top, Ab,* ${}_R Mod$ are well-powered with a natural choice, and so are the categories of groups and of rings.

10.6.3 Proposition. *A balanced category $\mathscr{C}$ with finite intersections of monomorphisms and with a generating set is well-powered.*

Proof. Let $\{G_\alpha\}$ be a generating set of $\mathscr{C}$. It may be assumed that $\{G_\alpha\}$ is not empty, since otherwise $\mathscr{C}$ would be a preordered class whose morphisms are isomorphisms. So, for $A \in |\mathscr{C}|$, we consider the set M of all morphisms $G_\alpha \to A$, i.e., $M = \bigcup [G_\alpha, A]$. To every monomorphism $m: A' \rightarrowtail A$ we assign the subset $N(m)$ of M which consists of the morphisms $G_\alpha \to M$ that factor through m (that is, those of the form $m f_\alpha$). We shall show that, if $m_1: A_1 \to A$ and $m_2: A_2 \to A$ are non-equivalent monomorphisms, then $N(m_1)$ and $N(m_2)$ are different. This implies the proposition. We consider the intersection

$$(1)\qquad \begin{array}{ccc} A_3 & \xrightarrow{n_2} & A_2 \\ {\scriptstyle n_1}\downarrow & & \downarrow{\scriptstyle m_2} \\ A_1 & \xrightarrow[m_1]{} & A \end{array}$$

(pullback). Here n_1 and n_2 are monomorphisms. If n_1 and n_2 are also epimorphisms, then n_1 and n_2 are isomorphic, since $\mathscr{C}$ is balanced, and so m_1 and m_2 are equivalent. Now suppose that n_2, for instance, is not an epimorphism. Then there are two morphisms $u, v: A_2 \to B$ with $u \neq v$ but $u\, n_2 = v\, n_2$. For a suitable G_α there is a morphism $f: G_\alpha \to A_2$ with $u f \neq v f$. Now, $m_2 f: G_\alpha \to A$ does not factor through m_1, since otherwise f would factor through n_2 by the definition of pullbacks. This is impossible because of $u\, n_2 = v\, n_2$.

10.6.4 Proposition. *Let $\mathscr{B}$ be a small and $\mathscr{C}$ a finitely complete, well-powered category. Then $[\mathscr{B}, \mathscr{C}]$ is well-powered. In particular, $[\mathscr{B}, Ens]$ is well-powered. In the additive case corresponding statements hold for $Add(\mathscr{B}, \mathscr{C})$ and $Add(\mathscr{B}, Ab)$.*

Proof. Let $T: \mathscr{B} \to \mathscr{C}$ be a functor. For every $A \in |\mathscr{B}|$, let a system $\mathfrak{M}_A$ of representatives for the monomorphisms with codomain $T(A)$ be chosen. If $\mu: S \to T$ is a monomorphism in $[\mathscr{B}, \mathscr{C}]$, then, by 10.1.4, $\mu_A: S(A) \to T(A)$ is a monomorphism for every $A \in |\mathscr{B}|$. Therefore, there is a $m_A \in \mathfrak{M}_A$ and an isomorphism ϱ_A such that $\mu_A = m_A\, \varrho_A$. For $f: A \to B$ in $\mathscr{B}$, let $S'(f) = \varrho_B\, S(f)\, \varrho_A^{-1}$ and let $S'(A)$ be the domain of m_A. Then S' is a functor, $\{\varrho_A\}: S \to S'$ is an isomorphism and $\{m_A\}: S' \to T$ a monomorphism. $\{m_A\}$ is a map of the set $|\mathscr{B}|$ into the set $\bigcup \mathfrak{M}_A$, and this proves the proposition. The additive case is treated similarly.

10.6.5 Special Representation Theorem. *Let $\mathscr{C}$ be a well-powered, complete category with a cogenerating set $\mathfrak{G}$. A functor $T: \mathscr{C} \to Ens$ is representable if and only if it preserves limits. If $\mathscr{C}$ is additive, then the same holds for an additive $T: \mathscr{C} \to Ab$.*

Proof. We assume that T preserves limits. It suffices, by 10.3.9, to show that T is proper. Let $\mathfrak{G} = \{G_\alpha\}$. We consider

$$(2)\qquad P = \prod_{G_\alpha} \prod_{x \in T(G_\alpha)} G_{\alpha, x} \quad \text{with} \quad G_{\alpha, x} = G_\alpha$$

and projections $pr_{\alpha,x}$ and, additionally, for $A \in |\mathscr{C}|$, corresponding to the dual of 10.5.4,

(3) $$Q = \prod_{e \in \bigcup [A, G_\alpha]} G_e\,, \qquad G_e \text{ is the codomain of } e,$$

with projections pr_e. We define $\Delta\colon A \to Q$ by

(4) $$pr_e\,\Delta = e\,.$$

Δ is a monomorphism (dual of 10.5.4). We may assume that $T(A) \neq \emptyset$. $a \in T(A)$ induces maps $[A, G_\alpha] \to T(G_\alpha)$ for all G_α by means of the rule $e \mapsto T(e)\,(a)$. With this,

(5) $$pr_e\,u = pr_{\alpha,\,T(e)\,(a)}\,, \qquad G_\alpha \text{ is the codomain of } e,$$

defines a morphism $u\colon P \to Q$. Now there is a pullback in $\mathscr{C}$, namely

(6) $$\begin{array}{ccc} M_a & \overset{m_a}{\rightarrowtail} & P \\ {\scriptstyle v_a}\downarrow & & \downarrow{\scriptstyle u} \\ A & \overset{\Delta}{\rightarrowtail} & Q \end{array}$$

Since Δ is a monomorphism, so is m_a. Application of T produces a (in general not canonically chosen) pullback in *Ens*, where $T(P)$ and $T(Q)$ are products with projections $T(pr_{\alpha,x})$ or, resp., $T(pr_e)$. In $T(P)$ there exists an element y with

(7) $$T(pr_{\alpha,x})\,(y) = x\,.$$

For $e\colon A \to G_\alpha$ and $x = T(e)\,(a)$, (4) and (5) thus imply

$$T(pr_e)\,T(\Delta)\,(a) = T(e)\,(a) = T(pr_{\alpha,T(e)(a)})\,(y) = T(pr_e)\,T(u)\,(y)\,,$$

and so $T(\Delta)\,(a) = T(u)\,(y)$. A comparison with (6) shows that there is an element z_a in $T(M_a)$ such that

(8) $$T(m_a)\,(z_a) = y \qquad \text{and} \qquad T(v_a)\,(z_a) = a\,.$$

u and Δ determine (6) only up to an isomorphism, so that only the equivalence class of the monomorphism m_a is determined. Since $\mathscr{C}$ is well-powered, the choice of representatives for the equivalence classes of monomorphisms with codomain P yields a dominating set for T.

10.6.6 Remarks. (a) This proposition and its proof are valid even if $\{G_\alpha\}$ is empty (compare 10.5.1). P and Q are then terminal objects.

(b) 10.6.5 implies that $\mathscr{C}$ (under the given assumptions) has an initial object: consider the constant functor $Z_{\mathscr{C}}$, where Z is a set of one element. In 16. 4. 8 it will turn out that $\mathscr{C}$ is even cocomplete.

(c) In the proof of 10.6.5 look at those monomorphisms $m\colon M \to P$ for which there is an element z in $T(M)$ with $T(m)\,(z) = y$. Let $n\colon N \to P$ be the intersection of a set of representatives (and thus of all) of these monomorphisms. Since $T(n)\colon T(N) \to T(P)$ is the intersection of the corresponding monomorphisms in *Ens*, a comparison with (8) shows that T is dominated by N alone. This then implies that a proposition corresponding to 10.6.5 is valid, if the assumption that $\mathscr{C}$ is well-powered is replaced by the following one: for every class of monomorphisms in $\mathscr{C}$ with the same codomain there exists an intersection and it is preserved by T. (For representable functors this is guaranteed by 7.7.7).

(d) If Δ in (4), (6) is always an equalizer for every $A \in |\mathscr{C}|$ (compare 10.2.1, 10.5.2 and later 17.2.1 for the dual case), then in 10.6.5 the condition that $\mathscr{C}$ is well-powered can be replaced by the requirement that for every $A \in |\mathscr{C}|$ the classes of equivalent equalizers with codomain A have a set as a system of representatives. One confirms easily that if Δ in (6) is an equalizer, then m_a is an equalizer (see also later 12.3.5).

10.6.7 A *category* $\mathscr{C}$ is called *co-well-powered* (also colocally small or locally cosmall) if $\mathscr{C}^0$ is well-powered. The categories in 10.6.2 are also co-well-powered. 10.6.3 and 10.6.4 can be dualized.

10.7 Problems

10.7.1 Carry out 10.1.9.

10.7.2 A functor $T: \mathscr{C} \to Ens$ is representable if and only if $\mathscr{C}^0/T$ (see 10.2.3) has a terminal object.

10.7.3 Construct the counterexample in 10.2.7.

10.7.4 Let $T: Ens \to Ens$ be a functor which maps $\emptyset$ into itself and every non-empty set into a set of one element. Show: If M is a non-empty set, then $\{M\}$ is a dominating set for T.

10.7.5 Assign to every group the set underlying its commutator group and get thus a setvalued functor K on the category of groups. K is proper with a dominating set $\{F_n \mid n \geq 1\}$, were F_n denotes a free group of rank n.

10.7.6 Let the category $\mathscr{C}$ be finitely complete. For the functor $T: \mathscr{C} \to Ens$ the following are equivalent:

(i) There is a small filtered category Σ and a functor $S: \Sigma \to \mathscr{C}^0$ such that T is a colimit object of H^*S.

(ii) T is proper and preserves finite limits.

Remark. Functors with the property (i) are sometimes found in the literature under the name of *prorepresentable functors*.

10.7.7 Carry out an elementary proof of the representation theorem 10.3.9 using the following steps:

Step 1. Let $\mathfrak{D} = \{D_i\}$ be a dominating set for T. We may assume that no $T(D_i)$ is empty. T is dominated by the individual object $D = \prod D_i$.

Step 2. For all $d \in T(D)$ let $D_d = D$. Let $B = \Pi D_d$ with projections pr_d. In $T(B)$ there is an element v with $T(pr_d)(v) = d$ for all $d \in T(D)$. For every $x \in T(X)$ there is an $f: B \to X$ with $T(f)(v) = x$.

Step 3. Let M be the set of all endomorphisms α of B with $T(\alpha)(v) = v$. For every pair $(\alpha, \beta) \in M \times M$ let $d_{\alpha,\beta}$ be an equalizer of α and β. There exists an intersection $m: A \to B$ of all monomorphisms $d_{\alpha,\beta}$. There is a $u \in T(A)$ with $T(m)(u) = v$.

Step 4. m is a coretraction. For $h: A \to A$ one shows: if $T(h)(u) = u$, then $h = 1_A$.

Step 5. (A, u) *represents* T.

10.7.8 Let $\mathscr{C}$ be a complete category and $T: \Sigma \to \mathscr{C}$ a diagram. For T, let there be a set $\mathfrak{M}$ of objects in $\mathscr{C}$ such that the following holds: If $\alpha: T \to A_\Sigma$ is a natural transformation, then there is an $M \in \mathfrak{M}$ and $\mu: T \to M_\Sigma$ such that α factors through μ; i.e., $\alpha = f_\Sigma \mu$ for a suitable $f: M \to A$. Then T has a colimit. (Hint: use 8.1.4 and 7.5.3 for $S: \mathscr{C} \to [\Sigma, \mathscr{C}]$).

10.7.9 Prove the statements in 10.3.11.

10.7.10 Let the category $\mathscr{C}$ have pushouts and a generator G. If $m: A \rightarrowtail B$ is a monomorphism, but not an epimorphism, then there is a morphism $G \to B$ that does not factor through m. If, conversely, an object G has this property, and if $\mathscr{C}$ has equalizers, then G is a generator. (Hint: Use the dual of 7.9.9 (a).)

11. Objects with an Algebraic Structure

11.1 Algebraic Structures

Algebraic structures on sets are created by "algebraic combinations" such as multiplication or addition of two elements, or formation of inverses and neutral elements with to respect to such a combination. We restrict ourselves here to operations that are defined everywhere rather than only for the elements of suitable subsets. This means in particular, that we leave out fields and division algebras, since for them inverses with respect to multiplication are not defined for all elements. With this restriction, algebraic combinations can be described by maps from products, and the usual laws satisfied by such combinations can be described by commutative diagrams. This, however, is possible in an arbitrary category, provided the necessary products and a terminal object exist, which will always be assumed in the following. Dualization produces co-algebraic structures, where coproducts and initial objects are required. This does not yield any interesting examples in *Ens*, but it does in other categories.

11.1.1 Definition. Let A be an object of the category $\mathscr{C}$. An *n-ary algebraic operation* on A is a morphism $t\colon \prod_{1\le j\le n} A_j \to A$, where $A_j = A$ for $1 \le j \le n$.

The case $n = 0$ is included: $\prod A_j$ is then understood to be a terminal object Z of $\mathscr{C}$. An n-ary co-algebraic operation on A is a morphism t: $A \to \coprod_{1\le j\le n} A_j$, where for $n = 0$ $\coprod A_j$ is understood to be an initial object J of $\mathscr{C}$.

11.1.2 An object A that is provided with a nullary operation n: $Z \to A$ (i.e., such a morphism is fixed) is called a *pointed object*. In *Ens* this yields exactly the pointed sets. In the same way a nullary operation $A \to J$ produces a copointed object; in *Ens* the only possibility for this is $A = \emptyset$.

11.1.3 We call an object provided with a binary operation $u\colon A \sqcap A \to A$ a *multiplicative object* (later also called additive) and u is called the corresponding multiplication (addition). In the case of $v\colon A \to A \sqcup A$ one has a co-multiplicative object.

11.1.4 We now introduce some notation. Let $X_1, X_2, \ldots, X_n, Y$ be objects, $f_j\colon Y \to X_j$ morphisms. For the morphism $f\colon Y \to \prod X_j$ given by $f_j = pr_j f$ (pr_j are the projections of the product) we write $(f_1, f_2, \ldots, f_n)'$. If $Y = \prod Y_j$ is also a product of n factors with projections q_j and $f_j = g_j q_j$ for $g_j\colon Y_j \to X_j$, then we write $g_1 \sqcap g_2 \sqcap \cdots \sqcap g_n$ or $\prod g_j$ for f. If $\{j_1, j_2, \ldots, j_s\}$ is a subset of $\{1, 2, \ldots, n\}$, then there is a morphism $(pr_{j_1}, pr_{j_2}, \ldots, pr_{j_s})'\colon \prod X_j \to \prod X_{j_k}$, which we call a *projection onto the partial product* determined by $\{j_1, j_2, \ldots, j_s\}$; we use the abbreviated notation $pr_{j_1 j_2 \ldots j_s}$ for it. For coproducts

$$(f_1, f_2, \ldots, f_n)\colon \coprod X_j \to Y, \coprod g_j = g_1 \sqcup g_2 \sqcup \cdots \sqcup g_n\colon \coprod X_j \to \coprod Y_j,$$

$$in_{j_1 j_2 \ldots j_s}\colon \coprod X_{j_k} \to \coprod X_j \text{ are defined analogously.}$$

11.1.5 Let A be provided with a multiplication $u\colon A \sqcap A \to A$ and pointed by means of $n\colon Z \to A$. We consider

(1)

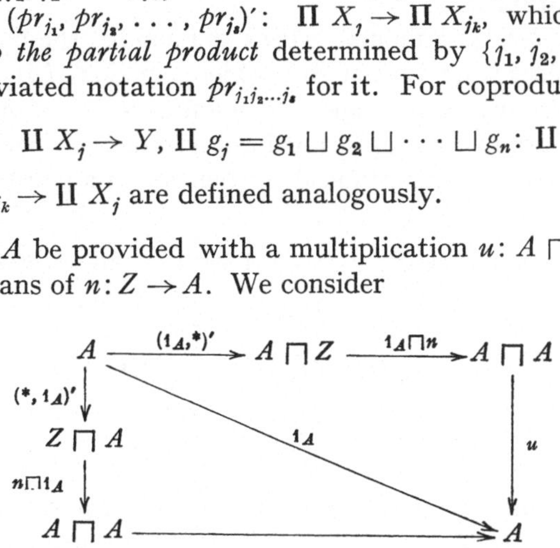

where $*: A \to Z$ is the only existing morphism. If in (1) the upper triangle is commutative, then n is called a *right unit* for u, similarly we say that n is a left unit if the lower triangle is commutative, and n is a unit if (1) is commutative.

A multiplicative object with a unit is called an *H-object* (H in honour of H. Hopf). A co-H-object is a co-multicative object with a counit.

11.1.6 A *multiplication* $u: A \sqcap A \to A$ is called *associative* if

$$\begin{array}{ccc} A \sqcap A \sqcap A & \xrightarrow{u \sqcap 1_A} & A \sqcap A \\ {\scriptstyle 1_A \sqcap u}\downarrow & & \downarrow{\scriptstyle u} \\ A \sqcap A & \xrightarrow{u} & A \end{array} \tag{2}$$

is commutative. Here $1_A \sqcap u$ is defined in the obvious way by $(pr_1, u\, pr_{23})'$ and correspondingly $u \sqcap 1_A = (u\, pr_{12}, pr_3)'$. Associativity of comultiplication is given by the diagram dual to (2). An H-object with an associative multiplication is called a *monoid* (*semigroup with unit*).

11.1.7 Let the H-object A be provided also with a unary operation $v: A \to A$. We consider

$$\begin{array}{ccc} A & \xrightarrow{(1_A, v)'} & A \sqcap A \\ {\scriptstyle (v, 1_A)'}\downarrow & \overset{*}{\searrow}\ Z\ \overset{n}{\searrow} & \downarrow{\scriptstyle u} \\ A \sqcap A & \xrightarrow{u} & A \end{array} \tag{3}$$

v is called a *right inversion* (left inversion) for u if the upper (lower) triangle is commutative. v is called an *inversion for u* if it is a right and left inversion for u.

A *group object* is a monoid with inversion for the multiplication. A co-group object is its dual. Group objects in *Ens*, *Top*, the category of differentiable, or, resp., algebraic manifolds are groups, topological groups, Lie groups or, resp., algebraic groups. In the pointed homotopy category 1.2.6 the (reduced) suspension of a pointed space is a co-group object.

11.1.8 A *multiplication* $u: A \sqcap A \to A$ is called *commutative* if

$$\begin{array}{ccc} A \sqcap A & & \\ & \overset{u}{\searrow} & \\ {\scriptstyle (pr_2, pr_1)'}\downarrow & & A \\ & \underset{u}{\nearrow} & \\ A \sqcap A & & \end{array} \tag{4}$$

is commutative.

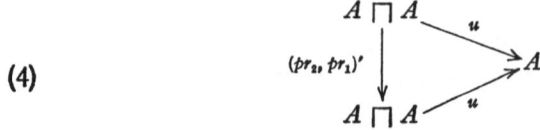

The meaning of ring object, Lie ring object and commutative ring object should be clear. Distributive laws are also commutative diagrams. Neutral operations corresponding to n-ary algebraic operations can be defined (with products in which $n - 1$ factors are terminal). Associativity conditions are produced through different choices of n factors from products with $2n - 1$ factors, commutativity conditions by permutations of the factors of n-fold products.

11.1.9 Definition. In a category $\mathscr{C}$ with finite products and a terminal object Z, an *object* A is said to have an *algebraic structure* if it is provided with a set $\{t_j\}$ of algebraic operations which satisfy certain relations of the following kind: commutativity conditions are required of a set of finite diagrams $D_\nu: \Sigma_\nu \to \mathscr{C}$ where the vertices of Σ_ν are mapped onto products of finitely many copies of A and Z; if further, for a vertex i of Σ_ν, $pr_1, pr_2, \ldots, pr_k$ denote the projections of $D_\nu(i)$ onto its factors, then an arrow $a: i' \to i$ of Σ_ν is to be mapped onto a morphism $D_\nu(a)$ where the $pr_i\, D_\nu(a)$ are to be composed of a projection of $D_\nu(i')$ onto a partial product and of an operation t_j or 1_A or of the canonical morphism "$*$" with codomain Z (this includes $*: A \to Z$ and 1_Z).

It ought to be clear what is meant by objects with an algebraic structure of the same type $\mathfrak{S}$, "$\mathfrak{S}$-objects", and by the type of an algebraic structure.

If an object A of the category $\mathscr{C}$ is provided with an algebraic structure of the type $\mathfrak{S}$, we say that A is the *carrier* of the thus created $\mathfrak{S}$-object.

11.2 Operations of an Object on Another

11.2.1 Definition. An *operation of the object K on the object A* is a morphism $w: K \sqcap A \to A$.

It may be that K or A are provided with algebraic operations and that w is subject to conditions described by commutative diagrams, whereby the descriptions of diagrams in 11.1.9 is to be extended in the obvious way. If K has a multiplication $u: K \sqcap K \to K$ and if

$$\text{(5)} \qquad \begin{array}{ccc} K \sqcap K \sqcap A & \xrightarrow{u \sqcap 1_A} & K \sqcap A \\ {\scriptstyle 1_K \sqcap w}\downarrow & & \downarrow{\scriptstyle w} \\ K \sqcap A & \xrightarrow{w} & A \end{array}$$

is commutative, then w is called a *left operation* of K on A. If in the top arrow of (5) u in $u \sqcap 1_A$ is replaced by $u\,(pr_2, pr_1)': K \sqcap K \to K$, and if the new diagram is commutative, then one has a *right operation*. It is usual to then consider the morphism

$$w(pr_2, pr_1)': A \sqcap K \to K \sqcap A \to A$$

instead of $w: K \sqcap A \to A$. If the multiplication on K is commutative, the two concepts coincide.

11.2.2 Examples. Left and right modules satisfy 11.2.1, as do algebras over a commutative ring and operations of a topological group on a topogical space. For objects A, B of a category $\mathcal{C}$, $[A, A]$ operates from the right and $[B, B]$ from the left on $[A, B]$ (compare 1.5.2).

11.2.3 It is clear that several objects may be operating on one given object, and that any compatibility conditions are commutative diagrams, e.g. for bimodules.

The dual of an operation is a co-operation $w\colon A \to K \sqcup A$, more precisely, a left or right co-operation. Examples for this are found in homotopy theory.

11.2.4 A terminal object Z caries every type of algebraic structure, since products of arbitrarily many factors Z are always isomorphic to Z. Z also allows every type of operations of objects on itself. This is well known for sets of one element, i.e., terminal objects of *Ens*.

11.3 Homomorphisms

11.3.1 Definition. Let the objects A and A' of the category $\mathcal{C}$ be provided with algebraic structures of the same type $\mathfrak{S}$. A morphism $f\colon A \to A'$ is called an *$\mathfrak{S}$-homomorphism* (just homomorphism if there is no doubt about $\mathfrak{S}$) if for every pair of corresponding algebraic operations $t\colon \prod A_j \to A$ and $t'\colon \prod A'_j \to A'$ the diagram

$$\begin{array}{ccc} \prod A_j & \xrightarrow{\prod f_j} & \prod A'_j \\ {\scriptstyle t}\downarrow & & \downarrow{\scriptstyle t'} \\ A & \xrightarrow{f} & A' \end{array} \tag{6}$$

with $f_j = f$ for all j is commutative.

For groups, monoids, rings (as objects with an algebraic structure in *Ens*) this yields the usual definition of homomorphism.

Note that in (6) $f_j\colon \prod A_j \to \prod A'_j$ includes the projections of the two products according to the definition of $\prod f_j$. The description of the diagrams belonging to an algebraic structure in 11.1.9 shows that a homomorphism induces natural transformations between corresponding diagrams.

11.3.2 Proposition. *For an algebraic structure of type $\mathfrak{S}$ the $\mathfrak{S}$-objects of a category $\mathcal{C}$ are the objects and the $\mathfrak{S}$-homomorphisms between them are the morphisms of a category $\mathcal{C}_{\mathfrak{S}}$. We call it the category of $\mathfrak{S}$-objects over $\mathcal{C}$.*

This proposition follows immediately from definition 11.3.1. Note that an object in $\mathcal{C}$ may be provided with different structures of type $\mathfrak{S}$, which produces different $\mathfrak{S}$-objects. For instance, a set can be provided with various group structures (there are exceptions). Thus *Ab*, the category of groups or of rings are based on *Ens*. $\mathscr{C}_{\mathfrak{S}}$ is not a subcategory of $\mathscr{C}$. However, there is a forgetful functor $\mathscr{C}_{\mathfrak{S}} \to \mathscr{C}$. There is also a forgetful functor $\mathscr{C}_{\mathfrak{S}} \to \mathscr{C}_{\mathfrak{S}'}$ if $\mathfrak{S}$ is the result of adding further algebraic operations or diagram conditions to $\mathfrak{S}'$. The examples of the forgetful functor from the category of abelian groups to the category of groups and from the category of rings to *Ab* are clear.

(6) implies immediately

11.3.3 Proposition. *The forgetful functor $\mathscr{C}_{\mathfrak{S}} \to \mathscr{C}$ is faithful and it reflects isomorphisms.*

11.3.4 Theorem. *Let the functor $T: \mathscr{C} \to \mathscr{D}$ preserve finite products (including terminal objects). T induces a functor $T_{\mathfrak{S}}: \mathscr{C}_{\mathfrak{S}} \to \mathscr{D}_{\mathfrak{S}}$ for every type of algebraic structure. If T is faithful and $T(A)$ provided with an $\mathfrak{S}$-structure, then there is at most one $\mathfrak{S}$-structure on A which is taken into the given one for $T(A)$ by T. There always is one provided T is fully faithful and $\mathscr{C}$ has finite products.*

Proof. The first statement is an immediate consequence of the fact that every functor preserves commutative diagrams. If T is faithful, then T takes non-commutative diagrams into non-commutative ones. So T reflects the commutativity of diagrams. Furthermore, there is at most one morphism $t: \prod A_j \to A$ with a given image $t': T(\prod A_j) \to \to T(A)$. If T is fully faithful and if $\mathscr{C}$ has finite products, then there is such a t.

Here we have made use of the fact that $T(\prod A_j)$ is a (not necessarily canonically chosen) product with projections $T(pr_j)$, and that this product is uniquely determined up to an isomorphism.

11.3.5 Corollary. *Let $\mathscr{C}$ be an arbitrary category. By means of the Yoneda embedding $A \mapsto H_A$, $f \mapsto H_f$ of $\mathscr{C}$ in $[\mathscr{C}^0, Ens]$ every $\mathfrak{S}$-structure for $A \in |\mathscr{C}|$ is transfered to an $\mathfrak{S}$-structure for H_A, and $\mathfrak{S}$-homomorphisms over $\mathscr{C}$ go into $\mathfrak{S}$-homomorphisms over $[\mathscr{C}^0, Ens]$. If $\mathscr{C}$ has finite products, one obtains in this way all $\mathfrak{S}$-structures for the contravariant functor H_A.*

Proof. $[\mathscr{C}^0, Ens]$ is complete. The embedding is fully faithful by 4.2.2 and preserves limits by 10.2.5.

11.3.6 For every type $\mathfrak{S}$ of an algebraic structure there is a dual type of a co-algebraic structure. We call it the co-structure of type $\mathfrak{S}$. If 11.3.5 is dualized by means of the embedding $\mathscr{C}^0 \to [\mathscr{C}, Ens]$ given by the rule $A \mapsto H^A$, $f \mapsto H^f$, one obtains corresponding statements for $\mathfrak{S}$-co-structures for objects of $\mathscr{C}$ and $\mathfrak{S}$-structures for the functors H^A.

Naturally, 11.3.2 through 11.3.4 can be dualized to give statements about co-algebraic structures. In 11.3.4 contravariant functors can be included in the way we just did above.

11.3.7 Analogous to 11.3.1, homomorphisms for operations w: $K \sqcap A \to A$ may be defined, where K and A are additionally provided with algebraic structures of type $\mathfrak{R}$ or, resp., $\mathfrak{S}$. If the same situation exists for $w'\colon K' \sqcap A' \to A'$, then an operations homomorphism compatible with the $\mathfrak{R}$- and $\mathfrak{S}$-structures consists of a pair (k, f) with $k\colon K \to K'$, $f\colon A \to A'$ such that

$$(7) \qquad \begin{array}{ccc} K \sqcap A & \xrightarrow{\;w\;} & A \\ {\scriptstyle k \sqcap f}\downarrow & & \downarrow{\scriptstyle f} \\ K' \sqcap A' & \xrightarrow{\;w'\;} & A' \end{array}$$

is commutative and k is an $\mathfrak{R}$-homomorphism, f an $\mathfrak{S}$-homomorphism.

11.3.2 through 11.3.6 carry over in an obvious way. We leave the explicit formulation to the reader.

A simple example for an operations homomorphism is provided by modules. Here K and K' are rings, A and A' additive groups, and w and w' define left module structures. One obtains "module homomorphisms with a change of rings".

11.3.8 11.3.5, 11.3.6 and their analogues for operations and co-operations may be used to weaken 11.1.9 and 11.2.1 and thus to talk about a weak $\mathfrak{S}$-structure, resp., $\mathfrak{S}$-co-structure, for $A \in |\mathscr{C}|$, if H_A, resp., H^A, is provided with an $\mathfrak{S}$-structure.

11.4 Reduction to Ens

11.3.5 and 11.3.6 allow us to reduce investigations of algebraic and co-algebraic structures to the subcategory of representable functors of $[\mathscr{C}^0, Ens]$, resp., $[\mathscr{C}, Ens]$. This even produces a reduction to *Ens*, since the following is true generally:

11.4.1 Proposition. *Let the category $\mathfrak{D}$ have finite products. If the functor $T\colon \mathscr{C} \to \mathfrak{D}$ as an object of $[\mathscr{C}, \mathfrak{D}]$ is provided with an algebraic structure of type $\mathfrak{S}$, then at every place $A \in |\mathscr{C}|$ there is an $\mathfrak{S}$-structure for $T(A)$ and for every morphism $f\colon A \to A'$, $T(f)$ is an $\mathfrak{S}$-homomorphism. If, conversely, for every $A \in |\mathscr{C}|$ an $\mathfrak{S}$-structure for $T(A)$ is fixed in such a way that, for any $f\colon A \to A'$, $T(f)\colon T(A) \to T(A')$ is always an $\mathfrak{S}$-homomorphism, then T has a uniquely determined $\mathfrak{S}$-structure which coincides for every $A \in |\mathscr{C}|$ with the previously given one.*

If $\mathfrak{D}$ has finite coproducts, then the same is true for co-algebraic structures.

Addition. The first statement of this proposition means that T is of the form $T = U\,T_{\mathfrak{S}}$, where $U\colon \mathfrak{D}_{\mathfrak{S}} \to \mathfrak{D}$ is the forgetful functor and $T_{\mathfrak{S}}$ a well determined functor $\mathscr{C} \to \mathfrak{D}_{\mathfrak{S}}$.

Proof. Morphisms in $[\mathscr{C}, \mathfrak{D}]$ are natural transformations. They and their compositions are defined "pointwise", as are products of functors (7.5.2). This implies immediately that every algebraic operation t for T is an operation of the same type at every place A and that for $T(f): T(A) \to T(A')$ condition (6) in 11.3.1 (with necessary changes in notation) is satisfied. A statement about the commutativity of a diagram says that certain morphisms are the same (compare 6.2), and, for natural transformations $\alpha, \beta: T_1 \to T_2$, $\alpha = \beta$ is equivalent to $\alpha_A = \beta_A$ for all $A \in |\mathscr{C}|$. This provides the first statement of the proposition and also its converse, since what is required here is exactly that the algebraic operations given at every place can be put together to form the corresponding natural transformations $\Pi\, T_j \to T$.

11.4.2 The remarks above enable us to transfer wellknown results about algebraic structures on sets and their homomorphisms to objects with corresponding algebraic structures in arbitrary categories and their homomorphisms. We mention the following:

(1) For a multiplication $u: A \sqcap A \to A$ there is at most one neutral operation; more precisely: if u has a left and a right unit, then these coincide (as pointings).

(2) For an associative multiplication the associativity is valid for any finite number of factors. The same then holds for commutativity.

(3) Foı an associative multiplication u with a unit there is at most one inversion; more precisely: if u has a left and a right inversion, then they coincide.

(4) If an associative multiplication has a left unit n and a left inversion v, then n is a unit and v an inversion. That is, one has a group object. Furthermore, $v\, v$ is the identity morphism.

(5) Let $\mathscr{C}_G, \mathscr{C}_H, \mathscr{C}_M$ be the categories of group objects, H-objects, or, resp., multiplicative objects over $\mathscr{C}$. And let $U_1: \mathscr{C}_G \to \mathscr{C}_H$, $U_2: \mathscr{C}_H \to \mathscr{C}_M$ be the corresponding forgetful functors. U_1 and $U_2\, U_1$ are fully faithful, while U_2 is faithful. This comes from the fact that for groups (over *Ens*) it suffices to require of a mapping of the carrier sets that it be compatible with the group multiplications in order for it to be a homomorphism. Also, different mappings for H-sets are still different if the H-sets are considered only as M-sets.

11.4.3 Naturally, 11.3.6 provides the dual statements for coalgebraic structures. Further, it is obvious that operations of an object on another (both possibly with an additional algebraic structure) as well as cooperations (with additional co-algebraic structures) can be reduced to operations on sets.

11.5 Limits and Filtered Colimits

11.5.1 Proposition. *If the category $\mathscr{C}$ has finite products, resp., products, or if $\mathscr{C}$ is finitely complete, resp., complete, then the same is true for* *the category $\mathscr{C}_{\mathfrak{S}}$ of $\mathfrak{S}$-objects over $\mathscr{C}$ for every type $\mathfrak{S}$ of an algebraic structure.*

Proof. What we show exactly, is this: If in $\mathscr{C}$ finite products and limits for all diagrams of a fixed type Σ exist, then these exist in $\mathscr{C}_{\mathfrak{S}}$. Since limits of type Σ commute with products and since natural transformations between diagrams of type Σ induce morphisms between the limit objects, it follows directly from the definitions that a limit for a diagram $D: \Sigma \to \mathscr{C}_{\mathfrak{S}}$ is obtained as follows: One forms a limit (L, λ) for $U D: \Sigma \to \mathscr{C}$, where $U: \mathscr{C}_{\mathfrak{S}} \to \mathscr{C}$ is the forgetful functor, and verifies that L becomes the carrier of a canonical $\mathfrak{S}$-structure with regard to which every $\lambda_i: L \to D(i)$ (i is a vertex of Σ) is a homomorphism. Here we have made use of the fact that for a terminal Z, Z_Σ has the limit $(Z, \{1_Z\})$ (7.1.7).

11.5.2 Example. Let A and B be sets that are provided with a multiplication $u: A \times A \to A$ and $v: B \times B \to B$. For $(a_1, a_2) \in A \times A$ we write $a_1 a_2$ for $u(a_1, a_2)$, and similarly $b_1 b_2$ for $v(b_1, b_2)$. The diagram

$$\begin{array}{ccc} (A \times B) \times (A \times B) & \xrightarrow{pr_1 \times pr_1} & A \times A \\ \Big\downarrow{\scriptstyle u''} & & \Big\downarrow{\scriptstyle u} \\ A \times B & \xrightarrow{pr_1} & A \end{array}$$

and the corresponding one for pr_2 and v then define the multiplication u'' for $A \times B$ that is induced by u and v. One has $(a_1, b_1) \times \times (a_2, b_2) = (a_1 a_2, b_1 b_2)$. If the multiplications u and v have units e and 0, then $(e, 0)$ is a unit for u''.

11.5.3 Corollary. *The forgetful functor $U: \mathscr{C}_{\mathfrak{S}} \to \mathscr{C}$ preserves and reflects finite products, or, resp., products, finite limits, limits if $\mathscr{C}$ has finite products, or, resp., products, finite limits, limits.*

Proof. The preserving follows from the construction, the reflecting from 11.3.3 and the definition of limits.

11.5.4 Corollary. *For $A \in |\mathscr{C}|$, let the contravariant functor H_A: $\mathscr{C} \to Ens$ be provided with an algebraic structure of type $\mathfrak{S}$, giving a contravariant functor $H_{A\mathfrak{S}}: \mathscr{C} \to Ens_{\mathfrak{S}}$.* If (L, λ) *is a colimit of the diagram $T: \Sigma \to \mathscr{C}$, then $(H_{A\mathfrak{S}}(L), H_{A\mathfrak{S}}\lambda)$ is a limit of the contravariant diagram $H_{A\mathfrak{S}} T$.*

In other words: The algebraic structure provided by H_A for $[L, A]_{\mathscr{C}}$ is the structure obtained for $[L, A]$ as a limit object of $H_A T$.

Proof. H_A takes colimits into limits (8.7.3), therefore the claim follows by 11.5.3.

11.5.5 Example. Let $u\colon H_A \sqcap H_A \to H_A$ be given and with it $u_X\colon [X, A] \times [X, A] \to [X, A]$ for every $X \in |\mathscr{C}|$. If the coproduct $X \sqcup Y$ exists in $\mathscr{C}$, then the isomorphism $[X \sqcup Y, A] \to [X, A] \times [Y, A]$ takes the map

$$u_{X \sqcup Y}\colon [X \sqcup Y, A] \times [X \sqcup Y, A] \to [X \sqcup Y, A]$$

into

$$[X, A] \times [Y, A] \times [X, A] \times [Y, A] \xrightarrow{1 \times \tau \times 1}$$
$$[X, A] \times [X, A] \times [Y, A] \times [Y, A] \xrightarrow{u_X \times u_Y} [X, A] \times [Y, A]$$

(compare 11.5.2), where $1 \times \tau \times 1$ interchanges the two middle factors of the first product. (This is a commutation of limits with limits, compare also 11.5.2). If multiplication at every place X is indicated by a point, one obtains

$$(x, y) \cdot (x', y') = (x \cdot x',\ y \cdot y')$$

for $x, x' \in [X, A]$ and $y, y' \in [Y, A]$.

11.5.6 Remarks. 11.5.1 makes it clear, why the restriction to algebraic operations that are defined everywhere (beginning of 11.1) was necessary: the category of fields does not have products. As is shown by the category of rings (with unit) for $\mathscr{C} = Ens$, $\mathscr{C}_{\mathfrak{S}}$ need not have zero-morphisms and kernels. The existence of coproducts can in general not be deduced, as might be expected by a comparison of coproducts in *Ens*, *Ab* and the category of groups, which have no relations to each other. Coequalizers are also not obtained. The following holds, however:

11.5.7 Theorem. *Let the category $\mathscr{C}$ be finitely complete and have filtered colimits which commute with finite limits. Then the same is true for the category $\mathscr{C}_{\mathfrak{S}}$ of $\mathfrak{S}$-objects over $\mathscr{C}$ for every type $\mathfrak{S}$ of an algebraic structure.*

Proof. For a small filtered category $\mathscr{X}$, we consider a functor $T\colon \mathscr{X} \to \mathscr{C}_{\mathfrak{S}}$. U is the forgetful functor $\mathscr{C}_{\mathfrak{S}} \to \mathscr{C}$. One forms the colimit (L, λ) of $U\,T$. By assumption, the formation of finite products commutes with the formation of filtered colimits in $\mathscr{C}$. The algebraic operations for the colimit object of $U\,T$ and the commutativity of the diagrams corresponding to the structure are implied by the fact that natural transformations of functors $\mathscr{X} \to \mathscr{C}$ induce uniquely determined morphisms for the colimit objects. The natural transformation $\lambda\colon U\,T \to L_{\mathscr{X}}$ gives an $\mathfrak{S}$-homomorphism for every $i \in |\mathscr{X}|$. Here we use the fact, that $Z_{\mathscr{X}}$ has the colimit $(Z, \{1_Z\})$ if $\mathscr{X}$ is filtered and therefore connected (9.1.3).

The explicit construction for $\mathscr{C} = Ens$ given in 9.4.7 and 9.4.8 thus turns out to be a special case.

11.5.8 Corollary. *Under the assumptions of* 11.5.7 *the forgetful functor $\mathscr{C}_{\mathfrak{S}} \to \mathscr{C}$ preserves and reflects filtered colimits.*

11.5.9 11.5.1 and 11.5.7 carry over to operations of an object on another. A classical result is obtained in the case of module homomorphisms with a change of rings (compare 11.3.7).

11.6 Homomorphically Compatible Structures

11.6.1 Theorem (Eckmann-Hilton). *Let the category $\mathscr{C}$ have finite products and let the object A in $\mathscr{C}$ be provided with two H-structures (u, n) and (u', n'). If $u'\colon A \sqcap A \to A$ is a homomorphism with respect to the multiplication u on A and the multiplication induced by u on $A \sqcap A$, then $u = u'$, $n = n'$ and u is associative and commutative.*

Proof. Because of 11.3.5 and 11.4.1, it is sufficient to supply a proof for $\mathscr{C} = Ens$. So let A be a set. For $(x, y) \in A \times A$ we write $x\,y$ instead of $u(x, y)$ and $x + y$ instead of $u'(x, y)$. n and n' each determine uniquely a unit e, or, resp., 0 of A, so that $x\,e = e\,x = x$ and $x + 0 = 0 + x = x$ for all $x \in A$. The multiplication induced by u in $A \times A$ we denote by u''. By 11.5.2, it is described by $u''(x_1, x_2, x_3, x_4) = (x_1\,x_3,\ x_2\,x_4)$. The assumption for u' means, because of 11.3.1 (6), that

$$(8)\qquad \begin{array}{ccc} (A\times A)\times(A\times A) & \xrightarrow{u'\times u'} & A\times A \\ \downarrow{\scriptstyle u''} & & \downarrow{\scriptstyle u} \\ A\times A & \xrightarrow{\quad u' \quad} & A \end{array}$$

is commutative, so that

$$(x_1 + x_2)\,(x_3 + x_4) = (x_1\,x_3) + (x_2\,x_4)$$

holds. This implies in succession

$$(1)\qquad e = e\,e = (e + 0)\,(0 + e) = (e\,0) + (0\,e) = 0 + 0 = 0\,,$$

$$(2)\quad x\,y = (x + 0)\,(0 + y) = (x + e)\,(e + y) = (x\,e) + (e\,y) = x + y\,,$$

$$(3)\quad x\,y = (0 + x)\,(y + 0) = (e + x)\,(y + e) = (e\,y) + (x\,e) = y + x\,,$$

$$(4)\quad (x + y) + z = (x + y) + (0 + z) = (x + y)\,(e + z) = (x\,e) + (y\,z) = x + (y + z)\,.$$

11.6.2 Corollary. *Let $\mathscr{C}$ be a category with finite products. The H-objects of the category $\mathscr{C}_H$ are those objects of $\mathscr{C}_H$ whose H-structure is commutative and associative. In particular: for the category of groups the H-objects are the abelian groups.*

11.6.3 Corollary. *Let A and B be objects of the category $\mathscr{C}$, and let H^A and H_B be provided with H-structures. If $A \sqcup A$ or $B \sqcap B$ exists in $\mathscr{C}$, then the H-structures provided by H^A and H_B on $[A, B]_{\mathscr{C}}$ coincide and they are commutative and associative.*

Proof. Suppose $A \sqcup A$ exists. Since $A \mapsto H^A$, $f \mapsto H^f$ gives a full embedding of $\mathscr{C}^0$ in $[\mathscr{C}, Ens]$, which preserves products, the multiplication for the H-structure of H_A comes from a uniquely determined comultiplication $v\colon A \to A \sqcup A$ (11.3.4). Let $u\colon H_B \sqcap H_B \to H_B$ be the multiplication for H_B.

$$\begin{array}{ccc} [A \sqcup A, B] \times [A \sqcup A, B] & \xrightarrow{[v,B]\times[v,B]} & [A, B] \times [A, B] \\ \downarrow{\scriptstyle u_{A \sqcup A}} & & \downarrow{\scriptstyle u_A} \\ [A \sqcup A, B] & \xrightarrow{[v,B]} & [A, B] \end{array}$$

is commutative, since u is a natural transformation. Taking into account the uniquely determined isomorphism $[A \sqcup A, B] \to [A, B] \times \times [A, B]$, the statement for the case considered follows from 11.6.1. The other case is its dual.

11.6.4 Corollary. *Let H_A and H^A be provided with H-structures for every $A \in |\mathscr{C}|$. If in $\mathscr{C}$ $A \sqcup A$ or $A \sqcap A$ always exists, then $\mathscr{C}$ is semi-additive in a uniquely determined way* (*compare* 1.5). *Addition in $[A, B]$ is the algebraic operation originating from the multiplication for H_B.*

Proof. By 11.6.3, every $[A, B]$ has a commutative monoid structure which is uniquely determined by H^A or by H_B and whose composition is regarded as addition. For $b\colon B \to B'$, H_b is a homomorphism since the structure is provided by H^A; for $a\colon A \to A'$ H^a is a homomorphism.

11.6.5 By 11.6.4, a category with finite products or finite coproducts can be made into a semi-additive category in at most one way. For, if $\mathscr{C}$ is semi-additive, then every H_A and H^A has an H-structure provided by this, by the definition of semi-additivity (1.5). If $\mathscr{C}$ could be provided with two different semi-additive structures, a contradiction with 11.6.4 would arise. The existence of these products or coproducts in 11.6.3 and 11.6.4 is necessary.

Counterexample. The rings $\boldsymbol{Z}$ and $\boldsymbol{Z}[x]$ (polynominal ring) have an isomorphic multiplicative structure because they have the same invertible elements ± 1, unique decomposition into prime factors and a countably infinite number of prime elements. This multiplicative structure can be regarded as the composition of morphisms in a category with only one object. This category, then, can be made into an additive one in different ways, for, $\boldsymbol{Z}$ and $\boldsymbol{Z}[x]$ as rings are not isomorphic. $\boldsymbol{Z}$ is a principal ideal ring, whereas $\boldsymbol{Z}[x]$ is not.

11.6.6 The following are well known special cases of 11.6.1:

(1) Fundamental groups of H-spaces (H-objects in *Top* or the corresponding homotopy category), in particular of topological groups, are commutative. The group composition for H-spaces can also be described by the H-structure of the space.

(2) Double suspensions in the pointed homotopy category are commutative co-groups (dual of 11.6.1). In particular, the sphere S^n for $n \geq 2$ is a commutative co-group. Therefore, apart from the fundamental group, homotopy groups are abelian.

11.7 Problems

11.7.1 Describe a ring by algebraic operations with arities 0, 1, and 2 together with appropriate commutative diagrams.

11.7.2 Let K be a ring (or field) and A a left K-module (or, resp., vector space). Express the laws for the operation of K on A by means of diagrams.

11.7.3 Let R be a ring. Then ${}_R\mathit{Mod} = \mathit{Ens}_{\mathfrak{S}}$ for a suitable type $\mathfrak{S}$ of an algebraic structure. (Regard the elements of R as unary algebraic operations for $\mathfrak{S}$ and find the corresponding commutative diagrams).

11.7.4 Transfer 11.3.2 through 11.3.5 to operations and their homomorphisms in the sense given in 11.2.

11.7.5 Let $\mathscr{C}$ be finitely complete. If there is exactly one nullary operation associated with an algebraic structure of type $\mathfrak{S}$, then $\mathscr{C}_{\mathfrak{S}}$ has a zero object. For $\mathscr{C} = \mathit{Ens}$ the converse also holds.

11.7.6 Carry out 11.5.9.

11.7.7 What H-objects are there in the category of rings? What ring objects are there in the category of groups?

11.7.8 The following is an example of a more general notion of algebraic structure, which uses arbitrary finite limits instead of finite products.

Let $\mathscr{C}$ be a finitely complete category and $A \in |\mathscr{C}|$. A *category structure* on A, or a *category-object* over $\mathscr{C}$ with *carrier* A, is given by a pullback

(1)
$$\begin{array}{ccc} B & \xrightarrow{c_0} & A \\ {\scriptstyle c_1}\downarrow & & \downarrow{\scriptstyle d_1} \\ A & \xrightarrow{d_0} & A \end{array}$$

and a morphism $m\colon B \to A$ such that conditions (2), (3), (5), (8) below are satisfied.

(2) $$d_1\, d_0 = d_0; \quad d_0\, d_1 = d_1,$$

(3) $$d_0\, m = d_0\, c_0; \; d_1\, m = d_1\, c_1.$$

By (1) and (2), there are uniquely determined morphisms $j_0, j_1: A \to B$ such that

(4)
$$c_1 j_0 = 1_A; \quad c_0 j_0 = d_0, \qquad c_0 j_1 = 1_A; \quad c_1 j_1 = d_1 .$$

m is required to satisfy

(5)
$$m j_0 = m j_1 = 1_A .$$

If a pullback

(6)
$$\begin{array}{ccc} C & \xrightarrow{p_0} & B \\ {\scriptstyle p_1}\downarrow & & \downarrow{\scriptstyle c_1} \\ B & \xrightarrow{c_0} & A \end{array}$$

has been chosen, then there are uniquely determined morphisms n_0, $n_1: C \to B$ such that

(7)
$$c_0 n_0 = m p_0; \quad c_1 n_0 = c_1 p_1, \qquad c_1 n_1 = m p_1; \quad c_0 n_1 = c_0 p_0 ,$$

and m is required to satisfy

(8)
$$m n_0 = m n_1 .$$

(a) Check the claims made for j_0, j_1, n_0, n_1. (Hint: $d_0 c_1 p_1 = d_1 m p_0$.)

(b) Different choices of (6) result in equivalent conditions (8).

(c) The following rules are consequences of (3), (4), (5):

(9)
$$d_0 d_0 = d_0; \quad d_1 d_1 = d_1 .$$

(d) Define $e: E \to A$ as a limit of the following diagram

$$A \overset{d_0}{\underset{d_1}{\overset{1_A}{\rightrightarrows}}} A .$$

Then e is a monomorphism, and there are uniquely determined morphisms $\Delta^0, \Delta^1: A \to E$ such that

(10)
$$d_0 = e \Delta^0; \quad d_1 = e \Delta^1 .$$

It follows that $1_E = \Delta^0 e = \Delta^1 e$ and that

$$\begin{array}{ccc} B & \xrightarrow{c_0} & A \\ {\scriptstyle c_1}\downarrow & & \downarrow{\scriptstyle \Delta^1} \\ A & \xrightarrow{\Delta^0} & E \end{array}$$

is a pullback.

(e) Let $\mathscr{C} = \mathit{Ens}$ and $A = \operatorname{Mor} \mathscr{X}$ for some small category $\mathscr{X}$. For $u: X \to Y$ as an element of A, set $d_0 u = 1_X$, $d_1 u = 1_Y$. Make the canonical choice for (1). Then the composition of morphisms in $\mathscr{X}$

defines m. Using this, interpret the above and show that, conversely, every category object over *Ens* uniquely determines a small category.

(f) Let $\mathcal{A} = (A, d_0, d_1, c_0, c_1, m)$ and $\mathcal{A}' = (A', d_0', d_1', c_0', c_1', m')$ be category objects over $\mathscr{C}$. Define a functorial morphism $\mathcal{A}' \to \mathcal{A}$ as a triple $(\mathcal{A}', \mathcal{A}, F)$, where F is a $\mathscr{C}$-morphism $F\colon A' \to A$ satisfying

(11) $$F\, d_0' = d_0\, F\,, \quad F\, d_1' = d_1\, F \quad \text{and}$$

(12) $$F\, m' = m\, \overline{F}\,.$$

Here $\overline{F}$ is the $\mathscr{C}$-morphism which is uniquely determined by

$$c_0\, \overline{F} = F\, c_0' \quad \text{and} \quad c_1\, \overline{F} = F\, c_1'\,.$$

Composition of functorial morphisms is defined as in $\mathscr{C}$. Using finite limits instead of finite products, transfer 11.3 through 11.5. (To be continued in 16.8.10.)

12. Abelian Categories

12.1 Survey

12.1.1 Definition. A *category* is called *abelian* if it satifies the following axioms:

A0 There is a zero object.
A1 There are finite products.
A1⁰ There are finite coproducts.
A2 Every morphism has a kernel.
A2⁰ Every morphism has a cokernel.
A3 Every monomorphism is a kernel
A3⁰ Every epimorphism is a cokernel.

12.1.2 This will be shown to have the following implications: An abelian category is in a unique way semi-additive and thus additive. It is therefore also finitely complete and finitely cocomplete (7.2.6, 7.4.2 and dual).

12.1.3 Further implications are: Finite products are also finite coproducts. If in_j are injections, pr_j projections, then for $\amalg A_j = \prod A_j$ with $j = 1, 2, \ldots, n$ and $n \geq 1$

(1) $$pr_k\, in_j = \delta_{kj} = \begin{cases} 0 & \text{for } j \neq k \\ 1_{A_k} & \text{for } j = k \end{cases}$$

(2) $$\sum_{j=1}^{n} in_j\, pr_j = 1_{\prod A_j}$$

hold.

12.1.4 There is a unique (up to isomorphisms) "natural" decomposition into an epimorphism followed by a monomorphism for every morphism. Abelian categories are the proper framework for the study of exact sequences, in fact, they are the foundation of homological algebra. Ab, ${}_R Mod$ and Mod_R are abelian categories.

12.1.5 The axioms in 12.1.1 are self-dual. The dual of an abelian category is abelian.

12.1.6 Proposition. *If $\mathcal{C}$ is an arbitrary, $\mathcal{A}$ an abelian category, then $[\mathcal{C}, \mathcal{A}]$ is abelian and so is Add $(\mathcal{C}, \mathcal{A})$, provided $\mathcal{C}$ is additive.*

Proof. This follows from the "pointwise" construction of limits and colimits in functor categories, if 10.1.4, 12.1.2 and 12.4.3 below are taken into account, whereby every monomorphism in $\mathcal{A}$ is the kernel of its cokernel or, resp., every epimorphism is the cokernel of its kernel.

12.1.7 The category of groups satisfies all the axioms except A 3, Ens_* all except A 3^0. This shows that a weakening of the definition results in a loss of essential properties. Nevertheless, we will occasionally use only parts of the system of axioms, sometimes combined with a semi-additive structure, to gain useful lemmas. But we will resist the temptation to lose ourselves in the game of axiomatic puzzles.

12.1.8 The system of axioms 12.1.1 can be reduced. A 1 or A 1^0 may be dropped (12.5). It should also be pointed out that there are equivalent systems.

12.2 Semi-additive Structure

12.2.1 Let $A = A_1 \sqcup A_2 \sqcup \ldots \sqcup A_m$ be a coproduct with injections in_j and $B = B_1 \sqcap B_2 \sqcap \cdots \sqcap B_n$ be a product with projections pr_k. If for every pair (k, j) a morphism f_{kj}: $A_j \to B_k$ is given, then there is exactly one morphism f: $A \to B$ with $pr_k\, f\, in_j = f_{kj}$. We use the matrix

$$(f_{kj}) = \begin{pmatrix} f_{11} & \cdots & f_{1m} \\ f_{n1} & \cdots & f_{nm} \end{pmatrix} \tag{3}$$

to denote f. Note the special cases $m = 1$ and $n = 1$.

12.2.2 Proposition. *Let the category $\mathcal{C}$ have a zero object and finite products and coproducts. We assume that for every pair (A, B) of objects a product and a coproduct has been chosen. Then*

$$\varrho_{A,B} = \begin{pmatrix} 1_A & 0 \\ 0 & 1_B \end{pmatrix} : A \sqcup B \to A \sqcap B \tag{4}$$

is a natural transformation of the bifunctors $\sqcup, \sqcap: \mathscr{C} \times \mathscr{C} \to \mathscr{C}$.

Proof. That $\sqcup$ and $\sqcap$ are bifunctors is implied by 7.3.3, 8.3.3. For $f: A \to X$, $g: B \to Y$ one has

$$(f \sqcap g)\, \varrho_{A,B} = \begin{pmatrix} f & 0 \\ 0 & g \end{pmatrix} = \varrho_{X,Y}\, (f \sqcup g)\,.$$

Here the zero morphisms are essential, as is made evident by *Ens* with $A = \emptyset$, $B \neq \emptyset$. In general $\varrho = \{\varrho_{A,B}\}$ is neither an epi- nor a monomorphism.

12.2.3 Proposition. *If under the assumptions of* 12.2.2 *the natural transformation* ϱ *is an isomorphism, then* $\mathscr{C}$ *has a uniquely determined semi-additive structure.*

Proof. Every $A \in |\mathscr{C}|$ has a multiplication

$$u: A \sqcap A \xrightarrow{\varrho^{-1}} A \sqcup A \xrightarrow{(1,1)} A$$

for which $0 \to A$ is a right unit, as is shown by

$$(5) \qquad \begin{array}{ccccccc} A & \xrightarrow{\binom{1}{0}} & A \sqcap 0 & \xrightarrow{\varrho^{-1}} & A \sqcup 0 & \xrightarrow{(1,0)} & A \\ & & \downarrow{\scriptstyle 1_A \sqcap 0} & & \downarrow{\scriptstyle 1_A \sqcup 0} & & \downarrow{\scriptstyle 1_A} \\ & & A \sqcap A & \xrightarrow{\varrho^{-1}} & A \sqcup A & \xrightarrow{(1,1)} & A \end{array}$$

The top consists of isomorphisms with inverses pr_1, ϱ, in_1 (7.3.5 and dual). Because of $pr_1\, \varrho\, in_1 = 1_A$, 1_A is the composite of the top line. The two rectangles are commutative. By 11.1.5, $0 \to A$ is a right unit for u, and similarly it is also a left unit. Thus, A, and also H_A, has an H-structure. Dually one obtains a co-H-structure for A and with it an H-structure for H^A. The proposition then follows from 11.6.4.

12.2.4 Lemma. *Let* $\mathscr{C}$ *be semi-additive.* $A_1 \xleftarrow{pr_1} A \xrightarrow{pr_2} A_2$ *is a product of* A_1 *and* A_2 *if and only if there are morphisms* $in_j: A_j \to A$ *for* $j = 1, 2$ *with* $pr_k\, in_j = \delta_{kj}$ *(i. e.,* 1_{A_k} *for* $k = j$, 0 *for* $k \neq j$*) and* $in_1\, pr_1 + in_2\, pr_2 = 1_A$.

Proof. Suppose that there is a product. Define $in_1: A_1 \to A$ by $\begin{pmatrix} 1 \\ 0 \end{pmatrix}$ and in_2 analogously. Then $pr_k\, in_j = \delta_{kj}$ hold. $pr_j\,(in_1\, pr_1 + in_2\, pr_2) = pr_j$ and with it $in_1\, pr_1 + in_2\, pr_2 = 1_A$ follows by the definition of products. Conversely, let in_1, in_2 have the required properties. If $f_j: B \to A_j$ are given, set $f = in_1 f_1 + in_2 f_2 \cdot pr_j\, f = f_j$ follows and this implies again $f = in_1 f_1 + in_2 f_2$. Thus one has a product.

12.2.5 Proposition. *A semi-additive category with a zero object has finite products if and only if it has finite coproducts. If this is the case, then the finite coproducts are also products, and the formulas* (1) *and* (2) *above are valid.*

Proof. 12.2.4 can obviously be modified for arbitrary finite products with at least one factor; also its dual, which proves the proposition. The case of an empty index set is trivial.

12.2.6 Convention. In a semi-additive category with finite products we always choose finite coproducts in such a way that their objects coincide with the corresponding products and formulas (1), (2) of 12.1.3 apply (compare 8.3.2). We then talk about *biproducts* and use the notation $\bigoplus A_j$.

12.2.1 now describes a morphism of biproducts. The composition of such morphisms is exactly the multiplication of matrices, their addition is the addition of matrices. Both are verified by means of (1) and (2). We need only consider the cases $n \leq 2$, $m \leq 2$. We use Δ for the diagonal morphism

$$\Delta = \begin{pmatrix} 1 \\ 1 \end{pmatrix} \colon A \to A \oplus A ,$$

∇ for the co-diagonal morphism $\nabla = (1,1) \colon A \oplus A \to A$.

With this, addition of f, g: $A \to B$ is described by each of the following three composites:

$$A \xrightarrow{\Delta} A \oplus A \xrightarrow{(f,g)} B , \tag{6}$$

$$A \xrightarrow{\binom{f}{g}} B \oplus B \xrightarrow{\nabla} B , \tag{7}$$

$$A \xrightarrow{\Delta} A \oplus A \xrightarrow{\left(\begin{smallmatrix} f & 0 \\ 0 & g \end{smallmatrix}\right)} B \oplus B \xrightarrow{\nabla} B . \tag{8}$$

12.2.7 Proposition. *Let $\mathcal{C}$, $\mathcal{D}$ be semi-additive categories. An additive functor $T\colon \mathcal{C} \to \mathcal{D}$ preserves products (including zero objects, if one exists). If $\mathcal{D}$ has a zero object and $\mathcal{C}$ finite products, then every functor $T\colon \mathcal{C} \to \mathcal{D}$ which preserves finite products, is additive.*

Proof. If T is additive, then T preserves zero morphisms and hence zero objects, since they are characterized by $1_0 = 0$. T preserves finite products because of 12.2.4, 12.2.5.

Conversely, let this be the case and let $\mathcal{C}$ have finite products. Then T preserves zero objects and thus zero morphisms. The proof of 12.2.4 shows that T preserves biproducts. T is additive because of (6).

12.3 Kernels and Cokernels

We shall assume the existence of a zero object throughout.

12.3.1 It makes sense here, to talk about kernels and cokernels. We designate a selected kernel $k\colon K \to A$ of $f\colon A \to B$ by $\ker f$, and similarly a cokernel by $\operatorname{coker} f$. Every kernel is a monomorphism (7.2.2). In the sense of the preordering 6.5.4 of monomorphisms with codomain A, $\ker f$ is maximal among the monomorphisms annihilated by f. $\ker f$ is also characterized by

(9)
$$\begin{array}{ccc} K & \longrightarrow & 0 \\ {\scriptstyle \ker f}\downarrow & & \downarrow \\ A & \xrightarrow{f} & B \end{array}$$

being a pullback. 1_A is always to be chosen as kernel of $A \to 0$ and as cokernel of $0 \to A$.

12.3.2 If $m: B \to C$ is a monomorphism, then $f: A \to B$ and $m f$ have the same kernels, because for $u: X \to A$, $m f u = 0$ is equivalent to $f u = 0$.

12.3.3 If m is a monomorphism, then $\ker m = 0$. In particular, $\ker(\ker f) = 0$. This follows from 12.3.2 with $f = 1_B$.

12.3.4 Theorem. *Let the square on the right in*

(10)
$$\begin{array}{ccccc} A_1 & \xrightarrow{a_1} & A_2 & \xrightarrow{a_2} & A_3 \\ \Downarrow & & \downarrow{\scriptstyle f_2} & & \downarrow{\scriptstyle f_3} \\ K & \xrightarrow{\ker g} & B_2 & \xrightarrow{g} & B_3 \end{array}$$

be commutative and $a_2 a_1 = 0$.

(a) *There is exactly one morphism* $f_1: A_1 \to K$ *which makes the left hand square commutative.*

(b) *If the square on the left becomes a pullback, then* a_1 *is a kernel of* a_2.

(c) *If* a_1 *is a kernel of* a_2 *and* f_3 *a monomorphism, then the left hand square is a pullback.*

(d) *If the right hand square is a pullback, then* a_1 *is a kernel of* a_2 *if and only if* f_1 *is an isomorphism.*

Proof. Since $g f_2 a_1 = 0$, (a) follows immediately from the definition of kernels.

(b) If $v: X \to A_2$ with $a_2 v = 0$ is given, then $g f_2 v = f_3 a_2 v = 0$, and there is exactly one morphism $u: X \to K$ with $(\ker g) u = f_2 v$. By the pullback property, there is exactly one morphism $w: X \to A_1$ with $a_1 w = v$ and $f_1 w = u$. By 7.8.2, a_1 is a monomorphism, and therefore, $a_1 w = v$ alone determines w uniquely. Thus a_1 is a kernel of a_2.

(c) If $u: X \to K$ and $v: X \to A_2$ are given with $(\ker g) u = f_2 v$, then $0 = g(\ker g) u = g f_2 v = f_3 a_2 v$. Since f_3 is a monomorphism, $a_2 v = 0$. Therefore, there is exactly one $w: X \to A_1$ with $a_1 w = v$. One has $(\ker g) f_1 w = f_2 a_1 w = f_2 v = (\ker g) u$ and thus $f_1 w = u$, since $\ker g$ is a monomorphism, which completes this proof.

(d) The morphism $0: K \to A_3$ uniquely determines a morphism $h: K \to A_2$ with $a_2 h = 0$ and $f_2 h = \ker g$. If one has $v: X \to A_2$ with $a_2 v = 0$, then $g f_2 v = 0$, and there is a unique $u: X \to K$ with

$f_2 v = (\ker g)\, u = f_2\, h\, u$. From $a_2 v = 0 = a_2\, h\, u$ the pullback property implies $v = h\, u$. So h is a kernel of a_2. For $v = a_1$ and $a_1 = h\, u$ in particular, it follows that $f_2\, a_1 = f_2\, h\, u = (\ker g)\, u$ and thus $u = f_1$ because of (a). $a_1 = h\, f_1$ then implies (d).

12.3.5 Remark. 12.3.4 (a) through (c) carry over to equalizers. Instead of a_2 and g one has two pairs of morphisms a_2, a_2' and g, g' with $a_2\, a_1 = a_2'\, a_1$, $f_3\, a_2 = g\, f_2$ and $f_3\, a_2' = g'\, f_2$. The place of $\ker g$ is taken by an equalizer of g and g'.

12.3.6 If $k\colon K \to A$ is a kernel of $f\colon A \to B$ and $p\colon A \to C$ a cokernel of k, then k is also a kernel of p. In the category of monomorphisms with codomain A (6.5.4, 6.5.6) $\ker f \cong \ker \operatorname{coker} \ker f$ is therefore valid.

Proof. Since $f\, k = 0$, there is exactly one morphism $q\colon C \to B$ with $q\, p = f$ by the definition of cokernel. Furthermore $p\, k = 0$. If one has $v\colon X \to A$ with $p\, v = 0$, then $q\, p\, v = f\, v = 0$ and there is exactly one morphism $w\colon X \to K$ with $v = k\, w$. So k is a kernel of p.

12.3.7 Definition. We assume that there are zero objects, kernels and cokernels. For $f\colon A \to B$ we set $\operatorname{im} f = \ker(\operatorname{coker} f)$ and $\operatorname{coim} f = \operatorname{coker}(\ker f)$ and call $\operatorname{im} f$ the *image*, and $\operatorname{coim} f$ the *coimage* of f.

12.3.8 Proposition. *Let the category $\mathscr{C}$ have a zero object, kernels and cokernels. For $f\colon A \to B$ there is a decomposition*

(11)
$$\begin{array}{ccc} \bar{A} & \xrightarrow{\bar{f}} & B' \\ {\scriptstyle \operatorname{coim} f}\uparrow & & \downarrow{\scriptstyle \operatorname{im} f} \\ A & \xrightarrow{f} & B \\ {\scriptstyle \ker f}\uparrow & & \downarrow{\scriptstyle \operatorname{coker} f} \\ \bullet & & \bullet \end{array}$$

with a uniquely determined $\bar{f}$, such that $f = (\operatorname{im} f)\, \bar{f}\, (\operatorname{coim} f)$. *If*

(12)
$$\begin{array}{ccc} A & \xrightarrow{f} & B \\ {\scriptstyle h_1}\downarrow & & \downarrow{\scriptstyle h_2} \\ C & \xrightarrow{g} & D \end{array}$$

is commutative, then (12) *can be extended in a unique way to a natural transformation of diagram* (11) *into the corresponding diagram for* $g\colon C \to D$.

Proof. Since $f\,(\ker f) = 0$, there is a $u\colon \bar{A} \to B$ with $f = u\,(\operatorname{coim} f)$ by the definition of $\operatorname{coim} f$. $(\operatorname{coker} f)\, u = 0$ follows, because $(\operatorname{coker} f)\, f = 0$ and $\operatorname{coim} f$ is an epimorphism. Therefore, there is an $\bar{f}$ with $u = (\operatorname{im} f)\, \bar{f}$. Since $\operatorname{coim} f$ is an epimorphism and $\operatorname{im} f$ a monomorphism, $\bar{f}$ is unique. To prove the second statement, start with omitting $\bar{f}$ from (11). The extension of (12) to a natural transformation is then provided by the definitions of kernel and cokernel and 12.3.7. If $\bar{f}$ and $\bar{g}$ are inserted again now, the commutativity of the new top rectangle follows from the fact that coim f is an epimorphism and im g a monomorphism. One adds coim f in front, im g at the end and makes use of the commutativity of the other faces of the cube which is obtained from the top part of (11) by means of (12) (compare 6.2.7).

12.3.9 Remarks. If in $\mathscr{C}$ kernels and cokernels are chosen, the second statement of 12.3.8 says that there is a functor from the category of $\mathscr{C}$-morphisms (6.5.1) to a category of commutative diagramms over $\mathscr{C}$ (natural factorization). This applies in particular to *Ab* with subgroups as kernels and factor groups as cokernels, and similarly to ${}_R Mod$, where $\bar{f}$ is an isomorphism.

Without additional conditions $\bar{f}$ need not be a monomorphism or an epimorphism, as is shown by Ens_*, Top_* and the category of groups. 12.3.7 and with it the notation in 12.3.8 are, in fact, not correct without additional conditions if one wants to have a general definition of image and coimage that is characterized by extremal conditions and that is natural.

12.4 Factorization of Morphisms

In the following lemmas the hypotheses are stated in the notation for the axioms in 12.1.1.

12.4.1 A0, A2, A3 imply: monomorphisms with the same codomain have finite intersections.

Proof. If $m: K \to B_2$ is a monomorphism, then m is a kernel of, let us say, $g: B_2 \to B_3$. If $f_2: A_2 \to B_2$ is also a monomorphism, then the claim follows from 12.3.4 (c) with $f_3 = 1_{B_3}$ and $a_2 = g f_2$ for the intersection of two monomorphisms, which is sufficient by 7.8.6.

12.4.2 A0, A1, A2, A3 imply finite completeness, in particular the existence of equalizers and pullbacks.

Proof. 12.4.1 and 7.8.8.

12.4.3 A0, A2, A2°, A3: if m is a monomorphism, then m is a kernel of coker m. If also coker $m = 0$, then m is an isomorphism. Every bimorphism is an isomorphism. (The category is balanced).

Proof. The first claim follows from A3 and 12.3.6. If coker $m = 0$ for $m: A \to B$, then m and 1_B are kernels of coker m, so m is an isomorphism. If m is a bimorphism, then coker $m = 0$ (dual of 12.3.3).

12.4.4 Lemma. *We assume* A0 *and* A2°. *The rule* $m \mapsto$ coker m (*with chosen cokernels*) *determines a functor* γ *from the category of mono-*

morphisms with codomain A (6.5.4) *to the category of epimorphisms with domain A. If* A2 *and* A3 *are also assumed, then* γ *induces an injective map if one goes over to equivalence classes of monomorphisms and of epimorphisms. This map is a bijection if in addition* A3^0 *is assumed.*

Proof. In

$$(13)\qquad \begin{array}{ccccc} K_1 & \xrightarrow{m_1} & A & \xrightarrow{p_1} & L_1 \\ \downarrow t & & \| 1_A & & \\ K_2 & \xrightarrow{m_2} & A & \xrightarrow{p_2} & L_2 \end{array}$$

let $m_1 = m_2 t$, and let m_1, m_2 be monomorphisms, so that t is also a monomorphism. Further, let $p_j = \operatorname{coker} m_j$ for $j = 1, 2$. $p_2 m_1 = 0$ and by the definition of cokernel there is exactly one morphism $\gamma(t)$: $L_1 \to L_2$ with $\gamma(t)\, p_1 = p_2$. This, obviously, yields a functor and in particular an isomorphism $\gamma(t)$ if t is an isomorphism. If A2 and A3 are assumed, m_j is a kernel of p_j (12.4.3) and any two kernels of p_j are equivalent. The rest follows from the dual of 12.4.3. The reverse map for equivalence classes is induced by $p \mapsto \ker p$.

12.4.5 A0, A2, A2°, A3: Suppose that f: $A \to B$ and a monomorphism m: $M \to B$ are given. f is of the form $f = m\, g$ if and only if $(\operatorname{coker} m)\, f = 0$. If this is the case, then $\operatorname{im} f$ is of the form $\operatorname{im} f = m\, h$; i.e., $\operatorname{im} f$ is the smallest monomorphism through which f can be factored.

Proof. $(\operatorname{coker} m)\, m = 0$ implies $(\operatorname{coker} m)\, m\, g = 0$. Now let $(\operatorname{coker} m)\, f = 0$. Then there is a g: $A \to M$ with $f = m\, g$, by 12.4.3 and the definition of kernels. By the definition of cokernels, there is a q with $q\,(\operatorname{coker} f) = \operatorname{coker} m$. By the definition of kernels and 12.4.3, there is an h with $\operatorname{im} f = m\, h$, and $\operatorname{im} f$ is smaller than m in the sense of 6.5.4.

12.4.6 Let A0, A2, A2°, A3 be satisfied and assume that there are equalizers (this is the case if, for instance, A1 is satisfied (12.4.2) or if the category is additive (7.2.6)). Then one has:

(a) The following are equivalent

(i) f: $A \to B$ is an epimorphism,

(ii) $\operatorname{im} f = 1_B$,

(iii) $\operatorname{coker} f = 0$.

(b) In the factorization $f = (\operatorname{im} f)\, f'$, f' is an epimorphism.

Proof. (a) By the dual of 12.3.6, $\operatorname{coker} f = \operatorname{coker}(\operatorname{im} f)$. Therefore (ii) and (iii) are equivalent by 12.4.3. (i) implies (iii) by the dual of 12.3.3. Now let $\operatorname{coker} f = 0$, and let u, v: $B \to Y$ be given with $u f = v f$. Let m: $M \to B$ be an equalizer of u and v. Then f is of the form $f = m\, g$. By 12.4.5 and (ii), 1_B is of the form $m\, h$. Therefore m is a retraction. Hence m is an isomorphism, and $u = v$ and thus (i) follows.

(b) We set $\operatorname{im} f\colon B' \to B$ and $f' = \bar{f}(\operatorname{coim} f)$ as in 12.3.8. Let $u, v\colon B' \to X$ with $u f' = v f'$ be given and let $m'\colon M \to B'$ be an equalizer of u and v. Then f' is of the form $m' g'$, m' is a monomorphism, and so is $m = (\operatorname{im} f)\, m'$. By 12.4.5, $\operatorname{im} f$ and m are equivalent monomorphisms with codomain B. So m' is an isomorphism, $u = v$ and therefore f' is an epimorphism.

12.4.7 A0, A2, A2°, A3: Let

$$(14)\qquad \begin{array}{ccc} A & \overset{a}{\twoheadrightarrow} & B \\ {\scriptstyle \bar a}\downarrow & & \downarrow{\scriptstyle b} \\ D & \underset{\bar b}{\rightarrowtail} & C \end{array}$$

be commutative, a an epimorphism and $\bar{b}$ a monomorphism. Then there is exactly one morphism $d\colon B \to D$ with $\bar{a} = da$ and $b = \bar{b}\, d$. Here d is an epimorphism (monomorphism) if $\bar{a}$ is an epimorphism (b a monomorphism). d is an isomorphism if it is a bimorphism.

Proof. Let $c\colon C \to E$ be a cokernel of $\bar{b}$. Then $0 = c\,\bar{b}\,\bar{a} = c\, b\, a$. Since a is an epimorphism, $c\, b = 0$ follows. By 12.4.3, $\bar{b}$ is a kernel of c. Therefore there is exactly one morphism d with $b = \bar{b}\, d$. $\bar{b}\, da = b\, a = \bar{b}\,\bar{a}$ and from this $da = \bar{a}$ follows, because $\bar{b}$ is a monomorphism. The next statement follows from 5.1.5⁰ or, resp., 5.1.5, the last one from 12.4.3.

12.4.8 Proposition. *Let* A0, A2, A2°, A3 *be satisfied. If a morphism can be factored into an epimorphism followed by a monomorphism, then the factorization is unique up to an isomorphism. If there is such a factorization for every morphism* (*compare* 12.4.6), *then it is natural. More precisely, let*

$$(15)\qquad \begin{array}{ccccc} A & \overset{a}{\twoheadrightarrow} & B & \overset{b}{\rightarrowtail} & C \\ & & \xrightarrow[\;c\;]{} & & \\ {\scriptstyle h}\downarrow & & & & \downarrow{\scriptstyle k} \\ & & \xrightarrow{\;z\;} & & \\ X & \underset{x}{\twoheadrightarrow} & Y & \underset{y}{\rightarrowtail} & Z \end{array}$$

be commutative, let a and x be epimorphisms, b and y monomorphisms. Then there is exactly one morphism $d\colon B \to Y$ with $d\, a = x\, h$ and $y\, d = k\, b$. Here d is an epimorphism (monomorphism) if h is an epimorphism (k a monomorphism). d is an isomorphism if it is a bimorphism.

These statements follow immediately from 12.4.7.

12.4.9 Proposition. *Let* A0, A2, A2°, A3, A3° *be satisfied. If there is a factorization $f = f'' f'$ for f with an epimorphic f' and a monomorphic*

f'', *then* f' *is a cokernel of* ker f *and* f'' *a kernel of* coker f. *In the factorization* 12.3.8 $\bar{f}$ *is an isomorphism.*

Remark. There is a factorization $f = f'' f'$ for an arbitrary f, provided there are equalizers or coequalizers, also, in particular, if in addition A1 or A1^0 is assumed, or if the category is additive.

Proof. By 12.3.2, ker f is also a kernel of f', and by the dual of 12.4.3, f' is a cokernel of ker f. Dually, then, f'' is a kernel of coker f. The last assertion is thus evident. The remark follows from 12.4.6 (b) and its dual.

12.4.10 Remark. Let us assume that in the category $\mathscr{C}$ every morphism can be factored uniquely up to an isomorphism into an epimorphism followed by a monomorphism (as is the case, e. g., in *Ens* and in every category [$\mathscr{C}$, *Ens*]). If (14) is commutative with an epimorphic a and a monomorphic $\bar{b}$, then there is exactly one morphism $d\colon B \to D$ with $\bar{a} = da$ and $b = d\bar{b}$. This follows by factoring $\bar{a}$ and b into epi- and monomorphisms and considering the factorization of $b\,a = \bar{b}\,\bar{a}$. 12.4.8 is accordingly valid including the fact that every bimorphism $f\colon A \to B$ is an isomorphism, which follows from 5.3.4 with $1_B f = f 1_A$.

If the factorization of morphisms is natural in the category $\mathscr{C}$, a natural factorization is implied for every functor category [$\mathscr{D}$, $\mathscr{C}$], resp. *Add* ($\mathscr{D}$, $\mathscr{C}$); compare 10.1.6, 10.1.9. The factorization into epi- and monomorphisms is guaranteed to be unique up to isomorphisms if 10.1.5 is valid.

12.5 The Additive Structure

We use Puppe's proof [7]. It is based on the fact that the diagonal map $\Delta\colon B \to B \oplus B$ has the cokernel $(1, -1)\colon B \oplus B \to B$ which provides a subtraction and addition for all $[A, B]$ and all $[B, C]$. However, the existence of coproducts is not a requirement for the existence of a cokernel of $\Delta\colon B \to B \oplus B$.

12.5.1 Proposition. *Let the category* $\mathscr{C}$ *satisfy axioms* A0, A2, A2°, A3, A3° *of* 12.1.1 *and in addition* A1 *or* A1°. *Then* $\mathscr{C}$ *has a uniquely determined additive structure and* $\mathscr{C}$ *is abelian.*

Proof. We assume A1, the other case is dual. The second claim follows from the first one by 12.2.5. By 11.6.5, there is at most one semi-additive structure. It remains to be shown that there exists an additive structure. For $B \in |\mathscr{C}|$ consider

(16)

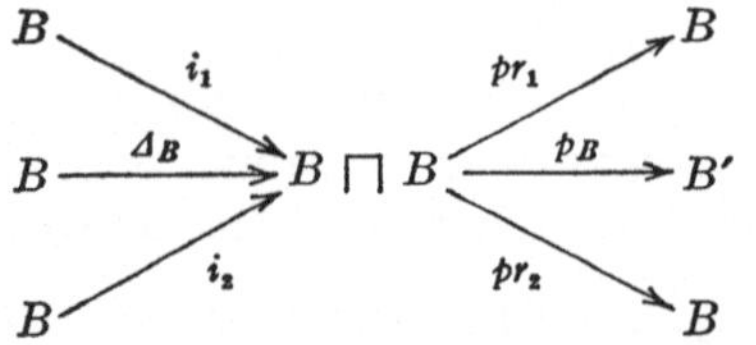

where $i_1 = \begin{pmatrix}1\\0\end{pmatrix}$, $i_2 = \begin{pmatrix}0\\1\end{pmatrix}$, $\Delta_B = \begin{pmatrix}1\\1\end{pmatrix}$, p_B is a cokernel of Δ_B and pr_1, pr_2 are the projections. The proof now consists of a chain of lemmas.

12.5.2 i_1 is a kernel of pr_2 and pr_2 is a cokernel of i_1, and correspondingly for i_2 and pr_1.

Proof. $pr_2\, i_1 = 0$. Since $pr_1\, i_1 = 1_B$, i_1 is a monomorphism and pr_1, pr_2 are epimorphisms. For $w = \begin{pmatrix}u\\v\end{pmatrix}: X \to B \sqcap B$, $pr_2\, w = 0$ is equivalent to $v = 0$. Since i_1 is a monomorphism, $\begin{pmatrix}u\\0\end{pmatrix} = i_1\, u$ implies that i_1 is a kernel of pr_2. pr_2 is a cokernel of i_1 by the dual of 12.4.3.

12.5.3 $p_B i_1$ and $p_B i_2$ are isomorphisms.

Proof. Since $pr_2\, \Delta_B = 1_B$, Δ_B is a monomorphism and by 12.4.3 it is a kernel of p_B. We consider (compare later 13.5.6)

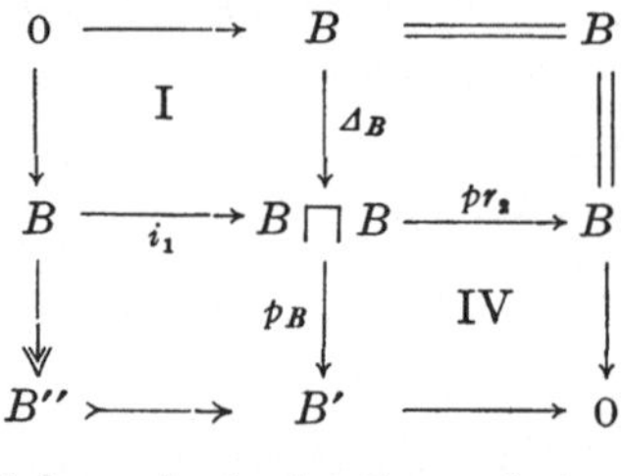

where $B \twoheadrightarrow B'' \rightarrowtail B'$ is a factorization of $p_B\, i_1$ into an epi- and a monomorphism which exists by 12.4.6 (b). By 12.5.2 and 12.3.4 (c), I is a pullback. From 12.3.4 (b), applied to the first two columns, it follows that $B \twoheadrightarrow B''$ has the kernel 0. Correspondingly, using the dual of 12.3.4, one shows that $B'' \rightarrowtail B'$ has the cokernel 0. By 12.4.3 and its dual, $B \twoheadrightarrow B''$ and $B'' \rightarrowtail B'$ are isomorphisms. Hence $p_B\, i_1$ is an isomorphism.

12.5.4 We set

(17) $$s_B = (p_B\, i_1)^{-1}\, p_B \colon\ B \sqcap B \to B\,,$$

(18) $$m_B = s_B\, i_2 = (p_B\, i_1)^{-1}\,(p_B\, i_2) \colon\ B \to B\,,$$

(19) $$a_B = s_B\,(1_B \sqcap m_B^{-1}) \colon\ B \sqcap B \to B\,.$$

a_B and $n_B\colon 0 \to B$ define an H-structure for B.

Proof. Since

$$a_B\begin{pmatrix}1\\0\end{pmatrix}\overset{(19)}{=}s_B\,(1_B\sqcap m_B^{-1})\begin{pmatrix}1\\0\end{pmatrix}=s_B\begin{pmatrix}1\\0\end{pmatrix}\overset{(17)}{=}1_B\,,$$

$$a_B\begin{pmatrix}0\\1\end{pmatrix}\overset{(19)}{=}s_B\,(1_B\sqcap m_B^{-1})\begin{pmatrix}0\\1\end{pmatrix}=s_B\begin{pmatrix}0\\m_B^{-1}\end{pmatrix}$$

$$=s_B\begin{pmatrix}0\\1\end{pmatrix}m_B^{-1}\overset{(18)}{=}1_B\,,$$

the claim follows from 11.1.5.

12.5.5 Let $f\colon B\to C$ be given. If as in (16) through (19) an H-structure for C is defined, then f is a homomorphism.

$$(20)\qquad\begin{array}{ccccc}B & \xrightarrow{\Delta_B} & B\sqcap B & \xrightarrow{s_B} & B\\ {\scriptstyle f}\downarrow & & \downarrow{\scriptstyle f\sqcap f} & & \downarrow{\scriptstyle g}\\ C & \xrightarrow[\Delta_C]{} & C\sqcap C & \xrightarrow[s_C]{} & C\end{array}$$

The left rectangle is commutative. By 12.5.3 and (17), s_B is a cokernel of Δ_B and s_C a cokernel of Δ_C. According to the dual of 12.3.4 (a), there exists a unique g such that the right rectangle is commutative. One can thus deduce the following

$$(21)\qquad g\overset{(17)}{=}g\,s_B\,i_1\overset{(20)}{=}s_C(f\sqcap f)\,i_1=s_C\begin{pmatrix}f\\0\end{pmatrix}=s_C\begin{pmatrix}1_C\\0\end{pmatrix}f\overset{(17)}{=}f\,,$$

$$(22)\qquad f\,m_B\overset{(21)}{=}g\,m_B\overset{(18)}{=}g\,s_B\,i_2\overset{(20)}{=}s_C\,(f\sqcap f)\,i_2=s_C\begin{pmatrix}0\\1_C\end{pmatrix}f=m_C\,f\,,$$

$$(23)\qquad f\,a_B\overset{(21)}{=}g\,a_B\overset{(19)}{=}g\,s_B\,(1_B\sqcap m_B^{-1})\overset{(20)}{=}s_C\,(f\sqcap f)\,(1_B\sqcap m_B^{-1})=$$
$$=s_C\,(f\sqcap f\,m_B^{-1})\overset{(22)}{=}s_C\,(f\sqcap m_C^{-1}\,f)=$$
$$=s_C\,(1\sqcap m_C^{-1})\,(f\sqcap f)\overset{(19)}{=}a_C\,(f\sqcap f)\,.$$

The assertion follows from (23) and $f\,0=0\colon 0\to B\to C$.

12.5.6 There exists a semi-additive structure for $\mathscr{C}$ such that, for $f_1, f_2\in[A,B]$, $f_1+f_2=a_B\begin{pmatrix}f_1\\f_2\end{pmatrix}$.

Proof. By 11.3.5, $H_B\colon\mathscr{C}^0\to Ens$ has an H-structure provided by B, where the binary operation is given by composing the isomorphism $[?,B]\times[?,B]\cong[?,B\sqcap B]$ with $[?,a_B]$ (11.3.4). Written as an addition, it has the given form at the "point" A. If for every object of $\mathscr{C}$ an H-structure as in (16) through (19) is fixed, then 12.5.5 and 11.4.1 imply, that $H^A\colon\mathscr{C}\to Ens$ has an H-structure. With this, the assertion follows from 11.6.4.

12.5.7 The semi-additive structure is even additive.

Proof. For $f\colon A \to B$ one has

$$f + m_B f = a_B \binom{f}{m_B f} = a_B (1 \sqcap m_B) \binom{f}{f} \overset{(19)}{=\!=} s_B \binom{f}{f} = s_B \Delta_B f \overset{(17)}{=\!=} 0\,.$$

12.6 Idempotents

12.6.1 Definition. An endomorphism $h\colon A \to A$ is called *idempotent* if $h\,h = h$.

One verifies immediately

12.6.2 If $r\colon A \to B$ is a retraction with $r\,i = 1_B$, then $i\,r\colon A \to A$ is idempotent. If in an additive category $h\colon A \to A$ is idempotent, then $1_A - h$ is also idempotent.

12.6.3 Proposition. *Let $\mathscr{C}$ be an additive category.*

(a) *If $r\colon A \to B$ is a retraction with $ri = 1_B$, then i is a kernel of $1_A - ir$.*

(b) *If $h\colon A \to A$ is idempotent, $i_1\colon A_1 \to A$ a kernel of h, and $i_2\colon A_2 \to A$ a kernel of $1_A - h$, then A is a biproduct of A_1 and A_2 with injections i_1, i_2.*

Proof. (a) $(1_A - i\,r)\,i = 0$. If $u\colon X \to A$ is given with $(1_A - i\,r)\,u = 0$, then $u = i(r\,u)$. Since i is a monomorphism, i is a kernel of $1_A - i\,r$.

(b) Since $h\,(1_A - h) = 0$, there is a $p_1\colon A \to A_1$ with $i_1\,p_1 = 1_A - h$ by the definition of kernels. There is also a $p_2\colon A \to A_2$ with $i_2\,p_2 = h$. It follows that $i_1\,p_1 + i_2\,p_2 = 1_A$, $i_1\,p_1\,i_1 = i_1 - h\,i_1 = i_1$ and from this that $p_1\,i_1 = 1_{A_1}$, because i_1 is a monomorphism; also $i_1\,p_1\,i_2 = (1_A - h)\,i_2 = 0$ and thus $p_1\,i_2 = 0$. One obtains $p_2\,i_2 = 1_{A_2}$ and $p_2\,i_1 = 0$ similarly and therefore (b) by the dual of 12.2.4.

12.7 Problems

12.7.1 Carry out the proof of 12.1.6.

12.7.2 A semi-additive category $\mathscr{C}$ can be enlarged to a category of matrices over $\mathscr{C}$ (in 12.2.1 A and B are to be replaced by m-tupels or, resp., n-tupels). The enlargement has finite products. Additive functors can be extended to the enlargements.

12.7.3 What is the analogue of 12.3.2 for equalizers in an arbitrary category?

12.7.4 Carry out 12.3.5.

12.7.5 If k is a kernel of $g\,f$ and if $f\,k = 0$, then k is also a kernel of f. Also formulate the analogous case for equalizers.

12.7.6 Fill in the details of 12.3.8.

12.7.7 Carry out 12.4.10.

12.7.8 A0, A2, A2° and A3 imply: if there exists $A \sqcup B$ with injections in_1, in_2, then

$$\begin{array}{ccc} 0 & \longrightarrow & B \\ \downarrow & & \downarrow{\scriptstyle in_2} \\ A & \xrightarrow{in_1} & A \sqcup B \end{array}$$

is a pullback and a pushout.

12.7.9 In a category with equalizers the converse of the first statement in 12.6.2 is true. It is also true if every morphism admits a factorization into an epimorphism and a monomorphism.

12.7.10 Let A, B be two objects in a finitely complete additive category such that $[A, B] = 0$. Show that any retract of $A \oplus B$ (i. e., the codomain of some retraction) is of the form $A' \oplus B'$, where A' is a retract of A and B' retract of B.

13. Exact Sequences

13.1 Exact Sequences in Exact Categories

13.1.1 Definition. In a category with a zero object and kernels a *sequence* of two morphisms

(1) $$A \xrightarrow{f} B \xrightarrow{g} C$$

is called *exact* if

(i) $$g f = 0\,,$$

(ii) in the factorization $f = (\ker g)\, f'$ guaranteed by (i), f' is an epimorphism.

A sequence of morphisms

$$\ldots \to A_{n-1} \xrightarrow{a_{n-1}} A_n \xrightarrow{a_n} A_{n+1} \xrightarrow{a_{n+1}} A_{n+2} \to \ldots$$

is called *exact at A_n* if a_{n-1} and a_n satisfy (i) and (ii). It is called *exact,* if it is exact at every place.

This definition is correct for groups also. In view of the intended applications however, we restrict ourselves to stronger conditions for the categories considered.

13.1.2 Definition. A *category* is called *exact* if it has a zero object and if every morphism f admits a factorization $f = f'' f'$ such that f' is a cokernel and f'' a kernel.

13.1.3 Proposition. *If $\mathscr{C}$ is an exact category, then it satisfies* A0, A2, A2⁰, A3, A3⁰ *of* 12.1.1. *Statements* 12.4.8 *and* 12.4.9 *are valid for the factorizations of morphisms.*

Proof. A0 is true by assumption. If f is a monomorphism, then f' is a monomorphic cokernel. By 8.2.5 and 8.2.2, f' is an isomorphism. Therefore f is a kernel and A3 is true. To prove A2, let f' be a cokernel of some g and let $g = g'' g'$ be a factorization of g as in 13.1.2. Since g' is an epimorphism, f' is also a cokernel of g'' (dual of 12.3.2), and since g'' is a kernel, g'' is a kernel of f' by 12.3.6. g'' is a kernel of f because of 12.3.2. Hence A2 is valid. $\mathscr{C}^0$ is exact by definition 13.1.2. Therefore A2⁰ and A3⁰ are valid in $\mathscr{C}$. The last claim of the proposition is evident.

13.1.4 Lemma. *In an exact category the following statements are equivalent:*

(E_1) (1) *is exact; i. e., there is a factorization* $f = (\ker g) f'$ *of* f *with an epimorphic* f'.
(E_1^0) *There is a factorization* $g = g''(\operatorname{coker} f)$ *of* g *with a monomorphic* g''.
(E_2) im f *is a kernel of* g.
(E_2^0) coim g *is a cokernel of* f.
(E_3) im f *is a kernel of* coim g.
(E_4) ker g *is a kernel of* coker f.

The dual statements of (E_3) *and* (E_4).

Proof. By 12.3.6, g and coim g have the same kernels, and im f is a kernel of coker f. By 12.4.9, (E_1), (E_2), (E_3), (E_4) all say that g, coim g and coker f have the same kernels. By 12.4.4, (E_3) is equivalent to coim g being a cokernel of im f. This is the dual of (E_3). The duals of (E_1) through (E_4) are also equivalent because the dual of an exact category is exact.

13.1.5 In an exact category the following are valid:

(a) $0 \to A \xrightarrow{m} B$ is exact if and only if m is a monomorphism.
(a⁰) $A \xrightarrow{p} B \to 0$ is exact if and only if p is an epimorphism.
(b) $0 \to A \xrightarrow{j} B \to 0$ is exact if and only if j is an isomorphism.
(c) $0 \to A \xrightarrow{m} B \xrightarrow{f} C$ is exact if and only if m is a kernel of f.
(c⁰) $A \xrightarrow{f} B \xrightarrow{p} C \to 0$ is exact if and only if p is a cokernel of f.

Proof. (a) 1_A is a cokernel of $0 \to A$. Thus (a) follows immediately from (E_1^0). (a⁰) is its dual. (b) follows from (a) and (a⁰). If in (c) m is a kernel of f, then exactness at A follows from (a), exactness at B

from (E_1). The converse follows also from (a) and (E_1). (c^0) is dual to (c).

13.1.6 Proposition. *Let*

(2) $$A \xrightarrow{f} B \xrightarrow{g} C \xrightarrow{h} D\,.$$

(a) *If h is a monomorphism, then* (2) *is exact at B if and only if $A \xrightarrow{f} B \xrightarrow{hg} D$ is exact.*

(b) *If g is a cokernel of f and if $A \xrightarrow{f} B \xrightarrow{hg} D$ is exact, then h is a monomorphism.*

(c) *If g is a monomorphism and if $A \xrightarrow{gf} C \xrightarrow{h}$ is exact, then $A \xrightarrow{f} B \xrightarrow{hg} D$ is exact.*

Proof. (a) follows immediately from (E_2) because of 12.3.2. (b) follows from (E_2^0) and 12.4.9. In (c) $f = (im\, f)\, f'$ with an epimorphic f'. We consider.

(3) $$\begin{array}{ccccc} B' & \xrightarrow{\operatorname{im} f} & B & \xrightarrow{hg} & D \\ \downarrow{\scriptstyle 1_{B'}} & & \downarrow{\scriptstyle g} & & \downarrow{\scriptstyle 1_D} \\ B' & \xrightarrow{g(\operatorname{im} f)} & C & \xrightarrow{h} & D \end{array}$$

The assumptions, together with (E_1), imply, that g (im f) is a kernel of h. The composite of the top line in (3) is 0, the right square is commutative and the left one is a pullback, since g is a monomorphism. Thus (c) follows from 12.3.4 (b).

13.1.7 Let the diagram

(4) $$\begin{array}{ccccc} A_1 & \xrightarrow{a_1} & A_2 & \xrightarrow{a_2} & A_3 \\ {\scriptstyle f_1}\downarrow & & {\scriptstyle f_2}\downarrow & & \downarrow{\scriptstyle f_3} \\ B_1 & \xrightarrow[b_1]{} & B_2 & \xrightarrow[b_2]{} & B_3 \end{array}$$

be commutative, assume f_1 to be an epimorphism and f_2 a monomorphism and the bottom line exact.

(a) If f_3 is a monomorphism, then the top line is exact.

(b) If a_2 is a cokernel of a_1, then f_3 is a monomorphism.

Proof. By the dual of 13.1.6 (a),

$$A_1 \to B_2 \xrightarrow{b_2} B_3$$

is also exact with $b_1 f_1 = f_2 a_1 \colon A_1 \to B_2$. By 13.1.6 (c),

$$A_1 \xrightarrow{a_1} A_2 \to B_3$$

is exact with $b_2 f_2 = f_3 a_2 \colon A_2 \to B_3$. Thus our statements above follow from 13.1.6 (a), (b).

13.2 Short Exact Sequences

13.2.1 A *short exact sequence* is an exact sequence of the form

(5) $$0 \to A \xrightarrow{m} B \xrightarrow{p} C \to 0\,.$$

Exact sequences are special diagrams. Thus natural transformations and isomorphisms of exact sequences are defined. In analogy to the case of Ab one also writes for (5), or for an isomorphic exact sequence,

$$0 \to A \to B \to B/A \to 0 .$$

13.2.2 There are two short exact sequences associated with every morphism $f\colon A \to B$, namely

$$0 \to K \xrightarrow{\ker f} A \xrightarrow{\operatorname{coim} f} \bar{A} \to 0 \quad \text{and} \tag{6}$$

$$0 \to B' \xrightarrow{\operatorname{im} f} B \xrightarrow{\operatorname{coker} f} K' \to 0 \tag{7}$$

with an isomorphism $\bar{f}\colon \bar{A} \to B'$ (12.4.9).

According to (E_2), (E_2^0), the exactness of $A \xrightarrow{f} B \xrightarrow{g} C$ can be described by an isomorphism between two of the short exact sequences belonging to f and g.

13.2.3 Longer exact sequences like

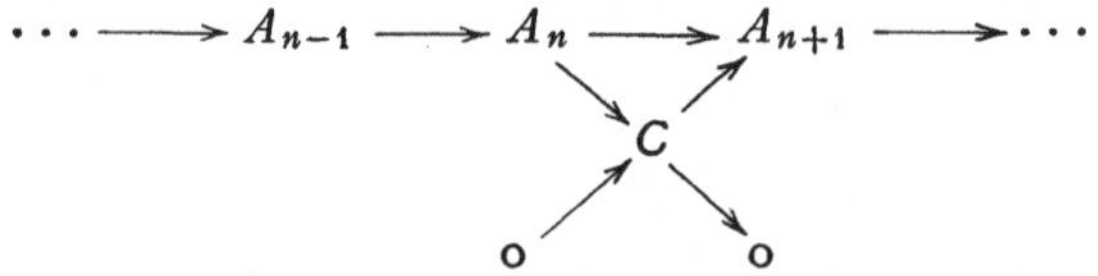

may be broken up into two exact sequences by factoring a morphism into an epimorphism followed by a monomorphism (13.1.5). Conversely, an exact sequence ending with $C \to 0$ can be coupled with one starting with $0 \to C$.

13.2.4 One says that the *short exact sequence* (5) *splits*, if p is a retraction. For exact additive categories, 12.6.3 says: If (5) splits, then B is a biproduct of A and C, where m is one of the injections and p is one of the projections. The injection belonging to C is in general not uniquely determined, as is seen, for instance, in the case of $B = \mathbf{Z} \oplus \mathbf{Z}$ in Ab.

13.2.5 Proposition. *In an exact additive category $\mathscr{C}$ the following are equivalent:*

(a) $0 \to A \xrightarrow{f} B \xrightarrow{g} C$ *is exact;*

(b) $0 \to [X, A] \to [X, B] \to [X, C]$ *is exact in Ab for every $X \in |\mathscr{C}|$.*

By 13.1.5 (c), this follows immediately from 7.7.3. The dual formulation is:

$A \to B \to C \to 0$ is exact if and only if $0 \to [C, X] \to [B, X] \to [A, X]$ is exact in Ab for all $X \in |\mathscr{C}|$.

In Ab $0 \to \mathbf{Z} \xrightarrow{f} \mathbf{Z} \xrightarrow{p} \mathbf{Z}_2 \to 0$ is exact if f is multiplication by 2 and p a cokernel of f.

$$0 \to [\mathbf{Z}_2, \mathbf{Z}] \to [\mathbf{Z}_2, \mathbf{Z}] \to [\mathbf{Z}_2, \mathbf{Z}_2] \to 0 \quad \text{and}$$
$$0 \to [\mathbf{Z}_2, \mathbf{Z}_2] \to [\mathbf{Z}, \mathbf{Z}_2] \to [\mathbf{Z}, \mathbf{Z}_2] \to 0$$

are not exact. Hence the proposition and its dual do not generalize to arbitrary exact sequences.

13.2.6 According to 13.1.5 (a^0) and (c) and 10.4.1, an object P of an exact additive category is projective if and only if for every short exact sequence $0 \to A \to B \to C \to 0$ the exact sequence $0 \to [P, A] \to \to [P, B] \to [P, C] \to 0$ is always exact in Ab.

13.2.7 Proposition. *In an exact additive category $\mathscr{C}$ the following are equivalent:*
(a) $0 \to A \xrightarrow{f} B \xrightarrow{g} C \to 0$ *is a split short exact sequence;*
(b) *for every* $X \in |\mathscr{C}|$
$0 \to [X, A] \to [X, B] \to [X, C] \to 0$ *is exact in* Ab.

Proof. Obviously (b) follows from (a). If (b) is assumed, then 13.2.5 implies first the exactness of $0 \to A \xrightarrow{f} B \xrightarrow{g} C$; if we then choose $X = C$ in (b), it follows that g is a retraction . Hence g is an epimorphism and the condition of 13.2.4 is satisfied.

13.3 Exact and Faithful Functors

13.3.1 Definition. A *functor* $T\colon \mathscr{C} \to \mathscr{D}$ between exact categories is called *left exact,* if it preserves kernels, *right exact,* if it preserves cokernels, and *exact,* if it is both right and left exact. T is called *half exact* if for every short exact sequence (5) in $\mathscr{C}$ $T(A) \to T(B) \to T(C)$ is exact.

13.3.2 Proposition. *Let $\mathscr{C}$ and $\mathscr{D}$ be exact categories and $T\colon \mathscr{C} \to \mathscr{D}$ a functor.*

(a) *T is left exact (right exact) if for every short exact sequence* (5)

$$0 \to T(A) \xrightarrow{T(m)} T(B) \xrightarrow{T(p)} T(C) \quad (T(A) \to T(B) \to T(C) \to 0)$$

is exact.

(b) *T is exact if and only if T takes exact sequences into exact sequences.*

(c) *If $\mathscr{C}$ is abelian, $\mathscr{D}$ exact additive and $T\colon \mathscr{C} \to \mathscr{D}$ additive, then T is left exact if and only if it preserves finite limits.*

(d) *If $\mathscr{C}$ is abelian and $\mathscr{D}$ exact additive, then every half exact functor $T\colon \mathscr{C} \to \mathscr{D}$ is additive.*

Proof. (a) If T is left exact, then the statement for short exact sequences follows from 13.1.5 (c). To prove the converse we consider for $f\colon A \to B$ the two short exact sequences (6), (7) belonging to f, together with the isomorphism $\bar{f}\colon \bar{A} \to B'$. Then $T(\operatorname{im} f)$ is a monomorphism, and hence also $T(\operatorname{im} f)\ T(\bar{f})$. Since $f = (\operatorname{im} f)\ \bar{f}\ (\operatorname{coim} f)$, $T(f)$ and $T(\operatorname{coim} f)$ have the same kernels (12.3.2). Thus (6) implies that $T(\ker f)$ is a kernel of $T(f)$. The right exact case is dual to the left exact one

(b) follows immediately from (a) and 13.2.3.

(c) follows from 7.4.5 because of 12.2.7.

(d) For an arbitrary $A \in |\mathscr{C}|$, $0 \to 0 \to A \xrightarrow{1_A} A \to 0$ is exact. Since T is half exact, $T(0) \to T(A)$ is a zero morphism. So T preserves zero morphisms. As zero objects 0 are characterized by $1_0 = 0$, T preserves zero objects, too. If $A \oplus B$ is a biproduct in $\mathscr{C}$ with injections in_1, in_2 and projections pr_1, pr_2, then $T(pr_k)\, T(in_j) = \delta_{kj}$ and therefore, $T(pr_k)$ is a retraction, and $T(in_j)$ is a coretraction. By 12.5.1 and the half exactness of T,

$$0 \to T(A) \xrightarrow{T(in_1)} T(A \oplus B) \xrightarrow{T(pr_2)} T(B) \to 0$$

is exact, because $T(in_1)$ is a monomorphism, and $T(pr_2)$ an epimorphism. By 13.2.4, $T(A \oplus B)$ is a biproduct with injections $T(in_j)$ and projections $T(pr_j)$. T is additive on account of 12.2.7.

13.3.3 For an exact additive category $\mathscr{C}$ the embedding $H_*\colon \mathscr{C} \to \to Add(\mathscr{C}^0, Ab)$ is left exact by 13.2.5, but not exact.

13.3.4 Proposition. *In abelian categories finite products of exact sequences are exact. Here we understand the product of two sequences*

$$\ldots A_{n-1} \xrightarrow{a_{n-1}} A_n \xrightarrow{a_n} A_{n+1} \ldots$$

and

$$\ldots B_{n-1} \xrightarrow{b_{n-1}} B_n \xrightarrow{b_n} B_{n+1} \ldots$$

to be the sequence

$$\ldots \to A_{n-1} \oplus B_{n-1} \xrightarrow{\begin{pmatrix} a_{n-1} & 0 \\ 0 & b_{n-1} \end{pmatrix}} A_n \oplus B_n \xrightarrow{\begin{pmatrix} a_n & 0 \\ 0 & b_n \end{pmatrix}} A_{n+1} \oplus B_{n+1} \to \ldots .$$

Proof. It obviously suffices to give a proof for the product of two short exact sequences, which is easily calculated using 13.1.5 (c), (c^0). It also can be done as follows: according to 7.6.3, the formation of products commutes with the formation of kernels and the dual applies to coproducts and cokernels. Furthermore, finite products and finite coproducts coincide.

13.3.5 Proposition. *Let $T\colon \mathscr{C} \to \mathscr{D}$ be a faithful functor.*

(a) *If $\mathscr{C}$, $\mathscr{D}$ are arbitrary categories, then T reflects mono- and epimorphisms and also commutative diagrams.*

(b) *If $\mathscr{C}$ and $\mathscr{D}$ have zero morphisms and $\mathscr{C}$ kernels, then T reflects zero morphisms and preserves them, if there exists a morphism f in $\mathscr{C}$ with $T(f) = 0$.*

(c) *If $\mathscr{C}$ and $\mathscr{D}$ are exact, then T takes non-exact sequences into non-exact ones and thus reflects exact sequences. In particular, T reflects kernels and cokernels.*

(d) *If $\mathscr{C}$ is abelian, $\mathscr{D}$ additive and T also additive, then T reflects finite limits and colimits.*

Proof. (a) For $m: B \to C$ in $\mathscr{C}$, suppose f and g satisfy $m f = m g$. If $T(m)$ is a monomorphism, then $T(f) = T(g)$. Since T is faithful, $f = g$ and m is a monomorphism. The epimorphic case is dual. A commutativity condition for a diagram fails if two morphisms with the same domain and codomain are different. After applying T they remain different.

(b) If $n: A \to B$ is a zero morphism in $\mathscr{C}$, then every arbitrary $X \xrightarrow{0} Y$ can be factored through n: $X \xrightarrow{0} A \xrightarrow{n} B \xrightarrow{0} Y$. This implies: If T preserves one zero morphism, then it preserves them all. Now let $T(f) = 0$ for $f: B \to C$. Then $f(\ker f) = 0$ and $T(f)\, T(\ker f) = 0$, and T preserves zero morphisms. For $0: B \to C$ we thus have $T(0) = 0$ and hence $f = 0$.

(c) For $g: A \to B$, $f: B \to C$ assume first that $f g \neq 0$. Then by (b) $T(f)\, T(g) \neq 0$. Now let $f g = 0$, let $k: K \to B$ be a kernel of f and $c: B \to D$ a cokernel of g. We consider

$$\begin{array}{ccccc} & & T(K) & & \\ & & \downarrow{\scriptstyle T(k)} & & \\ T(A) & \xrightarrow{T(g)} & T(B) & \xrightarrow{T(f)} & T(C)\,. \\ & & \downarrow{\scriptstyle T(c)} & & \\ & & T(D) & & \end{array}$$

If the row is exact, then T preserves zero morphisms by (b). Therefore, $T(f)\, T(k) = 0$ and $T(c)\, T(g) = 0$. Thus $T(k)$ factors through $\ker T(f)$ and $T(c)$ through $\operatorname{coker} T(g)$, and since $(\operatorname{coker} T(g))\, (\ker T(f)) = 0$, one has $T(c\, k) = 0$, which implies $c\, k = 0$. So k factors through $\operatorname{im} g = \ker c$. Since $g f = 0$, $\operatorname{im} g$ factors through k. So k and $\operatorname{im} g$ are equivalent and $A \xrightarrow{g} B \xrightarrow{f} C$ is exact. If $T(g)$ is a kernel of $T(f)$, then g is a monomorphism by (a) and therefore a kernel of f. Cokernels are the dual case.

(d) The proof of 12.2.7 shows that T preserves finite products as biproducts. For the finite diagram $D: \Sigma \to \mathscr{C}$ every natural transformation $\xi: A_\Sigma \to D$ induces a uniquely determined morphism $h: A \to \Pi\, D(i)$ with $\xi_i = pr_i\, h$, i ranges over the vertices of Σ. Let $c: \Pi\, D(i) \to C$ be a cokernel of h. If $(T(A), T\xi)$ is a limit of TD, then $T(h)$ is a monomorphism and so is h by (a). For the natural transformation $\beta: B_\Sigma \to D$, let $g: B \to \Pi\, D(i)$ be the induced morphism. If g does not factor through h, then $c\, g \neq 0$, since h is a kernel of c. $T(c\, g) \neq 0$ would follow. Because $T(c\, h) = 0$, $T(g)$ could then not factor through $T(h)$ and hence $(T(A), T\xi)$ not be a limit of TD. Therefore, g factors through h, and even uniquely, since h is a monomorphism. Thus is follows that (A, ξ) is a limit of D. The case of colimits is dual.

13.3.6 Remark. A faithful functor need not preserve zero morphisms. Let $A \neq 0$ in an abelian category $\mathscr{C}$. Set $T(f) = \begin{pmatrix} f & 0 \\ 0 & 1_A \end{pmatrix}$

13.3.7 Let $\mathscr{C}$ be an abelian category and $\mathscr{D}$ an exact additive category. If $T: \mathscr{C} \to \mathscr{D}$ is an exact functor, which reflects zero objects, then T is faithful.

Proof. By 13.3.2 (d), T is additive. It is therefore sufficient to show that if $f \neq 0$, then $T(f) \neq 0$. But for $f \neq 0$, $\operatorname{im} f \neq 0$ and $\operatorname{coim} f \neq 0$. The two short exact sequences belonging to f (13.2.2 (6), (7)) are taken into exact sequences by T. Since T reflects zero objects, $T(\operatorname{im} f)$ and $T(\operatorname{coim} f)$ and hence $T(f)$ are not 0.

13.4 Exact Squares

In this section the categories involved are always assumed to be abelian.

13.4.1 A square

(1)
$$\begin{array}{ccc} C & \xrightarrow{a} & A \\ {\scriptstyle b}\downarrow & & \downarrow{\scriptstyle \bar{b}} \\ B & \xrightarrow{\bar{a}} & D \end{array}$$

gives rise to morphisms

(2)
$$C \xrightarrow{\binom{a}{b}} A \oplus B \xrightarrow{(\bar{b}, -\bar{a})} D\,,$$

and (1) is commutative if and only if in (2) the composite morphism $\bar{b}\,a - \bar{a}\,b$ is 0. (1) is called *exact* if (2) is exact. Pullbacks and pushouts are special cases. By 7.8.5, (1) is a pullback if and only if $\begin{pmatrix} a \\ b \end{pmatrix}$ in (2) is a kernel of $(\bar{b}, -\bar{a})$; and, correspondingly, it is a pushout if $(\bar{b}, -\bar{a})$ is a cokernel of $\begin{pmatrix} a \\ b \end{pmatrix}$. If (1) is a pullback and a pushout, it is called *bicartesian* (Freyd calls it a Doolittle square). This is the case if and only if

$$0 \to C \xrightarrow{\binom{a}{b}} A \oplus B \xrightarrow{(\bar{b}, -\bar{a})} D \to 0$$

is exact.

13.4.2 Remarks. (a) Let (1) be commutative. If one forms a pullback of $\bar{a}$ and $\bar{b}$:

$$(3) \qquad \begin{array}{ccc} P & \xrightarrow{a'} & A \\ {\scriptstyle b'}\downarrow & & \downarrow{\scriptstyle \bar{b}} \\ B & \xrightarrow{\bar{a}} & D \end{array}$$

then there is exactly one morphism $c\colon C \to P$ with $a' c = a$ and $b' c = b$. (1) is exact if and only if c is an epimorphism. This follows immediately from (2) and the dual of 13.1.6 (a), (b).

(b) If from the pullback (3) one forms the pushout of a', b'

$$(4) \qquad \begin{array}{ccc} P & \xrightarrow{a'} & A \\ {\scriptstyle b'}\downarrow & & \downarrow{\scriptstyle \bar{b}'} \\ B & \xrightarrow{\bar{a}'} & Q \end{array}$$

then (4) is bicartesian and the uniquely determined morphism $d\colon Q \to D$ with $d\bar{a}' = \bar{a}$ and $d\bar{b}' = \bar{b}$ is a monomorphism. This follows from (2) and the dual of (a).

(c) If (1) is exact and if one forms first the pushout of a and b and then the pullback of the two new morphisms, the bicartesian square (4) results, up to an isomorphism. This follows from (2) and 13.1.4 (E_1), (E_1^0).

13.4.3 Proposition.

(a) *If* (1) *is exact and a or b a monomorphism, then* (1) *is a pullback.*

(b) *If* (1) *is a pullback, then b* (ker a) *is a kernel of $\bar{a}$. a is a monomorphism if and only if $\bar{a}$ is a monomorphism.*

(c) *If* (1) *is a pullback and $\bar{a}$ an epimorphism, then* (1) *is bicartesian and a is an epimorphism.*

Proof. (a) If a is a monomorphism, then so is $\begin{pmatrix} a \\ b \end{pmatrix}$, because $pr_1 \begin{pmatrix} a \\ b \end{pmatrix} = a$; (2) implies that (1) is a pullback.

(b) The first claim is 12.3.4 (d). This implies: if a is a monomorphism, then ker $\bar{a} = 0$, and therefore $\bar{a}$ is a monomorphism. The converse is 7.8.2.

(c) follows from the duals of (a) and (b).

13.4.4 Remarks. The analogue of 13.4.3 (c) in which a instead of $\bar{a}$ is assumed to be an epimorphism, does not hold. Counterexamples are provided by 7.8.9.

If (1) is a pullback (pushout) in *Ens* and $\bar{a}$ an epimorphism (a a monomorphism), then a is also an epimorphism ($\bar{a}$ a monomorphism). This follows from the fact that in *Ens* epimorphisms are retractions (and monomorphisms with non-empty domain are coretractions). By means of 10.1.4 and the "pointwise" construction of limits and colimits, corresponding statements for pullbacks and pushouts are also valid for every category $[\mathscr{C}, Ens]$.

13.4.5 Proposition. *If in*

(5)
$$\begin{array}{ccccc} A_1 & \xrightarrow{a_1} & A_2 & \xrightarrow{a_2} & A_3 \\ {\scriptstyle f_1}\downarrow & & \downarrow{\scriptstyle f_2} & & \downarrow{\scriptstyle f_3} \\ B_1 & \xrightarrow{b_1} & B_2 & \xrightarrow{b_2} & B_3 \end{array}$$

the left and right squares are exact, or, resp., pullbacks, pushouts, bicartesian, the same is true for the big rectangle.

Proof. The claim about pullbacks is contained in 7.8.4, the one for pushouts is its dual, and they together imply the bicartesian property. Now let the left and right squares in (5) be exact. We first form the pullback for b_2 and f_3 and obtain a_2', f_2' and the epimorphism c_2 according to 13.4.2 (a). Then we form the pullback for b_1 and f_2' obtaining

(6)
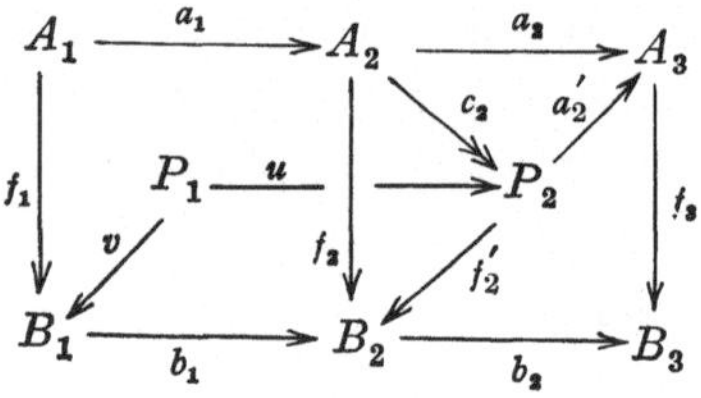

and, finally, the pullback for c_2 and u:

(7)
$$\begin{array}{ccc} T & \xrightarrow{w} & A_2 \\ {\scriptstyle c_1}\downarrow & & \downarrow{\scriptstyle c_2} \\ P_1 & \xrightarrow{u} & P_2 \end{array}$$

By 13.4.3 (c), (7) is bicartesian and c_1 is an epimorphism. By (6), one has $f_2'(c_2 a_1) = b_1 f_1$, and there is exactly one morphism $s\colon A_1 \to P_1$ with $v s = f_1$ and $u s = c_2 a_1$. By (7), this implies that there is exactly one morphism $t\colon A_1 \to T$ with $c_1 t = s$ and $w t = a_1$. Because of the statement about pullbacks that has already been proved, (6) and (7) now give a pullback

$$\begin{array}{ccc} T & \xrightarrow{w} & A_2 \\ {\scriptstyle v c_1}\downarrow & & \downarrow{\scriptstyle f_2' c_2} \\ B_1 & \xrightarrow{b_1} & B_2 \end{array}$$

It follows now from 13.4.2 (a) and the assumption, that t is an epimorphism. Thus $s = c_1 t$ is an epimorphism and the assertion follows from (6) again via 13.4.2 (a) and statement (5) for pullbacks.

13.4.6 Proposition. *Let* (5) *be commutative, let a_2 and b_2 be monomorphisms, and assume that the big rectangle is exact (a pullback). Then the left square is exact (a pullback).*

Proof. We consider

$$(8)\qquad \begin{array}{ccccc} A_1 & \xrightarrow{\binom{a_1}{f_1}} & A_2 \oplus B_1 & \xrightarrow{(f_2,\,-b_1)} & B_2 \\ \downarrow{\scriptstyle 1_{A_1}} & & \downarrow{\scriptstyle u} & & \downarrow{\scriptstyle b_2} \\ A_1 & \xrightarrow{\binom{a_2 a_1}{f_1}} & A_3 \oplus B_1 & \xrightarrow{(f_3,\,-b_2 b_1)} & B_3 \end{array}$$

with $u = \begin{pmatrix} a_2 & 0 \\ 0 & 1 \end{pmatrix}$. By assumption and by (2), the lower line in (8) is exact, b_2 is a monomorphism and so is u. Furthermore, (8) is commutative. By 13.1.7, the top line in (8) is exact, and hence so is the left square in (5). If the big rectangle in (5) is a pullback, then $\begin{pmatrix} a_2 a_1 \\ f_1 \end{pmatrix} = u \begin{pmatrix} a_1 \\ f_1 \end{pmatrix}$ is a monomorphism by 13.4.1 and thus the same is true for $\begin{pmatrix} a_1 \\ f_1 \end{pmatrix}$.

13.4.7 Lemma. *Let* (1) *be a pullback in an exact category and let a and $\bar a$ be monomorphisms. Form the cokernels u, $\bar u$ of a and $\bar a$ and the morphism v induced by them,*

$$(9)\qquad \begin{array}{ccccc} C & \overset{a}{\rightarrowtail} & A & \overset{u}{\twoheadrightarrow} & F \\ \downarrow{\scriptstyle b} & & \downarrow{\scriptstyle \bar b} & & \downarrow{\scriptstyle v} \\ B & \underset{\bar a}{\rightarrowtail} & D & \underset{\bar u}{\twoheadrightarrow} & G \end{array} .$$

Then v is a monomorphism and a is a kernel of $\bar u\, \bar b$.

Proof. The second claim follows from the first one and 13.1.6 (a). To prove the first claim, we factor $\bar u\, \bar b$ into an epimorphism u' followed by a monomorphism v'. By 12.3.4 (c), $\ker u'$ gives rise to a pullback for $\bar a$, $\bar b$. Therefore, a is a kernel of u' and u' a cokernel of a. Thus u und u', and hence v and v' also, differ only by an isomorphism.

13.4.8 Theorem. *In an abelian category let the square* (1) *be exact. Let a and $\bar a$ be factored into epimorphisms followed by monomorphisms. Then, using kernels k, $\bar k$ and cokernels u, $\bar u$ of a and $\bar a$ together with the morphism that exists according to* 12.4.8, *the following commutative diagram results*

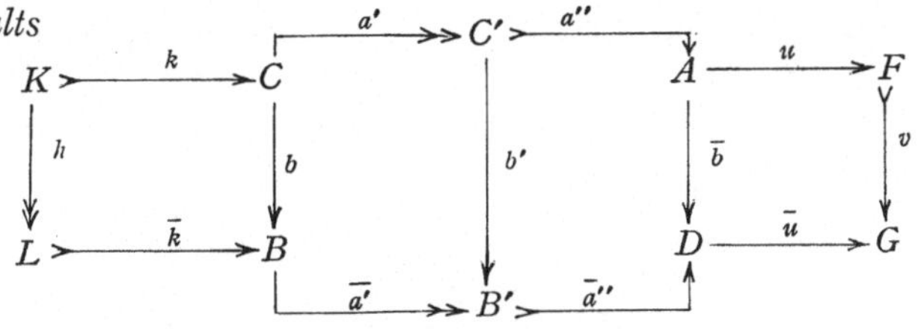

where $a'' a' = a$, $\bar a'' \bar a' = \bar a$. The following hold:

(a) *a', b, $\bar{a}'$, b' form a pushout.*
(b) *a'', b', $\bar{a}''$, $\bar{b}$ form a pullback.*
(c) *h is an epimorphism and $\bar{a}'$ is a cokernel of bk.*
(d) *v is a monomorphism and a'' is a kernel of $\bar{u}\bar{b}$.*
(e) *If* (1) *is a pullback, then the square in* (a) *is bicartesian and h is an isomorphism.*
(f) *If* (1) *is bicartesian, then the squares in* (a) *and* (b) *are also bicartesian and h and v are isomorphisms.*

Proof. (a) follows from 13.4.6 and the dual of 13.4.3 (a). (b) is its dual. (d) is 13.4.7, since a and a'' or, resp., $\bar{a}$ and a'' have the same cokernels (13.1.6 (a) dual). (c) is its dual. (e) follows from (a), 13.4.6 and 12.3.4 (d). Finally, (f) follows from (e) and its dual.

13.5 Some Diagram Lemmas

The categories involved are always assumed to be exact. Sometimes weaker assumptions would suffice.

13.5.1 Let the columns and the second and third rows in the commutative diagram

$$(1)\qquad \begin{array}{ccccccc} & & 0 & & 0 & & 0 \\ & & \downarrow & & \downarrow & & \downarrow \\ 0 & \to & K_1 & \to & K_2 & \to & K_3 \\ & & \downarrow & \mathrm{I} & \downarrow & & \downarrow \\ 0 & \to & A_1 & \to & A_2 & \to & A_3 \\ & & \downarrow & & \downarrow & & \\ 0 & \to & B_1 & \to & B_2 & & \end{array}$$

be exact. Then the square I is a pullback and the first row is exact.

Convention. By $B_1 \to B_2$ etc. here and below we mean the corresponding morphism in the diagram considered.

Proof. 12.3.4 (c) (applied to the first two columns) yields the fact that I is a pullback. Since $K_3 \to A_3$ is a monomorphism, $K_1 \to K_2 \to K_3 = 0$ and, by 12.3.4 (b) (applied to the first two rows), $K_1 \to K_2$ is a kernel of $K_2 \to K_3$, which is what was to be shown.

13.5.2 Addition. Let the columns and the middle row in (1) be exact. Assume further, that $A_1 \to B_1$ is an epimorphism. Then the first row is exact if and only if the third row is exact.

Proof. If the first row is exact, then I is again a pullback, by 12.3.4 (c). From

$$\begin{array}{ccccc} K_1 & \to & A_1 & \to & B_2 \\ \downarrow & \mathrm{I} & \downarrow & & \| \\ K_2 & \to & A_2 & \to & B_2 \end{array}$$

with $A_1 \to B_2 = A_1 \to A_2 \to B_2$ and from 12.3.4 (b), it follows that $K_1 \to A_1$ is a kernel of $A_1 \to B_2$. Then $B_1 \to B_2$ is a monomorphism, by 13.1.6 (b).

13.5.3 The Kernel Lemma. *Let*

$$
\begin{array}{ccccccc}
 & & 0 & & 0 & & 0 \\
 & & \downarrow & & \downarrow & & \downarrow \\
 & & K_1 & & K_2 & & K_3 \\
 & & \downarrow & & \downarrow & & \downarrow \\
0 & \to & A_1 & \to & A_2 & \to & A_3 \\
 & & \downarrow & & \downarrow & & \downarrow \\
0 & \to & B_1 & \to & B_2 & \to & B_3
\end{array} \tag{2}
$$

be commutative with exact columns and rows. Then there are uniquely determined morphisms $K_1 \to K_2$ and $K_2 \to K_3$ such that the completed diagram is commutative. Furthermore

$$0 \to K_1 \to K_2 \to K_3$$

is exact.

Proof. The existence follows from the definition of kernels and the rest follows from 13.5.1.

13.5.4 The Four Lemma. *Let the rows in the commutative diagram*

$$
\begin{array}{ccccccc}
A_1 & \xrightarrow{a_1} & A_2 & \xrightarrow{a_2} & A_3 & \xrightarrow{a_3} & A_4 \\
\downarrow f_1 & & \downarrow f_2 & & \downarrow f_3 & & \downarrow f_4 \\
B_1 & \xrightarrow[b_1]{} & B_2 & \xrightarrow[b_2]{} & B_3 & \xrightarrow[b_3]{} & B_4
\end{array} \tag{3}
$$

be exact, let f_1 be an epimorphism and let f_2 and f_4 be monomorphisms. Then f_3 is a monomorphism.

Proof. The rows are broken up at a_2 and b_2 as in 13.2.3. From 12.4.8, one gets

$$
\begin{array}{ccccc}
A_2 & \twoheadrightarrow & A_2' & \rightarrowtail & A_3 \\
\downarrow f_2 & & \downarrow f_2' & & \downarrow f_3 \\
B_2 & \twoheadrightarrow & B_2' & \rightarrowtail & B_3
\end{array} \tag{4}
$$

f_2' is a monomorphism, by 13.1.7 (b). This, together with $\ker f_4 = 0$, implies $\ker f_3 = 0$, by 13.5.3.

13.5.5 The Five Lemma. *Let*

$$
\begin{array}{ccccccccc}
A_1 & \longrightarrow & A_2 & \longrightarrow & A_3 & \longrightarrow & A_4 & \longrightarrow & A_5 \\
\downarrow f_1 & & \downarrow f_2 & & \downarrow f_3 & & \downarrow f_4 & & \downarrow f_5 \\
B_1 & \longrightarrow & B_2 & \longrightarrow & B_3 & \longrightarrow & B_4 & \longrightarrow & B_5
\end{array} \tag{5}
$$

be commutative, have exact rows, and assume that f_2 and f_4 are isomorphisms, that f_1 is an epimorphism, and that f_5 is a monomorphism. Then f_3 is an isomorphism.

This follows from 13.5.4 and its dual.

13.5.6 The Nine Lemma (3 × 3 *Lemma*).

$$
(6)\qquad
\begin{array}{ccccccccc}
 & & 0 & & 0 & & 0 & & \\
 & & \downarrow & & \downarrow & & \downarrow & & \\
0 & \longrightarrow & K_1 & \longrightarrow & K_2 & \longrightarrow & K_3 & \longrightarrow & 0 \\
 & & \downarrow & \mathrm{I} & \downarrow & \mathrm{II} & \downarrow & & \\
0 & \longrightarrow & A_1 & \longrightarrow & A_2 & \longrightarrow & A_3 & \longrightarrow & 0 \\
 & & \downarrow & \mathrm{III} & \downarrow & \mathrm{IV} & \downarrow & & \\
0 & \longrightarrow & B_1 & \longrightarrow & B_2 & \longrightarrow & B_3 & \longrightarrow & 0 \\
 & & \downarrow & & \downarrow & & \downarrow & & \\
 & & 0 & & 0 & & 0 & &
\end{array}
$$

(a) *Let* (6) *be commutative with exact columns and an exact middle row. Then the first row is exact if and only if the third row is exact.*

(b) *Let the middle row and the middle column be exact. Then* (6) *is commutative with exact rows and columns if and only if* I *is a pullback,* IV *is a pushout and if* $K_2 \to K_3 \to A_3$, $A_1 \to B_1 \to B_2$ *are factorizations of* $K_2 \to A_2 \to A_3$, $A_1 \to A_2 \to B_2$ *into an epimorphism and a monomorphism.*

Proof. (a) Let the first row be exact. Then $B_1 \to B_2 \to B_3 \to 0$ is exact, by the dual of 13.5.1. By 13.5.2, $0 \to B_1 \to B_2$ is also exact. The converse is dual.

(b) If (6) is commutative, with exact rows and columns, then I is a pullback by 13.5.1, and IV is a pushout by its dual; the claims about II and III are evident. If, conversely, II and III consist of the required factorizations, then a pullback at I can be constructed as in 12.3.4 (c) from a kernel of $K_2 \to K_3$. Since I is assumed to be a pullback, $K_1 \to K_2$ is a kernel of $K_2 \to K_3$. For the same reason $K_1 \to A_1$ is a kernel of $A_1 \to B_1$. The conclusion for IV is dual.

13.5.7 First Isomorphism Theorem. *Let* $N \rightarrowtail M$ and $M \rightarrowtail A$ *be monomorphisms. Then*

$$
(7)\qquad
\begin{array}{ccccccccc}
 & & 0 & & 0 & & 0 & & \\
 & & \downarrow & & \downarrow & & \downarrow & & \\
0 & \longrightarrow & N & \longrightarrow & M & \longrightarrow & M/N & \longrightarrow & 0 \\
 & & \downarrow & & \downarrow & & \downarrow & & \\
0 & \longrightarrow & N & \longrightarrow & A & \longrightarrow & A/N & \longrightarrow & 0 \\
 & & \downarrow & & \downarrow & & \downarrow & & \\
 & & 0 & \longrightarrow & A/M & \longrightarrow & A/N/M/N & \longrightarrow & 0 \\
 & & & & \downarrow & & \downarrow & & \\
 & & & & 0 & & 0 & &
\end{array}
$$

is commutative with exact rows and columns.

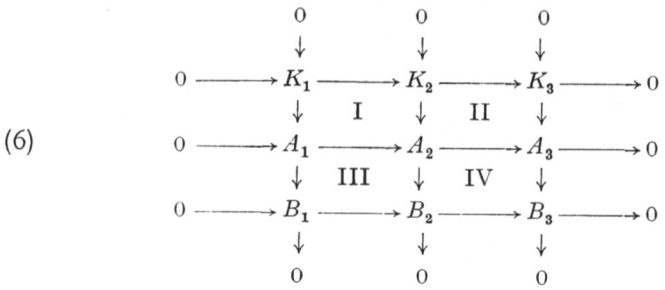

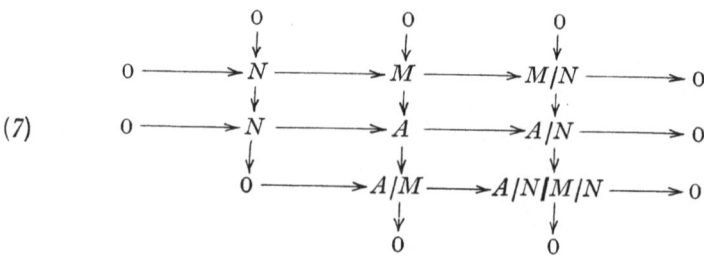

Proof. The two first rows and the first two columns are exact by the definition of cokernels, $M/N \to A/N$ exists also by the definition of cokernels, and it is a monomorphism by 13.1.7 (b). Thus the third column is exact. The morphism of the third row exist by the definition of cokernels. The exactness follows from 13.5.6 (a).

13.5.8 Remark. 13.5.7 gives the exact sequence

$$0 \to N \to M \to A/N \to A/M \to 0 \tag{8}$$

whose morphisms are uniquely determined by those of the middle row and of the middle column in (7).

13.5.9 Proposition (*Connecting morphism lemma*). *Let*

$$\begin{array}{ccccccccc} & & A_1 & \longrightarrow & A_2 & \longrightarrow & A_3 & \longrightarrow & 0 \\ & & \downarrow f_1 & & \downarrow f_2 & & \downarrow f_3 & & \\ 0 & \longrightarrow & B_1 & \longrightarrow & B_2 & \longrightarrow & B_3 & & \end{array} \tag{9}$$

be commutative with exact rows.

(a) *By adding kernels* k_i: $K_i \to A_i$ *and cokernels* c_i: $B_i \to C_i$ *for* f_i ($i = 1, 2, 3$) *one obtains an exact sequence*

$$K_1 \to K_2 \to K_3 \xrightarrow{\Delta} C_1 \to C_2 \to C_3 \tag{10}$$

with induced morphisms.

(b) *If* $A_1 \to A_2$ *is a monomorphism, then* $K_1 \to K_2$ *is also a monomorphism. If* $B_2 \to B_3$ *is an epimorphism, then so is* $C_2 \to C_3$.

(c) *The assignment of* (10) *to* (9) *is natural; i. e., a natural transformation of diagrams of type* (9) *induces a natural transformation for the corresponding exact sequences* (10).

Proof. We do not go into all the details. The essential point consists in the construction of Δ.

(a) We start by assuming that $A_1 \to A_2$ is a monomorphism, $B_2 \to B_3$ an epimorphism. Then, by 13.5.3 and its dual, there are exact sequences

$$0 \to K_1 \to K_2 \to K_3 \quad \text{and} \quad C_1 \to C_2 \to C_3 \to 0\,. \tag{11}$$

By factoring $K_2 \to K_3$, $C_1 \to C_2$ and $A_i \to B_i$ into epi- and monomorphisms one obtains

$$K_2 \twoheadrightarrow K \rightarrowtail K_3; \quad C_1 \twoheadrightarrow C \rightarrowtail C_2; \quad A_i \twoheadrightarrow D_i \rightarrowtail B_i\,. \tag{12}$$

Then, according to 13.5.2 and its dual, there are exact sequences

$$0 \to D_1 \to D_2 \to E \to 0\,, \quad 0 \to F \to D_2 \to D_3 \to 0\,. \tag{13}$$

Furthermore, $D_1 \to D_2 \to D_3$ is a zero morphism. By 13.5.6, there are exact sequences

$$0 \to K \to A_3 \to E \to 0\,, \quad 0 \to F \to B_1 \to C \to 0\,. \tag{14}$$

And by (8), (12), (13), (14), there are exact sequences

$$\begin{aligned} &K_2 \to K_3 \to E \to D_3 \to 0 \ \text{(because } D_3 = A_3/K_3)\,, \\ &0 \to D_1 \to F \to E \to D_3 \to 0\,, \\ &0 \to D_1 \to F \to C_1 \to C_2\,. \end{aligned} \tag{15}$$

By factoring the middle morphisms in (15) on gets (10) because of (11) for the special case considered here. The general case can be reduced to this one by factoring $A_1 \to A_2$ and $B_2 \to B_3$.

(b) is contained in the proof of (a).

(c) follows from the fact that the factorization of morphisms is natural, as is the formation of kernels and cokernels.

13.6 Problems

13.6.1 Let the category $\mathscr{C}$ have two objects o, A, where o is a zero object, but A is not.

(a) If $[A, A]$ consists of 1_A and 0 only, then $\mathscr{C}$ is an exact category with a uniquely determined additive structure.

(b) Let $[A, A]$ consist of 1_A, 0 and a morphism f. If $ff = f$, then $\mathscr{C}$ satisfies the axioms A0, A2, A2^0, A3, A3^0 of 12.1.1, but $\mathscr{C}$ is not exact. What can be said if $ff = 1_A$?

13.6.2 If one of two morphisms with the same codomain in an exact category is a monomorphism, then the pullback of the two morphisms exists. An exact category with pullbacks is abelian.

13.6.3 In the category Ens_* consider those maps of pointed sets for which every element different from the base point has at most one inverse image. These maps form an exact category. In contrast to Ens_*, it is not a category with products or coproducts.

13.6.4 Let

$$0 \to A_1 \xrightarrow{m_1} B \xrightarrow{p_1} C_1 \to 0\,,$$

$$0 \to A_2 \xrightarrow{m_2} B \xrightarrow{p_2} C_2 \to 0\,,$$

be short exact sequences in an exact category. If $p_2 m_1$ is a monomorphism (or, resp., an epimorphism, isomorphism), then the same is true for $p_1 m_2$. Kernels of retractions are coretractions.

13.6.5 Fill in the details of 13.5.9.

13.6.6 In an exact category $\mathscr{C}$ we shall call the square 13.4.1 (1) exact if it is commutative and if in the diagram 13.4.8 h is an epimorphism and v a monomorphism.

(a) If the square is exact, then 13.4.8 (a), (b) are valid.

(b) If the square is exact, then so is the one which one obtains by reflecting it on its diagonal $C - D$ (Hint: use 13.5.9).

(c) If the square is exact and if a is a monomorphism, then the square is a pullback and $\bar{a}$ is a monomorphism. The converse of this is also true.

(d) If the square is commutative and if 13.4.8 (a), (b) are satisfied, then the square is exact.

(e) If $\mathscr{C}$ is abelian, then the definition of an exact square used here is equivalent to the one used in 13.4.1.

Remark. It is an open question, whether in an exact category pullbacks are exact squares in the sense given here.

13.6.7 Let a_2 and g in 12.3.4 (10) be epimorphisms, let a_1 be a kernel of a_2 and let f_1 be an isomorphism. Then the square on the right is a bicartesian square provided the category is abelian.

13.6.8 In an exact category let $h = g f$ be given. Then morphisms g_*, f_*, Δ, g^*, f^* can be chosen in a unique way such that the diagram

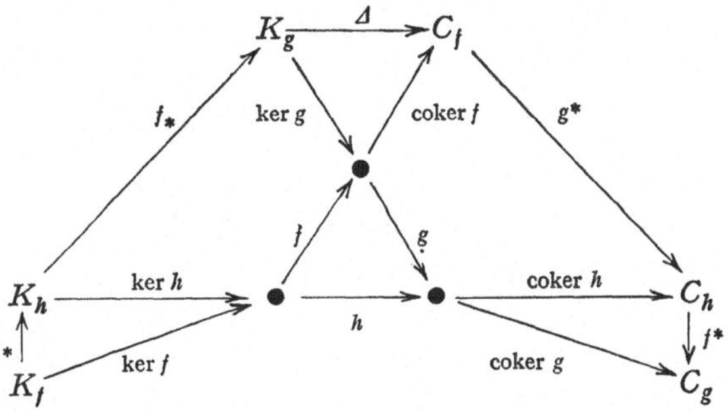

is a commutative. Further, the sequence

$$0 \to K_f \to K_h \to K_g \to C_f \to C_h \to C_g \to 0 \qquad \text{is exact .}$$

(Hint: Let $g = g'' g'$ be a factorization of g into a cokernel and a kernel. Use 13.5.9 with $f\colon A_2 \to B_2$ and $g'\colon B_2 \to B_3$. Also use the dual result).

14. Colimits of Monomorphisms

14.1 Preordered Classes

14.1.1 We start by recalling that a preordered class $\mathscr{K}$ is a category in which every set of morphisms $[A, B]$ has at most one element and where we write $A \leq B$ for $[A, B] \neq \emptyset$. Given a family $\{A_i\}$ of objects of $\mathscr{K}$, $C \in |\mathscr{K}|$ is called an *upper bound* if $A_i \leq C$ for all i. By a *directed class* we mean a preordered class in which any two objects have an upper bound. Obviously then, every non-empty finite family of objects has an upper bound. Every directed class is a special filtered category. For any set the finite subsets and their inclusions form a directed set.

14.1.2 In a preordered class equalizers and coequalizers exist trivially and are always isomorphisms (one can take identity morphisms).

Products, if they exist, are *infimums*; i.e., greatest common lower bounds of all the objects involved. Coproducts, analogously, are *supremums*.

Note that a preordered set can be complete. Conversely, the following holds:

14.1.3 Every complete or cocomplete small category $\mathcal{K}$ is a preordered set.

Proof. We assume that a set of morphisms, say $[A, B]$, has more than one element. Let I be an index set with higher cardinality than that of the morphism set of $\mathcal{K}$. The assumption that ΠB_i, with $B_i = B$ for all $i \in$ I, exists, then leads to a contradiction using $[A, \Pi B_i]$. A similar argument can be made using $[\amalg A_i, B]$.

14.1.4 If a preordered set is complete, then it has a smallest element, namely an infimum of all the objects. It also has a largest element, an infimum of the empty family. From this it follows that a complete preordered set is also cocomplete. Trivially there is an upper bound for every family of objects. An infimum of all upper bounds is a supremum of the family, since every member of the family is a lower bound for the set of upper bounds.

14.1.5 In a category $\mathcal{C}$ the colimit of a functor $T: \mathcal{Y} \to \mathcal{C}$, where $\mathcal{Y}$ is a directed set, is called a *directed colimit*. Every directed colimit is also filtered.

14.1.6 In a preordered class $\mathcal{K}$ a filtered colimit is also a directed one. More exactly: If $\mathcal{X}$ is a (small) filtered category and $T: \mathcal{X} \to \mathcal{K}$ a functor, then T is of the form $T = R\,P$, where $R: \mathcal{Y} \to \mathcal{K}$ is an embedding of the smallest subcategory of $\mathcal{K}$, which contains all morphisms $T(a)$ with $a \in \operatorname{Mor} \mathcal{X}$. $P: \mathcal{X} \to \mathcal{Y}$ is surjective on objects. Since $\mathcal{X}$ is filtered and $\mathcal{K}$ preordered, $\mathcal{Y}$ is a directed class (set) and $P: \mathcal{X} \to \mathcal{Y}$ is final. By 9.2.3, T has a colimit if and only if R has a colimit.

14.1.7 Let $\mathcal{K}$ be a preordered set. If objects A, B are said to be equivalent if $A \leq B$ and $B \leq A$, then the equivalence classes together with the preordering induced by $\mathcal{K}$, form an ordered set $\dot{\mathcal{K}}$. $\dot{\mathcal{K}}$ is directed or, resp., finitely complete, finitely cocomplete, complete, cocomplete, if and only if $\mathcal{K}$ has these properties. If $\mathcal{K}$ is a preordered class, then one obtains $\dot{\mathcal{K}}$ as a set of the higher universe $\mathfrak{B}$. Completeness or, resp., cocompleteness, is to be understood (as before) with respect to $\mathfrak{U}$-diagrams.

14.2 Unions of Monomorphisms

14.2.1 Let $\mathscr{C}$ be an arbitrary category and $A \in |\mathscr{C}|$. The monomorphisms with (fixed) codomain A form a preordered class $\mathscr{M}/A$, their equivalence classes an ordered class $\dot{\mathscr{M}}/A \cdot 1_A$ is a greatest object of $\mathscr{M}/A$. Infimums (i. e., products), and thus all limits, in $\mathscr{M}/A$ are intersections (7.8.6). The existence of finite or, resp., arbitrary intersections is guaranteed if $\mathscr{C}$ is finitely complete or, resp., complete. Compare also with 12.4.1. We write $\bigcap m_i : \bigcap M_i \to A$ for an intersection of a family $\{m_i : M_i \rightarrowtail A\}$ of monomorphisms (if it exists).

14.2.2 We call supremums in $\mathscr{M}/A$ *unions*. We write $\bigcup m_i$: $\bigcup M_i \to A$ for the union of the family $\{m_i: M_i \rightarrowtail A\}$ of monomorphisms (if it exists). The existence of (finite) unions does not follow from the (finite) cocompleteness of $\mathscr{C}$ without additional conditions.

14.2.1⁰—2⁰. We designate the category of epimorphisms with fixed domain A by $A/\mathscr{E}$. By passing from $\mathscr{C}$ to $\mathscr{C}^0$ epimorphisms are changed into monomorphisms with reversal of the preordering. 1_A is a smallest object of $A/\mathscr{E}$. Cointersections (8.8.4) are supremums. Infimums are called *co-unions.*

14.2.3 If $\mathscr{C}$ is well-powered (10.6.1) and $\mathscr{C}$ has intersections, then unions of monomorphisms with the same codomain A exist. This follows from 14.1.3, 14.1.4, and 14.1.7 if one uses $\mathscr{M}/A$.

14.2.4 Let the category $\mathscr{C}$ satisfy A0, A2, A2⁰, A3, A3⁰ of 12.1.1. By 12.4.4, $\bigcup m_i$ of the family $\{m_i: M_i \rightarrowtail A\}$ of monomorphisms exists if and only if the cointersection of $\{\text{coker } m_i\}$ exists, and then $\bigcup m_i$ is a kernel of this cointersection. By 12.4.1 and its dual, finite intersections and finite unions always exist.

14.2.5 Theorem. *Let every morphism in $\mathscr{C}$ factor uniquely up to an isomorphism into an epimorphism f' followed by a monomorphism f''. If (L, λ) is a colimit of the diagram $T: \Sigma \to \mathscr{C}$, $\alpha: T \to A_\Sigma$ a natural transformation and $f: L \to A$ the uniquely determined morphism with $f_\Sigma \lambda = \alpha$, then $f'' = \bigcup \alpha_i''$.*

Proof. Start by assuming Σ to be non-empty. The diagram

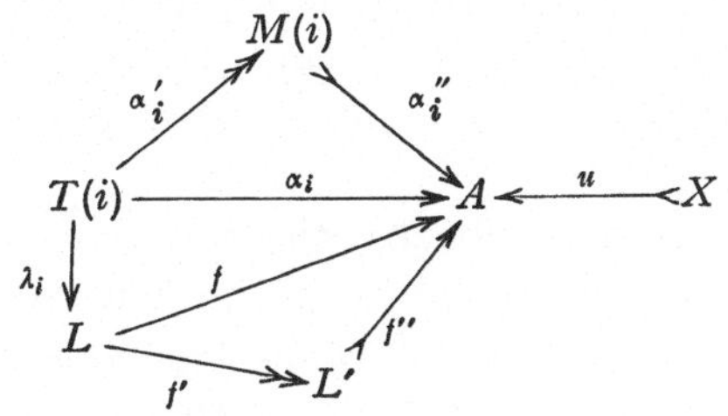

is commutative. By 12.4.10, there is exactly one morphism n_i: $M(i) \to L'$ with $f'' n_i = \alpha_i''$. Thus f'' is an upper bound for $\{\alpha_i''\}$. If $u: X \rightarrowtail A$ is an upper bound for $\{\alpha_i''\}$, then all α_i factor through u, and since u is a monomorphism, there is a natural transformation $T \to X_\Sigma$. Therefore, f factors through u by the colimit property and so does f'', again by 12.4.10. If Σ is empty, then L is initial in $\mathscr{C}$ and, for $f: L \to A$, f'' is initial in $\mathscr{M}/A$ as one verifies easily.

14.2.6 Corollary. *If $\mathscr{C}$ satisfies the condition in 14.2.5 and if $\mathscr{C}$ has (finite) coproducts, then (finite) unions of monomorphisms with the same codomain A exist.*

Set $(L, \lambda) = (\coprod M_i, \{in_i\})$ for $\{m_i: M_i \rightarrowtail A\}$.

14.2.7 Corollary. *For any two monomorphisms $m: M \rightarrowtail A$, n: $N \rightarrowtail A$ in an abelian category $\mathscr{C}$ the square* I *in*

(1) 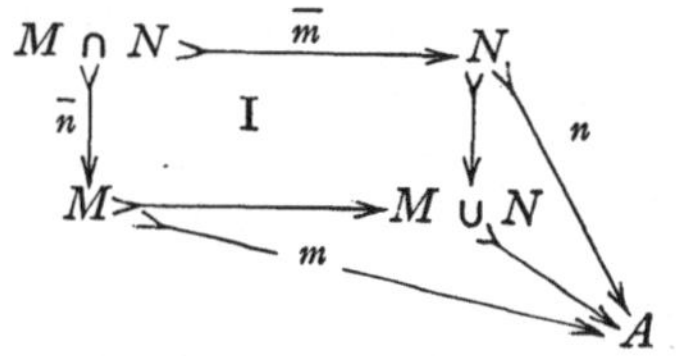

is bicartesian.

This follows by 14.2.5 from 13.4.2 (b).

14.2.8 Remark. If in an arbitrary category the union $\bigcup m_i$: $\bigcup M_i \rightarrowtail A$ of $\{m_i: M_i \rightarrowtail A\}$ exists, and if for the thus created monomorphisms $M_i \rightarrowtail \bigcup M_i$ the intersection exists, one obtains $\bigcap m_i$: $\bigcap M_i \to A$. This follows from the definition of intersections in 7.8.6, since $\bigcup m_i$ is a monomorphism.

14.3 Inverse Images of Monomorphisms

14.3.1 Suppose pullbacks exist. If $f: A \to B$ is given, then 7.8.3 enables us to construct for every monomorphism $n: N \to B$ an induced monomorphism $m: M \to A$. If there is no canonical choice of pullbacks, then m is only determined up to being preceded by an isomorphism. We then make a choice and call m an *inverse image* of n with respect to f. We write $m = f^{-1}(n)$. If f is also a monomorphism, then $f \circ f^{-1}(n) = n \circ n^{-1}(f)$ is the intersection of f and n by the definition of intersection.

By the properties of pullbacks, the transition to inverse images preserves the preordering of monomorphisms with fixed codomain, and in particular it takes equivalent ones into equivalent ones. As $f^{-1}(1_B)$ we always choose 1_A. By 7.8.4 the construction of inverse images is compatible with the composition of morphisms up to an isomorphism: If $f\,h$ is defined, then $(f\,h)^{-1}(n)$ and $h^{-1}(f^{-1}(n))$ are equivalent. Since limits commute with limits (7.6.3), the formation of finite intersections (or of arbitrary ones, if possible) commutes with the transition to inverse images (up to an isomorphism).

14.3.2 Proposition. *In Ens the transition to inverse images commutes with unions. The same holds for every functor category* $[\mathscr{C}, Ens]$.

Proof. The first part is known. Let $\{\mu_i\colon S_i \rightarrowtail T\}$ be a family of monomorphisms in $[\mathscr{C}, Ens]$. By 10.1.4, $\mu_{i,A}$ is a monomorphism for every $A \in |\mathscr{C}|$. We construct $\bigcup \mu_i$ as follows: For every $A \in |\mathscr{C}|$ we form the union $S(A)$ of the image sets $\mu_{i,A}(S_i(A)) \subset T(A)$. Then $T(f)$ induces a map $S(f)\colon S(A) \to S(B)$ for $f\colon A \to B$ in $\mathscr{C}$. By 10.1.6, this follows from the fact that set maps commute with the formation of unions of subsets. Thus one obtains a functor S with a monomorphism $\mu\colon S \to T$ which is at every place an inclusion. Obviously $\mu = \bigcup \mu_i$. The claim now follows from the "pointwise" construction of pullbacks in $[\mathscr{C}, Ens]$.

Remark. Another proof is based on the fact that colimits are universal in $[\mathscr{C}, Ens]$. One uses the construction 14.2.6 taking 7.8.4, 7.8.2 and 13.4.4 into account.

14.3.3 In *Ab* the transition to inverse images does not even commute with finite unions.

Counterexample. With respect to the diagonal map $\Delta\colon \mathbf{Z} \to \mathbf{Z} \oplus \mathbf{Z}$ both injections $in_1, in_2\colon \mathbf{Z} \to \mathbf{Z} \oplus \mathbf{Z}$ have the inverse image 0, and $1_{\mathbf{Z}\oplus\mathbf{Z}}$ is the union of in_1 and in_2.

14.4 Images of Monomorphisms

14.4.1 General assumption: Every morphism f has a unique factorization, up to an isomorphism, $f = f''f'$ with an epimorphic f' and a monomorphic f''.

14.4.2 If a monomorphism $m\colon M \rightarrowtail A$ and a morphism $f\colon A \to B$ is given, we choose a factorization $(f\,m) = (f\,m)''\,(f\,m)'$ and then call $(f\,m)''$ the *image* of m with respect to f. We write

$$(f\,m)'' = f(m)\colon f(M) \rightarrowtail B$$

(2)

$$\begin{array}{ccc} M & \twoheadrightarrow & f(M) \\ {\scriptstyle m}\downarrow & & \downarrow{\scriptstyle f(m)} \\ A & \xrightarrow{\;f\;} & B \end{array}$$

By 12.4.10, the transition to images preserves the preordering of monomorphisms; in particular, equivalent ones are taken into equivalent ones. If $g\,f$ is defined, then $(g\,f)(m)$ is equivalent to $g(f(m))$ as follows immediately from (2).

Remark. In an exact category $\operatorname{im} f = f(1_A)$. If an arbitrary category satisfies 14.4.1, then $\operatorname{im} f$ can be described this way. In categories not subject to 14.4.1 the correct definition of images requires the consideration of special classes of monomorphisms. For instance, in *Top* every morphism factors uniquely, up to an isomorphism, into an epimorphism followed by an equalizer. So in this case images are to be defined as equalizers. We content ourselves with this glimpse of the more general situation.

14.4.3 We assume that there are pullbacks and that 14.4.1 is satisfied. For $f\colon A \to B$ and monomorphisms $m\colon M \rightarrowtail A$ and $n\colon N \rightarrowtail B$ the following holds:

(a) $m \leq f^{-1}(f(m))$,

(b) $n \geq f(f^{-1}(n))$,

(c) $f(m)$ and $f(f^{-1}(f(m)))$ are equivalent,

(d) $f^{-1}(n)$ and $f^{-1}(f(f^{-1}(n)))$ are equivalent.

Proof. (a) follows from the definitions and the pullback property. (b) follows from the definitions, by 12.4.10. (c) follows from (a) and (b), if f is applied to (a) and if in (b) n is replaced by $f(m)$. (d) follows analogously.

14.4.4 If f in (2) is a monomorphism, then $f(m) = f\,m$ and the transition to images with respect to f induces an injective map $\mathcal{M}/A \to \mathcal{M}/B$. Here m and $f^{-1}(f(m))$ are equivalent, as is easily verified.

14.4.5 Let $f\colon A \to B$ be a morphism in an abelian category and $m\colon M \rightarrowtail A$ a monomorphism. m and $f^{-1}(f(m))$ are equivalent if and only if $m \geq \ker f$.

Proof. Since $f(m) \geq 0$, $f^{-1}(f(m)) \geq \ker f$. Now, let $m \geq \ker f$. By 14.4.3 (a), there is a commutative diagram

$$\begin{array}{ccccc}
K & \rightarrowtail & M & \twoheadrightarrow & f(M) \\
\| & & \downarrow{\scriptstyle m} & & \| \\
K & \rightarrowtail & \bullet & \twoheadrightarrow & f(M) \\
\| & & \downarrow{\scriptstyle f^{-1}(f(m))} & & \downarrow{\scriptstyle f(m)} \\
K & \underset{\ker f}{\rightarrowtail} & A & \underset{f}{\longrightarrow} & M
\end{array}$$

The first and the second rows are exact, by 13.1.7 (a). Thus the claim follows from 13.5.5.

14.4.6 Let $f\colon A \twoheadrightarrow B$ be an epimorphism in an abelian category or in *Ens* or in a functor category $[\mathscr{C}, \mathit{Ens}]$. If $n\colon N \rightarrowtail B$ is a monomorphism, then n and $f(f^{-1}(n))$ are equivalent.

For abelian categories this follows from 13.4.3 (c); for the other cases from 13.4.4.

14.4.7 Let condition 14.4.1 be satisfied and assume that there are (finite) coproducts. Then, for every (finite) family $\{m_i\colon M_i \rightarrowtail A\}$ of monomorphisms and for every $f\colon A \to B$, $f(\bigcup m_i)$ and $\bigcup f(m_i)$ are equivalent.

Proof. Let $f(m_i) = n_i\colon N_i \rightarrowtail B$ where $f\,m_i = n_i\,p_i$ with an epimorphic $p_i\colon M_i \twoheadrightarrow N_i$. By 8.3.3, $p = \amalg\, p_i\colon \amalg\, M_i \to \amalg\, N_i$ is an epimorphism. $\{m_i\}$ and $\{n_i\}$ determine morphisms $m\colon \amalg\, M_i \to A$ and $n\colon \amalg\, N_i \to B$ which satify $f\,m = n\,p$. If m, n and $f\,m$ are factored into epi- and monomorphisms,

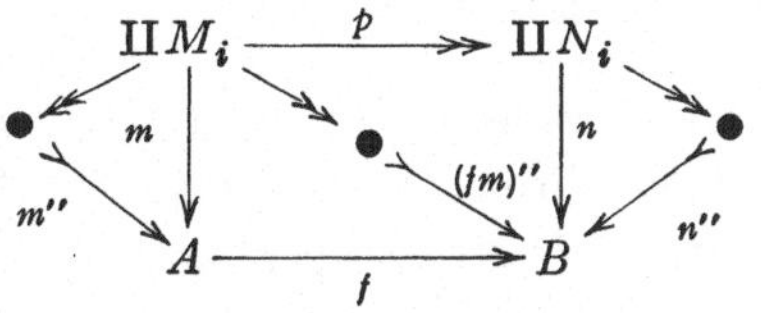

then n'' and $(f\,m)''$ are equivalent, since p is an epimorphism. The construction of $f(m'')$ shows that $f(m'')$ and $(f\,m)''$ are equivalent. Thus the claim follows from 14.2.6.

14.4.8 Remark. The image of an intersection of monomorphisms is a lower bound for the intersection of the images. As *Ab* and *Ens* show, there need not be an isomorphism.

14.4.9 Images of epimorphisms are constructed dually to 14.3.1 with pushouts, inverse images dually to 14.4.2 by factorization of morphisms. In an exact category inverse images of monomorphisms can be constructed, as in 12.3.4 (c), as kernels of the inverse images of their cokernels. Dually one gets images of epimorphisms as cokernels of the images of their kernels. Thus the required pullbacks and pushouts exist, and 14.4.5 is also valid for exact categories. Furthermore, 14.2.4 is valid for unions, and its dual is valid for intersections.

14.4.10 In a cocomplete abelian category let $\{n_i\colon N_i \rightarrowtail B\}$ be a family of monomorphisms and $p\colon A \twoheadrightarrow B$ an epimorphism. Then $p^{-1}(\bigcup n_i)$ and $\bigcup p^{-1}(n_i)$ are equivalent. (Compare with 14.3.3).

Proof. Obviously $\bigcup p^{-1}(n_i) \geq \ker p$. By 14.4.5, $\bigcup p^{-1}(n_i)$ is equivalent to $p^{-1}\big(p\big(\bigcup p^{-1}(n_i)\big)\big)$. The statement then follows from 14.4.7 and 14.4.6.

14.5 Constructions for Colimits

14.5.1 In a (finitely) complete category let $\Pi\, B_i$ be a (finite) product with projections pr_i. For $u, v\colon A \to \Pi\, B_i$ let h be an equalizer and h_i an equalizer of $pr_i\, u$ and $pr_i\, v$. Then h is an intersection of the familiy $\{h_i\}$.

Proof. For $w: X \to A$, $u\, w = v\, w$ is equivalent to $pr_i\, u\, w = pr_i\, v\, w$ for all i. Therefore, by the definition of equalizers, for a monomorphic w, $w \leq h$ is equivalent to $w \leq h_i$ for all i.

14.5.2 Let $T: \Sigma \to \mathscr{C}$ be a diagram for the complete category $\mathscr{C}$. The construction of a limit for T in 7.4.2 can be described by means of 14.5.1 as follows: For the vertex set Ve of Σ, one forms $\prod\limits_{i \in Ve} T(i)$ with projections pr_i. For every arrow a of Σ let $d_a: D_a \to \prod T(i)$ be an equalizer of $pr_{e(a)}$ and $T(a)\, pr_{o(a)}$. Then, using $\bigcap d_a: \bigcap D_a \to \prod T(i)$, $(\bigcap D_a, \{pr_i \circ (\bigcap d_a)\})$ is a limit of T.

14.5.3 Let $\mathscr{C}$ be a cocomplete abelian category and $T: \Sigma \to \mathscr{C}$ a diagram with colimit (L, λ). From $\coprod T(i)$ (with injections in_i) there is an epimorphism $c: \coprod T(i) \twoheadrightarrow L$ with $c\, in_i = \lambda_i$. $\bigcup$ im $\big(in_{o(a)} - in_{e(a)} T(a)\big)$ is a kernel of c.

Proof. By 8.2.6 and the dual of 14.5.2, c is the cointersection of coker $\big(in_{o(a)} - in_{e(a)}\, T(a)\big)$. Thus the claim follows from 14.2.4.

14.5.4 Proposition. *Let the category $\mathscr{C}$ have coproducts. Let J be a subset of the set I. For $\coprod A_i$ (with injections in_i) $in_j: A_j \to \coprod A_i$ defines a morphism*

$$in_J: \coprod_{j \in J} A_j \to \coprod_{i \in I} A_i\,.$$

(a) *If J ranges over the finite subsets of I, then $\coprod A_i$ is a directed colimit of $\{\coprod A_j\}$ by means of $\{in_J\}$, provided that, for $J \subset J'$, $in_{J,J'}: \coprod\limits_{j \in J} A_j \to$ $\to \coprod\limits_{j' \in J'} A_{j'}$ is defined correspondingly to in_J.*

(b) *If $\mathscr{C}$ has a zero object, then in_J is a coretraction and $1_{\coprod A_i} = \bigcup in_J = \bigcup in_i$.*

Proof. (a) is easily deduced from the definition of colimits. (b) Let $\coprod A_j$ have injections h_j. One defines $p_J: \coprod A_i \to \coprod A_j$ by $h_i: A_i \to \coprod A_j$ for $i \in J$ and $0: A_i \to \coprod A_j$ otherwise. p_J is a retraction for in_J. (This follows also from the duals of 7.3.4 and 7.7.7). Thus the last statement follows from (a).

14.5.5 Proposition. *If in a cocomplete abelian category finite limits commute with filtered colimits, then they also commute with pseudofiltered colimits.*

Proof. By 9.1.3, pseudofiltered colimits are coproducts of filtered ones. Every finite coproduct is a biproduct and commutes with limits. Thus the statement follows from 14.5.4.

14.6 Grothendieck Categories

14.6.1 Definition. A *Grothendieck category* is a cocomplete abelian category satisfying the following condition:

(AB5) If $m\colon A \rightarrowtail B$ is a monomorphism and $\{n_i\colon N_i \rightarrowtail B\}$ a directed family of monomorphisms, then

$$\bigcup (m \cap n_i) \cong m \cap \bigcup n_i\,. \tag{1}$$

The fact that $\{n_i\}$ is directed means, that the full subcategory determined by the objects n_i in $\mathcal{M}/B$ (monomorphisms with codomain B) is a directed set. The isomorphism exists in $\mathcal{M}/B$.

Some authors require additionally that there be a distinguished generating set in a Grothendieck category.

14.6.2 In a cocomplete abelian category, (AB5) is equivalent to the following: directed unions of monomorphisms commute with the transition to inverse images with respect to arbitrary morphisms $f\colon A \to B$ (up to an isomorphism):

$$f^{-1}\left(\bigcup n_i\right) \cong \bigcup f^{-1}(n_i)\,,\quad \{n_i\} \text{ is directed.} \tag{2}$$

Proof. For $m\colon A \rightarrowtail B$ and $n\colon N \rightarrowtail B$ we consider the pullback

$$\begin{array}{ccc} \bullet & \rightarrowtail & N \\ {\scriptstyle m^{-1}(n)}\downarrow & & \downarrow{\scriptstyle n} \\ A & \underset{m}{\rightarrowtail} & B \end{array} \tag{3}$$

The diagonal is $m \cap n = m\bigl(m^{-1}(n)\bigr)$. Thus (1) follows from (2) and 14.4.7. Conversely, (2) follows from (1) for monomorphic f by the definition of monomorphism; the general case then follows from 14.4.10 by factoring f.

14.6.3 Let $\mathcal{C}$ be a cocomplete abelian category. For $\mathcal{C}$, (AB5) is equivalent to

(AB5′) Every filtered colimit of monomorphisms with the same codomain is a monomorphism.

Proof. First, let (AB5) be satisfied, let $\mathcal{X}$ be a small filtered category and $T\colon \mathcal{X} \to \mathcal{C}/B$ (compare 6.5.3) a functor, where every

$$T(i) = n_i\colon N_i \rightarrowtail B \quad \text{for} \quad i \in |\mathcal{X}|$$

is a monomorphism with codomain B. The colimit of T is constructed "pointwise". Since $\mathcal{X}$ is filtered (and thus connected), $(B, \{1_B\})$ is a colimit of $B_{\mathcal{X}}$. Let $h\colon L \to B$ be colimit object of T and $k\colon K \to L$ a kernel of h. For every $i \in |\mathcal{X}|$ there is a commutative diagram

(4)

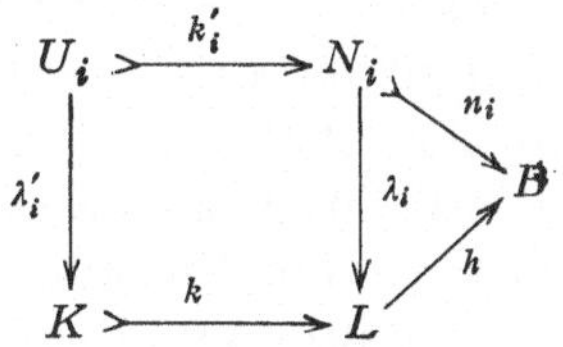

Here $(L, \{\lambda_i\})$ is a colimit of T at the place $0 \in |\mathbf{2}|$ and the left hand square in (4) is a pullback for k and λ_i. Hence $n_i k_i' = h k \lambda_i' = 0$. By 12.3.4 (b), k_i' is a kernel of n_i and therefore $U_i = 0$. Now, λ_i is a monomorphism because $h \lambda_i = n_i$, and, by 14.2.5 with $\lambda_i = \alpha_i$ and $f = 1_L$, one has $1_L = \bigcup \lambda_i$. By the definition of colimits, T induces a functor $S\colon \mathcal{X} \to \mathcal{M}/L$ with $S(i) = \lambda_i$, and by 14.1.6, $\{\lambda_i\}$ is a directed family. Now it follows from (1) that $k = k \cap 1_L = k \cap \bigcup \lambda_i \cong \bigcup (k \cap \lambda_i) = \bigcup \lambda_i k_i' = 0$. Therefore, h is a monomorphism.

Conversely, let (AB5') be satisfied. In the situation 14.6.1 we form for every i, as in 14.2.7, the commutative diagram

(5)

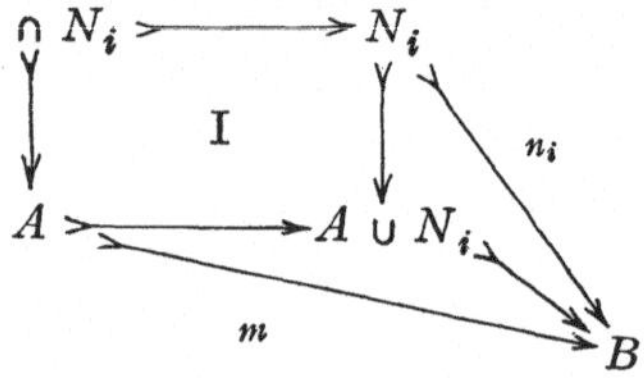

where all morphisms are monomorphisms and I is bicartesian. Every morphism $n_i \to n_j$ in $\mathcal{M}/B$ can be extended with 1_B and 1_A to a natural transformation of the corresponding diagrams (5). Since $\{n_i\}$ is directed, we obtain the directed colimits

(6)

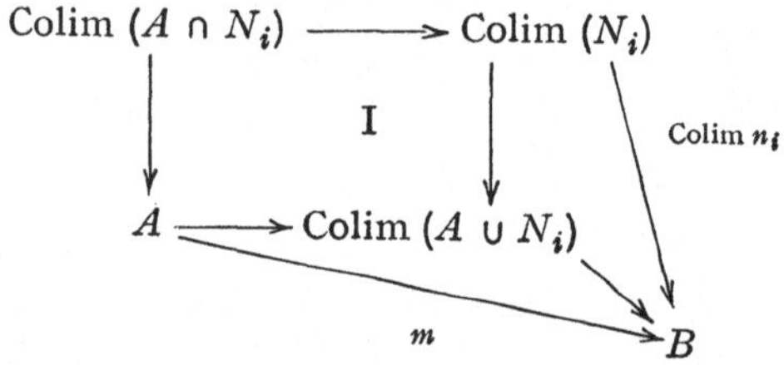

By assumption, the morphisms with codomain B are monomorphisms and, by 14.2.5, they are unions of the corresponding morphisms in the diagrams (5). It follows that all morphisms in (6) are monomorphisms. Furthermore, I is a pushout, since colimits commute with colimits. By 13.4.3 (c), I is bicartesian. Since $\operatorname{Colim}(A \cup N_i) \to B$ is a monomorphism, the outer contour of (6) is a pullback. Hence (1) is valid.

14.6.4 Lemma. *Let $\mathcal{X}$ be a small, filtered category and $\mathcal{C}$ a Grothendieck category. Let (L, λ) be a colimit of the functor $T\colon \mathcal{X} \to \mathcal{C}$. For $j \in |\mathcal{X}|$, let $k_j\colon K_j \to T(j)$ be a kernel of $\lambda_j\colon T(j) \to L$ and for every $u\colon j \to i$ in $\mathcal{X}$ let $k_u\colon K_u \to T(j)$ be a kernel of $T(u)\colon T(j) \to T(i)$. Then*

$$k_j \cong \bigcup k_u, \quad j \text{ is the domain of } u\,.$$

Proof. From $\lambda_j = \lambda_i T(u)$ for $u\colon j \to i$, $k_j \geq k_u$ follows and thus $k_j \geq \bigcup k_u$. We have to show that $\bigcup k_u \geq k_j$. For this, let $\mathcal{Y}$ be the full subcategory of $\mathcal{X}$ whose objects are all $i \in |\mathcal{X}|$ for which there is a $u\colon j \to i$. $\mathcal{Y}$ contains j and 1_j. Furthermore, $\mathcal{Y}$ is filtered and final in $\mathcal{X}$, as follows easily from definitions 9.3.4 and 9.4.9. By 9.2.3, $\mathcal{X}$ can be replaced by $\mathcal{Y}$. In order to avoid new notation, we assume that $\mathcal{Y} = \mathcal{X}$.

By 14.5.3 and the definition of kernels, there is the following commutative diagram

$$
(7) \qquad \begin{array}{ccccc}
K_j & \xrightarrow{k_j} & T(j) & \xrightarrow{\lambda_j} & L \\
\downarrow & \text{I} & \downarrow{\scriptstyle in_j} & & \|{\scriptstyle 1_L} \\
K & \xrightarrow{k} & \amalg\, T(i) & \xrightarrow{c} & L
\end{array}
$$

where $k = \bigcup_a \operatorname{im}\left(in_{o(a)} - in_{e(a)}\, T(a)\right)$ is a kernel of c and a ranges over all the morphisms of $\mathcal{X}$. in_j is a monomorphism (compare 14.5.4 (b)) and I is a pullback, by 12.3.4 (c). For every finite subset J of the morphism set of $\mathcal{X}$, we form

$$(8) \qquad v_J = \bigcup_{a \in J} \operatorname{im}\left(in_{o(a)} - in_{e(a)}\, T(a)\right).$$

$\{v_J\}$ is a directed family with respect to inclusions of subsets. Further, one also has $k = \bigcup v_J$. Thus it follows from (7) and (2), that

$$(9) \qquad k_j = in_j^{-1}(k) = \bigcup_J in_j^{-1}(v_J).$$

We shall show that for every J there is a $u: j \to i$ such that

$$(10) \qquad k_u \geq in_j^{-1}(v_J).$$

This, together with (9), yields $\bigcup k_u \geq k_j$ and thus completes the proof.

Let J consist of the morphisms $a_\nu: i_\nu \to i'_\nu$ with $\nu = 1, 2, \ldots, n$. For every ν we choose a $u_\nu: j \to i_\nu$. Since $\mathcal{X}$ is filtered, there is an $h \in |\mathcal{X}|$ with morphisms $v_\nu: i'_\nu \to h$. The objects $j, i_1, i_2, \ldots, i_n, i'_1, i'_2, \ldots, i'_n$ are not necessarily pairwise different and, for every one, there are one or more morphisms of the form $v_\nu\, a_\nu\, u_\nu$, $v_\nu\, a_\nu$, v_ν with codomain h. Repeated application of 9.3.4 (ii) shows, however, that h can be chosen in such a way that morphisms of the form $v_\nu\, a_\nu\, u_\nu$, $v_\nu\, a_\nu$, v_ν coincide whenever they have the same domain. Let this be the case, and set $u = v_\nu\, a_\nu\, u_\nu: j \to h$. We define $f: \amalg\, T(i) \to T(h)$ as follows: For $i = j$, or, resp., i_ν, i'_ν, let $f\, in_i = T(u)$ or, resp., $T(v_\nu\, a_\nu)$, $T(v_\nu)$; for all other i, let $f\, in_i = 0$. Then, for a_ν, $f\, in_{i_\nu} = T(v_\nu\, a_\nu) = f\, in_{i'_\nu}\, T(a_\nu)$. Thus, it follows from (8) and 14.4.7 that $f\, v_J = 0$ and therefore surely $f\, in_j\, in_j^{-1}(v_J) = 0$. (10) is valid since $f\, in_j = T(u)$.

14.6.5 Proposition. *In a Grothendieck category every filtered colimit of monomorphisms is a monomorphism.*

Proof. Let $\mathcal{X}$ be a small filtered category, $\mathcal{C}$ a Grothendieck category and $T: \mathcal{X} \to [\mathbf{2}, \mathcal{C}]$ a functor, where $T(i) = n_i: A_i \rightarrowtail B_i$ is a monomorphism for all $i \in |\mathcal{X}|$. T consists of two functors $R, S: \mathcal{X} \to \mathcal{C}$ and of a monomorphic natural transformation $n: R \to S$.

Let (L, λ) be a colimit of R, (M, μ) a colimit of S and $h\colon L \to M$ the morphism induced by n, i.e., $(h, \{(\lambda_i, \mu_i)\})$ is a colimit of T. For $u\colon j \to i$ in $\mathscr{X}$, we consider

$$(11)\qquad \begin{array}{ccccc} K_u & \xrightarrow{k_u} & A_j & \xrightarrow{R(u)} & A_i \\ \downarrow & \mathrm{I} & \downarrow{\scriptstyle n_j} & & \downarrow{\scriptstyle n_i} \\ K'_u & \xrightarrow[k'_u]{} & B_j & \xrightarrow[S(u)]{} & B_i \end{array}$$

with kernels k_u of $R(u)$ and k'_u of $S(u)$. By 12.3.4 (c), I is a pullback, so $k_u = n_j^{-1}(k'_u)$. It follows from 14.6.4 and (2) that $\ker \lambda_j = n_j^{-1}(\ker \mu_j)$. If we replace j again by i, we obtain, after factoring λ_i and μ_i into epi- and monomorphisms,

$$(12)\qquad \begin{array}{ccccccc} K_i & \xrightarrow{\ker \lambda_i} & A_i & \twoheadrightarrow & A'_i & \xrightarrow{\operatorname{im} \lambda_i} & L \\ \downarrow & \mathrm{I} & \downarrow{\scriptstyle n_i} & & \downarrow{\scriptstyle m_i} & & \downarrow{\scriptstyle h} \\ K'_i & \xrightarrow[\ker \mu_i]{} & B_i & \twoheadrightarrow & B'_i & \xrightarrow[\operatorname{im} \mu_i]{} & M \end{array}$$

Here I is a pullback from what was just said above. m_i is the induced morphism for cokernels. By 13.4.7, m_i is a monomorphism. For $a\colon i \to i'$, $T(a)$ can be extended to a natural transformation of the corresponding diagrams (12). Thus it follows from 14.6.3 that h, as a filtered colimit of the monomorphisms $(\operatorname{im} \mu_i)\, m_i$, is a monomorphism.

14.6.6 Theorem. *For a cocomplete abelian category the following are equivalent:*

(a) (AB 5) *is valid.*

(b) *Pseudofiltered colimits commute with finite limits.*

(c) *Pseudofiltered colimits of exact sequences are exact.*

The statements resulting from (b) *and* (c), *if pseudofiltered is replaced by directed, are also equivalent to* (a).

Proof. Since colimits commute with colimits, it follows from (a) by 14.6.5 that filtered colimits of short exact sequences are exact. Thus the corresponding statement for arbitrary exact sequences and with it the commutability of filtered colimits with kernels follows by factorization as in 13.2.3. Finite products are biproducts and hence they commute with all colimits. (b) follows from 14.5.5. (c) follows from (b). If in (b) or (c) pseudofiltered is replaced by directed, and, if one notes that every monomorphism is a kernel, one obtains the statement in 14.6.5 and the special case (AB 5′) for the directed case and thus again (a), by 14.6.3.

Warning. (b) does not imply (1) for pseudofiltered families (14.3.3 !).

14.6.7 Remark. 14.6.6 (c) implies

(AB 4) Coproducts $\amalg\, m_i\colon \amalg\, A_i \to \amalg\, B_i$ of monomorphisms are monomorphisms.

14.6.8 Proposition. *In the Grothendieck category* $\mathscr{C}$ *let there be a product* $\prod A_i$ *with projections* pr_i. *Then* $h\colon \coprod A_i \to \prod A_i$ *with* $pr_j\, h\, in_i = \delta_{ji}$ *is a monomorphism.*

Proof. Let $\{i_1, i_2, \ldots, i_n\}$ be a finite subset of the index set I. $h_\nu\colon A_{i_\nu} \to \prod A_i$ is defined by $pr_j\, h_\nu = \delta_{j i_\nu}$ and therefore $h_{i_1 i_2 \ldots i_n}\colon \coprod A_{i_\nu} \to \prod A_i$ with $h_\nu = h_{i_1 i_2 \ldots i_n}\, in_{i_\nu}$. Since $\coprod A_{i_\nu}$ is a biproduct, $h_{i_1 i_2 \ldots i_n}$ is a monomorphism and even a coretraction with $pr_{i_1 i_2 \ldots i_n}$ as a corresponding retraction. Thus the claim follows from 14.5.4 and 14.6.3.

14.6.9 *Ab* and with it $[\mathscr{C}, Ab]$ and $Add\,(\mathscr{C}, Ab)$ for every small or, resp., additive small category $\mathscr{C}$ are complete Grothendieck categories (10.2.1, 10.1.9). Since limits and colimits are constructed „pointwise", the dual of (AB 4), which is known for *Ab*, is also valid in these categories, i.e.:

Products of epimorphisms are epimorphisms.

On account of the completeness and the commutability of limits with limits it is equivalent to:

Products of exact sequences are exact.

14.7 Problems

14.7.1 Let $\mathscr{C}$ be a complete category and $A \in |\mathscr{C}|$. Show: The categories $\mathscr{C}/A$ (6.5.3) and $\mathscr{M}/A$ (14.2.1) are complete, and in $\mathscr{C}/A$, $\mathscr{M}/A$ is closed with respect to limits. If $\mathscr{C}$ is cocomplete, then $\mathscr{C}/A$ is cocomplete and $\Delta^\circ\colon \mathscr{C}/A \to \mathscr{C}$ (6.5.3) preserves and reflects colimits.

14.7.2 Prove: For any two monomorphisms $m\colon M \rightarrowtail A$, $n\colon N \rightarrowtail A$ in an exact category there is a commutative diagram

$$\text{(1)}\qquad \begin{array}{ccccccccc}
 & & 0 & & 0 & & 0 & & \\
 & & \downarrow & & \downarrow & & \downarrow & & \\
0 & \longrightarrow & N \cap M & \longrightarrow & N & \longrightarrow & N/N \cap M & \longrightarrow & 0\\
 & & \downarrow & & \downarrow & & \downarrow & & \\
0 & \longrightarrow & M & \longrightarrow & A & \longrightarrow & A/M & \longrightarrow & 0\\
 & & \downarrow & & \downarrow & & \downarrow & & \\
0 & \longrightarrow & M/M \cap N & \longrightarrow & A/N & \longrightarrow & A/N \cup M & \longrightarrow & 0\\
 & & \downarrow & & \downarrow & & \downarrow & & \\
 & & 0 & & 0 & & 0 & &
\end{array}$$

with exact rows and columns. (Hint: use 13.4.7 and 13.5.6 (b).)

14.7.3 Prove the following theorem (*Second Isomorphism Theorem*): For any two monomorphisms $M \rightarrowtail A$, $N \rightarrowtail A$ in an exact category there is an isomorphism $N/M \cap N \rightarrowtail\!\!\!\twoheadrightarrow M \cup N/N$ for which

$$\text{(2)}\qquad \begin{array}{ccccccccc}
0 & \longrightarrow & M \cap N & \longrightarrow & N & \longrightarrow & N/M \cap N & \longrightarrow & 0\\
 & & \downarrow & & \downarrow & & \downarrow & & \\
0 & \longrightarrow & M & \longrightarrow & M \cup N & \longrightarrow & M \cup N/N & \longrightarrow & 0
\end{array}$$

is commutative.

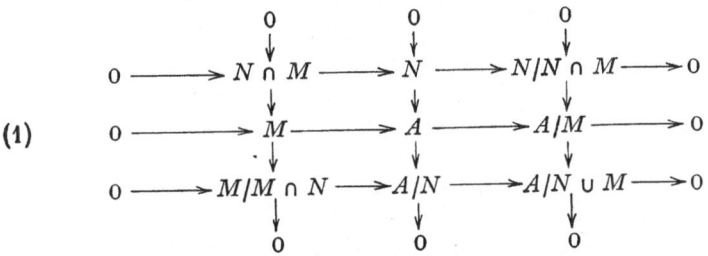

Remark. In an abelian category (2) follows also from 14.2.7 and 13.4.8.

14.7.4 Carry out the remark in 14.3.2.

14.7.5 Show that 14.4.6 is also valid in exact categories.

14.7.6 If $\mathscr{C}$ is a complete Grothendieck category for which the dual of 14.6.8 holds, then $\mathscr{C}$ has only zero objects. (Hint: For $A \in |\mathscr{C}|$, consider $\amalg A_n$ with $n \in \boldsymbol{N}$ and $A_n = A$ for all n. Define $\Delta\colon A \to \Pi A_n$ by $pr_n \Delta = 1_A$, $\nabla\colon \amalg A_n \to A$ by $\nabla in_n = 1_A$, and $h\colon \amalg A_n \to \Pi A_n$ by $pr_m h\, in_n = \delta_{mn}$. Then h is an isomorphism. Show that for $f = \nabla h^{-1} \Delta$, $1_A + f = f$.)

14.7.7 Let $\mathscr{K}$ be a preordered class which is cocomplete (as a category). Let $\mathscr{H}$ be a subclass (with the induced preordering) such that for every strongly ordered subset of $\mathscr{H}$ the supremum (formed in $\mathscr{K}$) is an element of $\mathscr{H}$ and such that in $\mathscr{K}$ $\mathscr{H}$ is closed with respect to isomorphisms (i.e., $\mathscr{H}$ is a union of equivalence classes). Show: If $\mathfrak{M}$ is a non-empty subset of $\mathscr{K}$ such that the supremum of $\mathfrak{M}$ is not in $\mathscr{H}$, then $\mathfrak{M}$ contains a finite subset with the same property. (Hint: Independent of the order of $\mathscr{K}$, consider all well orderings of the subsets of $\mathscr{K}$. Amongst the ordinals, determined by these well orderings, there is a smallest, say γ, for which the supremum of a corresponding subset of $\mathfrak{M}$ is not an element of $\mathscr{H}$. By the minimality of γ, γ is a limit ordinal or finite. Use the assumption to show that γ can not be a limit ordinal.)

14.7.8 For a cocomplete abelian category condition (AB 5) is equivalent to the condition arising from (AB 5) if one replaces "directed" by "strongly ordered". (Hint: Trivially, $\bigcup (m \cap n_i) \leq m \cap \bigcup n_i$. Set $\mathscr{K} = \mathscr{M}/B$ and let $\mathscr{H}$ consists of all monomorphisms $k\colon K \rightarrowtail B$ such that $m \cap k \leq \bigcup (m \cap n_i)$. Use 14.7.7.)

15. Injective Envelopes

15.1 Modules over Additive Categories

15.1.1 Definition. Let $\mathscr{C}$ be an additive category and R a ring. We consider R as an additive category with only one object $*$. An additive functor $F\colon R \to \mathscr{C}$ is an object $F(*) = A$ of $\mathscr{C}$ with a ring homomorphism $\varrho\colon R \to [A, A]_{\mathscr{C}}$. We say that A is provided with a *left R-module structure*, or that we have a *left R-module object* in $\mathscr{C}$. We designate it by ${}_{\varrho}A$ or by ${}_{R}A$, if there is no danger of confusion. Note, however, that there could be more than one left module structure on $A \in |\mathscr{C}|$. We call A the *underlying $\mathscr{C}$-object* of ${}_{R}A$.

We call $Add(R, \mathscr{C})$ the category, ${}_R\mathscr{C}$, of left R-module objects over $\mathscr{C}$. If ${}_\varrho A$, ${}_\sigma B$ are objects of ${}_R\mathscr{C}$, then a morphism $\bar{f}: {}_\varrho A \to {}_\sigma B$ in ${}_R\mathscr{C}$ is, by 2.6.1, a $\mathscr{C}$-morphism $f: A \to B$, for which

$$(1) \qquad f\,\varrho(r) = \sigma(r)\,f \qquad \text{for all} \quad r \in R\,.$$

We call $\bar{f}$ a (module) *homomorphism* and f the *underlying $\mathscr{C}$-morphism.*

15.1.2 Examples. For $\mathscr{C} = Ab$, ${}_RAb = {}_RMod$: for $A \in |{}_RMod|$, $r \in R$ induces an endomorphism of the additive group of A. If its effect on $a \in A$ is denoted (as usual) by ra, then $r\,(a_1 + a_2) = r\,a_1 + r\,a_2$. The fact, that one has a homomorphism of R into the ring of endomorphisms of the additive group of A is expressed by $1\,a = a$, $r_2(r_1\,a) = (r_2\,r_1)\,a$, $(r_1 + r_2)\,a = r_1\,a + r_2\,a$ for $r_1, r_2 \in R$ and $a \in A$.

For $\mathscr{C} = {}_SMod$, one correspondingly gets R-S-bimodules; i.e. ${}_R({}_SMod) = {}_{R,S}Mod$, where $r(s\,a) = s(r\,a)$ for $r \in R$, $s \in S$ and $a \in A \in |{}_{R,S}Mod|$.

For $R = \mathbf{Z}$, ${}_{\mathbf{Z}}\mathscr{C}$ is isomorphic to $\mathscr{C}$ in the evident, canonical way.

15.1.3 R^0 is the opposite ring of R (2.4.2). We call $Add\,(R^0, \mathscr{C})$ the category of *right R-modules* over $\mathscr{C}$. This is consistent with the canonical isomorphism ${}_{R^0}Mod = Mod_R$ which results from $r^0\,a \mapsto a\,r$, since $r_2^0\,r_1^0\,a = (r_1\,r_2)^0\,a$. For a commutative R, one thus has ${}_R\mathscr{C} = \mathscr{C}_R$.

We identify $({}_R\mathscr{C})^0$ with $(\mathscr{C}^0)_R$, because, for any categories $\mathscr{B}$, $\mathscr{C}$, $[\mathscr{B}^0, \mathscr{C}^0]$ can always be regarded as the dual category of $[\mathscr{B}, \mathscr{C}]$, and in the additive case $Add(\mathscr{B}^0, \mathscr{C}^0)$ as the dual of $Add(\mathscr{B}, \mathscr{C})$ (compare 4.5.6). The relation $({}_R\mathscr{C})^0 = (\mathscr{C}^0)_R$ allows the transfer of results about left-module categories to right-module categories by dualization.

15.1.4 There is the forgetful functor $U: {}_R\mathscr{C} \to \mathscr{C}$. It assigns to every object or, resp., morphism in ${}_R\mathscr{C}$ its underlying object or, resp., morphism in $\mathscr{C}$, and it coincides with the partial evaluation functor $E(?, *)$: $Add(R, \mathscr{C}) \to \mathscr{C}$ (3.7.1). The forgetful functor is faithful, and it reflects isomorphisms (compare 2.6.7).

The existence of limits and colimits (of a special kind or in general) is inherited by ${}_R\mathscr{C}$ from $\mathscr{C}$, and the forgetful functor preserves and reflects them. This follows from the "pointwise" construction in functor categories. Similarly, statements about commutability of finite limits with filtered or, resp., pseudofiltered limits (10.1.2) remain valid. If, in particular, $\mathscr{C}$ is exact or, resp., abelian, a Grothendieck category, then the same is true for ${}_R\mathscr{C}$, and the forgetful functor in particular is here exact and it also reflects exact sequences.

15.1.5 Let $\mathscr{C}$, $\mathscr{D}$ be additive categories and $T: \mathscr{C} \to \mathscr{D}$ an additive functor. T induces a "lifted" functor

$$(2) \qquad {}_RT\colon {}_R\mathscr{C} \to {}_R\mathscr{D} \qquad \text{with} \qquad U\,{}_RT = T\,U\,.$$

For $A \in |\mathscr{C}|$, there is the ring homomorphism

$$T_{A,A}\colon [A, A]_{\mathscr{C}} \to [T(A), T(A)]_{\mathscr{D}},$$

and $\varrho\colon R \to [A, A]$ yields $T_{A,A}\,\varrho\colon R \to [T(A), T(A)]$, so that one obtains

(3) $$_R T(_\varrho A) = {}_{T_{A,A}\varrho}(T(A)).$$

For $\bar{f}\colon {}_\varrho A \to {}_\sigma B$ one now gets $_R T(\bar{f})$ as the homomorphism $_R T(_\varrho A) \to$ $\to {}_R T(_\sigma B)$ with underlying $\mathscr{D}$-morphism $T(f)$. This constitutes a simple special case of 16.1.4.

15.1.6 For $A, B \in |\mathscr{C}|$, $[A, A]$ operates from the left on $[B, A]$ (1.5.2). Using $\varrho\colon R \to [A, A]$ a left R-module $[B, {}_\varrho A]$ is thus produced from $[B, A] \in |Ab|$, which is described by

(4) $$r f = \varrho(r) f \quad \text{for} \quad f\colon B \to A,\ r \in R \quad \text{and}\ \varrho\colon R \to [A, A].$$

Correspondingly, $[_\varrho A, B]_{\mathscr{C}}$ is a right R-module with

(4′) $$f r = f \varrho(r) \quad \text{for} \quad f\colon A \to B,\ r \in R \quad \text{and} \quad \varrho\colon R \to [A, A].$$

The partial Hom-functors for $\mathscr{C}$ give rise to the functors

(5) $$[\mathrm{Op}\,?, {}_\varrho A]_{\mathscr{C}}\colon \mathscr{C}^0 \to {}_R Mod; \qquad [_\varrho A, ?]_{\mathscr{C}}\colon \mathscr{C} \to Mod_R.$$

This applies in particular to the case of $R = [A, A]$ and $\varrho = 1_R$.

From (5) one gets the bifunctors $\mathscr{C}^0 \times {}_R\mathscr{C} \to {}_R Mod$ and $(_R\mathscr{C}^0) \times$ $\times\, \mathscr{C} \to Mod_R$, and by applying the forgetful functor $U\colon {}_R Mod \to Ab$ or, resp., $Mod_R \to Ab$ one obtains from (4), (4′) and (5)

(6) $$U[?, {}_\varrho A]_{\mathscr{C}} = [?, A]_{\mathscr{C}}; \qquad U[_\varrho A, ?]_{\mathscr{C}} = [A, ?]_{\mathscr{C}}.$$

For $_\varrho A \in |_R\mathscr{C}|$ and $_\sigma B \in |_S\mathscr{C}|$, $[A, B]_{\mathscr{C}}$ becomes a right R - left S - bimodule, by (4) and (4′). For $_R A$, $_R B \in |_R\mathscr{C}|$ the additive group of $_R\mathscr{C}$-morphisms from $_R A$ to $_R B$ is frequently designated by

(7) $$\mathrm{Hom}_R(_R A, {}_R B)$$

in order to avoid confusion with the twosided R-module $[_R A, {}_R B]_{\mathscr{C}}$.

If R is commutative, then (7) can be provided with an R-module structure such that (7) becomes a contra-covariant functor with codomain $_R Mod$. Using the notation in (1), one sets $r f = f \varrho(r) = \sigma(r) f$ for $r \in R$ and $\bar{f}\colon {}_\varrho A \to {}_\sigma B$. Since R is commutative, $r f$ is again a homomorphism $_\varrho A \to {}_\sigma B$, and, for $g\colon {}_\sigma B \to {}_\tau C$, one has $r(g f) = (r g) f =$ $= g\,(r f)$.

15.1.7 For additive categories $\mathscr{C}$, $\mathscr{D}$ there exists, according to the additive version of 3.6.3, the canonical isomorphism

$$Add(\mathscr{C}, {}_R\mathscr{D}) = Add(\mathscr{C}, Add(R, \mathscr{D})) \cong Add(R, Add(\mathscr{C}, \mathscr{D})) = {}_R Add(\mathscr{C}, \mathscr{D}).$$

This means, in particular, that additive functors $\mathscr{C} \to {}_R\mathscr{D}$ can be regarded as left R-module objects over $Add(\mathscr{C}, \mathscr{D})$.

The Yoneda embedding $H_*: \mathscr{C} \to Add(\mathscr{C}^0, Ab)$ provides a bijection between the module structures on $A \in |\mathscr{C}|$ and on $H_A \in |Add(\mathscr{C}^0, Ab)|$. Lifting of H_* or, resp., H^* results therefore in a full embedding

$$(8) \qquad {}_RH_*: {}_R\mathscr{C} \to {}_RAdd(\mathscr{C}^0, Ab) \cong Add(\mathscr{C}^0, {}_RMod)\,,$$
$$\qquad {}_{R^0}H^*: ({}_R\mathscr{C})^0 = (\mathscr{C}^0)_R \to Add(\mathscr{C}, Ab)_R \cong Add(\mathscr{C}, Mod_R)\,.$$

At $A \in |\mathscr{C}|$, one obtains in this way (5) from (8); and (6) from (8) and (2).

If one notes that an isomorphism $\gamma: F \to G$ provides, for additive functors $\mathscr{C} \to \mathscr{D}$, a bijection between the module structures on F and G, then (8) together with what was said above implies immediately the

Proposition. *Let $\mathscr{C} \to Mod_R$ be an additive functor. If $UT: \mathscr{C} \to Ab$ (with the forgetful functor $U: Mod_R \to Ab$) is representable, then T is isomorphic to a functor $[{}_\varrho A, ?]_\mathscr{C}$ for an appropriate ${}_\varrho A \in |{}_R\mathscr{C}|$.*

Remarks. In combination with the additive versions of 10.3.9 and 10.6.5 one notes that, by 15.1.4, T preserves limits if and only if this is the case for $U\,T$. If $\mathscr{C} = Mod_R$, then ${}_\varrho A$ is a bimodule, even if R is commutative.

15.1.8 Let R^+ be the additive group of the ring R. Every multiplication with a ring element from the left is an endomorphism of R^+ and thus the left R-module ${}_RR$ in ${}_RMod$ is created. (Note that in general R is not the endomorphism ring of R^+ as evidenced, for instance, by the complex numbers.) Left ideals of R are (by definition) submodules of ${}_RR$. The right R-module R_R and the twosided module ${}_RR_R$ are obtained analogously with right ideals or, resp., two sided ideals as submodules. ${}_RR$, R_R, ${}_RR_R$ are always supposed to have the meaning given above.

For $A \in |{}_RMod|$ a homomorphism $f: {}_RR \to A$ is already completely determined by $f(1)$, and $f \mapsto f(1)$ produces the canonical isomorphism

$$(9) \qquad \mathrm{Hom}_R({}_RR, ?) \overset{\cong}{\Rightarrow} U(?)$$

with $U: {}_RMod \to Ab$ the forgetful functor. We mention that this implies that ${}_RR$ is a projective generator of ${}_RMod$ (10.4.1, 10.5.1, 15.1.4). By the dual of (5), (9) provides a canonical isomorphism

$$(10) \qquad \mathrm{Hom}_R({}_RR_R, ?) \overset{\cong}{\Rightarrow} 1_{RMod}(?)\,.$$

For Mod_R, (9) and (10) are correspondingly valid. (9) and (10) also imply that R^0 or, resp., R is the endomorphism ring of ${}_RR$ or, resp., R_R. By (6), this implies further that

$$(11) \qquad \mathrm{Hom}_R(R_R, [{}_RA, B]_\mathscr{C}) \cong [A, B]_\mathscr{C}$$

for ${}_R A \in |{}_R\mathscr{C}|$ and $B \in |\mathscr{C}|$, and that there exists an isomorphism of contra-co-variant functors, which we shall later extend to an isomorphism of trifunctors (tensor product, 17.7). However, we mention already here the following isomorphism of contra-co-variant functors

$$(12) \qquad \psi\colon \operatorname{Hom}_R(M, \operatorname{Hom}_{\mathbf{Z}}(R_R, G)) \xrightarrow{\approx} [U(M), G]_{Ab}$$

for $M \in |{}_R Mod|$ and $G \in |Ab|$. One can prove (12) by a calculation. For $f\colon M \to \operatorname{Hom}_{\mathbf{Z}}(R_R, G)$ and $m \in M$, ψ is given by $(\psi(f))(m) = (f(m))(1)$ and the inverse isomorphism by $((\varphi(h)(m))(r) = h(r\,m)$ for $h\colon U(M) \to G$, $m \in M$ and $r \in R$.

15.1.9 The non-additive case. If $\mathscr{C}$ is an arbitrary category and R a category with only one object, then one can regard the functor category $[R, \mathscr{C}]$ as the category of R-objects over $\mathscr{C}$ as in 15.1.1. The preceding considerations carry over to this case (with *Ens* instead of *Ab*) without any trouble. If, in particular, R is terminal in *Cat*, then there is the trivial isomorphism $[R, \mathscr{C}] \cong \mathscr{C}$. Here one can take for R specifically a set of one element with its identity map.

15.2 Essential Extensions

This section is preparatory for 15.3. The category $\mathscr{C}$ is, in the following, always assumed to be abelian.

15.2.1 Definition. An *extension* of the object $A \in |\mathscr{C}|$ is a monomorphism $m\colon A \rightarrowtail B$. It is called *proper* if m is not an isomorphism. If $0 \to A \xrightarrow{m} B \xrightarrow{p} C \to 0$ is exact, then this short exact sequence (occasionally also only B) is called an extension of A by C. (This is also done similarly in other, not necessarily abelian categories as, e.g., the category of groups.) B, and thus m and p, are not determined by A and C, not even up to an isomorphism, as is well known for *Ab*.

An *extension* $m\colon A \rightarrowtail B$ is called *essential* if the following holds: If $n\colon N \rightarrowtail B$ is a monomorphism with $n \cap m = 0$, then $n = 0$. ("Every non-trivial subobject of B meets A".) Note, that $m \cap n = 0$ is equivalent to $m^{-1}(n) = 0$ and also to $n^{-1}(m) = 0$.

In *Ab* every endomorphism of $\mathbf{Z}$, that is different from 0, is essential. The sequence of the additive groups $\mathbf{Z}_p, \mathbf{Z}_{p^2}, \ldots, \mathbf{Z}_{p^n}, \ldots$ produces successive essential extensions.

15.2.2 If $m\colon A \rightarrowtail B$ and $n\colon B \rightarrowtail C$ are extensions, then $n\,m$ is essential if and only if m and n are essential.

Proof. If m and n are essential, then $n\,m$ is essential by 7.8.4 and by definition. If m is not essential, then there is a monomorphism $q\colon D \rightarrowtail B$ with $q \neq 0$ and $m \cap q = 0$, and since $n\,m \cap n\,q = n\,(m \cap q)$, $n\,m$ is not essential. If n is not essential, then $n\,m$ is all the more not essential.

15.2.3 Let $m\colon A \rightarrowtail B$ be a monomorphism. Then the following are equivalent:

(a) m is essential.

(b) If $g\colon D \to B$ is any morphism different from 0, then $g \circ g^{-1}(m) \neq 0$.

(c) Every morphism $f\colon B \to C$, for which $f\,m$ is a monomorphism, is itself a monomorphism.

Proof. (b) follows from (a) if g is factored into an epi- and a monomorphism and if pullbacks are formed. Take note of 14.4.6 here. (a) follows from (b) by restriction to a monomorphic g. To prove the equivalence of (a) and (c) let $k\colon K \to B$ be a kernel of $f\colon B \to C$. A comparison of

(1)
$$\begin{array}{ccccc} \bullet & \xrightarrow{\ker fm} & A & \xrightarrow{fm} & C \\ & & \downarrow{\scriptstyle m} & & \| \\ K & \xrightarrow{k} & B & \xrightarrow{f} & C \end{array}$$

with 12.3.4 shows: $f\,m$ is a monomorphism if and only if $k \cap m = 0$. If m is essential, (c) follows. If m is not essential, then there is a monomorphism $k \neq 0$ with $k \cap m = 0$ and, for $f = \operatorname{coker} k$, $f\,m$ is a monomorphism, but f is not.

15.2.4 Let $\mathscr{C}$ be a Grothendieck category. Every filtered colimit of essential extensions of the object A is then an essential extension of A.

Proof. Let $\mathscr{X}$ be a small filtered category and $T\colon \mathscr{X} \to A/\mathscr{C}$ (morphisms with domain A, dual of 6.5.3) a functor such that $T(i)\colon A \rightarrowtail B_i$ is an essential extension for every $i \in |\mathscr{X}|$. By 14.6.5 and 9.1.3, the colimit of T consists of a monomorphism $m\colon A \rightarrowtail B$ and of morphisms $\mu_i\colon B_i \to B$ with $\mu_i\, T(i) = m$ for all i. By 15.2.3 (c), every μ_i is a monomorphism. If $n\colon N \rightarrowtail B$ is a monomorphism and $m \cap n = 0$, then $\mu_i \cap$ $\cap\, n = 0$ for all i, because $T(i)$ is essential. Since $\{\mu_i\}$ is a filtered family of monomorphisms with $\bigcup \mu_i = 1_B$, $n = 0$ follows from (AB 5) in 14.6.1. Therefore, m is essential.

Remark. In general, the essential extensions of an object do not form a filtered category. Consider, e.g., the essential extension $\mathbf{Z}_2 \to \mathbf{Z}_4$ in Ab. The two automorphisms of $\mathbf{Z}_4$ show that 9.3.4 (ii) is not satisfied for the essential extensions of $\mathbf{Z}_2$.

15.2.5 Let $m\colon A \rightarrowtail B$ be a monomorphism in a well-powered Grothendieck category. Then, there is an epimorphism $p\colon B \twoheadrightarrow C$ such that $p\,m$ is a monomorphism and essential.

Proof. There is a set $\mathfrak{M}$ of representatives for the equivalence classes of monomorphisms with codomain B. Let $\mathfrak{D}$ be the subset of those monomorphisms whose intersection with m is 0. For $\mathfrak{D} = \{0\}$, m is essential and we set $p = 1_B$. Now let $\mathfrak{D} \neq \{0\}$. $\mathfrak{D}$ is ordered (as a set in $\mathscr{M}/B$). Let $\{d_i\}$ be a strongly ordered subset. By (AB 5)

$m \cap \bigcup d_i = \bigcup (m \cap d_i) = 0$. Hence $\bigcup d_i$ is equivalent to an element of $\mathfrak{D}$, and by Zorn's lemma, there is a maximal element in $\mathfrak{D}$; suppose it is $k: K \rightarrowtail B$. Let $p = \operatorname{coker} k: B \twoheadrightarrow C$. Then, by (1) and 12.3.4, $p\,m$ is a monomorphism. If $n: N \rightarrowtail C$ is a monomorphism, then $(p\,m)^{-1}(n) = m^{-1}(p^{-1}(n))$. Here $p^{-1}(n) \geq k$. Since k is maximal, $m^{-1}(p^{-1}(n)) = 0$ implies that $p^{-1}(n)$ is equivalent to k. $n = 0$ follows then from 14.4.6. Therefore, $p\,m$ is essential.

15.2.6 An object $Q \in |\mathscr{C}|$ is injective if and only if every extension of Q is a coretraction. If $\mathscr{C}$ is a well-powered Grothendieck category, then Q is injective if and only if Q does not possess a proper essential extension.

Proof. By the dual of 10.4.6, every extension $m: Q \rightarrowtail B$ of an injective object Q is a coretraction. If m is a proper extension, then m is not essential by 15.2.3 (c).

For the converse of the first statement, let a monomorphism $n: N \rightarrowtail L$ and a morphism $f: N \to Q$ be given. We form the pushout

(2)
$$\begin{array}{ccc} N & \overset{n}{\rightarrowtail} & L \\ {\scriptstyle f}\downarrow & & \downarrow{\scriptstyle \bar{f}} \\ Q & \overset{\bar{n}}{\longrightarrow} & P \end{array}$$

By the dual of 14.4.3 (c), $\bar{n}$ is a monomorphism. If $\bar{n}$ is a coretraction, then there is an $r: P \to Q$ with $r\,\bar{n} = 1_Q$, which implies $(r\,\bar{f})\,n = r\,\bar{n}\,f = f$. Now the first claim follows by the definition of injective (10.4.1°). The converse of the second statement follows from what has just been proved by means of (2) and 15.2.5: If Q does not have a proper essential extension, then $\bar{n}$ is a coretraction.

15.2.7 Definition. An *injective envelope* for the object A is an essential extension $m: A \rightarrowtail Q$ with an injective Q.

15.2.8 In a well-powered Grothendieck category, injective envelopes, if they exist, are maximal essential extensions. Here the following holds: If $m: A \rightarrowtail Q$ and $m': A \rightarrowtail Q'$ are injective envelopes, then there is an isomorphism $h: Q \to Q'$ with $h\,m = m'$.

Proof. The first statement follows immediately from 15.2.6. For the second one, one sees that h with $h\,m = m'$ exists, since Q' is injective. h is a monomorphism by 15.2.3 (c), and it is essential by 15.2.2. 15.2.6 shows it to be an isomorphism.

15.3 Existence of Injectives

First, we are concerned with some results for modules. We make use of the fact that ${}_R\mathit{Mod}$ is a well-powered Grothendieck category (15.1.4, 10.6.3).

15.3.1 Proposition. *Let R be a ring. A left R-module A is injective if and only if for every left ideal L of R and every module homomorphism f: $L \to A$ there is an $a \in A$ with $f(r) = r\,a$ for all $r \in L$.*

Proof. If A is injective, then f can be extended to a homomorphism f': ${}_RR \to A$. $a = f'(1)$ has the desired property. Now let the condition above be satisfied and let m: $A \rightarrowtail B$ be a proper extension. We may assume that m is an inclusion. Let b be an element of B that is not in A. The set of elements r of R, for which $r\,b$ is in A, is a left ideal L. By assumption, there is an $a \in A$ with $r\,b = r\,a$ for all $r \in L$. The submodule of B generated by $b - a$ shows that m is not an essential extension. A is injective by 15.2.6.

15.3.2 Proposition. *In Ab, $T = \mathbf{Q}/\mathbf{Z}$ is an injective cogenerator.*

Proof. An additive group A is called *divisible*, if for every $a \in A$ and every $n \in \mathbf{Z}$, $n \neq 0$, there is always an $a' \in A$ with $n\,a' = a$. 15.3.1 shows for $R = \mathbf{Z}$ that in Ab the divisible groups are exactly the injective ones. In particular, T is injective. For $B \in |Ab|$ and $b \in B$ with $b \neq 0$, there is a homomorphism f': $B' \to T$ with $f'(b) \neq 0$, where B' is the cyclic subgroup of B generated by b. Since T is injective, f' can be extended to a homomorphism f: $B \to T$. If g: $A \to B$ is a homomorphism different from 0, then there is an $a \in A$ with $g(a) \neq 0$ and therefore f: $B \to T$ with $f\,g \neq 0$, which completes the proof.

The last conclusion is a special case of the following proposition.

15.3.3 Proposition. *Let U be an injective object in an abelian category $\mathcal{C}$. U is an injective cogenerator if and only if $[A, U]_{\mathcal{C}} \neq 0$ for every A that is not a zero object.*

Proof. The condition is necessary by definition 10.5.1°, as 1_A and 0: $A \to A$ show. If it is satisfied, then H_U: $\mathcal{C} \to Ab$ reflects zero objects. Furthermore, H_U is exact by the dual of 13.2.6. By 13.3.7, H_U is faithful, and hence it is a cogenerator.

15.3.4 Proposition. *We assume that for the functors T: $\mathcal{C} \to \mathcal{D}$ and S: $\mathcal{D} \to \mathcal{C}$ there is an isomorphism*

$$(1) \qquad \varphi: [S(?), ??]_{\mathcal{C}} \to [?, T(??)]_{\mathcal{D}}$$

of contra-co-variant functors. (This is a pair of adjoint functors, see later 16.4.1.)

(a) *If S is faithful and $A \in |\mathcal{C}|$ a cogenerator, then $T(A)$ is a cogenerator in $\mathcal{D}$.*

(b) *If S preserves monomorphisms and if $A \in |\mathcal{C}|$ is injective, then $T(A)$ is injective.*

Proof. (a) The assumptions together with (1) imply immediately that $[?, T(A)]_{\mathcal{D}}$ is a faithful contravariant functor.

(b) By assumption and definition 10.1.4⁰, $[S(?), A]$ takes monomorphisms into epimorphisms. $T(A)$ is injective by (1) and 10.1.4⁰.

15.3.5 Proposition. *${}_R Mod$ has an injective cogenerator and injectives.*

Proof. Taking into account the dual of 10.5.5, one only has to prove the existence of an injective cogenerator. This, however, follows from 15.3.2 by 15.1.8 (12) and 15.3.4 because the forgetful functor U: ${}_R Mod \to Ab$ is faithful and exact (15.1.4):

$\mathrm{Hom}_{\mathbf{Z}}(R_R, \mathbf{Q}/\mathbf{Z})$ is an injective cogenerator of ${}_R Mod$.

15.3.6 Proposition. *Let $\mathscr{C}$ be an abelian category with injectives and a generator G. If $\mathscr{C}$ is also complete or cocomplete, then $\mathscr{C}$ has an injective cogenerator.*

Proof. According to 10.6.3, $\mathscr{C}$ is well-powered, and, by 12.4.4, it is also co-well-powered. Let $\{g_e\colon G \twoheadrightarrow G_e\}$ be a set of representatives for the equivalence classes of epimorphisms with domain G. Now let $\mathscr{C}$ be complete. We consider $P = \prod G_e$ and an extension $m\colon P \rightarrowtail Q$ with an injective Q. By 15.3.3, Q is a cogenerator if $[A, Q] \neq 0$ for all A that are not zero objects. For $A \neq 0$ there is a $g\colon G \to A$ with $g \neq 0$, since G is a generator. g factors through some G_e, call it G_d, with $g = g'' g_d$, so that $g''\colon G_d \to A$ is a monomorphism. Since $g \neq 0$, $G_d \neq 0$. Now, pr_d is a retraction (7.3.4). Let $pr_d i_d = 1_{G_d}$. Since Q is injective, there is an $h\colon A \to Q$ with $h g'' = m i_d$. Since m and i_d are monomorphisms and $G_d \neq 0$, $m i_d \neq 0$ follows and hence $h \neq 0$. So $[A, Q] \neq 0$. If $\mathscr{C}$ is cocomplete, one sets $P = \coprod G_e$.

Remark. From 16.4.8 and its dual it will follow that here $\mathscr{C}$ is both complete and cocomplete.

15.3.7 Theorem. *If $\mathscr{C}$ is a Grothendieck category with a generator G, then $\mathscr{C}$ has an injective cogenerator and every object has an injective envelope. $\mathscr{C}$ is also complete.*

Proof. $R = [G, G]$ is a ring which operates on $[G, A]$ from the right (15.1.6). We therefore consider $H^G = [G, ?]$ as a functor $T\colon \mathscr{C} \to Mod_R$. $H^G\colon \mathscr{C} \to Ab$ and hence T is an embedding (10.5.1), and T preserves limits because the forgetful functor $Mod_R \to Ab$ reflects limits (15.1.4). We now prove two lemmas.

Lemma 1. *T preserves and reflects essential extensions.*

Proof. Let $m\colon A \rightarrowtail B$ be a monomorphism in $\mathscr{C}$. Then $T(m)$ is a monomorphism, since T preserves limits. Now, let m be essential, let $M \neq 0$ be a submodule of $T(B)$ and $g \neq 0$ an element of M. g is a $\mathscr{C}$-morphism $G \to B$. We consider the pullback

$$\begin{array}{ccc} C & \xrightarrow{h} & A \\ {\scriptstyle j}\downarrow & & \downarrow{\scriptstyle m} \\ G & \xrightarrow[g]{} & B \end{array}$$

By 15.2.3 (b), $g\,j = m\,h \neq 0$. Since G is a generator, there is an f: $G \to C$ with $g\,j\,f = m\,h\,f \neq 0$. $j\,f \in R$ and $g \in M$ imply $g\,j\,f \in M$. Further one has $h\,f \in T(A)$ and $T(m)\,(h\,f) = m\,h\,f = g\,j\,f$. Thus the inverse image of M with respect to $T(m)$ does not consist of 0 only, so that $T(m)$ is essential. Assume now, conversely, that $T(m)$ is essential and that n: $N \rightarrowtail B$ is a monomorphism with $m \cap n = 0$. Since T preserves limits, $T(n)$ is a monomorphism and $T(m) \cap T(n) = 0$. $T(m)$ being essential, $T(n) = 0$ follows. Since T is faithful and additive, $n = 0$. Therefore, m is essential.

Remark. So far we have only made use of the fact that $\mathscr{C}$ is an abelian category with a generator.

Lemma 2. *T is a full embedding.*

Proof. It remains to be shown that T is full. Let u: $T(A) \to T(B)$ be a morphism in Mod_R. As in 10.5.4, we consider the epimorphism

$$(2) \qquad p\colon \coprod_{e \in T(A)} G_e \twoheadrightarrow A \ \text{ with } G_e = G \text{ for all } e \text{ and } p\ in_e = e.$$

Now, $u(e)$ is a $\mathscr{C}$-morphism $G \to B$ and there is the morphism

$$(3) \qquad q\colon \coprod_{e \in T(A)} G_e \to B \qquad \text{with} \quad q\ in_e = u(e)\,.$$

Let k: $K \to \coprod G_e$ be a kernel of p. We show that $q\,k = 0$. Since p is a cokernel of k, there is an f: $A \to B$ with $q = f\,p$. (2) and (3) then imply that $f\,e = u(e)$, and thus $T(f) = u$ by the definition of T (compare 2.2.5).

By 14.5.4, there exists, for every finite subset D of $T(A)$, the inclusion

$$in_D\colon \coprod_{d \in D} G_d \to \coprod_{e \in T(A)} G_e$$

and thus $(\coprod G_e, \{in_D\})$ is a filtered colimit. Therefore, p is a filtered colimit of the morphisms $p\ in_D$. Let $k_D: K_D \to \coprod G_d$ be a kernel of $p\ in_D$. By 14.6.6 (b), k is a colimit of the kernels k_D and it suffices to show that $q\,in_D\,k_D = 0$. Since G is a generator, this is equivalent to $q\,in_D\,k_D\,h = 0$ whenever h: $G \to K_D$. Let in'_d and pr'_d be the injections and projections of the biproduct $\coprod G_d$. We set $r_d = pr'_d\,k_D\,h$: $G \to G$. Since $in_D\,in'_d = in_d$ and

$$(4) \qquad k_D\,h = \sum in'_d\,pr'_d\,k_D\,h = \sum in'_d\,r_d\,,$$

(2) implies

$$0 = p\,in_D\,k_D\,h = \sum_{d \in D} p\,in_D\,in'_d\,r_d = \sum_{d \in D} d r_d\,.$$

From this, (3) and (4)

$$\begin{aligned} q\,in_D\,k_D\,h &= \sum_{d \in D} q\,in_D\,in'_d\,r_d = \sum_{d \in D} q\,in_d\,r_d \\ &= \sum_{d \in D} u(d)\,r_d = u\,\Big(\sum_{d \in D} d\,r_d\Big) = 0 \end{aligned}$$

follows, the latter because $r_a \in R$ and u is a module homomorphism. Thus lemma 2 is proved.

Proof of the theorem.[1] By 15.3.5, given $A \in |\mathscr{C}|$, there is a monomorphism in Mod_R $\alpha: T(A) \rightarrowtail J$ with an injective J. If $m: A \rightarrowtail B$ is an essential extension, then $T(m)$ is an essential extension by lemma 1. Since J is injective, there is a $\beta: T(B) \to J$ with $\alpha = \beta\, T(m)$, and by 15.2.3 (c), β is a monomorphism. B and m are determined uniquely by β because T is an embedding. If $\beta: T(B) \rightarrowtail J$ and $\beta': T(B') \rightarrowtail J$ are equivalent monomorphisms, then there is exactly one isomorphism f: $B \to B'$ with $\beta'\, T(f) = \beta$ because T is fully faithful; and m and $f\, m$ are isomorphic essential extensions in $\mathscr{C}$.

Now let a fixed A be chosen. We consider monomorphisms in Mod_R which are of the form $\beta: T(B) \to J$ and such that there is an essential extension $\mu: T(A) \to T(B)$ with $\beta\, \mu = \alpha$. From the class of all monomorphisms with codomain J one obtains by restriction an equivalence and a preordering for the ones considered here. By choosing representatives for the equivalence classes one gets an ordered set $\mathscr{Y}$. By lemma 1 and 2, there exists a functor W from $\mathscr{Y}$ into the category of essential extensions of A; namely: $\beta \in |\mathscr{Y}|$ determines uniquely $\mu: T(A) \rightarrowtail T(B)$ with $\beta\, \mu = \alpha$ and thus uniquely $W(\beta) = m$: $A \rightarrowtail B$ with $T(m) = \mu$ (m exists by lemma 2, is a monomorphism by 13.3.5, and is essential by lemma 1). For $\beta' \in |\mathscr{Y}|$, $\beta': T(B') \to J$, there is at most one morphism $\nu: \beta \to \beta'$; i.e., $\nu: T(B) \to T(B')$ with $\beta'\, \nu = \beta$. Here ν is an essential extension by 15.2.2 (since, for a suitable μ', $\alpha = \beta'\, \mu'$ and since this implies $\mu' = \nu\, \mu$). $W(\nu)$ is the essential extension of B with $T(W(\nu)) = \nu$. Thus one has W as a functor; and, by what was said above, every essential extension of A is isomorphic to one of the form $W(\beta)$.

We show that every strongly ordered subset $\mathscr{X}$ of $\mathscr{Y}$ has an upper bound (in $\mathscr{Y}$). $W(\mathscr{X})$ is a strongly ordered set of essential extensions of A. By 15.2.4 one obtains as a colimit an extension $g: A \rightarrowtail C$ with monomorphisms $n_\beta: B_\beta \to C$, where $\beta \in |\mathscr{X}|$ and $n_\beta\, W(\beta) = g$. If to every $\beta \in |\mathscr{X}|$ the diagram

$$T(A) \xrightarrow{TW(\beta)} T(B_\beta) \xrightarrow{\beta} J$$

is assigned, one obtains a strongly ordered set of diagrams which have a colimit in Mod_R

$$T(A) \xrightarrow{\nu} L \xrightarrow{\gamma} J$$

with morphisms $\lambda_\beta: T(B_\beta) \to L$. Here $\nu = \lambda_\beta\, TW(\beta)$, $\beta = \gamma\, \lambda_\beta$ for all $\beta \in |\mathscr{X}|$; ν and γ are monomorphisms by 14.6.5 and ν is essential by 15.2.4. Since the colimit in $\mathscr{C}$ is transferred to a natural transformation

[1] This proof may be omitted, since 19.8.7 will yield it as a corollary.

by T, there is exactly one morphism $\varrho: L \to T(C)$ with $\varrho\,\lambda_\beta = T(n_\beta)$ for all $\beta \in |\mathcal{X}|$. Now one has $\varrho\,\nu = \varrho\,\lambda_\beta\, TW(\beta) = T(n_\beta)\, TW(\beta) = T(g)$. Since ν is essential, ϱ is a monomorphism (15.2.3); $\varrho: L \to T(C)$ is essential because $T(g)$ is essential (15.2.2). Therefore (compare above), there is a monomorphism $\sigma: T(C) \rightarrowtail J$ with $\sigma\,\varrho = \gamma$, where $\sigma\, T(g) = \alpha$. So σ is equivalent to an element of $\mathcal{Y}$. This is an upper bound for $\mathcal{X}$ in $\mathcal{Y}$, since $\beta = \gamma\,\lambda_\beta = \sigma\,\varrho\,\lambda_\beta$.

By Zorn's lemma, there is a maximal element in $\mathcal{Y}$, let us call it δ: $T(D) \rightarrowtail J$. Let $s: D \rightarrowtail E$ be an essential extension of D in $\mathcal{C}$. Since $T(s)$ is essential, there is (see above) a monomorphism $\tau: T(E) \rightarrowtail J$ with $\delta = \tau\, T(s)$. Since $s\, W(\delta): A \rightarrowtail E$ is essential (15.2.2), τ is equivalent to an element ε of $\mathcal{Y}$. Then $\varepsilon = \delta$ because δ is maximal. Thus $T(s)$ and also s are isomorphisms. By 15.2.6, D is injective, and by 15.2.7, $W(\delta): A \rightarrowtail D$ is an injective envelope of A.

The existence of an injective cogenerator for $\mathcal{C}$ now follows from 15.3.6. The last statement of the theorem is a special case of the dual of 16.4.8.

15.3.8 Remarks. From the preceding proof one obtains by simplifying ($\mathcal{C}$, $1_\mathcal{C}$ instead of Mod_R, T):

If $\mathcal{C}$ is a well-powered Grothendieck category and if the object $A \in |\mathcal{C}|$ has an extension $a: A \to J$ with an injective J, then A has an injective envelope. However, a well-powered Grothendieck category need not have injectives (see Freyd [13]).

The use of injective envelopes is not always suitable. If there is an injective cogenerator Q, then, by the dual of 10.5.4, there is a monomorphism $m_A: A \rightarrowtail \Pi\, Q_e$ with $e \in [A, Q]$ and $Q_e = Q$ for all e. Here $\Pi\, Q_e$ is injective by the dual of 10.4.4. $f: A \to B$ induces, by means of $[f, Q]$, a morphism

$$f_*: \prod_{e \in [A,Q]} Q_e \to \prod_{d \in [B,Q]} Q_d \qquad \text{with } pr_d\, f_* = pr_{df}.$$

In this way a functor $\mathcal{C} \to [\mathbf{2}, \mathcal{C}]$ is created that assigns to every object of $\mathcal{C}$ an extension with an injective codomain.

15.3.9 We add the remark that under the assumptions of 15.3.6 an additive functor $\mathcal{C} \to Ab$ is representable if and only if it preserves limits. This follows from 10.6.5 because of 10.6.3. It is true, in particular, for $\mathcal{C} = {}_R Mod$.

15.4 An Embedding Theorem

15.4.1 Proposition. *Let $\mathcal{C}$ be a small, exact, additive category. Then in $Add(\mathcal{C}, Ab)$ $G = \coprod_{A \in |\mathcal{C}|} H^A$ is a projective generator which as a functor G: $\mathcal{C} \to Ab$ is left exact.*

Proof. Every H^A is projective by 10.4.3 and left exact by 13.2.5. All the H^A together form a generating set by 10.5.2. By 10.5.3 and 10.4.4, G is a projective generator. Since $\amalg H^A$ is constructed "pointwise" and since Ab is a Grothendieck category, G is left exact by 14.6.6.

15.4.2 Proposition. *Let $\mathscr{C}$ be an exact additive category. If an additive functor $T: \mathscr{C} \to Ab$ is an injective object of $Add(\mathscr{C}, Ab)$, then T is right exact.*

Proof. If $A \to B \to C \to 0$ is an exact sequence in $\mathscr{C}$, then $0 \to \to H^C \to H^B \to H^A$ is exact in $Add(\mathscr{C}, Ab)$ by 10.2.5, 10.2.7. If T is injective, then $[H^A, T] \to [H^B, T] \to [H^C, T] \to 0$ is exact, where $[H^A, T]$ etc. are the morphism groups for $Add(\mathscr{C}, Ab)$. By the Yoneda lemma (4.3.1), $T(A) \to T(B) \to T(C) \to 0$ is exact.

15.4.3 Definition. A functor is called a *monofunctor* if it preserves monomorphisms.

A right exact monofunctor between exact categories is exact.

15.4.4 Lemma. *Let $\mathscr{C}$ be an abelian category and $M: \mathscr{C} \to Ab$ an additive monofunctor. If $\mu: M \rightarrowtail N$ is an essential extension in $Add(\mathscr{C}, Ab)$, then N is also a monofunctor.*

We prove this indirectly. If N is not a monofunctor, then there is a monomorphism $f: A \rightarrowtail B$ in $\mathscr{C}$ such that $N(f)$ is not a monomorphism, i.e., in $N(A)$ there is an $x \neq 0$ with $N(f)(x) = 0$. By the Yoneda lemma, there is a natural transformation $\xi: H^A \to N$ with $Y(\xi) = = \xi_A(1_A) = x$. With objects and morphisms that are yet to be defined, we consider the following diagram

(1)

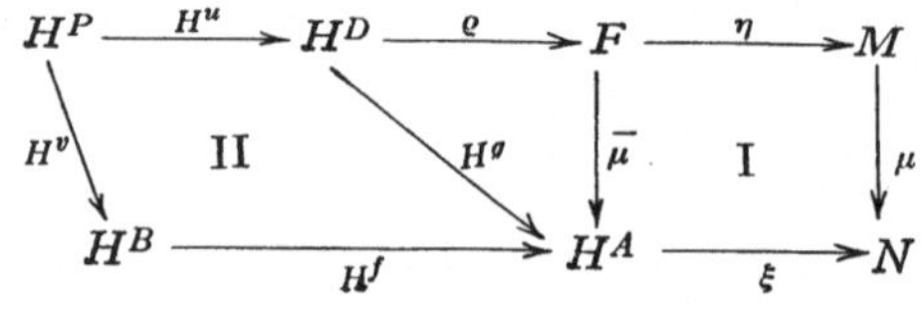

Here I is a pullback. Since $\xi \neq 0$, $\xi\,\bar{\mu} = \mu\,\eta \neq 0$ by 15.2.3 (b). So there is a $D \in |\mathscr{C}|$ with $y \in F(D)$ such that $(\mu\,\eta)_D\,(y) \neq 0$. y determines $\varrho: H^D \to F$ with $Y(\varrho) = y$ and thus $g: A \to D$ with $H^g = \bar{\mu}\varrho$. We form the pushout for f and g in $\mathscr{C}$. It gets transformed by means of the Yoneda embedding $\mathscr{C}^0 \to Add(\mathscr{C}, Ab)$ into the pullback II (10.2.5, 10.2.7). By 13.4.3 (b), if f is a monomorphism, then so is u. Since M is a monofunctor, $M(u)$ is monomorphic and hence by 4.2.4 so is $[H^u, M]: [H^D, M] \to [H^P, M]$. Since $0 \neq \eta\,\varrho \in [H^D, M]$, one gets $\mu\,\eta\,\varrho\,H^u \neq 0$. This constitutes a contradiction with the commutativity of (1), because $N(f)\,(x) = 0$ and the Yoneda lemma imply $\xi\,H^f = = [H^f, N]\,(\xi) = 0$.

15.4.5 Theorem. *Every small abelian category $\mathscr{C}$ has an exact embedding in Ab.*

Proof. $Add(\mathscr{C}, Ab)$ is a Grothendieck category (10.1.2, 14.6.6). By 15.4.1, it possesses a generator G that is a monofunctor. The injective envelope $\mu: G \rightarrowtail Q$ of G exists by 15.3.7. By 15.4.2 and 15.4.4, Q is exact. $G = \coprod H^A$ preserves and reflects zero morphisms. The same is true for Q, because μ is monomorphic, even "pointwise" on account of 10.1.4. Thus Q is an exact faithful functor. Anticipating 16.2.7, one can use this to obtain an embedding in Ab.

15.5 Problems

15.5.1 Give an example illustrating the last remark in 15.1.7.

15.5.2 Check 15.1.9. What is the analogue of ${}_R Mod$?

15.5.3 Determine the injective envelope of $\boldsymbol{Z}_p \in |Ab|$ as a subgroup of $\boldsymbol{Q}/\boldsymbol{Z}$.

15.5.4 Check the last remark in 15.3.8.

15.5.5 Let $\mathscr{C}$ be a co-well-powered cocomplete category in which filtered colimits of monomorphisms are monomorphisms. Further, let every morphism have a factorization into an epi- and a monomorphism, and let pushouts preserve monomorphisms; i. e., if 15.2 (2) is a pushout with a monomorphic n, then $\bar{n}$ is a monomorphism. Define essential extensions as monomorphisms which satisfy condition 15.2.3 (c), and prove that the analogues of 15.2.2 and 15.2.4 through 15.2.8 are then valid.

15.5.6 Let $\mathscr{C}$ be an abelian category and $T: \mathscr{C} \to Ab$ an additive functor. Call an element $a \in T(A)$ *effaceable* if there is a monomorphism $m: A \rightarrowtail B$ such that $T(m)(a) = 0$. Prove the following:

(a) There is an additive functor $E_T: \mathscr{C} \to Ab$ such that, for all $A \in |\mathscr{C}|$, $E_T(A)$ consists of the effaceable elements of $T(A)$ and the inclusions $E_T(A) \subset T(A)$ form a natural transformation. (Hint: use the dual of 12.4.3 (c).)

(b) If, for $S \in |Add(\mathscr{C}, Ab)|$, $E_S: \mathscr{C} \to Ab$ is defined correspondingly, and if $\mu: S \to T$ is a monomorphic natural transformation, then there is a pullback

$$\begin{array}{ccc} E_S & \subset & S \\ \downarrow & & \downarrow \mu \\ E_T & \subset & T \end{array}$$

(c) 15.4.4 also follows from (b).

15.5.7 If $\mathscr{C}$ is an abelian category and $\mathscr{M}$ a set of objects in $\mathscr{C}$, then there is a full small abelian subcategory of $\mathscr{C}$ which contains all the objects in $\mathscr{M}$.

15.5.8 Let $m: A \rightarrowtail B$ be a monomorphism in an abelian category. Show that the following two conditions are equivalent:

(a) There exists a maximal monomorphism $m': A' \rightarrowtail B$ such that $m \cap m' = 0$.

(b) There exists an epimorphism $c: B \to C$ such that $cm: A \to C$ is an essential extension.

If A has an injective envelope, then (a) and (b) are valid for every monomorphism with domain A.

16. Adjoint Functors

16.1 Composition of Functors and Natural Transformations

16.1.1 Rules. If $U: \mathscr{D} \to \mathscr{E}$ is a functor and $\xi: T \to T'$ a natural transformation of functors $T, T': \mathscr{C} \to \mathscr{D}$, then the rule $C \mapsto U(\xi_C)$ determines a natural transformation $U\,T \to U\,T'$, which we denote by $U * \xi$ or $U\,\xi$. If $S: \mathscr{B} \to \mathscr{C}$ is a functor, then the rule $B \mapsto \xi_{S(B)}$, for $B \in |\mathscr{B}|$, determines a natural transformation $\xi * S = \xi\,S: T\,S \to$ $\to T\,S'$. If further $R: \mathscr{A} \to \mathscr{B}$ and $V: \mathscr{E} \to \mathscr{F}$ are functors, the following rules obviously hold:

(1) $$(V\,U) * \xi = V * (U * \xi); \qquad C \mapsto VU(\xi_C)\,.$$

(2) $$\xi * (S\,R) = (\xi * S) * R; \qquad A \mapsto \xi_{SR(A)}\,.$$

(3) $$(U * \xi) * S = U * (\xi * S); \qquad B \to U(\xi_{S(B)})\,.$$

Thus one can write $V\,U\,\xi$, $\xi\,S\,R$, $U * \xi * S$, $U\,\xi\,S$. If, in addition, $\xi': T' \to T''$ and $\beta: S \to S'$ are natural transformations of functors $T', T'': \mathscr{C} \to \mathscr{D}$ or, resp., $S, S': \mathscr{B} \to \mathscr{C}$, then the following rules are also easily verified:

(4) $$U * (\xi'\,\xi) * S = (U * \xi' * S)\,(U * \xi * S)\,,$$

(5) $$(\xi * S')\,(T * \beta) = (T' * \beta)\,(\xi * S): \; T\,S \to T'\,S'\,.$$

16.1.2 If ξ is an isomorphism, then $U * \xi * S$ is also an isomorphism. For $\xi * S$ this follows from the fact that a natural transformation is an isomorphism if and only if it is one at every place; and $U * \xi$ is an isomorphism, because every functor preserves isomorphisms.

16.1.3 The rules (1) through (4) mean that the Hom-functor of *cat* may be regarded as a contra-co-variant functor with values in *cat* (and correspondingly for the category CAT of small $\mathfrak{V}$-categories).

For arbitrary categories,

$$[\mathscr{C}, U]\colon [\mathscr{C}, \mathscr{D}] \to [\mathscr{C}, \mathscr{E}]$$

is the functor for which $T \mapsto U\,T$, $\xi \mapsto U * \xi$, by (4) and 16.1.2. (1) becomes in this case

$$(1') \qquad [\mathscr{C}, V]\,[\mathscr{C}, U] = [\mathscr{C}, V\,U]\,.$$

In the same way, $[S, \mathscr{D}]\colon [\mathscr{C}, \mathscr{D}] \to [\mathscr{B}, \mathscr{D}]$ is the functor for which $T \mapsto T\,S$, $\xi \mapsto \xi * S$, and (2) becomes

$$(2') \qquad [R, \mathscr{D}]\,[S, \mathscr{D}] = [S\,R, \mathscr{D}]\,.$$

Together with $(U\,T)\,S = U\,(T\,S)$, (3) turns into

$$(3') \qquad [S, U]_{Cat} = [S, \mathscr{E}]\,[\mathscr{C}, U] = [\mathscr{B}, U]\,[S, \mathscr{D}]$$

so that this determines in fact a contra-co-variant functor.

(5) provides an additional structure. First, for each $\xi\colon T \to T'$ one obtains a natural transformation

$$(5_1') \qquad [\mathscr{B}, \xi]\colon [\mathscr{B}, T] \to [\mathscr{B}, T']$$

of functors from $[\mathscr{B}, \mathscr{C}]$ to $[\mathscr{B}, \mathscr{D}]$ whose component at $S \in [\mathscr{B}, \mathscr{C}]$ is $\xi\,S\colon T\,S \to T'\,S$; and for each $\beta\colon S \to S'$ one obtains a natural transformation

$$(5_2') \qquad [\beta, \mathscr{D}]\colon [S, \mathscr{D}] \to [S', \mathscr{D}]$$

between functors from $[\mathscr{C}, \mathscr{D}]$ to $[\mathscr{B}, \mathscr{D}]$ whose component at $T \in [\mathscr{C}, \mathscr{D}]$ is $T\,\beta\colon T\,S \to T\,S'$. Repeated applications of (5) lead to natural transformations $[\beta, U]_{Cat}$, $[R, \xi]_{Cat}$ and, finally, $[\alpha, \beta]_{Cat}$ with $\alpha\colon R \to R'$.

Note that in $(5_2')$ the order of S and S' in $[\beta, \mathscr{D}]$ is not reversed. From $(5_1')$ and $(5_2')$ it follows that there are functors

$$[\mathscr{B}, ?]\colon [\mathscr{C}, \mathscr{D}] \to [[\mathscr{B}, \mathscr{C}], [\mathscr{B}, \mathscr{D}]]\,,$$

$$[?, \mathscr{D}]\colon [\mathscr{B}, \mathscr{C}] \to [[\mathscr{C}, \mathscr{D}], [\mathscr{B}, \mathscr{D}]]\,.$$

16.1.4 If all categories and functors involved are additive, the above yields

$$[\mathscr{C}, U]\colon Add(\mathscr{C}, \mathscr{D}) \to Add(\mathscr{C}, \mathscr{E})\,,$$

$$[S, \mathscr{D}]\colon Add(\mathscr{C}, \mathscr{D}) \to Add(\mathscr{B}, \mathscr{D})$$

and the properties of 16.1.3 carry over without any difficulties.

16.2 Equivalences of Categories

16.2.1 Definition. A functor $T\colon \mathscr{C} \to \mathscr{D}$ is called an *equivalence* if there are a functor $S\colon \mathscr{D} \to \mathscr{C}$ and isomorphisms $\Psi\colon S\,T \to 1_{\mathscr{C}}$, $\Phi\colon 1_{\mathscr{D}} \to T\,S$. S is called *equivalence-inverse* to T. The categories $\mathscr{C}$ and $\mathscr{D}$ are said to be equivalent if there is an equivalence $T\colon \mathscr{C} \to \mathscr{D}$.

16.2.2 Examples and remarks. An equivalence is weaker than an isomorphism. (There is an analogy with the concept of homotopy equivalence in *Top*, where natural isomorphisms of functors correspond to the homotopies.) In applications, equivalences of categories are more frequent than isomorphisms.

If $\mathcal{C}$ is the category of finite dimensional vector spaces over a field K, then, for the contravariant functor $D: \mathcal{C} \to \mathcal{C}$ defined by the rules vectorspace $\mapsto$ dual vectorspace, linear transformation $\mapsto$ transposed transformation, Op D: $\mathcal{C} \to \mathcal{C}^0$ is an equivalence with an equivalence-inverse D Op: $\mathcal{C}^0 \to \mathcal{C}$.

In the theory of Lie groups the equivalence of the category of simply connected Lie groups with the category of finite dimensional Lie algebras is fundamental.

If in an exact category one chooses a cokernel for every monomorphism with codomain A and a kernel for every epimorphism with domain A, then the transition to cokernels is an equivalence $\mathcal{M}/A \to$ $\to A/\mathcal{E}$ by 12.4.4.

16.2.3 Proposition. *Let* $V: \mathcal{B} \to \mathcal{C}$ and $T: \mathcal{C} \to \mathcal{D}$ *be equivalences with equivalence-inverses* $U: \mathcal{C} \to \mathcal{B}$ *and* $S: \mathcal{D} \to \mathcal{C}$. *Then*

(a) Op T Op: $\mathcal{C}^0 \to \mathcal{D}^0$ *is an equivalence with equivalence-inverse* Op S Op.

(b) $T\ V$ *is an equivalence with equivalence-inverse* $U\ S$.

(c) *If* $\mathcal{A}$ *is an arbitrary category, then* $[\mathcal{A}, T]: [\mathcal{A}, \mathcal{C}] \to [\mathcal{A}, \mathcal{D}]$ *(compare* 16.1.3) *is an equivalence with equivalence-inverse* $[\mathcal{A}, S]$.

(d) $[T, \mathcal{A}]: [\mathcal{D}, \mathcal{A}] \to [\mathcal{C}, \mathcal{A}]$ *is an equivalence with equivalence-inverse* $[S, \mathcal{A}]$.

(e) T': $\mathcal{C} \to \mathcal{D}$ *is isomorphic to* T *if and only if* S *is equivalence-inverse to* T'. *This is the case if only* $T'\ S$ *is isomorphic to* $1_{\mathcal{D}}$ *or* $S\ T'$ *isomorphic to* $1_{\mathcal{C}}$.

Remark. Two equivalences $T, T': \mathcal{C} \to \mathcal{D}$ need not be isomorphic to each other: for $\mathcal{C} = \mathcal{D} = \mathcal{A} \sqcup \mathcal{A}$ consider $1_{\mathcal{C}}$ and interchange the two cofactors $\mathcal{A}$.

Proof. (a) follows from Op Op $= 1$ and the definition.

(b) From isomorphisms Ψ: $S\ T \to 1_{\mathcal{C}}$ and X: $U\ V \to 1_{\mathcal{B}}$ one obtains, by 16.1.1 and 16.1.2, the isomorphism

$$X\,(U * \Psi * V):\ U\,S\,T\,V \to U\,V \to 1_{\mathcal{B}}$$

and correspondingly $1_{\mathcal{D}} \to T\ V\ U\ S$.

(c) follows from 16.1.3 (1′) and ($5_1'$), (d) from 16.1.3 (2′) and ($5_2'$).

(e) If T is isomorphic to T', then $T\ S$ and $T'\ S$ are isomorphic by 16.1.2, and the same holds for $S\ T$ and $S\ T'$. Therefore, S is also equivalence-inverse to T'. If $T'\ S$ is isomorphic to $1_{\mathcal{D}}$, then $T'\ S\ T$ is isomorphic to T and to T', once more by 16.1.2. The case where $S\,T'$ is isomorphic to $1_{\mathcal{C}}$ is treated analogously.

16.2.4 Proposition. *Let $T\colon \mathcal{C} \to \mathcal{D}$ be an equivalence with equivalence-inverse $S\colon \mathcal{D} \to \mathcal{C}$. Then*

(a) *T is fully faithful and every object of $\mathcal{D}$ is isomorphic to an object of the form $T(A)$.*

(b) *T preserves and reflects limits and colimits including terminal and initial objects. In particular, T preserves and reflects monomorphisms and epimorphisms.*

(c) *If $\mathcal{C}$ or $\mathcal{D}$ is (semi-)additive, then there is exactly one (semi-)additive structure on the other category with respect to which T is additive. Thus S is also additive.*

(d) *If $\mathcal{C}$ is (finitely) complete or (finitely) cocomplete or exact or abelian or a Grothendieck category, then the same is the case for $\mathcal{D}$. If $\mathcal{C}$ is exact, then T preserves and reflects exact sequences.*

The above list of properties could be lengthened.

Proof. (a) For $A, B \in |\mathcal{C}|$, $(S\,T)_{A,B}\colon [A, B] \to [ST(A), ST(B)]$ is bijective. Since $(S\,T)_{A,B} = S_{T(A),T(B)}\,T_{A,B}$, $S_{T(A),T(B)}$ is surjective and $T_{A,B}$ is injective. Considering $(T\,S)_{T(A),T(B)}$ one finds that $S_{T(A),T(B)}$ is injective and thus bijective. Therefore, $T_{A,B}$ is bijective and T is fully faithful. $X \in |\mathcal{D}|$ is isomorphic to $TS(X)$.

(b) T reflects limits by (a) and 7.7.6. Let (L, λ) be a limit of $D\colon \Sigma \to \mathcal{C}$. D is isomorphic to $S\,T\,D$ and L is isomorphic to $ST(L)$ by an isomorphism $\Psi^{-1}\colon 1_{\mathcal{C}} \to S\,T$. Therefore, $(ST(L), ST\lambda)$ is a limit of $S\,T\,D$. Since S is fully faithful, $(T(L), T\lambda)$ is a limit of $T\,D$. The statement about monomorphisms follows from 7.8.9. The statements concerning colimits are dual to those about limits.

(c) Let $\mathcal{D}$ have a (semi-) additive structure. By (a) there is exactly one such structure on $\mathcal{C}$ for which T is additive. An isomorphism $\Phi\colon 1_{\mathcal{D}} \to T\,S$ then shows S to be additive.

(d) follows easily from (b) and (c).

16.2.5 Remark. If the additive functor $T\colon \mathcal{C} \to \mathcal{D}$ is an equivalence between additive categories and if $\mathcal{A}$ is an additive category, then

$$[\mathcal{A}, T]\colon \mathit{Add}(\mathcal{A}, \mathcal{C}) \to \mathit{Add}(\mathcal{A}, \mathcal{D})$$

and

$$[T, \mathcal{A}]\colon \mathit{Add}(\mathcal{D}, \mathcal{A}) \to \mathit{Add}(\mathcal{C}, \mathcal{A})$$

are equivalences. Using 16.2.4 (c), this follows as in 16.2.3 (c), (d) from 16.1.3 and 16.1.4.

16.2.6 Proposition. *Let $T\colon \mathcal{C} \to \mathcal{D}$ be a functor between arbitrary categories. There is a category $\mathcal{D}'$ with a full embedding $S\colon \mathcal{D} \to \mathcal{D}'$ such that S is an equivalence and ST is isomorphic to a functor $T'\colon \mathcal{C} \to \mathcal{D}'$ that is injective for the classes of objects.*

Proof. We may assume that $\mathscr{C}$ is not empty. Let $\mathscr{D}'$ have as objects pairs (A, X) with $A \in |\mathscr{C}|$ and $X \in |\mathscr{D}|$, and as morphisms from (A, X) to (A', X') triples (A, A', u) with $u\colon X \to X'$. Composition of morphisms is defined by $(A', A'', u')\,(A, A', u) = (A, A'', u'\,u)$ for $u'\colon X' \to X''$. Let B be a fixed object of $\mathscr{C}$. $X \mapsto (B, X), u \mapsto (B, B, u)$ defines a full embedding $S\colon \mathscr{D} \to \mathscr{D}'$. We define $V\colon \mathscr{D}' \to \mathscr{D}$ by $(A, X) \mapsto X, (A, A', u) \mapsto u$. Then $V\,S = 1_{\mathscr{D}}$; and $(A, X) \mapsto (A, B, 1_X)$ yields an isomorphism $\Phi\colon 1_{\mathscr{D}'} \to S\,V$. Now let $T'\colon \mathscr{C} \to \mathscr{D}'$ be defined by $T'(A) = (A, T(A))$ and $T'(f) = (A, C, T(f))$ for $f\colon A \to C$. Then $T = V\,T'$, $S\,T = S\,V\,T'$, and, by 16.1.2, $S\,T$ is isomorphic to T' by means of $\Phi\,T'$.

16.2.7 Remark. For $\mathscr{D} = Ens$ or $\mathscr{D} = Ab$, S in 16.2.6 can be chosen to be a functor $Ens \to Ens$ or, resp., $Ab \to Ab$. One replaces (A, X) by the set of pairs (A, x) with $x \in X$; in Ab with the evident group structure.

16.3 Skeletons

16.3.1 Definition. A *category* $\mathscr{K}$ is called *reduced* if any two isomorphic objects are identical. A subcategory $\mathscr{K}$ of $\mathscr{C}$ is called a *skeleton* of $\mathscr{C}$ if $\mathscr{K}$ is reduced and if the inclusion $\mathscr{K} \subset \mathscr{C}$ is an equivalence.

16.3.2 Proposition. *Let $\mathscr{H}$ and $\mathscr{K}$ be reduced categories. For a functor $T\colon \mathscr{H} \to \mathscr{K}$, the following are equivalent:*

(a) *T is an isomorphism,*

(b) *T is an equivalence,*

(c) *T is fully faithful and surjective for the classes of objects.*

Proof. (b) follows from (a), and (c) follows from (b) by 16.2.4 (a).

Now let (c) be satisfied. If $A, B \in |\mathscr{H}|$ have the same image $X = T(A) = T(B)$ under T, then there is a morphism $u\colon A \to B$ with $T(u) = 1_X$, because T is fully faithful. u is an isomorphism by 4.1.5. Since $\mathscr{H}$ is reduced, T is bijective on the classes of objects. It follows, that T is a bijection Mor $\mathscr{H} \to$ Mor $\mathscr{K}$ and thus an isomorphism.

16.3.3 Remarks concerning the axiom of choice. We assume that the axiom of choice applies to the universe $\mathfrak{U}$. Without this assumption many results could be obtained only for small categories, others would require elaborate formulations. If $\mathfrak{U}$ is the universal class for 3.1.1, and if for a subclass of $\mathfrak{U}$ there is an equivalence relation, then Gödel's axiom of choice allows the choice of representatives for the equivalence classes. Our assumption is motivated by this.

16.3.4 Proposition. *Every category has a skeleton.*

Proof. Isomorphism is an equivalence relation for the objects of $\mathscr{C}$. Choose an object from every equivalence class. Let $\mathscr{K}$ be the full subcategory of $\mathscr{C}$ whose objects are these chosen ones, and let $S: \mathscr{K} \to \mathscr{C}$ be the inclusion. $V: \mathscr{C} \to \mathscr{K}$ is constructed as follows: given $A \in |\mathscr{K}|$, choose an isomorphism $w_{A'}: A' \to A$ for every object A' in $\mathscr{C}$ that is isomorphic to A. Let $w_A = 1_A$. Set $V(A') = A$. If A, B are objects of $\mathscr{K}$ and $w_{A'}: A' \to A$, $w_{B'}: B' \to B$ chosen isomorphisms, set $V(f) = w_{B'} f w_{A'}^{-1}$ for $f: A' \to B'$. This defines a functor V. Since $w_A = 1_A$, $V S = 1_{\mathfrak{K}}$; and $\{w_{A'}\}$ is an isomorphism $1_{\mathscr{C}} \to S V$ by construction.

16.3.5 Remark. 16.3.4 makes it possible in many cases to reduce a category to a small category. This is the case, for instance, for finitely generated modules, in particular for finitely generated additive groups and for finite dimensional vector spaces. The same holds for Lie groups and for finite dimensional Lie algebras. In the cases named above there is a canonical choice for the objects of an equivalent small category.

16.3.6 Theorem. *A functor $T: \mathscr{C} \to \mathscr{D}$ is an equivalence if and only if T is fully faithful and if every object of $\mathscr{D}$ is isomorphic to an object of the form $T(A)$.*

Proof. 16.2.4 (a) furnishes one part. Now, conversely, let the other condition be satisfied, and let $S: \mathscr{K} \to \mathscr{C}$ and $R: \mathscr{L} \to \mathscr{D}$ be inclusions of skeletons with equivalence-inverses $V: \mathscr{C} \to \mathscr{K}$, $U: \mathscr{D} \to \mathscr{L}$. Every one of the functors U, T, S satisfies the corresponding conditions, and hence $U T S$ also. By 16.3.2, $U T S$ is an isomorphism, and by 16.2.3 (b), $R U T S V$ is an equivalence. It is evidently isomorphic to T by 16.1.2. By 16.2.3 (e), T is an equivalence.

16.3.7 Corollary. *Two categories are equivalent if and only if their skeletons are isomorphic.*

16.3.8 Corollary. *Every fully faithful functor can be factored into an equivalence and an inclusion of a full subcategory.*

16.3.9 Proposition. *Let $\mathscr{C}$, $\mathscr{D}$ be additive categories with finite products, and let $\mathscr{B}$ be a full subcategory of $\mathscr{C}$ such that every object of $\mathscr{C}$ is a finite product of objects of $\mathscr{B}$. The restriction to $\mathscr{B}$ of additive functors $\mathscr{C} \to \mathscr{D}$ and their natural transformations is an equivalence $\tilde{R}: Add(\mathscr{C}, \mathscr{D}) \to Add(\mathscr{B}, \mathscr{D})$ with an equivalence-inverse Q such that $\tilde{R} Q = 1_{Add(\mathscr{B}, \mathscr{D})}$.*

Proof. Let $R: \mathscr{B} \to \mathscr{C}$ be the inclusion. We show:

(a) Every additive functor $F': \mathscr{B} \to \mathscr{D}$ can be extended (in at least one way) to an additive functor $F: \mathscr{C} \to \mathscr{D}$.
(b) If $F, G: \mathscr{C} \to \mathscr{D}$ are additive, then every natural transformation ξ: $F R \to G R$ extends uniquely to a natural transformation $\alpha: F \to G$.

Q is defined on objects by choosing extensions as in (a) and is then completely determined by (b). From (a) and (b) it follows that $\tilde{R}$ is an equivalence by 16.3.6, and Q is equivalence-inverse to $\tilde{R}$ by 16.2.3 (e).

(a) For every object A of $\mathcal{C}$, choose a representation $A = \bigoplus X_j$ of A as a finite biproduct of objects of $\mathcal{B}$ with projections pr_j and injections in_j, where the objects of $\mathcal{B}$ are biproducts with only one factor and identity morphisms as projections and injections. Now every morphism in $\mathcal{C}$ is described by exactly one matrix whose elements are morphisms in $\mathcal{B}$ (compare 12.2.1). If $\bigoplus X_j$ is the representation chosen for A, set $F(A) = \bigoplus F'(A_j)$ (with choice of a biproduct in $\mathcal{D}$), where $F(A) = F'(A)$, provided A belongs to $\mathcal{B}$. One obtains F for morphisms by applying F' to the elements of the matrices describing the morphisms in $\mathcal{C}$. That F is an additive functor follows from the multiplication and addition of matrices, and $F' = F\,R$.

(b) Choose a representation as a biproduct as in (a) for the objects of $\mathcal{C}$. If $\alpha_A\colon F(A) \to G(A)$ for $A = \bigoplus X_j$ exists, then

(1) $$\alpha_A\, F(in_j) = G(in_j)\, \xi_{X_j}$$

is valid. Since $F(\bigoplus X_j)$ is a coproduct with injections $F(in_j)$ (12.2.7), α_A is uniquely determined by (1). Now we define α_A by (1). Here $\alpha_A = \xi_A$ for $A \in |\mathcal{B}|$ by the choice of A as a biproduct. Then let $B = \bigoplus Y_k$, with injections in_k and projections pr_k, be the representation chosen for $B \in |\mathcal{C}|$ and let $f\colon A \to B$ be a morphism in $\mathcal{C}$. For $g = f\, in_j\colon X_j \to B$

(2) $$g = \sum_k in_k\, pr_k\, g\colon X_j \to B$$

holds. Now $\xi_{Y_k}\, F(pr_k\, g) = G(pr_k\, g)\, \xi_{X_j}$. By (1) for B

$$\alpha_B\, F(in_k\, pr_k\, g) = G(in_k\, pr_k\, g)\, \xi_{X_j}$$

follows, and since F and G are additive,

$$\alpha_B\, F(f)\, F(in_j) = \alpha_B\, F(g) = G(g)\, \xi_{X_j} = G(f)\, \alpha_A\, F(in_j)$$

follows by (2), the last equation being implied by (1). Since $F(A)$ is a coproduct with injections $F(in_j)$, it follows further that $\alpha_B\, F(f) = G(f)\, \alpha_A$, which proves the statement in (b).

16.3.10 Remarks. If $\mathcal{C}$ is an additive category with finite biproducts and $\mathcal{B}$ a full small subcategory, then $\mathcal{B}$ can be completed to a full small subcategory with finite biproducts by choosing for any finite number of objects of $\mathcal{B}$ a biproduct in $\mathcal{C}$. The subcategory that one obtains in this way is again small; and it has finite biproducts by 7.7.7. A (small) additive category $\mathcal{B}$ can always be completed to a (small) additive category $\mathcal{C}$ with finite biproducts: one applies the above described procedure to the Yoneda embedding $H_*\colon \mathcal{B} \to Add(\mathcal{B}^0, Ab)$.

16.4 Adjoint Functors

16.4.1 Definition. Let $S\colon \mathcal{D} \to \mathcal{C}$ and $T\colon \mathcal{C} \to \mathcal{D}$ be functors. (S, T) is called a pair of *adjoint functors* if there is an isomorphism

(1) $$\bar{\varphi}\colon [S\,\mathrm{Op}(?),\ ??]_{\mathcal{C}} \cong\!\!> [\mathrm{Op}(?),\ T(??)]_{\mathcal{D}}$$

of bifunctors $\mathcal{D}^0 \times \mathcal{C} \to \mathit{Ens}$. $\bar{\varphi}$ can also be considered as an isomorphism

(2) $$\varphi\colon [S(?),\ ??]_{\mathcal{C}} \cong\!\!> [?,\ T(??)]_{\mathcal{D}}$$

of contra-co-variant functors. In this case T is called *right adjoint* to S (by means of φ), S is called *left adjoint* to T and φ is called an *adjunction isomorphism* for (S, T). We also say that φ makes T right adjoint to S (and S left adjoint to T) and that $(\varphi, S, T, \mathcal{C}, \mathcal{D})$ is an *adjunction* or an adjoint situation.

First examples are 4.5.2, 7.5.3, 8.5.2 and 15.1.8 (12). The terms "adjoint, coadjoint" are also in use, where adjoint means right adjoint or left adjoint depending on the author. Left and right adjoint refer to the position in the Hom-functor.

16.4.2 Proposition. *Let* $(\varphi, S, T, \mathcal{C}, \mathcal{D})$ *and* $(\chi, R, U, \mathcal{B}, \mathcal{C})$ *be adjunctions. Then the following pairs of functors are adjoint pairs:*

(a) $(\mathrm{Op}\,T\,\mathrm{Op},\ \mathrm{Op}\,S\,\mathrm{Op})$,

(b) $(R\,S,\ T\,U)$.

Proof. (a) follows immediately from the definition. (b) results from the isomorphisms given by χ and φ:

$$[R\,S(?),\ ??]_{\mathcal{B}} \cong\!\!> [S(?),\ U(??)]_{\mathcal{C}} \cong\!\!> [?,\ T\,U(??)]_{\mathcal{D}}\,.$$

Remark. (a) permits statements about adjoint pairs of functors to be dualized.

16.4.3 Let $(\varphi, S, T, \mathcal{C}, \mathcal{D})$ and $(\chi, R, U, \mathcal{C}, \mathcal{D})$ be adjunctions. A natural transformation $\tau\colon T \to U$ induces a uniquely determined natural transformation $\varrho\colon R \to S$ (reversed direction!) so that

$$\begin{array}{ccc} [S(X),\ A] & \xrightarrow{\ \varphi_{X,A}\ } & [X,\ T(A)] \\ {\scriptstyle [\varrho_X,\ A]}\downarrow & & \downarrow{\scriptstyle [X,\ \tau_A]} \\ [R(X),\ A] & \xrightarrow{\ \chi_{X,A}\ } & [X,\ U(A)] \end{array}$$

is always commutative. The *transformations* ϱ and τ are called *conjugate* to each other. If τ is an isomorphism, then ϱ is also an isomorphism.

Proof. $(X, A) \mapsto [X, \tau_A]$ is a natural transformation σ between contra-co-variant functors (compare 4.5.3) and thus so is $\chi^{-1}\,\sigma\,\varphi\colon [S(?),\ ??] \to [R(?),\ ??]$. Therefore, the first statement follows from 4.5.4, the second one from 4.1.5.

16.4.4 Corollary. *In a pair of adjoint functors (S, T), each of the two functors determines the other uniquely up to an isomorphism.*

16.4.5 Proposition. *The functor $T: \mathscr{C} \to \mathscr{D}$ has a left adjoint $S: \mathscr{D} \to \mathscr{C}$ if and only if $[X, T(?)]_{\mathscr{D}} : \mathscr{C} \to Ens$ is representable for every $X \in |\mathscr{D}|$.*

Proof. If $(\varphi, S, T, \mathscr{C}, \mathscr{D})$ is an adjunction, then $\varphi_X: [S(X), ?] \to$ $\to [X, T(?)]$ is a representation. The converse follows from 4.5.1.

Remark. If $S(X) \in |\mathscr{C}|$ and $\eta_X: X \to T\,S(X)$ are given, then $[X, T(?)]_{\mathscr{D}}$ is represented by $(S(X), \eta_X)$ if and only if, for every $A \in |\mathscr{C}|$ and every $u: X \to T(A)$ in $\mathscr{D}$, there is exactly one $f: S(X) \to A$ in $\mathscr{C}$ with $u = T(f)\, \eta_X$.

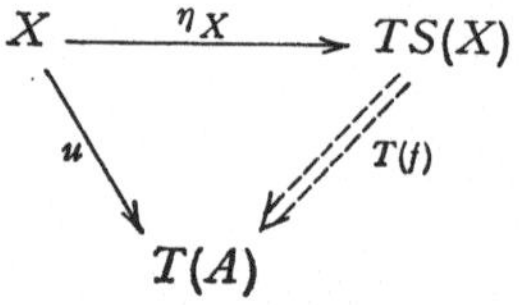

Since $u \in [X, T(A)]$ and $T(f)\,\eta_X = [X, T(f)]\,(\eta_X)$, this follows immediately from 4.4.2.

16.4.6 Proposition. *If $(\varphi, S, T, \mathscr{C}, \mathscr{D})$ is an adjunction, then T preserves all limits in $\mathscr{C}$ (including large ones), in particular monomorphisms, and S colimits, in particular epimorphisms.*

A right adjoint functor preserves limits, a left adjoint one colimits.

Proof. $[X, T(?)]$ is representable and preserves limits for every $X \in |\mathscr{D}|$ by 7.7.4. By 7.7.5, T preserves limits. The statement for S is dual (16.4.2 (a)).

16.4.7 Proposition. *Let $\mathscr{C}$ be a complete category. A functor $T: \mathscr{C} \to \mathscr{D}$ has a left adjoint if and only if it preserves limits and if $[X, T(?)]_{\mathscr{D}}$ is proper for every $X \in |\mathscr{D}|$. If $\mathscr{C}$ is in addition well-powered and possesses a cogenerating set, then T has a left adjoint if and only if it preserves limits.*

Proof. By 7.7.5, T preserves limits if and only if this is the case for all $H^X T$ with $X \in |\mathscr{D}|$. Thus the first statement follows from 16.4.5 and 10.3.9, the second one from 10.6.5.

16.4.8 Proposition. *Let $\mathscr{C}$ be a well-powered complete category with a cogenerating set. Then $\mathscr{C}$ is cocomplete.*

Proof. Let Σ be any small category, and let $T: \mathscr{C} \to [\Sigma, \mathscr{C}]$ be the functor $T(A) = A_\Sigma$, $T(f) = f_\Sigma$. Since limits are constructed "pointwise" in $[\Sigma, \mathscr{C}]$, T preserves limits, also for $\Sigma = \emptyset$. By 16.4.7, T has a left adjoint $S: [\Sigma, \mathscr{C}] \to \mathscr{C}$, which is determined uniquely up to an isomorphism by 16.4.4. A comparison with 8.5.2 shows that every functor $F: \Sigma \to \mathscr{C}$ has a colimit with colimit object $S(F)$.

16.4.9 Remark. By 12.4.4, an abelian category is well-powered if and only if it is co-well-powered. It is balanced. If it has a cogenerating set, then it is co-well-powered by the duals of 12.4.1 and 10.6.3. Therefore, well-powered can be replaced by abelian in 16.4.8.

16.5 Quasi-inverse Adjunction Transformations

16.5.1 For an adjunction $(\varphi, S, T, \mathcal{C}, \mathcal{D})$, the representation φ_X: $[S(X), ?]_{\mathcal{C}} \to [X, T(?)]_{\mathcal{D}}$ is described by $(S(X), \eta_X)$ as in 4.4.1, where

(1) $$\eta_X = \varphi_{X, S(X)}(1_{S(X)}) \colon X \to T\,S(X)\,.$$

In this case 4.2 (2) states: for $f \in [S(X), A]$, $\varphi_{X,A}(f) = [X, T(f)]\,(\eta_X)$, and this is $T(f)\,\eta_X = [\eta_X, T(A)]\;T(f)$. Therefore,

(2) $$\begin{array}{ccc} [S(X), A] & \xrightarrow{T_{S(X),A}} & [TS(X), T(A)] \\ & {\scriptstyle \varphi_{X,A}} \searrow \quad \swarrow {\scriptstyle [\eta_X, T(A)]} & \\ & [X, T(A)] & \end{array} \qquad \begin{array}{ccc} f & \longmapsto & T(f) \\ & \searrow \quad \swarrow & \\ & \varphi_{X,A}\,(f) = T(f)\,\eta_X & \end{array}$$

is commutative. $T_{S(X),A}$ is injective, because $\varphi_{X,A}$ is bijective.

16.5.2 Proposition. *For the adjunction* $(\varphi, S, T, \mathcal{C}, \mathcal{D})$ *the universal elements* η_X *of the representations* φ_Y: $[S(X), ?] \to [X, T(?)]$ *form a natural transformation*

$$\eta\colon 1_{\mathcal{D}} \to T\,S\,.$$

Proof. For an arbitrary $u\colon X \to Y$ in $\mathcal{D}$

$$S(u) = [S(X), S(u)]\,(1_{S(X)}) = [S(u), S(Y)]\,(1_{S(Y)})\,.$$

Applying φ, one gets $[X, T\,S(u)]\,(\eta_X) = [u, T\,S(Y)]\,(\eta_X)$; i.e. $T\,S(u)\,\eta_X = \eta_Y\,u\colon X \to T\,S(Y)$, and this is the statement. It is equivalent to

$$\begin{array}{ccc} [X, Y]_{\mathcal{D}} & \xrightarrow{(TS)_{X,Y}} & [TS(X), TS(Y)]_{\mathcal{D}} \\ & {\scriptstyle [X, \eta_Y]} \searrow \quad \swarrow {\scriptstyle [\eta_X, TS(Y)]} & \\ & [X, TS(Y)]_{\mathcal{D}} & \end{array}$$

being commutative for all $X, Y \in |\mathcal{D}|$. Factoring $T\,S$ yields, by (2),

(3) $$\begin{array}{ccc} [X, Y]_{\mathcal{D}} & \xrightarrow{S_{X,Y}} & [S(X), S(Y)]_{\mathcal{C}} \\ {\scriptstyle [X,\eta_Y]} \downarrow & \swarrow {\scriptstyle \varphi_{X,S(Y)}} & \downarrow {\scriptstyle T_{S(X),S(Y)}} \\ [X, TS(Y)]_{\mathcal{D}} & \xleftarrow[{[\eta_X, TS(Y)]}]{} & [TS(X), TS(Y)]_{\mathcal{D}} \end{array}$$

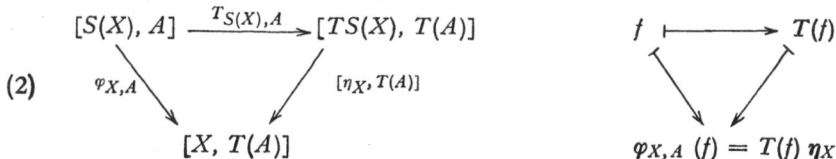

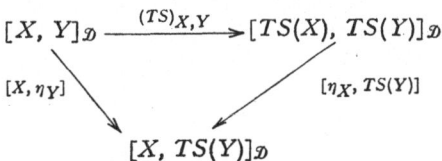

Remark. As the proofs show, (2) and (3) are already valid if φ: $[S(?), ??]_{\mathcal{C}} \to [?, T(??)]_{\mathcal{D}}$ is a natural transformation. If, conversely, a natural transformation $\eta\colon 1_{\mathcal{D}} \to T\,S$ is given, then φ can be defined by (2) as a composite of two natural transformations between contra-co-variant functors. Here (1) is valid again.

16.5.2⁰ Dualizing yields the result that the functor $S\colon \mathcal{D} \to \mathcal{C}$ has a right adjoint if and only if the contravariant functor $[S(?), A]_{\mathcal{C}}$ is representable for every $A \in |\mathcal{C}|$. Taking $\psi = \varphi^{-1}$, such a representation $\psi_A\colon [?, T(A)]_{\mathcal{D}} \to [S(?), A]_{\mathcal{C}}$ is described by $(T(A), \varepsilon_A)$, where

$$(1^0) \qquad \varepsilon_A = \psi_{A,\,T(A)}(1_{T(A)})\colon S\,T(A) \to A\,.$$

These universal elements form a natural transformation $\varepsilon\colon S\,T \to 1_{\mathcal{C}}$, and the two diagrams below are commutative.

$$(2^0) \qquad \begin{array}{ccc} [X, T(A)]_{\mathcal{D}} & \xrightarrow{S_{X,T(A)}} & [S(X), ST(A)]_{\mathcal{C}} \\ & {\scriptstyle\psi_{A,X}} \searrow \quad \swarrow {\scriptstyle [S(X),\varepsilon_A]} & \\ & [S(X), A]_{\mathcal{C}} & \end{array} \qquad \begin{array}{ccc} u & \longmapsto & S(u) \\ & \searrow \quad \swarrow & \\ & \psi_{A,X}(u) = \varepsilon_A\, S(u) & \end{array}$$

$$(3^0) \qquad \begin{array}{ccc} [A, B]_{\mathcal{C}} & \xrightarrow{T_{A,B}} & [T(A), T(B)]_{\mathcal{D}} \\ {\scriptstyle [\varepsilon_A, B]}\downarrow & \swarrow {\scriptstyle \psi_{B,T(A)}} & \downarrow {\scriptstyle S_{T(A),T(B)}} \\ [ST(A), B]_{\mathcal{C}} & \xleftarrow{[ST(A),\varepsilon_B]} & [ST(A), ST(B)]_{\mathcal{C}} \end{array}$$

16.5.3 Proposition. *Using the notation from above, the following statements are equivalent:*

(a) *T is faithful.*
(b) *T reflects epimorphisms.*
(c) *ψ preserves epimorphisms.*
(d) *Every ε_A is an epimorphism.*

Proof. (b) follows from (a) by 13.3.5. Let (b) be satisfied and let $u\colon X \to T(A)$ be an epimorphism. Then by (2), $u = T(\varphi^{-1}_{X,A}(u))\, \eta_X$. Therefore, $T(\varphi^{-1}_{X,A}(u))$, and thus $\varphi^{-1}_{X,A}(u) = \psi_{A,X}(u)$ are epimorphisms. (d) follows from (c) by (1⁰). (a) follows from (d) by (3⁰) and 5.1.4⁰ since ψ is an isomorphism.

16.5.4 Proposition. *With the same notation, the following hold:*

(a) *T is full if and only if all ε_A are coretractions.*
(b) *Let T be full and $X \in |\mathcal{D}|$. If $\eta_X\colon X \to TS(X)$ is a coretraction, then η_X is an isomorphism.*
(c) *If T is full and every η_X a monomorphism, then all η_X are bimorphisms.*
(d) *T is fully faithful if and only if ε is an isomorphism.*

Proof. (a) follows from (3⁰) and the dual of 5.2.4.

(b) Let $r\,\eta_X = 1_X$. Then $\eta_X\, r\, \eta_X = \eta_X$. Since T is full, there is an $f\colon S(X) \to S(X)$ with $T(f) = \eta_X\, r$. By (2) and $T(f)\,\eta_X = \eta_X =$

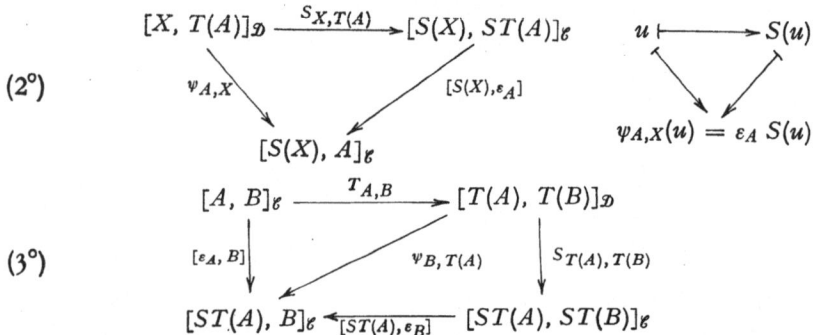

$= T(1_{S(X)})\,\eta_X,\ f = 1_{S(X)}$. Therefore, $\eta_X\, r = 1_{TS(X)}$, and r is inverse to η_X.

(c) For $u, v\colon T\,S(X) \to Y$, let $u\,\eta_X = v\,\eta_X$. $\eta_Y\, u\,\eta_X = \eta_Y\, v\,\eta_X$ follows. Since T is full, there are $f, g\colon S(X) \to S(Y)$ with $T(f) = \eta_Y\, u$, $T(g) = \eta_Y\, v$. Thus (2) implies $\varphi_{X, S(Y)}(f) = \varphi_{X, S(Y)}(g)$ and further $f = g$, hence $\eta_Y\, u = \eta_Y\, v$. Since η_Y is a monomorphism, $u = v$, and thus η_X is an epimorphism.

(d) follows from (a), 16.5.3 and 5.3.4.

Remarks. By the dual of 16.5.3, all η_X are monomorphic if and only if S is faithful. We shall make frequent use of statement (d).

16.5.5 Proposition. *If $(\varphi, S, T, \mathscr{C}, \mathscr{D})$ is an adjunction, then the corresponding natural transformations $\eta\colon 1_{\mathscr{D}} \to T\,S$, $\varepsilon\colon S\,T \to 1_{\mathscr{C}}$ satisfy*

(4) $$(T * \varepsilon)\,(\eta * T) = 1_T\,,$$

(4°) $$(\varepsilon * S)\,(S * \eta) = 1_S\,.$$

If S or T is full, then $\eta * T$ and $S * \eta$ are isomorphisms with inverses $T * \varepsilon$ or, resp., $\varepsilon * S$.

Proof. At $A \in |\mathscr{C}|$, the left side of (4) is $T(\varepsilon_A)\,\eta_{T(A)}$. By (2) and (1°), this is $\varphi_{T(A), A}(\varepsilon_A) = 1_{T(A)}$. (4°) is dual to (4). Now let S be full. Then, by the dual of 16.5.4 (a) $\eta_{T(A)}$ is a retraction, by (4) it is a monomorphism (even a coretraction), so that it is an isomorphism by 5.3.4. Similarly, $S(\eta_X)$ is a monomorphic retraction by (4°). The case where T is full is dual.

16.5.6 Remark. If S or T is full and $X \in |\mathscr{D}|$, then η_X is an isomorphism if and only if X is isomorphic to an object $T(A)$. For, by 16.5.5, $\eta_{T(A)}$ is an isomorphism, and for an isomorphism $u\colon X \to T(A)$, $\eta_X = TS(u^{-1})\,\eta_{T(A)}\,u\colon\ X \to TS(X)$ is an isomorphism. Conversely, $TS(X)$ is of the form $T(A)$. There is a dual statement for ε_A.

16.5.7 Proposition. *Let the functors $T\colon \mathscr{C} \to \mathscr{D}$ and $S\colon \mathscr{D} \to \mathscr{C}$ be given with a natural transformation $\eta\colon 1_{\mathscr{D}} \to T\,S$.* (2) *then defines a natural transformation $\varphi\colon [S(?), ??]_{\mathscr{C}} \to [?, T(??)]_{\mathscr{D}}$. It is an isomorphism if and only if there is a natural transformation $\varepsilon\colon S\,T \to 1_{\mathscr{C}}$ such that* (4) *and* (4°) *are valid.*

Proof. Suppose that η and ε exist. As the remark in 16.5.2 points out, φ and ψ are defined by (2) and (2°). Furthermore, (3) and (3°) are valid. If, in addition (4) holds, then $\varphi\,\psi\colon [?, T(??)]_{\mathscr{D}} \to [?, T(??)]_{\mathscr{D}}$ is an isomorphism, since for $u\colon X \to T(A)$ one has

$$\varphi_{X,A}\big(\psi_{A,X}(u)\big) \overset{(2^\circ)}{=\!=} \varphi_{X,A}\big(\varepsilon_A\, S(u)\big) \overset{(2)}{=\!=}$$
$$T(\varepsilon_A)\; TS(u)\,\eta_X \overset{(3)}{=\!=} T(\varepsilon_A)\,\eta_{T(A)}\,u \overset{(4)}{=\!=} 1_{T(A)}\,u\,.$$

From (4°) it follows dually that $\psi_{A,X}\big(\varphi_{X,A}(f)\big) = f$ for $f\colon S(X) \to A$. This together with 16.5.5 completes the proof.

16.5.8 Definition. 16.5.5 and 16.5.7 are the reason why ε and η are called *quasi-inverse adjunction transformations.*

16.5.9 Remark. Equivalence-inverse pairs of functors are a special case of adjoint functors; more exactly: $S\colon \mathcal{D} \to \mathcal{C}$ and $T\colon \mathcal{C} \to \mathcal{D}$ are equivalence-inverse if and only if (S, T) is a pair of adjoint functors in which T is fully faithful and where $\eta\colon 1_{\mathcal{D}} \to T\,S$ is an isomorphism.

Proof. If S, T are equivalence-inverse, then T is fully faihtful by 16.2.4 (a), and there is an isomorphism $\eta\colon 1_{\mathcal{D}} \to T\,S$. (2) then defines an adjunction isomorphism. (Note that the ε given by 16.5.2^0 need not be the Ψ specified in 16.2.1.) The converse follows immediately from 16.3.6 or from 16.5.4 (d).

16.5.10 Proposition. *Let $(\varphi, S, T, \mathcal{C}, \mathcal{D})$ be an adjunction, where $\mathcal{C}$ and $\mathcal{D}$ are additive categories.*

(a) *T is additive if and only if S is additive, and this is the case if and only if φ preserves addition, i.e., if it is an isomorphism of Ab-valued functors.*
(b) *If $\mathcal{C}$ and $\mathcal{D}$ have zero objects and if $\mathcal{C}$ or $\mathcal{D}$ have finite products, then T is additive.*

Proof. (a) If T is additive, then $[X, T(?)]_{\mathcal{D}}$ is additive. By 4.3, every φ_X preserves addition. By (3), this is equivalent to S being additive. The converse of this is its dual.

(b) If $\mathcal{C}$ has finite products, then T is additive by 16.4.6 and 12.2.7 The case where $\mathcal{D}$ has finite products is dual by (a).

16.5.11 Proposition. *Let $(\varphi, S, T, \mathcal{C}, \mathcal{D})$ be an adjunction with quasi-inverse adjunction transformations η, ε.*

(a) *For every category $\mathcal{A}$, $[\mathcal{A}, T]\colon [\mathcal{A}, \mathcal{C}] \to [\mathcal{A}, \mathcal{D}]$ is right adjoint to $[\mathcal{A}, S]\colon [\mathcal{A}, \mathcal{D}] \to [\mathcal{A}, \mathcal{C}]$ with quasi-inverse adjunction transformations*

$$[\mathcal{A}, \eta]\colon 1_{[\mathcal{A}, \mathcal{D}]} \to [\mathcal{A}, T]\,[\mathcal{A}, S] = [\mathcal{A}, T\,S]\,,$$

$$[\mathcal{A}, \varepsilon]\colon [\mathcal{A}, S]\,[\mathcal{A}, T] = [\mathcal{A}, S\,T] \to 1_{[\mathcal{A}, \mathcal{C}]}\,.$$

(b) *For every category $\mathcal{E}$, $[T, \mathcal{E}]\colon [\mathcal{D}, \mathcal{E}] \to [\mathcal{C}, \mathcal{E}]$ is left adjoint to $[S, \mathcal{E}]\colon [\mathcal{C}, \mathcal{E}] \to [\mathcal{D}, \mathcal{E}]$ with quasi-inverse adjunction transformations*

$$[\varepsilon, \mathcal{E}]\colon [T, \mathcal{E}]\,[S, \mathcal{E}] = [S\,T, \mathcal{E}] \to 1_{[\mathcal{C}, \mathcal{E}]}\,,$$

$$[\eta, \mathcal{E}]\colon 1_{[\mathcal{D}, \mathcal{E}]} \to [S, \mathcal{E}]\,[T, \mathcal{E}] = [T\,S, \mathcal{E}]\,.$$

Remark. If $\mathcal{C}$, $\mathcal{D}$, T, and thus also S, are additive, then the statements analogous to (a) and (b) are of course also valid for additive categories $\mathcal{A}$, $\mathcal{E}$ and the corresponding categories of additive functors (compare 16.1.4).

Proof. We restrict ourselves to the verification of (4) in case (a); one has the following equations:

$$([\mathcal{A}, T] * [\mathcal{A}, \varepsilon])\,([\mathcal{A}, \eta] * [\mathcal{A}, T]) =$$

$$[\mathcal{A}, T * \varepsilon]\,[\mathcal{A}, \eta * T] = [\mathcal{A}, (T * \varepsilon)\,(\eta * T)] = [\mathcal{A}, 1_T] = 1_{[\mathcal{A}, T]}\,.$$

16.5.12 Note. An adjunction $(\varphi, S, T, \mathcal{C}, \mathcal{D})$ leads to the functor $R = T\,S\colon \mathcal{D} \to \mathcal{D}$ with natural transformations

(1) $$\eta\colon 1_{\mathcal{D}} \to R \quad \text{and} \quad \mu\colon R\,R \to R\,,$$

where $\mu = T * \varepsilon * S$. They satisfy the following conditions:

(2) $$\mu\,(\eta * R) = \mu\,(R * \eta) = 1_R\,,$$

(3) $$\mu\,(\mu * R) = \mu\,(R * \mu)\,,$$

(4) $$(\eta * R) * \eta = (R * \eta) * \eta\,.$$

Here (2) follows from 16.5.5 by 16.1(4). 16.1(5) implies (4), since $\eta = \eta * 1_{\mathcal{D}} = 1_{\mathcal{D}} * \eta$. Similarly, $\varepsilon\,(\varepsilon * ST) = \varepsilon\,(ST * \varepsilon)$ which implies (3) by 16.1 (4).

If S or T is full, then μ is an isomorphism by 16.5.5.

In chapter 21 we shall discuss such "triples" (R, η, μ) where for a functor $R\colon \mathcal{D} \to \mathcal{D}$ and for natural transformations (1) equations (2) and (3) are satisfied.

16.6 Fully Faithful Adjoints

16.6.1 Theorem. *Let* $(\varphi, S, T, \mathcal{C}, \mathcal{D})$ *be an adjunction and* $T\colon \mathcal{C} \to \mathcal{D}$ *fully faithful.*

(a) *If* $R\colon \Sigma \to \mathcal{C}$ *is a diagram and* (L, λ) *a limit or, resp., colimit of* TR, *then* $\big(S(L), (\varepsilon * R)\,(S * \lambda)\big)$ *is a limit or, resp.,* $\big(S(L), (S * \lambda)\,(\varepsilon^{-1} * R)\big)$ *a colimit of* R.

(b) *If* $\mathcal{D}$ *is complete or, resp., finitely complete, cocomplete, finitely cocomplete, then the same is true for* $\mathcal{C}$.

Warning. (a) does not mean that S preserves limits.

Proof. (a) If (L, λ) is a colimit of TR, then $(S(L), S * \lambda)$ is a colimit of STR, since S preserves colimits by 16.4.6. By 16.5.4, ε is an isomorphism and, therefore, $\big(S(L), (S * \lambda)\,(\varepsilon^{-1} * R)\big)$ is a colimit of R.

Now let (L, λ) be a limit of TR. By 16.5.2 (3),

(1) $$\begin{array}{ccc} L_\Sigma & \xrightarrow{\ \lambda\ } & TR \\ {\scriptstyle (\eta_L)_\Sigma}\downarrow & & \downarrow{\scriptstyle \eta * TR} \\ TS(L)_\Sigma & \xrightarrow{(TS)*\lambda} & TSTR \end{array}$$

is commutative. By 16.5.5, $\eta * (T\,R) = (\eta * T) * R$ is an isomorphism and, by (1), $\big(L, ((T\,S * \lambda)\,(\eta_L)_\Sigma\big)$ is a limit of $T\,S\,T\,R$. By the definition of limits, there is exactly one morphism $u\colon\ TS(L) \to L$ with $TS * \lambda = ((TS * \lambda))\,(\eta_L)_\Sigma\, u_\Sigma$. Composing on the right with $(\eta_L)_\Sigma$ produces, again by the definition of limits, $u\,\eta_L = 1_L$, so that $S(u) \circ S(\eta_L) = 1_{S(L)}$ certainly holds. By 16.5.5, $S(\eta_L)$ is an isomorphism and $S(u) = \varepsilon_{S(L)}$. By 16.5.2 (3), $\eta_L\, u = TS(u)\,\eta_{TS(L)}$, and $\eta_L\, u = T(\varepsilon_{S(L)})\,\eta_{TS(L)} = 1_{TS(L)}$ follows from 16.5.5. Thus η_L is inverse to u and therefore, by (1), $(T\,S(L),\ T\,S * \lambda)$ is a limit of $TSTR$. Since T reflects limits (7.7.6), $(S(L),\ S * \lambda)$ is a limit of STR. By 16.5.4, ε is an isomorphism, which proves the statement in (a). (b) follows immediately from (a).

16.6.2 Corollary. *Let $(\varphi, S, T, \mathscr{C}, \mathscr{D})$ be an adjunction, T fully faithful and let S preserve finite limits. Further, let $\mathscr{D}$ be finitely complete.*

(a) *If $\mathscr{D}$ has filtered colimits and if these commute with finite limits, then the same is true for $\mathscr{C}$.*

(b) *If $\mathscr{D}$ has colimits and if colimits in $\mathscr{D}$ are universal* (9.5.5), *then the same is true for $\mathscr{C}$.*

Proof. (a) By 16.6.1, $\mathscr{C}$ is finitely complete and has filtered colimits. Then, by 10.1.2, it is sufficient to show that filtered colimits commute with pullbacks. If

$$\begin{array}{ccc} P & \to & R \\ \downarrow & & \downarrow \\ M & \to & N \end{array} \tag{2}$$

is a pullback in $[\mathscr{X}, \mathscr{C}]$, where $\mathscr{X}$ is a small filtered category, then, by applying T, one gets a pullback in $[\mathscr{X}, \mathscr{D}]$ and, by applying $S\,T$, again a pullback in $[\mathscr{X}, \mathscr{C}]$ by the assumption about S. ε^{-1} gives rise to an isomorphism between (2) and the pullback created by applying $S\,T$. This, together with 16.6.1, completes the proof, again because S preserves finite limits. (b) follows similarly.

16.6.3 Proposition. *Let $T\colon \mathscr{C} \to \mathscr{D}$ be a fully faithful functor.*

(a) *The following two statements are equivalent:*

 (i) *T has a left adjoint S, and $\eta_X \to TS(X)$ is an epimorphism for all $X \in |\mathscr{D}|$.*

 (ii) *For every $X \in |\mathscr{D}|$, there is an epimorphism $\eta_X\colon X \to T(B_X)$ with a suitable $B_X \in |\mathscr{C}|$ such that every morphism $u\colon X \to T(A)$ for an arbitrary $A \in |\mathscr{C}|$ can be factored through η_X.*

(b) *If $\mathscr{C}$ has products and if $\mathscr{D}$ is co-well-powered, then the following is equivalent to* (i):

 (iii) *T preserves products and every morphism of the form $u\colon X \to T(A)$ factors through an epimorphism $u'\colon X \to T(A_u)$ for a suitable $A_u \in |\mathscr{C}|$.*

(c) *Let the assumptions in* (b) *be satisfied. Further, let $\mathfrak{D}$ have equalizers and let every morphism in $\mathfrak{D}$ factor into an epimorphism followed by an equalizer. Then the following is equivalent to* (i):

(iv) *T preserves products and if $K \to T(A)$ is an equalizer in $\mathfrak{D}$, then K is isomorphic to an object of the form $T(C)$.*

Proof. (a) (ii) follows from (i) with $B_X = S(X)$ since $u = T(\varphi_{X,A}^{-1}(u)) \circ \circ\, \eta_X$, by 16.5.1. Now let (ii) be valid. Set $u = \bar{u}\,\eta_X$: $X \to T(B_X) \to \to T(A)$. $\bar{u}$ is determined uniquely by u, since η_X is an epimorphism. As T is fully faithful, there is exactly one f: $B_X \to A$ with $T(f) = \bar{u}$. By 16.4.5, (B_X, η_X) is a representation of $[X, T(?)]_{\mathfrak{D}}$, and (i) is true.

(b) (iii) follows immediately from (i) and (ii). Now let (iii) be valid. We consider epimorphisms of the form $X \to T(A)$ for $X \in |\mathfrak{D}|$. By assumption, there is among these a set $\{p_i\colon X \twoheadrightarrow T(A_i)\}$ such that every other one is equivalent to some p_i. Since $\mathscr{C}$ has products which are preserved by T, there is a morphism p: $X \to T(\prod A_i)$ with $T(pr_i)\, p = p_i$. It factors through an epimorphism p': $X \to T(A_p)$. If u: $X \to T(A)$ is any morphism, then u factors through some p_i and thus also through p and through p'. Thus (ii) and therefore (i) follows.

(c) (iv) implies (iii) immediately and thus (i) also. If (i) is satisfied, then T preserves products and the rest of the statement is a special case of the following proposition.

16.6.4 Proposition. *Let $(\varphi, S, T, \mathscr{C}, \mathfrak{D})$ be an adjunction and (L, λ) a limit of a diagram R: $\Sigma \to \mathfrak{D}$ with the following property:*

If, for $i \in |\Sigma|$, $R(i)$ is not isomorphic to an object of the form $T(A_i)$, then there is an arrow $j \to i$ in Σ such that $R(j)$ is isomorphic to an object of the form $T(A_j)$.

Further, let η_L: $L \to TS(L)$ be an epimorphism. Then η_L is an isomorphism

Proof. If Σ is empty, then L is terminal and η_L is an epimorphic coretraction and as such an isomorphism. Now let Σ be non-empty. One may assume that R satisfies the following condition:

If $R(i)$ does not have the form $T(A_i)$, then there is an arrow a: $j \to i$ such that $R(j)$ is of the form $T(A_j)$.

This can be achieved through replacing R by an isomorphic diagram and through changing λ: $L_\Sigma \to R$ accordingly.

We define morphisms λ_i': $TS(L) \to R(i)$ with $\lambda_i = \lambda_i'\,\eta_L$ as follows:

(a) If $R(i) = T(A_i)$ for a suitable $A_i \in |\mathscr{C}|$, then, by 16.5.1, there is exactly one morphism f_i: $S(L) \to A_i$ with $T(f_i)\,\eta_L = \lambda_i$. We set $\lambda_i' = T(f_i)$.

(b) If $R(i)$ does not have the form $T(A)$, we choose an arrow a: $j \to i$ such that $R(j) = T(A_j)$ for a suitable A_j and we set $\lambda_i' = R(a)\,\lambda_j'$, where λ'_j is determined as in (a). Then $\lambda'_i \eta_L = \lambda_i$ because $R(a)\,\lambda'_j \eta_L = R(a)\,\lambda_j = \lambda_i$.

Now $\{\lambda'_i\}$ is a natural transformation λ': $TS(L) \to R$. For, if b: $i \to k$ is any arrow in Σ, then $\lambda'_k \eta_L = \lambda_k = R(b)\,\lambda_i = R(b)\,\lambda'_i \eta_L$ and thus $\lambda'_k = R(b)\,\lambda'_i$, since η_L is an epimorphism.

Since (L, λ) is a limit of R, there is a morphism u: $TS(L) \to L$ with $\lambda' = \lambda\, u_\Sigma$. $\lambda = \lambda'(\eta_L)_\Sigma = \lambda(u\,\eta_L)_\Sigma$ and thus $u\,\eta_L = 1_L$ follows, again by the property of limits. Thus η_L is an epimorphic coretraction and thus an isomorphism.

Remarks. If T is the inclusion of a subcategory and η: $1_{\mathcal{D}} \to T\,S$ an epimorphism at every place $X \in |\mathcal{D}|$, then what this proposition says is: if a diagram of $\mathcal{D}$ starts from $\mathcal{C}$, and if it has a limit, then there is a limit object in $\mathcal{C}$. Note also 16.5.6. In the proof of the proposition, the only use that was made of the adjunction was the representation of $[L, T(\,?)]_{\mathcal{D}}$ by $(S(L), \eta_L)$. Note also that the proposition is not a special case of the dual of 9.2.3.

16.6.5 In applications, the inclusion T: $\mathcal{C} \to \mathcal{D}$ of a subcategory is often considered. The investigation of an arbitrary functor T: $\mathcal{C} \to \mathcal{D}$ can be reduced to this special case by 16.2.6 through composition with an equivalence. If T is faithful or, resp., full, fully faithful, and if T preserves (finite) limits or colimits, then corresponding statements hold after composition with an equivalence (16.2.4). (Finite) completeness, cocompleteness and commutativity statements for $\mathcal{D}$ are preserved by the transition to an equivalent category. For a fully faithful functor, 16.3.8 can also be applied. We shall return to this case in detail in 19.4.

Let $\mathcal{C}$ be a full subcategory of $\mathcal{D}$ and T: $\mathcal{C} \to \mathcal{D}$ the inclusion. If T has a left adjoint S: $\mathcal{D} \to \mathcal{C}$, then S can be chosen such that S is the identity on $\mathcal{C}$; i.e., such that $S\,T = 1_{\mathcal{C}}$. By 16.5.4 (d), this is a special case of the following proposition.

16.6.6 Proposition. *Let R: $\mathcal{C} \to \mathcal{E}$ and S: $\mathcal{D} \to \mathcal{E}$ be functors and T: $\mathcal{C} \to \mathcal{D}$ a functor which is injective on the object classes. If $S\,T$ is isomorphic to R, then there is a functor S' which is isomorphic to S and such that $S'\,T = R$.*

Proof. Let ξ: $S\,T \to R$ be an isomorphism. We set $S'(T(A)) = R(A)$, $\alpha_{T(A)} = \xi_A$ for all $A \in |\mathcal{C}|$ and $S'(X) = S(X)$, $\alpha_X = 1_{S(X)}$ for all those $X \in |\mathcal{D}|$ that are not of the form $T(A)$. For u: $X \to Y$ in $\mathcal{D}$, set $S'(u) = \alpha_Y\, S(u)\, \alpha_X^{-1}$. Then S' is a functor, α: $S \to S'$ an isomorphism and $S'\,T = R$.

16.6.7 If T: $\mathcal{C} \to \mathcal{D}$ is an inclusion, then a left adjoint S : $\mathcal{D} \to \mathcal{C}$ is called a *coreflection* (Mitchell, for instance) or a *reflection* (Freyd, for instance) depending on the author. We call a subcategory $\mathscr{C}$ of $\mathscr{D}$ *epireflective* if it is full and if for the inclusion $T: \mathscr{C} \to \mathscr{D}$ 16.6.3 (i) is valid. As shown above, a left adjoint can be chosen such that $S\,T = 1_{\mathscr{C}}$. We call it an *epireflector*.

For a complete category $\mathscr{D}$, every epireflective subcategory is complete by 16.6.1. If in addition every morphism in $\mathscr{D}$ factors into an epimorphism followed by an equalizer, and if $\mathscr{D}$ is co-well-powered, then the epireflective subcategories are characterized by 16.6.3 (c).

16.6.8 Applications. Let a full subcategory in the category *Top* of topological spaces and continuous maps be defined by properties that are inherited by products and subspaces. Every such subcategory is epireflective. Thus epireflective subcategories are defined by the separation axioms T_0, T_1, the Hausdorff separation axiom and also by the axioms for regularity and for complete regularity. Furthermore, if a class $\mathfrak{K}$ of spaces is given, then there is a smallest epireflective subcategory which includes $\mathfrak{K}$ and with any object all the ones isomorphic to it. Its objects are those spaces which are homeomorphic to a subspace of a product of spaces of $\mathfrak{K}$.

In the category of Hausdorff spaces analogous statements hold. Properties that are inherited by products and by closed subspaces define epireflective subcategories. This is true, e.g., for compactness. The corresponding epireflection is the Stone-Čech compactification.

In the category of uniform or, resp., separated uniform spaces (with uniformly continuous maps) epireflective subcategories are obtained in the same way. For example, in the category of uniform spaces, the subcategory of separated spaces is epireflective, as is the subcategory of complete spaces in the category of seperated spaces.

The case of locally convex vector spaces is analogous.

There are numerous other examples. To name one, in the category of groups the subcategory of abelian groups is epireflective and so is the subcategory of torsion-free groups in *Ab*.

16.6.9 A dual situation to 16.6.3 is given by the subcategory of Kelley spaces in the category of Hausdorff spaces. A Kelley space is a Hausdorff space in which a subset is closed if and only if its intersection with all compact subspaces is closed. This is the case, e.g., for locally compact spaces and for (CW-) cell complexes. Every Hausdorff topology can be refined to a Kelley topology by taking as closed sets those whose intersections with the compact sets of the existing topology are closed.

In the category of locally convex vector spaces the subcategory of bornological spaces plays the corresponding role.

Note that the dual of 16.6.1 is valid.

16.7 Tensor Products

16.7.1 The concept of an adjoint pair of functors admits the following generalization: let $\mathscr{C}, \mathscr{D}, \mathscr{M}$ be categories and $S\colon \mathscr{D} \times \mathscr{M} \to \mathscr{C}$, $T'\colon \mathscr{M}^0 \times \mathscr{C} \to \mathscr{D}$ bifunctors, where instead of T' the corresponding contra-co-variant functor T is considered. T is called *right adjoint* to S, and S *left adjoint* to T, if there is an isomorphism

$$(1) \qquad \varphi\colon [S(?, ??), ???]_{\mathscr{C}} \cong\!\!> [?, T(??, ???)]_{\mathscr{D}}$$

which is an isomorphism of trifunctors $\mathscr{D}^0 \times \mathscr{M}^0 \times \mathscr{C} \to Ens$. For every object $M \in |\mathscr{M}|$ one then has in particular a pair of adjoint functors $(S(?, M), T(M, ??))$ in the ordinary sense as in 16.4.1. 16.4.2 through 16.4.5 carry over without trouble. We formulate in particular the

Proposition. *The contra-co-variant functor T associated with the bifunctor $T'\colon \mathscr{M}^0 \times \mathscr{C} \to \mathscr{D}$ has a left adjoint $S\colon \mathscr{D} \times \mathscr{M} \to \mathscr{C}$ if and only if the functor $[X, T(M, ?)]_{\mathscr{D}}$ is representable for every pair (X, M) of objects $X \in |\mathscr{D}|$, $M \in |\mathscr{M}|$.*

16.7.2 An important special case results from the assumptions that there is a forgetful functor for $\mathscr{D}$, say $U\colon \mathscr{D} \to Ens$, that $\mathscr{M} = \mathscr{C}$, and that UT is the Hom-functor of $\mathscr{C}$. We then call the left adjoint S, if it exists, a *tensor product* and we write $X \otimes M$, $u \otimes f$ for $S(X, M)$ or, resp., $S(u, f)$. The condition that U is a forgetful functor can be weakened. Furthermore, one also talks about tensor products in more general situations — see example 2 below and later 17.7.

Example 1. In *Ens* there is the isomorphism

$$(2) \qquad [A \times B, C] \cong\!\!> [A, [B, C]]$$

as an isomorphism of trifunctors. According to 3.4.4 and 16.1.3, (2) is also valid in *cat* and analogously in *CAT* (small $\mathfrak{V}$-categories). There is also an analogue for the additive case. However, the tensor product of additive categories is not the product (see 16.7.4 below).

Example 2. With the usual tensor product for modules, there is the isomorphism

$$(3) \qquad \mathrm{Hom}_S(M \otimes_R N, G) \cong\!\!> \mathrm{Hom}_R(N, \mathrm{Hom}_S(M, G))$$

with $N \in |_R Mod|$, $G \in |_S Mod|$ and $M \in |_S Mod_R|$. There are explicit calculations to verify this. However, a proof will be furnished by 17.4.4. Note the special cases $R = \mathbf{Z}$ or $S = \mathbf{Z}$ and, in particular, the case where $R = S$ is a commutative ring so that right and left modules coincide and every module is also a bimodule. In this case all the Hom-functors in (3) have the codomain $_R Mod$.

Example 3. If X and Y are topological spaces, then one obtains from the set $[X, Y]$ of continuous maps $X \to Y$ by means of the compact-open topology a topological space ${}_{co}[X, Y]$ and from it by refinement (compare 16.6.9) the Kelly space ${}_{kco}[X, Y]$. Using the corresponding refinement $X \times_k Y$ of the topological product one obtains isomorphisms

$$(4) \qquad {}_{co}[X \times_k Y, Z] \xrightarrow{\cong} {}_{co}[X, {}_{co}[Y, Z]] = {}_{co}[X, {}_{kco}[Y, Z]]\,,$$

$$(4') \qquad {}_{kco}[X \times_k Y, Z] \xrightarrow{\cong} {}_{kco}[X, {}_{kco}[Y, Z]]\,,$$

where X and Y are Kelley spaces and Z is an arbitrary Hausdorff space.

Example 4. Locally convex vector spaces (with continuous linear maps) furnish another example analogous to example 3. In the place of the compact-open topology one has here the bounded-open topology and Kelley-spaces and -refinements are replaced by the bornological ones. One obtains

$$(5) \qquad {}_{bbo}[X \otimes_b Y, Z] \xrightarrow{\cong} {}_{bbo}[X, {}_{bbo}[Y, Z]]\,,$$

where X and Y are bornological vector spaces. *bbo* denotes the bornological refinement of the bounded-open topology. $X \otimes_b Y$ is a filtered colimit of the tensor products $A \otimes B$, where A and B range over algebraic subspaces of X and Y spanned by bounded, absolutely convex sets, and provided with the corresponding seminorm.

16.7.3 In the examples 1, 3, 4, as well as in example 2 provided that $R = S$ is a commutative ring, one has tensor products in the narrow sense (also called a *closed symmetric monoidal structure* on the category). Such a tensor product for a category $\mathscr{C}$ is a bifunktor $\otimes: \mathscr{C} \times \mathscr{C} \to \mathscr{C}$ which is a tensor product in the sense given above and for which in addition the following conditions are satisfied

(i) There is an isomorphism $\otimes\,(\otimes \times 1_{\mathscr{C}}) \to \otimes\,(1_{\mathscr{C}} \times \otimes)$; i.e. $(A \otimes B) \otimes C \cong A \otimes (B \otimes C)$ (associativity).

(ii) There is an isomorphism $\otimes\,\tau \to \otimes$, where τ interchanges the factors; i.e. $A \otimes B \cong B \otimes A$ (commutativity).

(iii) There is an object J and an isomorphm $J \otimes\ ? \to 1_{\mathscr{C}}$ (neutral object).

(iv) The isomorphisms (i), (ii), (iii) are compatible in a sense that can be made precise (coherence). This means, roughly speaking, that in combining them they can be treated as if they were algebraic identities.

The object J postulated in (iii) has the following meanings: in example 1 it is a terminal set or, resp., category; in example 2 for $R = S$ commutative it is the ring R as a module over itself; in example 3 it is a one-point space and in example 4 a one-dimensional space.

We restrict ourselves to these remarks and refer to MacLane [60, 61], Eilenberg-Kelly [36], Kelley [50], Linton [55], Bénabou [27] and Dubuc [11] for further details.

16.7.4 If $\mathcal{B}$ and $\mathcal{C}$ are additive categories, one obtains the tensor product $\mathcal{B} \otimes \mathcal{C}$ as follows: objects are pairs (B, C) with $B \in |\mathcal{B}|$, $C \in |\mathcal{C}|$, and we set $[(B, C), (B', C')] = [B, B'] \otimes [C, C']$, where this tensor product is taken in Ab. Its elements can be represented (not uniquely) in the form $\sum_{i=1}^{n} u_i \otimes f_i$. Composition with $\sum_{j=1}^{m} u'_j \otimes f'_j \in [B', B''] \otimes [C', C'']$ is defined by $\sum_{i=1}^{n} \sum_{j=1}^{m} u'_j u_i \otimes f'_j f_i$, where it can be verified that the composite is uniquely determined. $\mathcal{B} \otimes \mathcal{C}$ is again an additive category. If $\mathcal{D}$ is also additive, then there is an isomorphism between the category of biadditive functors from $\mathcal{B} \times \mathcal{C}$ to $\mathcal{D}$ with $Add\,(\mathcal{B} \otimes \mathcal{C}, \mathcal{D})$, and thus $Add(\mathcal{B} \otimes \mathcal{C}, \mathcal{D}) \overset{\cong}{\Rightarrow} Add(\mathcal{B}, Add(\mathcal{C}, \mathcal{D}))$ as in 3.8.1. The tensor product of (small) additive categories is a tensor product in the category of (small) additive categories with additive functors.

16.8 Problems

16.8.1 Check 16.1.

16.8.2 What are the skeletons of a preordered class (as a category)?.

16.8.3 Describe adjunctions $(\varphi, S, T, \mathcal{C}, \mathcal{D})$, where $\mathcal{C}$ and $\mathcal{D}$ are preordered classes. What do 16.5.5 and 16.6.1 say in this case? Show that $T\,S\,T \cong T$ and $S\,T\,S \cong S$.

Remark. When considering Op S, T Op, one often speaks of a *Galois correspondence* in this case.

16.8.4 Let $\mathcal{D}$ be an ordered class and $\mathcal{C}$ a full reflective subcategory with reflector $S: \mathcal{D} \to \mathcal{C}$. Further, let $K: \mathcal{D} \to \mathcal{D}$ be a contravariant functor with $K\,K = 1_{\mathcal{D}}$. What can be said about S, $K\,S\,K$, and about their restrictions to $\mathcal{C}$ and $K(\mathcal{C})$? In particular, consider the case where $\mathcal{D}$ consists of the subspaces of a topological space and their inclusions and where S is the closure operator and K the transition to complements.

16.8.5 Let $h: A \to B$ be a morphism in the category $\mathcal{C}$.

(a) The rule $u \mapsto h\,u$ defines a faithful functor $S: \mathcal{C}/A \to \mathcal{C}/B$ (objects over A or, resp., B, see 6.5.3).

(b) Let $\mathcal{C}$ have pullbacks. For every morphism $v: Y \to B$ define $T(v)$ by choosing a pullback

$$\begin{array}{ccc} X & \longrightarrow & Y \\ {\scriptstyle T(v)}\big\downarrow & & \big\downarrow{\scriptstyle v} \\ A & \xrightarrow{\;h\;} & B \end{array} .$$

Then the rule $v \mapsto T(v)$ extends to a functor $T\colon \mathcal{C}/B \to \mathcal{C}/A$, and $(S, T)$is a pair of adjoint functors.

(c) If h is a monomorphism, then S is a full embedding, $\varepsilon\colon S\,T \to 1$ is a monomorphism and $\eta\colon 1 \to T\,S$ an isomorphism. Furthermore, one gets an adjunction by restricting S and T to $\mathcal{M}/A$ and $\mathcal{M}/B$ (objects are monomorphisms with codomain A or, resp., B).

16.8.6 In the category $\mathcal{C}$, let $\mathfrak{E}$ and $\mathfrak{M}$ be two classes of morphisms such that $\mathfrak{E} \cap \mathfrak{M}$ is the class of isomorphisms. Assume that every morphism f in $\mathcal{C}$ has a factorization $f = f''\,f'$, where $f' \in \mathfrak{E}$ and $f'' \in \mathfrak{M}$, and that this factorization is natural in the sense of 12.4.8, 12.4.10. Let $\mathfrak{M}/A$ be the full subcategory of $\mathcal{C}/A$ whose objects are morphisms in $\mathfrak{M}$. Then $\mathfrak{M}/A$ is a full reflective subcategory of $\mathcal{C}/A$. Taking 14.7.1 into account, generalize 14.2.5.

16.8.7 Let $(\varphi, S, T, \mathcal{C}, \mathcal{D})$ be an adjunction.

(a) For $X \in |\mathcal{D}|$, φ induces an isomorphism between $S(X)/\mathcal{C}$ (objects under $S(X)$) and X/T (see 9.2.1). Dually, $\mathcal{D}/T(A)$ and S/A are isomorphic for $A \in \mathcal{C}$.

(b) T induces a functor $T'\colon \mathcal{C}/A \to \mathcal{D}/T(A)$ which has a left adjoint S'. (Hint: for $u\colon X \to T(A)$ consider $\varepsilon_A\, S(u) = \varphi^{-1}(u)$.) Discuss also the dual case.

16.8.8 Let $\mathcal{E}$ be a fixed small category. Then $[?, \mathcal{E}]_{cat}$ can be considered as a contravariant functor $M\colon cat \to cat$. M is right adjoint to itself; i. e., there is an isomorphism

$$\varphi\colon [?, M(??)]_{cat} \xrightarrow{\cong} [??, M(?)]_{cat}$$

between contravariant bifunctors. φ can be chosen in such a way that $\varphi = \varphi^{-1}$. What does 16.8.7 become in this case? Note that the examples in 16.7.2 give rise to analogous considerations.

16.8.9 Let the functor $T\colon \mathcal{C} \to \mathcal{D}$ have a left adjoint S and a right adjoint U. Then $S\,T$ is left adjoint to $U\,T$, and $T\,S$ is left adjoint to $T\,U$. S is fully faithful if and only if U is fully faithful.

16.8.10 (Continuation of 11.7.8). (a) Let Ens_{cat} be the category of category-objects over Ens. There is an equivalence $Ens_{cat} \to cat$, which becomes an isomorphism when restricted to the full subcategory for whose objects the pullback 11.7.8 (1) is a canonically chosen one.

(b) The forgetful functor $U\colon Ens_{cat} \to Ens$ has a left adjoint. (Hint: first consider terminal objects of Ens).

(c) Let $\mathcal{C}$ be a finitely complete category, and let $(\mathcal{A}', \mathcal{A}, F_0)$, $(\mathcal{A}', \mathcal{A}, F_1)\colon \mathcal{A}' \to \mathcal{A}$ be functorial morphisms between the category objects $\mathcal{A}'$, $\mathcal{A}$ over $\mathcal{C}$. Define a natural transformation

$(\mathcal{A}', \mathcal{A}, F_0) \to (\mathcal{A}', \mathcal{A}, F_1)$ as a triple (F_0, F_1, α), where α is a $\mathcal{C}$-morphism $\alpha\colon A' \to A$ satisfying

(13) $$d_0 \alpha = d_0 F_0, \qquad d_1 \alpha = d_1 F_1$$

and

(14) $$\alpha m' = m \beta_0 = m \beta_1 .$$

Here β_0, β_1 are uniquely defined by

$$c_0 \beta_0 = F_0 c_0'; \qquad c_1 \beta_0 = \alpha c_1',$$
$$c_1 \beta_1 = F_1 c_1'; \qquad c_0 \beta_1 = \alpha c_0',$$

provided (13) is valid. (Compare 2.7.6). One thus gets a small category $[\mathcal{A}', \mathcal{A}]$ whose objects are functorial morphisms and whose morphisms are natural transformations.

(d) The category $\mathcal{C}_{cat}$ of category-objects over $\mathcal{C}$, together with the categories described in (c), provides an example of a bicategory; i.e., a category whose morphism sets are given the structure of a small category subject to certain axioms. Using 16.1, list the axioms that one is inclined to require. Every ordinary category can be looked at as a bicategory in two different ways.

Remark. Other examples suggest a more general notion of bicategory, where the laws of associativity and distributivity are valid only up to specified isomorphisms. (As an example of this: Using 16.8.7 (b), try to construct a contravariant functor $\mathcal{C} \to Cat$ whose value at $A \in |\mathcal{C}|$ is $\mathcal{C}/A$). Coherence conditions are then involved as in 16.7.3. However, this general case needs intricate definitions, so that a smooth theory is not be expected.

17. Pairs of Adjoint Functors between Functor Categories

17.1 The Kan Construction

17.1.1 Diagramm categories. Let $\mathcal{C}$ be a category. We assign to it the following diagram category $Dg(\mathcal{C})$: objects are functors $T\colon \Sigma \to \mathcal{C}$, where Σ is a small category. If $T\colon \Sigma \to \mathcal{C}$ and $T'\colon \Sigma' \to \mathcal{C}$ are objects, then a morphism $(R, \varrho)\colon T \to T'$ consists of a functor $R\colon \Sigma \to \Sigma'$ and a natural transformation $\varrho\colon T \to T' R$.

(1)

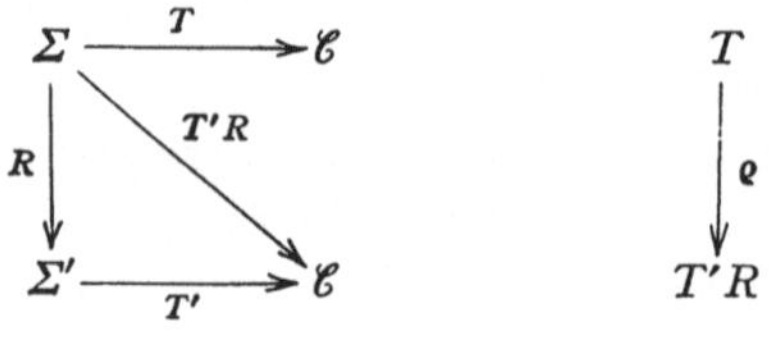

If $\mathscr{C}$ is cocomplete and if for every diagram in $\mathscr{C}$ a colimit is chosen, then a functor Colim: $Dg(\mathscr{C}) \to \mathscr{C}$ is constructed as follows. Colim $T = L$, where (L, λ) is the chosen colimit of T. Let $(R, \varrho): T \to T'$ be a $Dg(\mathscr{C})$-morphism and (L, λ), (L', λ') the colimits of T, T'. Now $\lambda' * R: T' R \to L'_\Sigma$ is a natural transformation (with $(\lambda' * R)_i = \lambda'_{R(i)}$ for $i \in |\Sigma|$) and thus so is $(\lambda' * R)\,\varrho: T \to L'_\Sigma$. By the definition of colimits, there is exactly one morphism $f: L \to L'$ with

(1) $$f_\Sigma \lambda = (\lambda' * R)\,\varrho\,.$$

We set $f =$ Colim (R, ϱ), and one verifies that this makes Colim: $Dg(\mathscr{C}) \to \mathscr{C}$ a functor.

Notice that $[\Sigma, \mathscr{C}]$ is a subcategory of $Dg(\mathscr{C})$ (always with 1_Σ for R) and that 8.6.1 also fits in here. A functor $F: \mathscr{C} \to \mathscr{D}$ induces a functor $Dg(\mathscr{C}) \to Dg(\mathscr{D})$ by the rule $T \mapsto F\,T$, $(R, \varrho) \mapsto (R, F\varrho)$. We shall call it $Dg(F)$.

17.1.1⁰ The dual situation is based on the category $Dg'(\mathscr{C})$, whose objects are again functors $T: \Sigma \to \mathscr{C}$ with a small category Σ, but where a morphism $(R, \varrho'): T \to T'$ consists of a functor $R: \Sigma \to \Sigma'$ and a natural transformation $\varrho': T' R \to T$. If $\mathscr{C}$ is complete and if limits are chosen for the diagrams, then there exists the contravariant functor Lim: $Dg'(\mathscr{C}) \to \mathscr{C}$ (compare 7.6.1).

If $F: \mathscr{C} \to \mathscr{D}$ is a contravariant functor, then F induces (covariant!) functors $Dg(\mathscr{C}) \to Dg'(\mathscr{D})$ and $Dg'(\mathscr{C}) \to Dg(\mathscr{D})$.

17.1.2 For the category $[\mathbf{2}, \mathscr{C}]$ (see 6.5.1, 6.5.2) there are the two functors $\Delta^0, \Delta^1: [\mathbf{2}, \mathscr{C}] \to \mathscr{C}$, which assign to every morphism of $\mathscr{C}$ its domain or, resp., its codomain and to every natural transformation of morphisms the corresponding $\mathscr{C}$-morphism for the domains or, resp., codomains. If $U: \mathscr{B} \to \mathscr{C}$ and $V: \mathscr{E} \to \mathscr{C}$ are functors, then the following diagram exists in Cat:

(2)
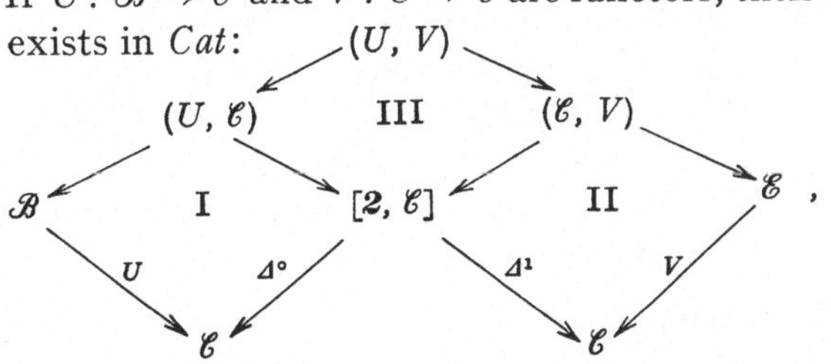

where I, II, III are pullbacks (Lawvere's Comma construction). The objects of the category (U, V) are triples (X, Y, a) with $X \in |\mathscr{B}|$, $Y \in |\mathscr{E}|$ and $a: U(X) \to V(Y)$ in $\mathscr{C}$. Morphisms from (X, Y, a) to (X', Y', a') are commutative squares

(3) $$\begin{array}{ccc} U(X) & \xrightarrow{\;a\;} & V(Y) \\ {\scriptstyle U(u)}\downarrow & & \downarrow{\scriptstyle V(v)} \\ U(X') & \xrightarrow{\;a'\;} & V(Y') \end{array}$$

with $u: X \to X'$ in $\mathcal{B}$ and $v: Y \to Y'$ in $\mathcal{C}$. Only special cases are needed here (as in 9.2.1 and in 10.2).

17.1.3 Let $U: \mathcal{B} \to \mathcal{C}$ be a fixed, chosen functor. For every $A \in |\mathcal{C}|$, let U/A be the following (possibly empty) category: objects are pairs (X, a) with $X \in |\mathcal{B}|$ and $a: U(X) \to A$, morphisms from (X, a) to (X', a') are triples (a, a', w) with $w: X \to X'$ such that $a' \, U(w) = a$. That is, $U/A = (U, \ulcorner A \urcorner)$ where $\ulcorner A \urcorner = A_1: \mathbf{1} \to \mathcal{C}$ is the functor whose value is A. ($\mathbf{1}$ is a terminal category.) A morphism $f: A \to B$ in $\mathcal{C}$ induces a functor $U/f: U/A \to U/B$ which takes (X, a) into $(X, f\, a)$ and (a, a', w) into $(f\, a, f\, a', w)$. Notice that U/f is faithful.

Now let $\mathcal{B}$ be a small category. Then there is a functor $Q: \mathcal{C} \to$ $\to Dg(\mathcal{B})$ described as follows: $Q_A: U/A \to \mathcal{B}$ is the functor given by the rule $(X, a) \mapsto X$, $(a, a', w) \mapsto w$, (i.e., Q_A is the projection of $(U, \ulcorner A \urcorner)$ on $\mathcal{B}$ in (2)), and

(4) $$Q_f = (U/f, 1_{Q_A}) \,,$$

which makes sense, since

(5) $$Q_B \, U/f = Q_A \,.$$

Besides the functor $Dg(U)\, Q$: $\mathcal{C} \to Dg(\mathcal{C})$, there is the functor Z: $\mathcal{C} \to Dg(\mathcal{C})$ with $Z_A = Z(A) = A_{U/A}$ and $Z_f = Z(f) = (U/f, f_{U/A})$ for f: $A \to B$. Here $(X, a) \mapsto a$ is a natural transformation $\gamma_A: U\, Q_A \to Z(A)$, and thus $(1_{U/A}, \gamma_A)$ is a $Dg(\mathcal{C})$-morphism, which gives rise to a natural transformation $\gamma: Dg(U)\, Q \to Z$.

17.1.4 If this construction is based on the functor $1_{\mathcal{B}}$ instead of on U: $\mathcal{B} \to \mathcal{C}$, one obtains for $Y \in |\mathcal{B}|$ the category $\mathcal{B}/Y$ of $\mathcal{B}$-morphisms with codomain Y taking the place of U/A. Analogous to U/f, $v: Y \to Y'$ gives rise to a functor $\mathcal{B}/v: \mathcal{B}/Y \to \mathcal{B}/Y'$. In the place of Q there is P: $\mathcal{B} \to Dg(\mathcal{B})$. Furthermore, there is a natural transformation β: $P \to Q\, U$. For, for $Y \in |\mathcal{B}|$ one has the functor $\Xi_Y: \mathcal{B}/Y \to U/U(Y)$, which takes $z: X \to Y$ into $(X, U(z))$ and $(z, z' w)$ (with $z' w = z$) into $(U(z), U(z'), w)$. Obviously $P_Y = Q_{U(Y)}\, \Xi_Y$, so that (Ξ_Y, id): $P_Y \to$ $\to Q_{U(Y)}$ is a $Dg(\mathcal{B})$-morphism and thus finally a natural transformation $\beta: P \to Q\, U$ is defined.

17.1.5 Remarks

(a) 1_Y is a terminal object of $\mathcal{B}/Y$.

(b) $(\Xi_Y, \gamma_{U(Y)}\, (U * \beta_Y))$ is a $Dg(\mathcal{C})$-morphism $U\, P_Y \to U\, Q_{U(Y)} \to$ $\to Z_{U(Y)}$. By definition of β and γ, the terminal object 1_Y of $\mathcal{B}/Y$ is here assigned the morphism $1_{U(Y)}$.

(c) U is faithful if and only if Ξ_Y: $\mathcal{B}/Y \to U/U(Y)$ is an embedding for every Y.

(d) U is fully faithful if and only if Ξ_Y is an isomorphism of categories for every Y. In this case $\beta: P \to Q\, U$ is an isomorphism.

(e) If $\mathscr{B}$ is finitely cocomplete and if U preserves finite colimits, then every U/A is filtered. This follows immediately from the definitions.

17.1.6 Theorem. *Let $\mathscr{B}$ be a small category, $U: \mathscr{B} \to \mathscr{C}$ a functor and $\mathscr{D}$ a cocomplete category.*

(a) *The functor $\tilde{U} = [U, \mathscr{D}]: [\mathscr{C}, \mathscr{D}] \to [\mathscr{B}, \mathscr{D}]$ has a left adjoint V: $[\mathscr{B}, \mathscr{D}] \to [\mathscr{C}, \mathscr{D}]$.*

(b) *$\tilde{U} = [U, \mathscr{D}]$ preserves limits and colimits.*

(c) *If U is fully faithful, then V is also fully faithful (however, not $\tilde{U}$ in general), and $F: \mathscr{B} \to \mathscr{D}$ is isomorphic to $V(F)\,U$.*

(d) *If every object of $\mathscr{C}$ has the form $U(X)$, then $\tilde{U}$ is faithful.*

(e) *If $\mathscr{B}$ is finitely cocomplete, $\mathscr{D}$ finitely complete, if U preserves finite colimits and if in $\mathscr{D}$ finite limits commute with filtered colimits, then V preserves finite limits.*

(f) *If $\mathscr{B}, \mathscr{C}, \mathscr{D}$ are additive categories and if U is additive, then* (a) *through* (e) *are correspondingly valid for the categories $Add(\mathscr{B}, \mathscr{D})$ and $Add(\mathscr{C}, \mathscr{D})$. Here $\tilde{U}$ and V are additive.*

Proof. We use the notations introduced above.

(a) We shall show: for every functor F: $\mathscr{B} \to \mathscr{D}$ the functor $[F, \tilde{U}(?)]_{[\mathscr{B}, \mathscr{D}]}$: $[\mathscr{C}, \mathscr{D}] \to ENS$ is representable. The claim of (a) then follows from 16.4.5.

Now let F be fixed. 17.1.3 and 17.1.1 yield a functor $V(F): \mathscr{C} \to \mathscr{D}$ with

(6) $$V(F)\,(A) = \operatorname{Colim} F\,Q_A = \operatorname{Colim}\left((U, \ulcorner A \urcorner) \xrightarrow{Q_A} \mathscr{B} \xrightarrow{F} \mathscr{D}\right),$$
$$V(F)\,(f) = \operatorname{Colim} F\,Q_f$$

for all $A \in |\mathscr{C}|$ and all $f \in \operatorname{Mor} \mathscr{C}$. By 17.1.5 (a) and the dual of 7.1.8, one has $F = \operatorname{Colim} Dg(F)\,P$ (if the colimits are chosen according to the dual of 7.1.8). By 17.1.4 and 17.1.1, $Dg(F) * \beta: Dg(F)\,P \to Dg(F)\,QU$ induces a natural transformation

$$\eta_F: F \to V(F)\,U.$$

For $A \in |\mathscr{C}|$, let $(V(F)\,(A), \lambda_A)$ be the chosen colimit. Considering $\lambda_{U(X)}$ at $(X, 1_{U(X)}) \in |U/U(X)|$, one gets

(7) $$\eta_{F,X} = \lambda_{U(X),(X, U(1_X))},$$

by the definition of η_F. (We could define η_F by (7), because for w: $X \to X'$ in $\mathscr{B}$,

$$V(F)\,\big(U(w)\big)\,\lambda_{U(X),(X, U(1_X))} = \lambda_{U(X'),(X', U(1_{X'}))}\,F(w)$$

follows from (1), (4), (5), (6), (7).)

$(V(F), \eta_F)$ will turn out to be a representation of $[F, U(?)]$. By the remark in 16.4.5, one has to prove the following: if $G: \mathscr{C} \to \mathscr{D}$ and a natural transformation $\xi: F \to G\,U$ are given, then there is exactly one natural transformation $\alpha: V(F) \to G$ such that

$$(8) \qquad \xi = (\alpha * U)\,\eta_F\,.$$

Since $(G * \gamma_A)\,(\xi * Q_A)$: $F\,Q_A \to G\,U\,Q_A \to G\,Z_A$ is a natural transformation and since $G\,Z_A = G(A)_{U/A}$, there is a unique $\alpha_A: V(F)\,(A) \to G(A)$ such that

$$(9) \qquad (\alpha_A)_{U/A}\,\lambda_A = (G * \gamma_A)\,(\xi * Q_A)\,;$$

i.e., at $(X, a) \in |U/A|$ one has

$$(9') \qquad \alpha_A\,\lambda_{A,\,(X,\,a)} = G(a)\,\xi_X\,.$$

Using this twice, one gets

$$(9'') \qquad \alpha_B\,\lambda_{B,\,(X,\,fa)} = G(f\,a)\,\xi_X = G(f)\,\alpha_A\,\lambda_{A,\,(X,\,a)}$$

for $f: A \to B$ in $\mathscr{C}$. By (1), (4), (5) and (6),

$$(10) \qquad \lambda_{B,\,(X,\,fa)} = V(F)(f)\,\lambda_{A,\,(X,\,a)}\,.$$

Since $(V(F)\,(A), \lambda_A)$ is a colimit,

$$(11) \qquad \alpha_B\,V(F)(f) = G(f)\,\alpha_A$$

follows from (9'') and (10), provided U/A is not empty. If, however, U/A is empty, then $V(F)\,(A)$ is initial in $\mathscr{D}$, so that (11) is true in any case. Therefore, $\alpha = \{\alpha_A\}: V(F) \to G$ is a natural transformation, and setting $A = U(X)$, $a = U(1_X)$ in (9'), one gets (8) by (7).

It remains to be shown that there is only one such α. For $a: U(X) \to A$,

$$(12) \qquad V(F)(a)\,\eta_{F,X} = \lambda_{A,\,(X,\,a)}\,,$$

by (7) and (10). If $\alpha: V(F) \to G$ is any natural transformation for which (8) holds, then

$$\alpha_A\,\lambda_{A,(X,a)} = \alpha_A\,V(F)(a)\,\eta_{F,X} = G(a)\,\alpha_{U(X)}\,\eta_{F,X} = G(a)\,\xi_X\,.$$

Hence (9') and (9) are valid. Thus, by the definition of colimits, α is determined uniquely and the proof of (a) is complete.

Before proving (b), we add some remarks. For $G = V(F)$ and $\xi = \eta_F$ one has $\alpha = 1_{V(F)}$ by (8). Therefore, $\eta = \{\eta_F\}: 1 \to \widetilde{U}\,V$ is an adjunction transformation. The quasi-inverse transformation $\varepsilon: V\,\widetilde{U} \to 1$ is given by the morphisms

$$(13) \qquad \varepsilon_{G,A}: \operatorname{Colim} G\,U\,Q_A \to G(A)$$

that are induced by $Dg(G) * \gamma$. This follows immediately from (6) and (9) if one sets $F = G\,U$ and $\xi = 1_{GU}$.

(b) As a right adjoint of V, $\widetilde{U}$ preserves limits. Since $\mathscr{D}$ is cocomplete, colimits in $[\mathscr{B}, \mathscr{D}]$ and $[\mathscr{C}, \mathscr{D}]$ can be formed "pointwise". A colimit in $[\mathscr{C}, \mathscr{D}]$ is therefore in particular a colimit at every place $U(X)$, and so $\tilde{U}$ preserves colimits.

(c) If U is fully faithful, then $\beta\colon P \to Q\,U$ is an isomorphism by 17.1.5 (d). Therefore, $Dg(F) * \beta\colon Dg(F)\,P \to Dg(F)\,Q\,U$ is an isomorphism. Transferring to colimits, one obtains the isomorphism η_F: $F \to \tilde{U}\,V(F) = V(F)\,U$. V is fully faithful by the dual of 16.5.4 (d).

(d) If $G_1, G_2\colon \mathscr{C} \to \mathscr{D}$ are functors and if $\alpha = \{\alpha_A\}\colon G_1 \to G_2$ is a natural transformation, then $\tilde{U}(\alpha) = \alpha * U = \{\alpha_{U(X)}\}$, which implies the claim made in (d) immediately.

(e) Under these conditions, $\mathscr{B}$ has an initial object which is preserved by U. Therefore, no U/A is empty, and every U/A is filtered by 17.1.5 (e). Finite limits in $[\mathscr{B}, \mathscr{D}]$ and $[\mathscr{C}, \mathscr{D}]$ can be formed "pointwise". Furthermore, the construction of V at the place A as a filtered colimit of type U/A commutes with finite limits in $\mathscr{D}$.

(f) To start with, we shall assume that $\mathscr{B}$ has finite products. Let $F\colon \mathscr{B} \to \mathscr{D}$ be additive. In $\mathscr{B}$ there exists the biproduct $X \oplus X$ with injections in_1, in_2 and diagonal map $\Delta\colon X \to X \oplus X$. By 12.2.7, U and F preserve biproducts and diagonal maps. Now, in $\mathscr{C}$ there is the following commutative diagram for $f, g\colon A \to B$:

(14) 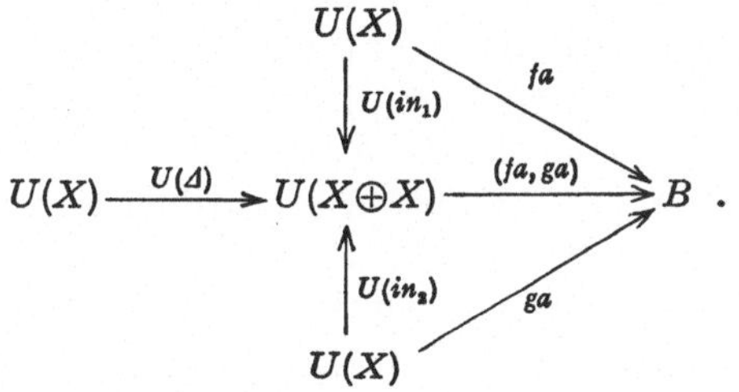

In $\mathscr{D}$ one obtains, by (6),

(14') 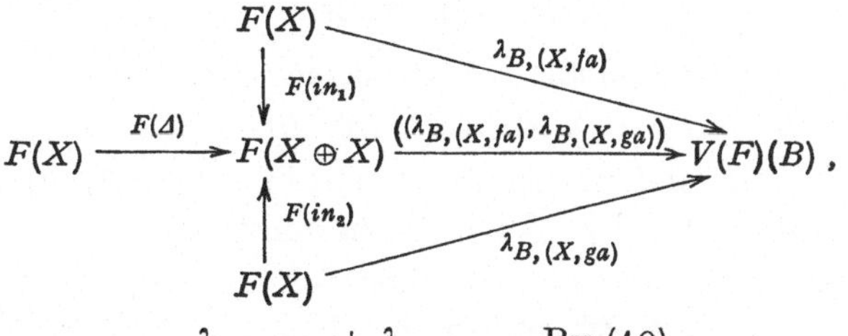

so that $\lambda_{B,(X, fa+ga)} = \lambda_{B,(X, fa)} + \lambda_{B,(X, ga)}$. By (10),

$$V(F)(f+g)\,\lambda_{A,(X,a)} = \lambda_{B,(X,(f+g)a)} = \big(V(F)(f) + V(F)(g)\big)\,\lambda_{A,(X,a)}$$

follows. By this and the definition of colimits, $V(F)(f+g) = V(F)(f) + V(F)(g)$ follows; thus $V(F)$ is additive. If $G\colon \mathscr{C} \to \mathscr{D}$ is additive, $\tilde{U}\,G = G\,U$ is obviously additive. Thus from V and $\tilde{U}$ one gets functors between $Add(\mathscr{B}, \mathscr{D})$ and $Add(\mathscr{C}, \mathscr{D})$ and, with $\tilde{U}$, V is also additive by 16.5.10. By 8.5.3, (b) holds again and (c), (d), (e) follow as before.

If $\mathcal{B}$ does not have finite products, then $\mathcal{B}$ can be completed to a small additive category $\mathcal{B}'$ with finite products as in 16.3.10. If $\mathcal{C}$ now has finite products, then U can be extended to an additive functor $U'\colon \mathcal{B}' \to \mathcal{C}$ as in 16.3.9, and for U' one is now dealing with the same case as above. If $R\colon \mathcal{B} \to \mathcal{B}'$ is the inclusion, then $U = U' R$ and $\tilde{U} = \tilde{R}\,\tilde{U}'$. By 16.3.9, $\tilde{R}$ is an equivalence, so that (f) is implied by 16.4.2 (b) in this case also. Note however, that in general (6) through (13) are not valid.

If neither $\mathcal{B}$ nor $\mathcal{C}$ have finite products, complete $\mathcal{C}$ by adding finite products to $\mathcal{C}'$. Let $S\colon \mathcal{C} \to \mathcal{C}'$ be the inclusion. By what was said above, $(\widetilde{SU})\colon Add(\mathcal{C}', \mathcal{D}) \to Add(\mathcal{B}, \mathcal{D})$ has a left adjoint V'. Let Q be equivalence-inverse to $\tilde{S}$. Then $V = \tilde{S}\, V'$ is left adjoint to $(\widetilde{SU})\, Q = \tilde{U}\,\tilde{S}\, Q$ and thus also to $\tilde{U}$, since $\tilde{U}$ and $\tilde{U}\,\tilde{S}\, Q$ are isomorphic. Thus (f) is proved in the general case.

17.1.6⁰ Let $\mathcal{B}$ again be a small category, $U\colon \mathcal{B} \to \mathcal{C}$ a functor and $\mathcal{D}$ a complete category. Then $U = [U, \mathcal{D}]\colon [\mathcal{C}, \mathcal{D}] \to [\mathcal{B}, \mathcal{D}]$ has a right adjoint V^+. U again preserves limits and colimits, and V^+ is fully faithful provided U is fully faithful.

This is what results from 17.1.6 if $\mathcal{B}, \mathcal{C}, \mathcal{D}$ are replaced by their dual categories. Here $V^+(F)\,(A) = \operatorname{Lim} F\, Q'_A$ with $Q'_A\colon A/U \to \mathcal{B}$, where the objects of A/U are pairs (X, a) with $a\colon A \to U(X)$ and where Q'_A takes the object (X, a) back into X.

Now let $\mathcal{D}$ again be cocomplete. According to earlier conventions, $[\mathcal{B}^0, \mathcal{D}]$, $[\mathcal{C}^0, \mathcal{D}]$ are to be regarded as categories of contravariant functors $\mathcal{B} \to \mathcal{D}$ or, resp., $\mathcal{C} \to \mathcal{D}$. $\mathrm{Op}\, U\, \mathrm{Op}\colon \mathcal{B}^0 \to \mathcal{C}^0$ corresponds to $U\colon \mathcal{B} \to \mathcal{C}$. For $\tilde{U} = [\mathrm{Op}\, U\, \mathrm{Op}, \mathcal{D}]$ there is a left adjoint given by colimits in $\mathcal{D}$, again by 17.1.6. In 17.1.6 (e) the assumptions about $\mathcal{B}$ and $U\colon \mathcal{B} \to \mathcal{C}$ are to be replaced by their duals.

17.1.7 Let $\mathcal{B}$ be small and $\mathcal{D}$ cocomplete. If $U\colon \mathcal{B} \to \mathcal{C}$ has a right adjoint $W\colon \mathcal{C} \to \mathcal{B}$, then $\tilde{W}\colon [\mathcal{B}, \mathcal{D}] \to [\mathcal{C}, \mathcal{D}]$ is left adjoint to $\tilde{U}$ by 16.5.11 (b) and, therefore, it is isomorphic to the functor V constructed in 17.1.6, by 16.4.4. If, in addition, $\mathcal{D}$ is complete, then $\tilde{W}$ preserves limits and colimits (on account of the pointwise construction in $[\mathcal{B}, \mathcal{D}]$ and $[\mathcal{C}, \mathcal{D}]$) and so does V.

17.1.8 The set-valued case. Let $\mathcal{B}$ again be small and $U\colon \mathcal{B} \to \mathcal{C}$ given. We set $U^0 = \mathrm{Op}\, U\, \mathrm{Op}\colon \mathcal{B}^0 \to \mathcal{C}^0$. Let further $H^*\colon \mathcal{B}^0 \to [\mathcal{B}, Ens]$ and $H^*\colon \mathcal{C}^0 \to [\mathcal{C}, Ens]$ be the Yoneda embeddings 4.2.2. For $G\colon \mathcal{C} \to Ens$ and $X \in |\mathcal{B}|$, a double application of 4.2.4 yields the isomorphisms

$$(15) \qquad [H^X, G\, U]_{[\mathcal{B}, Ens]} \cong G(U(X)) \cong [H^{U(X)}, G]_{[\mathcal{C}, Ens]}$$

as isomorphisms of bifunctors $\mathscr{B}^0 \times [\mathscr{C}, Ens] \to Ens$. An arbitrary functor $F\colon \mathscr{B} \to Ens$ is, by 10.2.1, a colimit object for a functor $\mathscr{B}^0/F \to [\mathscr{B}, Ens]$ which is of the form $H^* Q_F$ with respect to H^* (this is analogous to $U Q_A$ in 17.1.3, here with $\mathscr{B}^0/F$ instead of U/A). By applying 8.7.3 twice to (15), a functor V that is left adjoint to $\tilde{U}\colon [\mathscr{C}, Ens] \to [\mathscr{B}, Ens]$, is obtained through

$$(16) \qquad V(F) = \operatorname{Colim} H^* U^0 Q_F \quad \text{with} \quad Q_F\colon \mathscr{B}^0/F \to \mathscr{B}^0\,,$$

where

$$(17) \qquad \begin{array}{ccc} \mathscr{B}^0 & \xrightarrow{U^0} & \mathscr{C}^0 \\ {\scriptstyle H^*}\downarrow & & \downarrow{\scriptstyle H^*} \\ [\mathscr{B}, Ens] & \xrightarrow{V} & [\mathscr{C}, Ens] \end{array}$$

is commutative; i.e., $V(H^?) = H^{U(?)}$ for ? in $\mathscr{B}$. By 16.4.4, the functor V given here is isomorphic to the one constructed in 17.1.6. For $F = H^Y$, in particular, there has to be a natural transformation

$$(18) \qquad \varrho_A(?)\colon [Y, Q_A(?)] \to [U(Y), A]_{U/A} \qquad (? \in |U/A|)$$

which is a colimit. Conversely, an interchange of colimits shows the V considered here to be isomorphic to the one in 17.1.6.

(18) can be proved directly by constructing the colimit as in 8.4.4 with equivalence classes of the coproduct $\coprod [Y, X]_{(X,a)}$. It is not empty if and only if $[U(Y), A] \neq \emptyset$, and in this case the desired bijection is obtained by assigning to $b \in [U(Y), A]$ the equivalence class of $1_Y \in [Y, Y]_{(Y,b)}$.

17.1.9 The Ab-valued case. If $\mathscr{B}$, $\mathscr{C}$ and $U\colon \mathscr{B} \to \mathscr{C}$ are additive, then (15) holds for $Add(\mathscr{B}, Ab)$ and $Add(\mathscr{C}, Ab)$, for an additive $G\colon \mathscr{C} \to Ab$. If $\mathscr{B}$ has finite products, then $V(F)$ is additive since it is a colimit of representable and thus additive functors. The additive version of 10.2.1 needed for (16) requires additivity in 10.2.1 (3). It is provided by (14). For this, as in (14), it is sufficient to require that there be an $X \oplus X$ for every $X \in |\mathscr{B}|$.

The case where $\mathscr{C}$, but not $\mathscr{B}$, has finite products can be reduced to the one just treated (compare the end of the proof of 17.1.6) by completing $\mathscr{B}$ to a small additive category $\mathscr{B}'$ with finite products, and by also extending U. If $R\colon \mathscr{B} \to \mathscr{B}'$ is the inclusion, then $\tilde{R}\colon Add(\mathscr{B}', Ab) \to Add(\mathscr{B}, Ab)$ is an equivalence. As shown in the proof of 16.3.9, an equivalence-inverse S can be obtained by extending every additive functor $\mathscr{B} \to Ab$ to a functor $\mathscr{B}' \to Ab$. This can obviously be done in such a way that

$$\begin{array}{ccc} \mathscr{B}^0 & \xrightarrow{R^0} & \mathscr{B}^{0\prime} \\ {\scriptstyle H^*}\downarrow & & \downarrow{\scriptstyle H^*} \\ Add(\mathscr{B}, Ab) & \xrightarrow{S} & Add(\mathscr{B}', Ab) \end{array}$$

is commutative. Thus commutativity for (17) is achieved in this case also. However, (16) and (18) need not be true any more.

17.2 Dense Functors

17.2.1 If $\mathscr{B}$ and $\mathscr{C}$ are arbitrary categories and if $U: \mathscr{B} \to \mathscr{C}$ is a functor, then the construction in 17.1.3 makes sense. $U: \mathscr{B} \to \mathscr{C}$ is called *dense* (*left adequate*) if (A, γ_A) is a colimit of $U\,Q_A$ for every $A \in |\mathscr{C}|$. A *subcategory* $\mathscr{B}$ of $\mathscr{C}$ is called *dense* (in $\mathscr{C}$) if the inclusion is dense.

Here the categories U/A need not be small. 10.2.1 states that the Yoneda embedding $H^*: \mathscr{C}^0 \to [\mathscr{C}, Ens]$ of 4.2.2 is always dense. 10.3.8 shows that for dense functors between non-small categories colimits of small diagrams might be of interest. For a dense functor the codomain category need not be cocomplete, as evidenced by $1_{\mathscr{B}}$ for an arbitrary category $\mathscr{B}$; and colimits need not be preserved, as is shown again by the Yoneda embedding.

17.2.2 Examples of dense subcategories. Here $\mathscr{B}$ is a full subcategory of $\mathscr{C}$.

(a) $\mathscr{C} = Ens$, $\mathscr{B}$ has as its only object a set of one element (or any non-empty set).

(b) $\mathscr{C} = {}_R Mod$, $\mathscr{B}$ has as its only object $R \oplus R$ (as a biproduct of two left modules ${}_R R$).

(c) $\mathscr{C} = {}_R Mod$, $\mathscr{B}$ is the category of finitely generated modules (or finitely presentable modules).

(d) $\mathscr{C}$ is the category of rings (with unit), $\mathscr{B}$ has as only object the free associative $\boldsymbol{Z}$-algebra with two free generators.

(e) $\mathscr{C}$ is the category of commutative rings (with unit), $\mathscr{B}$ has as only object the polynomial ring $\boldsymbol{Z}[X, Y]$.

Examples (b) and (c) are valid in particular for $R = \boldsymbol{Z}$, that is for $\mathscr{C} = Ab$. $R \oplus R$ in (b) may not be replaced by the generator R, since in this subcategory there are not enough morphisms: if for $\mathscr{C} = Ab$ the subcategory $\mathscr{B}$ has only the object $\boldsymbol{Z}$, then $A = \boldsymbol{Z} \oplus \boldsymbol{Z}$ is not a colimit of $U\,Q_A$, instead of A one obtains a direct sum of countably many summands as a colimit object. In all cases (a) through (e), definition 17.2.1 can be verified directly (with some effort). The dual situation is represented by the category of (Hausdorff) compact spaces. The full subcategory whose only object is the unit square is codense.

17.2.3 Proposition. *The functor $U\colon \mathscr{B} \to \mathscr{C}$ is dense if and only if the functor $\check{U}\colon \mathscr{C} \to [\mathscr{B}^0, Ens]$ defined by the rule $A \mapsto [U(?), A]_{\mathscr{C}}$, $f \mapsto [U(?), f]_{\mathscr{C}}$ is fully faithful.*

Proof. There is a bijection

$$\theta_{A,B}\colon [U\,Q_A, B_{U/A}] \to [\check{U}(A), \check{U}(B)] \tag{1}$$

which is described as follows: Let $\alpha\colon U\,Q_A \to B_{U/A}$ be a natural transformation. For $(X, a) \in |U/A|$, one has $\alpha_{(X,a)}\colon U(X) \to B$, and for fixed $X \in |\mathscr{B}|$, the rule $a \mapsto \alpha_{(X,a)}$ then gives $\alpha_X\colon [U(X), A] \to [U(X), B]$. With α, $\{\alpha_X\}\colon [U(?), A] \to [U(?), B]$ is also a natural transformation and one verifies that $\alpha \mapsto \{\alpha_X\}$ is a bijection. Using $\gamma_{A,(X,a)} = a$ (compare 17.1.3), there is a map

$$\Omega_{A,B}\colon [A, B] \to [U\,Q_A, B_{U/A}] \tag{2}$$

given by the rule $f \mapsto f_{U/A}\,\gamma_A$.

Now $\theta_{A,B}\,\Omega_{A,B} = \check{U}_{A,B}\colon [A, B] \to [\check{U}(A), \check{U}(B)]$, as is shown by $f \mapsto \{a \mapsto f\,a\}$. For fixed A, $\{\Omega_{A,B}\}$ is a natural transformation $H^A \to N_{UQ_A}$ (compare 8.1.3) and it is an isomorphism if and only if (A, γ_A) is a colimit of $U\,Q_A$. After what we just proved, this is seen to be the case if and only if $\check{U}_{A,B}$ is an isomorphism for all B. This completes the proof.

17.2.4 Remarks. $\check{U}\colon \mathscr{C} \to [\mathscr{B}^0, Ens]$ is the composite of the Yoneda embedding $H_*\colon \mathscr{C} \to [\mathscr{C}^0, Ens]$ with $\widetilde{U}^0 = [\mathrm{Op}\,U\,\mathrm{Op}, Ens]\colon [\mathscr{C}^0, Ens] \to$ $\to [\mathscr{B}^0, Ens]$. $\check{U}$ preserves limits, because limits in functor categories are constructed pointwise (compare the discussion in the proof of 17.1.6 (b) for colimits) and because of 10.2.5.

For composites of functors one has further

$$(U_2\,U_1)^{\vee} = \widetilde{U}_1^0\,\widetilde{U}_2^0\,H_* = \widetilde{U}_1^0\,\check{U}_2 \tag{3}$$

and, therefore, by a double application of 17.2.3 one gets:

If $U_2\,U_1$ is dense, then $\check{U}_2$ is faithful. If $\widetilde{U}_1^0 = [\mathrm{Op}\,U_1\,\mathrm{Op}, Ens]$ is also faithful, then U_2 is dense. $\widetilde{U}_1^0$ is guaranteed to be faithful if U_1 is surjective for object classes, as is shown by 17.1.6 (d).

If, in particular, $U\colon \mathscr{B} \to \mathscr{C}$ is dense, and if U_1 is the functor induced by U whose codomain is the smallest subcategory of $\mathscr{C}$ containing all morphisms of the form $U(f)$ (the "image" of U), and if U_2 is the embedding of this subcategory in $\mathscr{C}$, then U_2 is dense.

17.2.5 $U\colon \mathscr{B} \to \mathscr{C}$ is fully faithful if and only if $\check{U}\,U$ is isomorphic to the Yoneda embedding $H_*\colon \mathscr{B} \to [\mathscr{B}^0, Ens]$. For, U gives rise to a natural transformation $[?, ??]_{\mathscr{B}} \to [U(?), U(??)]_{\mathscr{C}}$ of contra-covariant functors and the statement follows from $\check{U}\,U(X) = [U(?), U(X)]$.

For $H_*: \mathcal{B} \to [\mathcal{B}^0, Ens]$, $\check{H}_* \cong 1_{[\mathcal{B}^0, Ens]}$. Since $\check{H}^*(T) = [H_?, T]_{[\mathcal{B}^0, Ens]}$, this follows from 4.2.4. Therefore, 10.2.1 is a special case of 17.2.3.

17.2.6 Proposition. *Let $(\varphi, S, T, \mathcal{C}, \mathcal{D})$ be an adjunction. The following statements are equivalent:*

(a) *T is fully faithful,*

(b) *S is dense,*

(c) *$\varepsilon: S\,T \to 1_{\mathcal{C}}$ is an isomorphism.*

If further $U: \mathcal{B} \to \mathcal{D}$ is dense, then the following is also equivalent:

(d) *$S\,U$ is dense.*

Proof. (a) and (c) are equivalent by 16.5.4 (d). (b) is contained in (d) by setting $U = 1_{\mathcal{D}}$. Now, $\check{U}\,T: \mathcal{C} \to [\mathcal{B}^0, Ens]$ operates via $A \mapsto [U(?), T(A)]_{\mathcal{D}}$ and $(S\,U)^{\vee}: \mathcal{C} \to [\mathcal{B}^0, Ens]$ via $A \mapsto [S\,U(?), A]_{\mathcal{C}}$, so $\check{U}\,T$ and $(S\,U)^{\vee}$ are isomorphic. By assumption and 17.2.3, $\check{U}$ is fully faithful. Therefore, (a) and (d) are equivalent.

17.2.7 Theorem. *Let $\mathcal{B}$ be a small category, $\mathcal{D}$ a cocomplete category and $U: \mathcal{B} \to \mathcal{C}$ a dense functor. $\tilde{U} = [U, \mathcal{D}]: [\mathcal{C}, \mathcal{D}] \to [\mathcal{B}, \mathcal{D}]$ gives the maps* (2.2.7)

$$\tilde{U}_{G,G'}: [G, G']_{[\mathcal{C},\mathcal{D}]} \to [\tilde{U}(G), \tilde{U}(G')]_{[\mathcal{B},\mathcal{D}]} = [G\,U, G'\,U]_{[\mathcal{B},\mathcal{D}]}\,.$$

If G preserves colimits, then $\tilde{U}_{G,G'}$ is a bijection. In particular, two functors $G, G': \mathcal{C} \to \mathcal{D}$ that preserve colimits are isomorphic if and only if $G\,U$ and $G'\,U$ are isomorphic.

Proof. We make use of the notation in 17.1.3 and 17.1.6 again. Since U is dense, Colim $U\,Q = 1_{\mathcal{C}}$, provided (A, γ_A) is always chosen as a colimit of $U\,Q_A$. Since G preserves colimits, $(G(A), G\,\gamma_A)$ is a colimit of $G\,U\,Q_A$. By 17.1.6 (6) and (13), we may assume $V(G\,U) = G$ and $\varepsilon_G = 1_G$. With this, 16.5.5 (4) implies $\eta_{GU} = (\eta * U)_G = 1_{GU}$, and, by 16.5.1 (2), $\tilde{U}_{G,G'} = \tilde{U}_{V(GU),G'}$ is an isomorphism. The rest of the statement follows by a repeated application of this result.

17.2.8 Remarks. Consider the special case where U is an inclusion and $\tilde{U}$ thus a restriction. 17.2.7 asserts then that natural transformations of colimit preserving functors restricted to $\mathcal{B}$ can be extended in exactly one way.

One can dispense with the condition that $\mathcal{D}$ be cocomplete. The application of 17.1.6 in the proof will then have to be replaced by calculations as in the proof of 17.1.6 (a); Colim $G\,U\,Q$ exists by the assumptions made for U and G. The assumption that $\mathcal{B}$ is small can be replaced by the requirement that G preserves all colimits in $\mathcal{C}$, even large ones.

If in addition to $U: \mathcal{B} \to \mathcal{C}$ a functor $F: \mathcal{B} \to \mathcal{D}$ is given, then one can consider the question of the existence of a functor $G: \mathcal{C} \to \mathcal{D}$ which preserves colimits, and for which F and $G\,U$ are isomorphic. As $U = 1_{\mathcal{B}}$ shows, such a functor need not exist. The following lemma (with $W = 1_{\mathcal{C}}$) guarantees the existence of such a functor; here G is uniquely determined up to an isomorphism, this follows from what we just proved, since G is an "extension" of F that preserves colimits. Universal extension theorems will follow in 17.3.1 and 17.3.2.

17.2.9 Lemma. *We assume that there are the following functors*

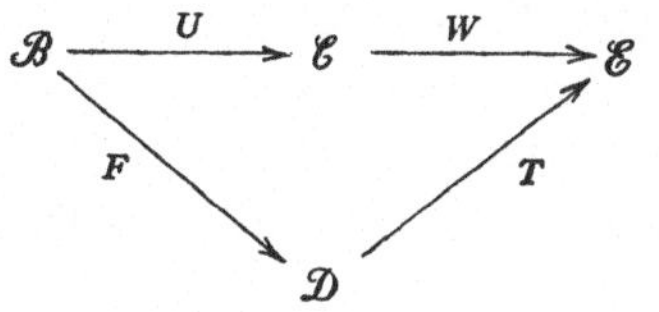

(*no commutativity condition*) *satisfying*:

(i) *For every object $A \in |\mathcal{C}|$, there is a category Σ_A and a functor Q_A: $\Sigma_A \to \mathcal{B}$ such that A is a colimit object of $U\,Q_A$.*

(ii) *W preserves the colimits in* (i).

(iii) *There is a bifunctor isomorphism*

$$\varphi\colon [F(?),\, ??]_{\mathcal{D}} \to [W\,U(?),\, T(??)]_{\mathcal{E}}\,. \tag{4}$$

Then the following statements are equivalent:

(a) *For all $A \in |\mathcal{C}|$, $F\,Q_A$ has a colimit in $\mathcal{D}$.*

(b) *There is a functor $G: \mathcal{C} \to \mathcal{D}$ with a bifunctor isomorphism*

$$\chi\colon [G(?),\, ??]_{\mathcal{D}} \to [W(?),\, T(??)]_{\mathcal{E}}\,. \tag{5}$$

In that case F and $G\,U$ are isomorphic and G preserves all colimits that are preserved by W.

Proof. First, let (a) be satisfied. We construct G analogously to $V(F)$ in 17.1.6. So let $(G(A), \varrho_A)$ be a (chosen) colimit of $F\,Q_A$. For $D \in |\mathcal{D}|$, one then has

$$\begin{aligned}
[G(A), D] &= [\operatorname{Colim} F\,Q_A, D] \cong \operatorname{Lim}\,[F\,Q_A, D] \\
&\cong \operatorname{Lim}\,[W\,U\,Q_A, T(D)] \cong [\operatorname{Colim} W\,U\,Q_A, T(D)] \\
&\cong [W(\operatorname{Colim} U\,Q_A), T(D)] = [W(A), T(D)]\,,
\end{aligned}$$

by assumption and 8.7.3 (possibly with a change of universe). In any case, the isomorphisms are natural with respect to the second argument. Therefore, one has a representation of $[W(A), T(??)]$ with representing object $G(A)$, and (b) follows from 4.5.1.

We now prove the additional assertion. Comparison of (4) and (5) together with 4.5.4 shows F and $G\,U$ to be isomorphic. The rest of the assertion follows from (5) by a double application of 8.7.3.

Now let (b) be satisfied. By what we just proved, it follows from (i) and (ii) that $G(A)$ is a colimit object of $G\,U\,Q_A$, and $F\,Q_A \cong G\,U\,Q_A$. Hence (a) holds.

17.2.10 The additive case. If $\mathcal{B}$, $\mathcal{C}$ are additive categories, if $U\colon \mathcal{B} \to \mathcal{C}$ is additive and if $\mathcal{B}$ has finite products, then 17.2.3 is valid for $\check{U}\colon \mathcal{C} \to Add(\mathcal{B}^0, Ab)$. It suffices to verify that (1) is additive (for (2) this is evident), and this follows as in 17.1.6 (14). Thus 17.2.4 through 17.2.6 carry over smoothly. In particular, the additive version of 10.2.1 is in this way obtained as a special case. The additive version of 17.2.7 with $\tilde{U}\colon Add(\mathcal{C}, \mathcal{D}) \to Add(\mathcal{B}, \mathcal{D})$ is valid by 16.3.9 and 17.1.6 (f), if U extends to a dense functor when $\mathcal{B}$ is completed with finite products. If in 17.2.9 all categories, the given functors and φ in (4) are additive, then G and χ in (5) are additive; this follows from (6) and 4.5.7.

17.3 Characterization of the Yoneda Embedding

17.3.1 Theorem. *Let $\mathcal{B}$ be a small category and $\mathcal{D}$ a cocomplete category.*

(a) *The functor $\tilde{H}_* = [H_*, \mathcal{D}]\colon [[\mathcal{B}^0, Ens], \mathcal{D}] \to [\mathcal{B}, \mathcal{D}]$ induced by the Yoneda embedding $H_*\colon \mathcal{B} \to [\mathcal{B}^0, Ens]$ has a left adjoint K and K is fully faithful.*

(b) *For $F\colon \mathcal{B} \to \mathcal{D}$, $K(F)$ is left adjoint to $\check{F}\colon \mathcal{D} \to [\mathcal{B}^0, Ens]$, and $K(F)\,H_* = F$.*

(c) *$G\colon [\mathcal{B}^0, Ens] \to \mathcal{D}$ has a right adjoint if and only if G preserves colimits. If this is the case, then $(G\,H_*)^\vee$ is right adjoint to G.*

(d) *K induces an equivalence between $[\mathcal{B}, \mathcal{D}]$ and the full subcategory of $[[\mathcal{B}^0, Ens], \mathcal{D}]$ whose objects are those functors $[\mathcal{B}^0, Ens] \to \mathcal{D}$ that preserve colimits.*

Proof. (a) is a special case of 17.1.6 (a), (c) where we have written K instead of V.

(b) We consider 17.2.9 for the following situation:

(1)

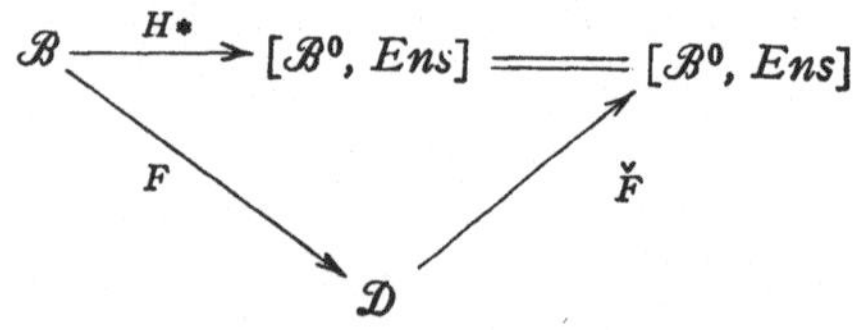

Here H_* is dense by 10.2.1, so that (i) of 17.2.9 is satisfied. (ii) is void, (iii) follows from the definition of $\check{F}$ (17. 2.3) and the Yoneda lemma:

(2) $$[F(X), A]_{\mathcal{D}} = \check{F}(A)\,(X) \cong [H_X, \check{F}(A)]_{[\mathcal{B}^0, Ens]}$$

and, by 4.2.4, these isomorphisms yield a bifunctor isomorphism

(3) $$[F(?), ??]_{\mathcal{D}} \cong [H_?, F(??)]_{[\mathcal{B}^0, Ens]}\,.$$

By lemma 17.2.9, $\check{F}$ has a left adjoint G and $G\,H_*$ is isomorphic to F. Therefore, $K(F)$ and $K(GH_*)$ are isomorphic. Being left adjoint to $\check{F}$, G preserves colimits (16.4.6), and H_* is dense. Thus it follows from 17.1.6 (13) that $K(G\,H_*) = K(\widetilde{H}_*(G))$ is isomorphic to G. So $K(F)$ is left adjoint to $\check{F}$ and $K(F)\,H_* \cong G\,H_* \cong F$. By 16.6.6, $K(F)$ can be changed by an isomorphism in such a way that $K(F)\,H_* = F$. If this is done for all $F: \mathcal{B} \to \mathcal{D}$, then, by 16.4.4, a functor K is created that is still left adjoint to $\widetilde{H}_*$.

(c) If G is to have a right adjoint, then G has to preserve colimits. Let this be the case. We set $F = G\,H_*$. By (b), $K(F)\,H_* = G\,H_*$, and by 17.2.7, $K(F)$ and G are isomorphic. Therefore, $\check{F} = (G\,H_*)^\vee$ is right adjoint to G.

(d) follows immediately from 16.3.8 by (a), (b), (c).

17.3.2 Theorem. *Let $\mathcal{B}$ be a small category, $\mathcal{C}$ a cocomplete category and $U: \mathcal{B} \to \mathcal{C}$ a functor. The following statements are equivalent:*

(a) *There is an equivalence $T: \mathcal{C} \to [\mathcal{B}^0, Ens]$ such that $T\,U$ is isomorphic to the Yoneda embedding $H_*: \mathcal{B} \to [\mathcal{B}^0, Ens]$.*

(b) *For every functor $F: \mathcal{B} \to \mathcal{D}$ into a cocomplete category $\mathcal{D}$, there is a functor $G: \mathcal{C} \to \mathcal{D}$ which preserves colimits and which is such that $G\,U$ is isomorphic to F. Furthermore, any two such functors are isomorphic.*

(c) *$\check{U}: \mathcal{C} \to [\mathcal{B}^0, Ens]$ is an equivalence.*

(d) *U is dense, $\check{U}$ preserves colimits, and $\check{U}\,U \cong H_*$.*

(e) *U is fully faithful and dense. Furthermore, for every $X \in |\mathcal{B}|$, $H^{U(X)}$: $\mathcal{C} \to Ens$ is a functor which preserves colimits.*

Proof. (a) $\Longrightarrow$ (b). Let S be equivalence-inverse to T. By 17.3.1 (b), $K(F)$ is a functor $[\mathcal{B}^0, Ens] \to \mathcal{D}$ which preserves colimits, so that $K(F)\,H_* = F$. Set $G = K(F)\,T$. If $G': \mathcal{C} \to \mathcal{D}$ preserves colimits, and if $G'\,U \cong F$, then $G'\,S$ preserves colimits, and $G'\,S\,H_* \cong G'\,S\,T\,U \cong$ $\cong G'\,U \cong F = K(F)\,H_*$ holds. Since H_* is dense, $G'\,S \cong K(F)$ follows from 17.2.7, and thus one has $G' \cong G'\,ST \cong K(F)\,T = G$.

(b) $\Longrightarrow$ (c). There is a functor $T: \mathcal{C} \to [\mathcal{B}^0, Ens]$ which preserves colimits such that $T\,U \cong H_*$. Now set $S = K(U)$ as in 17.3.1 (b). $T\,S$ and $S\,T$ preserve colimits, since S and T do. Now, $T\,S\,H_* =$ $= T\,K(U)\,H_* = T\,U \cong H_*$. By 17.2.7, $T\,S \cong 1_{[\mathcal{B}^0, Ens]}$. Further, $S\,T\,U \cong S\,H_* = K(U)\,H_* = U$. By (b) for $F = U$, $S\,T \cong 1_{\mathcal{C}}$. Therefore, T is equivalence-inverse to $K(U)$. By 17.3.1 (b) and 16.4.4, $\check{U}$ is isomorphic to T. One notices that (a) follows from (c).

(c) $\Longrightarrow$ (d). U is dense by 17.2.3. By 17.3.1 (b), $K(U)$ is equivalence-inverse to $\check{U}$. Thus $\check{U}\,U = \check{U}\,K(U)\,H_* \cong H_*$ follows.

(d) $\Longrightarrow$ (a). Since $\check{U}$ and $K(U)$ preserve colimits and since H_* and U are dense, 17.2.7 implies

$$\check{U}\,K(U) \cong 1_{[\mathscr{B}^0, Ens]}\,, \qquad \text{since} \qquad \check{U}\,K(U)\,H_* = \check{U}\,U \cong H_*\,,$$

$$K(U)\,\check{U} \cong 1_{\mathscr{C}}\,, \qquad \text{since} \qquad K(U)\,\check{U}\,U \cong K(U)\,H_* = U\,.$$

Taking $T = \check{U}$, (a) follows.

(d) $\Longleftrightarrow$ (e). By 17.2.5, U is fully faithful if and only if $\check{U}\,U \cong H_*$. For a diagram $D\colon \Sigma \to \mathscr{C}$, $\check{U}\,D = [U(?), D]_{\mathscr{C}}$ is a diagram in $[\mathscr{B}^0, Ens]$. It follows from the "pointwise" construction of colimits in functor categories that $\check{U}$ preserves colimits if and only if this is the case for all $H^{U(X)}$.

17.3.3 Theorem. *For the category $\mathscr{C}$ the following statements are equivalent:*

(a) *$\mathscr{C}$ is cocomplete and there is a dense small subcategory $\mathscr{C}'$ in $\mathscr{C}$.*

(b) *There is a small category $\mathscr{B}$ and a fully faithful functor $T\colon \mathscr{C} \to \to [\mathscr{B}^0, Ens]$ with a left adjoint S.*

Furthermore, if (a), and thus (b), *is satisfied, then $\mathscr{C}$ is also complete.*

Proof. Let (a) be valid. We set $\mathscr{B} = \mathscr{C}'$ and let $F\colon \mathscr{B} \to \mathscr{C}$ denote the inclusion. Then $T = \check{F}$ is fully faithful by 17.2.3, and right adjoint to $K(F)$ by 17.3.1 (b). Now let (b) be valid. Let $\mathscr{C}'$ be the full subcategory of $\mathscr{C}$ with objects $S(H_X)$ for $X \in |\mathscr{B}|$. Since $\mathscr{B}$ is small, so is $\mathscr{C}'$. Since T is fully faithful and $H_*\colon \mathscr{B} \to [\mathscr{B}^0, Ens]$ is dense, $S\,H_*$ is dense by 17.2.6 (d). By 17.2.4, $\mathscr{C}'$ is dense in $\mathscr{C}$. Also, $[\mathscr{B}^0, Ens]$ is complete and cocomplete. By 16.6.1 (b), the same is true for $\mathscr{C}$.

17.3.4 The additive case. By 17.1.6 and 17.2.10, the additive version of 17.3.1 is valid for $Add(\mathscr{B}^0, Ab)$ and $Add(Add(\mathscr{B}^0, Ab), \mathscr{D})$ instead of $[\mathscr{B}^0, Ens]$ and $[[\mathscr{B}^0, Ens], \mathscr{D}]$, at least if $\mathscr{B}$ has finite products. If this is not the case, $\mathscr{B}$ has to be completed to $\mathscr{B}'$ as in 16.3.10. If $R\colon \mathscr{B} \to \mathscr{B}'$ is the inclusion, one finds oneself in the following situation:

(4)

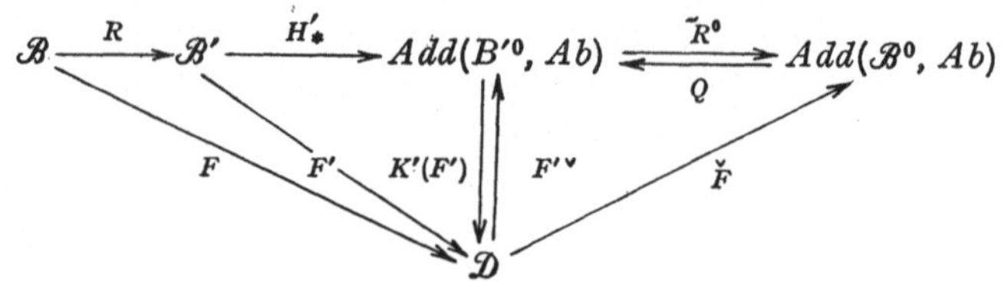

Here H'_* is the Yoneda embedding for $\mathscr{B}'$ and $\tilde{R}^0\,H'_*\,R = H_*$. F' is an extension of F. Then $\tilde{R}^0\,F'^{\vee} = \check{F}$. Let Q be equivalence-inverse to $\tilde{R}^0$. Taking $K(F) = K'(F')\,Q$, one has returned to the situation in 17.1.6 (f). One obtains thus the additive version of 17.3.1 in this case also.

The additive version of 17.3.2 is valid if $\mathcal{B}$ has finite products. If this is not the case, one completes $\mathcal{B}$ again with the result, according to (4), that here too the additive versions of 17.3.2 (a), (b), (c) are equivalent. In (d), (e), U is fully faithful and, by 16.3.8, $\mathcal{B}$ is equivalent to the subcategory of $\mathcal{C}$ which is the image of U. If, in 17.3.2 (d), (e), the requirement that U is dense is changed to requiring that completing the image by means of finite biproducts yields a dense subcategory of $\mathcal{C}$, then one obtains the statements which are equivalent to (a), (b), (c). We will return to this situation in 17.4.9.

The additive version of 17.3.3 is valid if, in (a), $\mathcal{C}'$ is a small subcategory which, completed by finite products, becomes dense. We choose this round about formulation because, by (4), it is still true that T in (b) is produced by the inclusion $F: \mathcal{B} \to \mathcal{C}$ ($\mathcal{B} = \mathcal{C}'$) as $\check{F}$; we will make use of this later.

17.4 Small Projective Objects

From 17.3.1 and 17.3.2 one concludes that in the category $[\mathcal{B}^0, Ens]$ the objects H_X (with $X \in |\mathcal{B}|$) are "independent with respect to colimits". The corresponding situation in the additive case leads to the formation of a concept which is adapted to cocomplete abelian categories.

17.4.1 Definition. An *object* P of the additive category $\mathcal{C}$ is called *small* if the *Ab*-valued functor $H^P = [P, ?]_\mathcal{C}$ preserves coproducts.

17.4.2 Proposition. *Let $\mathcal{C}$ be a cocomplete abelian category. The object $P \in |\mathcal{C}|$ is a small projective if and only if the Ab-valued functor H^P preserves colimits.*

Proof. If P is projective, then H^P is exact by 13.2.6, and it therefore preserves cokernels. If P is also small, then H^P preserves colimits by the dual of 7.4.5, since $\mathcal{C}$ is cocomplete. The converse follows from the definitions 17.4.1 and 10.4.1 by the dual of 7.8.9.

17.4.3 Proposition. *Let $\mathcal{B}$ be a small additive category. The objects H_X with $X \in |\mathcal{B}|$ form a generating set of small projective objects in $Add(\mathcal{B}^0, Ab)$.*

Proof. The objects H_X form a generating set by 10.5.2. Therefore, the statement follows from 17.4.2 and 10.4.3.

17.4.4 In ${}_R Mod$, ${}_R R$ is a small projective generator.

Proof. $[{}_R R_R, ?]$ is isomorphic to the identity functor of ${}_R Mod$ (15.1.8). The forgetful functor U: ${}_R Mod \to Ab$ preserves colimits (15.1.4). Since $U[{}_R R_R, ?] = [{}_R R, ?]$ (15.1.8) and by 17.4.2, ${}_R R$ is a small projective. Furthermore, ${}_R R$ is a generator (15.1.8).

17.4.5 Proposition. *Let $\mathscr{C}$ be an additive category with coproducts. The object P is small if and only if every morphism of P into a coproduct factors through the injection of a finite subcoproduct. I.e., for $f: P \to \coprod_{i\in I} A_i$, there is a finite subset J of I such that f is of the form $in_J f'$ with $f': P \to \coprod_{j\in J} A_j$ and $in_J: \coprod_{j\in J} A_j \to \coprod_{i\in I} A_i$ (compare 14.5.4).*

Proof. If P is any object, then there is the morphism

$$(1) \qquad h: \coprod [P, A_i] \to [P, \coprod A_i] \quad \text{with} \quad h\, in_i' = [P, in_i]$$

in Ab, where the injections of the coproduct in Ab or, resp., $\mathscr{C}$ have been denoted by in_i' or, resp., in_i. Every in_i is a coretraction and there is a corresponding retraction p_i with

$$(2) \qquad p_i\, in_i = 1; \qquad p_i\, in_j = 0 \quad \text{for} \quad j \neq i\,.$$

There are analogous retractions p_i' in Ab. An element a of $\coprod [P, A_i]$ determines uniquely a family $\{a_i \mid a_i \in [P, A_i]\}$ so that only a finite number of a_i are different from 0, and one has

$$(3) \qquad a = \Sigma\, in_i'(a_i) = \Sigma\, in_i'\, p_i'(a); \qquad a_i = p_i'(a)\,,$$

$$(4) \qquad h(a) = \Sigma\, h\, in_i'(a_i) = \Sigma\, [P, in_i]\, (a_i)\,,$$

$$(5) \qquad [P, p_i]\, \big(h(a)\big) = a_i\,.$$

This implies always that h is a monomorphism. h is an epimorphism if and only if, given any $f \in [P, \coprod A_i]$, there is an $a \in \coprod [P, A_i]$ with $h(a) = f$, and by (4), (5) this is the case if and only if f can be represented in the form

$$(6) \qquad f = in_J\, f'\,,$$

where J is the finite set of those indices for which $a_i = [P, p_i]\,(f)$ is not 0. This completes the proof.

Remark. By (2) and 12.2.5, $f: P \to \coprod A_i$ factors through a finite subcoproduct of $\coprod A_i$ if and only if

$$(7) \qquad f = \Sigma\, in_i\, p_i\, f$$

with only a finite number of summands different from 0.

17.4.6 Proposition. *Let $\mathscr{C}$ be a cocomplete abelian category.*

(a) *If P is a small projective, then it satisfies the following condition:*

(S) *The union of a filtered family $\{m_i: A_i \rightarrowtail P\}$ of monomorphisms with codomain P is equivalent to 1_P only if this is already so for one monomorphism of the family.*

(b) *If $\mathscr{C}$ is a Grothendieck category, then every object with the property* (S) *is small.*

Proof. (a) We use 14.2.5 with $A = P$, $T(i) = M(i) = A_i$, $\alpha_i = \alpha_i'' = m_i$, where L is a colimit object of the filtered family $\{A_i\}$. By assumption, we can choose $f'' = 1_P$, and f is an epimorphism. By 14.5.3, there is an epimorphism $c\colon \amalg A_i \twoheadrightarrow L$, and $f\,c$ is an epimorphism. Since P is projective, there is a $g\colon P \to \amalg A_i$ with $f\,c\,g = 1_P$. Since P is small, g factors through a finite subcoproduct (17.4.5) and, by (7), it is of the form $g = \sum in_i\,p_i\,g$, where only finitely many $p_i\,g\colon P \to A_i$ are not zero. Let those be $i = i_1, i_2, \ldots, i_n$. Since the index category is filtered, there is an index j with arrows $a_\nu\colon i_\nu \to j$ for $\nu = 1, 2, \ldots, n$ and, for the morphism $g' = \sum T(a_\nu)\,p_{i_\nu}\,g\colon P \to A_j$, $c\,in_j\,g' = c\,g$; this follows from $c\,(in_j - in_{i_\nu}\,T(a_\nu)) = 0$ by 14.5.3. Now (again using the notation from 14.2.5), $m_j = \alpha_j = f\,\lambda_j = f\,c\,in_j$ and, therefore, $m_j\,g' = f\,c\,in_j\,g' = f\,c\,g = 1_P$. This makes m_j a monomorphic retraction and thus an isomorphism.

(b) Set $A = \amalg A_i$ and let $f\colon P \to A$ be a morphism. By 14.5.4, $1_A = \bigcup in_J$, where in_J ranges over the injections of the finite subcoproducts of $\amalg A_i$. $\{in_J\}$ is filtered and, by 14.6.2, $1_P = f^{-1}(\bigcup in_J) = \bigcup f^{-1}(in_J)$. By (S), $f^{-1}(in_J) = 1_P$ for a suitable J. The corresponding pullback shows that f factors through in_J. P is small by 17.4.5.

17.4.7 Proposition. *Let $\mathcal{B}$ be a small additive category with finite biproducts. We assume further that in $\mathcal{B}$ every idempotent morphism h (12.6.1) is representable in the form $h = i\,r$, where i is a coretraction and r a corresponding retraction (idempotents split). An object T of $Add(\mathcal{B}, Ab)$ is a small projective if and only if T is a representable functor.*

Remark. By 12.6.3, idempotents in $\mathcal{B}$ do split if $\mathcal{B}$ has kernels or cokernels.

Proof. Let T be a small projective. By 17.2.10, T is a colimit of representable functors. Therefore, there is a coproduct $U = \amalg[A_i, ?]_{\mathcal{B}}$ with an epimorphism $c\colon U \twoheadrightarrow T$ in $Add(\mathcal{B}, Ab)$. Since T is a small projective, there is a g with $c\,g = 1_T$, and g can be represented in the form $in_J\,g'$, where in_J is the injection of a finite subcoproduct in U. This is a biproduct and isomorphic to $V = [\oplus A_j, ?]$ with $j \in J$ (8.3.4). Since g is a coretraction, so is g', and, taking $B = \oplus A_j$, one obtains a coretraction $\varphi\colon T \to H^B$. Let χ be a corresponding retraction. $\varphi\,\chi\colon H^B \to T \to H^B$ is then idempotent and, by the Yoneda lemma, there is an $h\colon B \to B$ with $\varphi\,\chi = H^h$. Since $H^h\,H^h = \varphi\,\chi = H^h$, h is idempotent. By assumption, there is a retraction $r\colon B \to C$ with a coretraction $i\colon C \to B$ such that $i\,r = h$. Now, $\varphi\,\chi = H^r\,H^i$, where φ and H^r are coretractions and χ, H^i retractions. By 12.6.2, $\psi = 1 - \varphi\,\chi = 1 - H^r\,H^i\colon H^B \to B^B$ is idempotent and, by 12.6.3, φ and H^r are both kernels of ψ. Therefore, T is isomorphic to H^C. The converse is true by 17.4.3.

17.4.8 Lemma. *Let $\mathcal{C}$ be a cocomplete abelian category and $\mathcal{B}$ a small full subcategory with finite biproducts. $\mathcal{B}$ is dense in $\mathcal{C}$ if one of the following assumptions holds:*

(a) $|\mathcal{B}|$ *is a generating set of small objects in* $\mathcal{C}$.
(b) $|\mathcal{B}|$ *is a generating set for* $\mathcal{C}$ *and* $\mathcal{C}$ *is a Grothendieck category.*

Proof. Let $U: \mathcal{B} \to \mathcal{C}$ be the inclusion. For $A \in |\mathcal{C}|$, we form U/A and $\gamma_A: U\,Q_A \to A_{U/A}$ as in 17.1.3. We have to show that (A, γ_A) is a colimit of $U\,Q_A$.

Since U is an inclusion, any object (P, a) of U/A is uniquely determined by the morphism $a: U(P) \to A$, where $U(P) = P$. So instead of $\gamma_A = \{\gamma_{A,(P,a)}\}$ we set $\gamma_A = \{i\}$ with $i: P_i \to A$, where i ranges over all morphisms with codomain A and domain in $\mathcal{B}$. Since the objects of $\mathcal{B}$ form a generating set for $\mathcal{C}$, 10.5.4 states that there is an epimorphism

$$g: \amalg P_i \twoheadrightarrow A \qquad \text{with} \quad g\, in_i = i\,.$$

A natural transformation $\mu: U\,Q_A \to B_{U/A}$ induces a morphism

$$q: \amalg P_i \to B \qquad \text{with } q\, in_i = \mu_i\,.$$

There is at most one $f: A \to B$ with $f\,g = q$, since g is an epimorphism. We show that such an f exists. The statement will then follow from the definition of colimits.

First, let (a) be satisfied and let $k: K \to \amalg P_i$ be a kernel of g. Since g is an epimorphism, g is a cokernel of k. The existence of f follows from the definition of cokernels, provided $q\,k = 0$. By the definition of a generating set, this is the case if and only if $q\,k\,u = 0$ for every morphism $u: Q \to K$ with $Q \in |\mathcal{B}|$.

For $v = k\,u: Q \to \amalg P_i$ we have to show that $q\,v = 0$. Since Q is small, 17.4.5 applies to v and, by (7), v can be represented in the form

$$v = \sum in_i\, v_i \qquad \text{with} \qquad v_i = p_i\, v: Q \to P_i\,,$$

where only finitely many v_i are not 0. Let those be $i = i_1, i_2, \ldots, i_n$. There exists an object M in $\mathcal{B}$ which is a biproduct of $P_{i_1}, P_{i_2}, \ldots, P_{i_n}$ with injections h_ν, and there exists the injection $in_J: M \to \amalg A_i$ with $in_J\, h_\nu = in_{i_\nu}$. (This follows from 12.2.7, since U is additive.) Further, the biproduct $\oplus\, Q_\nu$ with $Q_\nu = Q$ for $\nu = 1, 2, \ldots, n$ also exists in $\mathcal{B}$. Let its injections be k_ν. Let $\Delta: Q \to \oplus\, Q_\nu$ be the diagonal map. Then let $w_\nu = h_\nu\, v_{i_\nu}$, let $w': \oplus\, Q_\nu \to M$ be the morphism defined by $w'\, k_\nu = w_\nu: Q \to M$; let $w = w'\,\Delta$ and $g\, in_J = m: M \to A$. The following diagram is then commutative:

(8),

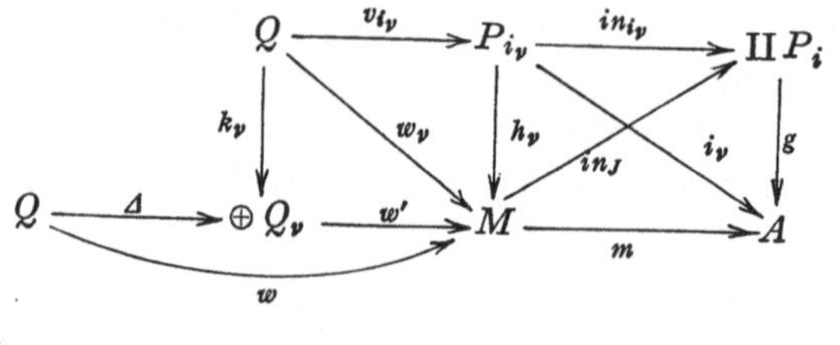

By 12.2.6 (6),

$$m\,w = m \sum w_\nu = g\, in_J \sum h_\nu\, v_{i_\nu} = g \sum in_{i_\nu}\, v_{i_\nu} = g\,v = g\,k\,u = 0$$

holds. Since $M \in |\mathcal{B}|$, m is one of the morphisms in the set $\{i\} = \gamma_A$, and so is $m\,w$. $m\,w = 0$ implies $\mu_{mw} = 0$, since for $0: Q \to Q$, $m\,w = m\,w\,0$ and thus $\mu_{mw} = \mu_{mw}0 = 0$, since $\mu: U\,Q_A \to B_{A/U}$ is a natural transformation. Taking into account that U is a full embedding, the commutativity of (8) implies now

$$0 = \mu_{mw} = \mu_m\,w = \mu_m \sum w_\nu = \sum \mu_m\,h_\nu\,v_{i_\nu} = \sum \mu_{i_\nu}\,v_{i_\nu}$$
$$= \sum q\,in_{i_\nu}\,v_{i_\nu} = q\,v\,,$$

which is what remained to be shown in case (a).

Now let (b) be satisfied, let $\amalg P_j$ be a finite subcoproduct of $\amalg P_i$ belonging to the index set J and let $in_J: \amalg P_j \to \amalg P_i$ be the inclusion. We may assume $\amalg P_j$ to be an object of $\mathcal{B}$. Let $k_J: K_J \to \amalg P_j$ be a kernel of $g\,in_J$ and again $k: K \to \amalg P_i$ a kernel of g. Since $\mathcal{C}$ is now a Grothendieck category, so that kernels commute with filtered colimits, and since g is a filtered colimit of the morphisms $g\,in_J$ by 14.5.4 (if J ranges over all finite subsets of the index set of $\amalg P_i$), k is a filtered colimit of the kernels $k_J \cdot q\,k = 0$ follows, provided $q\,in_J\,k_J = 0$ always, which is the case if and only if for every morphism $u: Q \to K_J$ with $Q \in |\mathcal{B}|$, $q\,in_J\,k_J\,u = 0$. Setting $v = in_J\,k_J\,u$, the statement follows as before.

17.4.9 Theorem. *A cocomplete abelian category $\mathcal{C}$ is equivalent to a category of the form $Add(\mathcal{B}^0, Ab)$ for a suitable small additive category $\mathcal{B}$ if and only if $\mathcal{C}$ has a generating set of small projectives. If this is the case, and if $\mathcal{B}$ is the full subcategory of $\mathcal{C}$ determined by these objects, then the Yoneda embedding $H_*: \mathcal{B} \to Add(\mathcal{B}^0, Ab)$ has an equivalence $\mathcal{C} \to Add(\mathcal{B}^0, Ab)$ as an extension.*

Proof. Let $\mathcal{B}$ be a subcategory possessing the properties described above. If $\mathcal{B}$ does not yet have finite biproducts, it is to be completed to $\mathcal{B}'$ as in 16.3.10. For the finite biproducts $\oplus P_j$ one has $[\oplus P_j, ?] \cong \oplus [P_j, ?]$. By 17.4.2, the objects of $\mathcal{B}'$ are also small projectives and they, of course, also constitute a generating set for $\mathcal{C}$. By 17.4.8, the inclusion $U: \mathcal{B}' \to \mathcal{C}$ is dense. The equivalence of $\mathcal{C}$ and $Add(\mathcal{B}'^0, Ab)$ stems from the additive version of 17.3.2, as in 17.3.4. From the factorization of H_*, as in 17.3.4 (4), the equivalence with $Add(\mathcal{B}^0, Ab)$ follows. The converse is implied by 17.4.3 and the fact that generating sets of small projective objects are preserved by equivalences.

17.4.10 Theorem. *An additive category is equivalent to a module category ${}_R Mod$ if and only if it is abelian and cocomplete and if in addition it has a small projective generator.*

Proof. ${}_R Mod$ has the required properties by 15.1.4 and 14.4.4. Now let $\mathcal{C}$ be cocomplete and abelian, and let G be a small projective generator. By 17.4.9, $\mathcal{C}$ is equivalent to $Add(\mathcal{B}^0, Ab)$, where $\mathcal{B}$ is the full subcategory of $\mathscr{C}$ with G as its only object. But, if $R = [G, G]_{\mathscr{C}}$, then $Add(\mathscr{B}^0, Ab) = {}_{R^0}Mod$ by 15.1.2.

17.4.11 Remarks. According to lemma 2 of 15.3.7, every Grothendieck category with a generator has a full embedding in a module category. By 15.1.4 and 13.2.6, this embedding is exact if and only if the generator is projective. We note here that, by 17.3.4, lemma 2 of 15.3.7 follows from 17.4.8 (b) and the additive version of 17.3.3. The direct proof given in 15.3.7 is a variation of the proof of 17.5.5 according to the proof of 17.4.8.

The proof of 17.4.9 shows in particular: If $\mathscr{B}$ is a small, additive category, then $Add(\mathscr{B}^0, Ab)$ has a small, dense full subcategory whose objects are finite biproducts of objects H_A.

17.5 Finitely Generated Objects

17.5.1 Definition. Let a generating set $\mathfrak{G}$ be fixed in the category $\mathscr{C}$. An *object* A is called *finitely generated* (with respect to $\mathfrak{G}$) if there is an epimorphism $\coprod G_j \twoheadrightarrow A$ from a finite coproduct of objects of $\mathfrak{G}$ to A.

In ${}_RMod$, the fixed generating set is always chosen to be the one object ${}_RR$ (in Ab it is $\mathbf{Z}$). "Finitely generated" then agrees with the usual definition, where a finitely generated module is generated by finitely many of its elements.

17.5.2 Proposition. *Every finitely generated module is small, every small projective module is finitely generated.*

Proof. The first statement follows immediately from 17.4.5 by what has been said above. Since ${}_RR$ is a generator of ${}_RMod$, every module M is a quotient of a free module. If M is a small projective, then 17.4.5 implies that M is a retract of a finitely generated free module, and that it is thus finitely generated itself.

17.5.3 Proposition. *Let $\mathscr{C}$ be a Grothendieck category with a distinguished generating set $\mathfrak{G}$. For every $A \in |\mathscr{C}|$, the monomorphisms with codomain A and finitely generated domain form a filtered class with union 1_A. There are final sets in this class, and A is a colimit object of the corresponding filtered families of domains.*

Every object is a filtered colimit of its finitely generated subobjects.

Proof. By 10.5.4, there is an epimorphism $u\colon \coprod G_e \twoheadrightarrow A$, where all G_e belong to $\mathfrak{G}$. For $G = \coprod G_e$, 1_G is the union of injections in_J of finite subcoproducts (14.5.4). Since u is an epimorphism, $u(1_G) = 1_A$ (14.4.2), and by 14.4.7, $1_A = \bigcup u(in_J)$.

By 14.2.6, the union of two monomorphisms with finitely generated domain has again a finitely generated domain. From this the first statement follows, by 14.1.2. The second statement follows from the fact that $\mathscr{C}$ is well-powered (10.6.3). The third one is implied by the second, because $\mathscr{C}$ is a Grothendieck category (14.6.3).

17.5.4 Definition. Let $\mathscr{C}$ be a finitely cocomplete category with a distinguished generating set $\mathfrak{G}$. An *object* A is called *finitely presentable* if there are finite coproducts M, N of objects of $\mathfrak{G}$ and morphisms f, g: $M \to N$ such that A is the codomain of a coequalizer c of f and g. One then calls

$$M \underset{g}{\overset{f}{\rightrightarrows}} N \xrightarrow{c} A$$

a finite presentation of A. In the additive case one can take $g = 0$; i.e., $c = \operatorname{coker} f$.

In ${}_R Mod$ and in the category of groups, this means that one is dealing with a representation by finitely many generators (N) and finitely many relations (M).

17.5.5 Proposition. *Let $\mathscr{C}$ be an abelian category with a projective generator G. The right operation of the ring $R = [G, G]_{\mathscr{C}}$ on the groups $[G, A]_{\mathscr{C}}$ determines a functor $T = [{}_R G, ?]_{\mathscr{C}}$: $\mathscr{C} \to Mod_R$. T is an exact embedding and, for every finitely generated $A \in |\mathscr{C}|$, $T_{A,B}$: $[A, B]_{\mathscr{C}} \to$ $\to \operatorname{Hom}_R(T(A), T(B))$ is an isomorphism.*

Proof. Since G is a projective generator, H^G: $\mathscr{C} \to Ab$ is an exact embedding (13.2.6). By 15.1.4, the same is true for T. If A is finitely generated, then there is an epimorphism p: $\bigoplus G_j \twoheadrightarrow A$, where $\bigoplus G_j$ is a finite biproduct with factors $G_j = G$, injections in_j and projections pr_j. Let φ: $T(A) \to T(B)$ be a module homomorphism. Now, $p\, in_j$: $G \to A$ is an element of $T(A)$ and thus $\varphi(p\, in_j): G \to B$ is defined. We consider the morphism $q = \sum \varphi(p\, in_j)\, pr_j$: $\bigoplus G_j \to B$ and show that there is a $g: A \to B$ in $\mathscr{C}$ such that $q = g\, p$. To do this, let $k: K \to \bigoplus P_j$ be a kernel of p, so that p is a cokernel of k. The existence of g follows from $q\, k = 0$, by the definition of cokernels. Since G is a generator, it suffices to show that for every morphism u: $G \to K$, $q\, k\, u = 0$. But, since φ is a module homomorphism and $pr_j\, k\, u$: $G \to G$ an element of R, one has

$$\begin{aligned} q\, k\, u &= \textstyle\sum \varphi(p\, in_j)\, pr_j\, k\, u = \varphi\, (\sum p\, in_j\, pr_j\, k\, u) \\ &= \varphi(p\, k\, u) = 0\,. \end{aligned}$$

Now we show that φ and $T(g)$ are the same map $T(A) \to T(B)$. Let $a: G \to A$ be an element of $T(A)$. Since G is projective and p an epimorphism, there is an $f: G \to \bigoplus G_j$ with $p\, f = a$. Since $pr_j\, f: G \to G$ is an element of R,

$$\begin{aligned} T(g)\, (a) &= g\, a = g\, p\, f = q\, f = \textstyle\sum \varphi(p\, in_j)\, pr_j\, f \\ &= \varphi(\textstyle\sum p\, in_j\, pr_j\, f) = \varphi(p\, f) = \varphi(a) \end{aligned}$$

follows.

Remark. Since $\mathcal{C}$ is finitely cocomplete, since T preserves finite biproducts and cokernels, and since $T(G) = R_R$, every finitely presentable right R-module is isomorphic to one of the form $T(A)$.

17.6 Natural Transformations with Parameters

17.6.1 Let $F, G: \mathcal{M} \to \mathcal{C}$ be functors. By 2.6.1, a natural transformation $\alpha: F \to G$ is a family $\{\alpha_M\}_{M \in |\mathcal{M}|}$ of $\mathcal{C}$-morphisms satisfying the following conditions

$$(1) \qquad \begin{aligned} &\alpha_M \in [F(M), G(M)]_{\mathcal{C}}\,, \\ &[F(p), G(N)]\,(\alpha_N) = [F(M), G(p)]\,(\alpha_M) \end{aligned}$$

for an arbitrary $p: M \to N$ in $\mathcal{M}$.

17.6.2 Let $P: \mathcal{A} \times \mathcal{B} \to \mathcal{C}$ be a bifunctor. The categories $\mathcal{B}$ and $\mathcal{C}$ can be replaced by $[\mathcal{M}, \mathcal{B}]$ and $[\mathcal{M}, \mathcal{C}]$ for any $\mathcal{M}$, giving the bifunctor

$$(2) \qquad \overline{P}: \mathcal{A} \times [\mathcal{M}, \mathcal{B}] \to [\mathcal{M}, \mathcal{C}]$$

described by

$$(2a) \qquad (A, U) \mapsto P(A, U(?))\,,$$

$$(2b) \qquad (f, U) \mapsto P(f, U(?))\,,$$

$$(2c) \qquad (A, \xi) \mapsto \{P(A, \xi_X)\}_{X \in |\mathcal{M}|}\,.$$

For example, in this way one obtains from the Hom-functor for $\mathcal{B}$ a contra-co-variant functor which is described by $(A, U) \mapsto [A, U(?)]_{\mathcal{B}}$ for objects. The dual of 4.5.3 follows from this using the canonical isomorphisms

$$[\mathcal{B}^0 \times [\mathcal{M}, \mathcal{B}], [\mathcal{M}, Ens]] \cong [[\mathcal{M}, \mathcal{B}], [\mathcal{B}^0, [\mathcal{M}, Ens]]]$$

and

$$[\mathcal{B}^0, [\mathcal{M}, Ens]] \cong [\mathcal{B}^0 \times \mathcal{M}, Ens]\,.$$

A corresponding replacement in the contravariant argument of a contra-co-variant functor results from the above by means of a partial dualization. Note that, by 4.5.6, $[\mathcal{M}^0, \mathcal{B}^0]$ may be regarded as the category which is dual to $[\mathcal{M}, \mathcal{B}]$. We make use of the conventions 2.4.5, 2.6.6 in the following.

17.6.3 We assume that besides $P: \mathcal{A} \times \mathcal{B} \to \mathcal{C}$ there is also a bifunctor $Q: \mathcal{X} \times \mathcal{Y} \to \mathcal{C}$. Through a replacement in Q a bifunctor

$$\overline{Q}: \mathcal{X} \times [\mathcal{M}, \mathcal{Y}] \to [\mathcal{M}, \mathcal{C}]$$

is given. For $U: \mathcal{M} \to \mathcal{B}$ and $V: \mathcal{M} \to \mathcal{Y}$, one can consider the $\mathfrak{V}$-set of natural transformations $P(A, U(?)) \to Q(X, V(?))$ for fixed $A \in |\mathcal{A}|$ and $X \in |\mathcal{X}|$. To be clear, we use the notation

$$(3) \qquad \mathrm{Nat}(P(A, U), Q(X, V)) \quad \text{or} \quad \mathrm{Nat}(P(A, U(?)), Q(X, V(?))$$

for it. (2), (2a), (2b), (2c) show (3) to be a functor

$$\mathcal{A}^0 \times [\mathcal{M}, \mathcal{B}]^0 \times \mathcal{X} \times [\mathcal{M}, \mathcal{Y}] \to ENS\,.$$

It is produced by composing the contra-co-variant Hom-functor of $[\mathcal{M}, \mathcal{C}]$ with Op $\overline{P}$ Op $\times\, \overline{Q}$.

17.6.4 Lemma. *Let* $P\colon \mathcal{A} \times \mathcal{B} \to \mathcal{C}$, $Q\colon \mathcal{X} \times \mathcal{Y} \to \mathcal{C}$, $R\colon \mathcal{X}^0 \times \mathcal{B} \to \mathcal{D}$, $S\colon \mathcal{A}^0 \times \mathcal{Y} \to \mathcal{D}$ *be bifunctors, and let there be a natural transformation*

(4) $\varphi_{A,B,X,Y}\colon [P(A, B), Q(X, Y)]_{\mathcal{C}} \to [R(X, B), S(A, Y)]_{\mathcal{D}}$

of functors $\mathcal{A}^0 \times \mathcal{B}^0 \times \mathcal{X} \times \mathcal{Y} \to Ens$. *Then there is a natural transformation*

(5) $\chi_{A,U,X,V}\colon \mathrm{Nat}\,(P(A, U), Q(X, V)) \to \mathrm{Nat}\,(R(X, U), S(A, V))$

of functors $\mathcal{A}^0 \times [\mathcal{M}, \mathcal{B}]^0 \times \mathcal{X} \times [\mathcal{M}, \mathcal{Y}] \to ENS$, *which for* $\alpha\colon (P(A, U) \to Q(X, V)$ *and* $M \in |\mathcal{M}|$ *is described by*

(6) $$(\chi_{A,U,X,V}(\alpha))_M = \varphi_{A,U(M),X,V(M)}(\alpha_M)\,.$$

If (4) *is an isomorphism, then so is* (5).

Proof. It first has to be shown that the left side of (6) really is a natural transformation $R(X, U) \to S(A, V)$. Suppose that one has $p\colon M \to N$ in $\mathcal{M}$. Since (4) is a natural transformation, (1) implies

$$\begin{aligned}[R(X, U(p), S(A, V(N))]\, \varphi_{A,U(N),X,V(N)}(\alpha_N) &= \\ = \varphi_{A,U(M),X,V(N)}\,[P(A, U(p), Q(X, V(N))]\,(\alpha_N) &= \\ = \varphi_{A,U(M),X,V(N)}\,[P(A, U(M)), Q(X, V(p))]\,(\alpha_M) &= \\ = [R(X, U(M)), S(A, V(p))]\, \varphi_{A,U(M),X,V(M)}(\alpha_M)&\end{aligned}$$

and this constitutes (1) for $\chi_{A,U,X,V}(\alpha)$.

By 2.6.8, the verification that (5) is a natural transformation can be carried out seperately for every argument. Suppose that there is an $f\colon A' \to A$ in $\mathcal{A}$. Then $\{[P(f, U(M)), Q(X, V(M))]\,(\alpha_M)\}$ is a natural transformation $\alpha'\colon P(A', U) \to Q(X, V)$, and it follows that $\{\chi_{A,U,X,V}\}$ is natural with respect to A again by a simple calculation based on (6) and (4). The conclusions for the other arguments are analogous. The last claim follows immediately from (6).

17.6.5 Remarks. (a) For functors $P\colon \mathcal{A} \times \mathcal{B} \to \mathcal{C}$, $Q\colon \mathcal{X} \times \mathcal{Y} \to \mathcal{C}$, $R'\colon \mathcal{X} \times \mathcal{B}^0 \to \mathcal{D}$, $S'\colon \mathcal{A} \times \mathcal{Y}^0 \to \mathcal{D}$, a natural transformation (isomorphism)

(4′) $\varphi_{A,B,X,Y}\colon [P(A, B), Q(X, Y)]_{\mathcal{C}} \to [S'(A, Y), R'(X, B)]_{\mathcal{D}}$

gives a natural transformation (isomorphism)

(5′) $\chi_{A,U,X,V}\colon \mathrm{Nat}\,(P(A, U), Q(X, V)) \to \mathrm{Nat}\,(S'(A, V), R'(X, U))$

for which (6) is also valid. This follows from 17.6.4 if $\mathcal{D}$ is replaced by $\mathcal{D}^0$ and if convention 2.4.5 is used.

(b) If $\mathcal{A}$ is a terminal category, then there is the canonical isomorphism $pr_2\colon \mathcal{A} \times \mathcal{B} \to \mathcal{B}$, and P, S may be regarded simply as functors $\mathcal{B} \to \mathcal{C}$ or, resp., $\mathcal{Y} \to \mathcal{D}$. Corresponding statements about $\mathcal{X}$, Q, R can be made. Observe, on the other hand, that $\mathcal{A}$ and $\mathcal{X}$ can be products of categories.

17.6.6 Example. For $U\colon \mathcal{M} \to \mathcal{D}$, $V\colon \mathcal{M} \to \mathcal{C}$, one obtains from the generalized adjunction 16.7.1

(7) $$\varphi_{D,A,C}\colon [S(D,A), C]_{\mathcal{C}} \xrightarrow{\cong} [D, T(A,C)]_{\mathcal{D}}$$

the generalized adjunction

(8) $$\chi_{U,A,V}\colon \mathrm{Nat}\,(S(U,A), V) \xrightarrow{\cong} \mathrm{Nat}\,\big(U, T(A,V)\big)\,.$$

17.6.7 The above obviously carries over to additive categories and additive or, resp., multiadditive functors, whereby *Ab* and *AB* take the place of *Ens* and *ENS*.

17.7 Tensor Products over Small Categories

17.7.1 Definition. Let $\mathcal{B}$ be a small additive category and $\mathcal{C}$ a cocomplete, additive category. For the additive functor $F\colon \mathcal{B} \to \mathcal{C}$, the additive version of 17.3.1 (b) yields as in 17.3.4 the adjunction isomorphism

(1) $$\varphi_F\colon [K(F)(?), ??]_{\mathcal{C}} \xrightarrow{\cong} [?, \check{F}(??)]_{Add\,(\mathcal{B}^0, Ab)}\,.$$

By the additive version of 17.6.2, the rule $(F, A) \mapsto \check{F}(A) = [F(?), A]_{\mathcal{C}}$ extends to a biadditive contra-co-variant functor $\langle ?, ??\rangle$ with corresponding bifunctor

(2) $$\langle \mathrm{Op}(?), ??\rangle\colon Add\,(\mathcal{B}, \mathcal{C})^0 \times \mathcal{C} \to Add\,(\mathcal{B}^0, Ab)\,.$$

Now, (1) states in particular that $[M, \langle F, ??\rangle]$ is always representable. By 16.7.1, there exists a bifunctor *tensor product*

(3) $$? \otimes_{\mathcal{B}} ??\colon Add\,(\mathcal{B}^0, Ab) \times Add\,(\mathcal{B}, \mathcal{C}) \to \mathcal{C}$$

with an isomorphism of trifunctors

(4) $$\varphi_{M,F,A}\colon [M \otimes_{\mathcal{B}} F, A]_{\mathcal{C}} \xrightarrow{\cong} [M, \langle F, A\rangle]_{Add\,(\mathcal{B}^0, Ab)}\,.$$

17.7.2 Theorem. (a) *For the Yoneda embedding* $H_*\colon \mathcal{B} \to Add(\mathcal{B}^0, Ab)$ *and the evaluation functor* $E\colon Add\,(\mathcal{B}, \mathcal{C}) \times \mathcal{B} \to \mathcal{C}$, *there is a bifunctor isomorphism*

(5) $$\chi_{X,F}\colon H_X \otimes_{\mathcal{B}} F \xrightarrow{\cong} E(F, X) = F(X)\,.$$

(b) *The tensor product* (3) *preserves colimits in* $Add\,(\mathcal{B}^0, Ab)$ *for every fixed* $F \in |Add\,(\mathcal{B}, \mathcal{C})|$.

(c) (a) *and* (b) *determine the tensor product uniquely up to an isomorphism.*

(d) *The functor described by* $F \mapsto ? \otimes_{\mathscr{B}} F, \alpha \mapsto ? \otimes_{\mathscr{B}} \alpha$ *is isomorphic to* K *and thus left adjoint to*

$$[H_*, \mathscr{C}]: Add\,(Add\,(\mathscr{B}^0, Ab), \mathscr{C}) \to Add\,(\mathscr{B}, \mathscr{C})\,.$$

It induces an equivalence between $Add\,(\mathscr{B}, \mathscr{C})$ *and the full subcategory of* $Add\,(Add\,(\mathscr{B}^0, Ab), \mathscr{C})$ *whose objects are the additive functors* $Add\,(\mathscr{B}^0, Ab) \to \mathscr{C}$ *which preserve colimits.*

(e) *The tensor product preserves colimits in* $Add\,(\mathscr{B}, \mathscr{C})$ *for every fixed* $M \in |Add\,(\mathscr{B}^0, Ab)|$.

Proof. (a) Using the Yoneda isomorphism 4.3.3, one obtains from (4)

(b) $$[H_X \otimes_{\mathscr{B}} F, A]_{\mathscr{C}} \xrightarrow{\cong} [H_X, \langle F, A\rangle] \xrightarrow{\cong} [F(X), A]$$

as an isomorphism of trifunctors. Thus (a) follows from the additive version 4.5.7 of 4.5.4.

(b) follows immediately from (4) and 16.4.6.

(c) By the additive version of 3.6.3, (3) corresponds uniquely to an additive functor

$$T: Add\,(\mathscr{B}^0, Ab) \to Add\,(Add\,(\mathscr{B}, \mathscr{C}), \mathscr{C})\,.$$

By the "pointwise" construction of colimits in functor categories, (b) is equivalent to T preserving colimits. TH_* is determined by (5). Using the additive versions of 17.2.1 and 17.2.7, (c) follows at least if $\mathscr{B}$ has finite biproducts. If this is not yet the case, the completion of $\mathscr{B}$ as in 16.3.10 makes a dense functor out of H_* (compare 17.4.11).

(d) By the additive version of 17.3.1 (a) and the dual of 16.5.4 (d), $\tilde{H}_* K$ is isomorphic to $1_{Add(\mathscr{B},\mathscr{C})}$. $E(??, ?): Add\,(\mathscr{B}, \mathscr{C}) \times \mathscr{B} \to \mathscr{C}$ is thus isomorphic to $E(\tilde{H}_*\, K(??), ?) = E(K(??)\, H_*, ?) = E'(K(??), H_?)$ where E' is the evaluation functor

$$E': Add\,(Add\,(\mathscr{B}^0, Ab), \mathscr{C}) \times Add\,(\mathscr{B}^0, Ab) \to \mathscr{C}\,.$$

Furthermore, $K(F)$, being a left adjoint of $\check{F}$, preserves colimits and thus so does $E'(K(F), ?)$. In this way, (c) implies that $? \otimes_{\mathscr{B}} ??$ is isomorphic to $E'(K(??), ?)$. Thus the functor given in (d) is isomorphic to K, and the remaining claims follow from the additive version of 17.3.1.

(e) follows from (d) by the "pointwise" construction of colimits in functor categories.

17.7.3 Remarks. (a) The proof of 17.7.2 (d) yielded the fact that, for a proper choice of φ_F, (1) is a trifunctor isomorphism, which is not evident, since both sides originate from trifunctors that are already fixed. If $\mathcal{B}$ has finite biproducts, i. e., if H_* is dense, it is simpler to deduce 17.7.2 (d) from 17.1.6 (13).

(b) 17.7.2 (b) and (e) do not imply that the tensor product preserves colimits in the product category on the left side of (3). The usual tensor product $Ab \times Ab \to Ab$ will be shown to constitute a special case. It preserves neither finite coproducts nor coequalizers in $Ab \times Ab$.

(c) Since $H_* \times 1_{Add(\mathcal{B},\mathcal{C})}$ is injective on objects, the tensor product can be normed in such a way (16.6.6) that (5) is the identity morphism.

(d) From (4) one obtains corresponding isomorphisms of trifunctors if one or more of the categories $Add(\mathcal{B}^0, Ab)$, $Add(\mathcal{B}, \mathcal{C})$, $\mathcal{C}$ are replaced by a functor category with the corresponding codomain (17.6.2); as an example, $Add(\mathcal{A}, Add(\mathcal{B}^0, Ab))$ produces in this way an isomorphism of trifunctors, namely

$$Add(\mathcal{A}, Add(\mathcal{B}^0, Ab))^0 \times Add(\mathcal{B}, \mathcal{C})^0 \times \mathcal{C} \to Add(\mathcal{A}^0, Ab) .$$

(e) As in 17.6.6, one obtains from (4) for $Add(\mathcal{A}, Add(\mathcal{B}^0, Ab))$, $Add(\mathcal{B}, \mathcal{C})$, $Add(\mathcal{A}, \mathcal{C})$ the adjunction

(7) $$\chi_{U,F,V}\colon [U \otimes_{\mathcal{B}} F, V]_{Add(\mathcal{A},\mathcal{C})} \xrightarrow{\cong} [U, \langle F, V\rangle]_{Add(\mathcal{A}, Add(\mathcal{B}^0, Ab))}$$

which, by 3.6.3, can be followed by the canonical isomorphism

(8) $$Add(\mathcal{A}, Add(\mathcal{B}^0, Ab)) \cong Add(\mathcal{B}^0, Add(\mathcal{A}, Ab)) .$$

17.7.4 Special cases. If $\mathcal{B}$ has only one object, then $\mathcal{B}$ is a ring R, $Add(\mathcal{B}^0, Ab) = Mod_R$, $Add(\mathcal{B}, \mathcal{C}) = {}_R\mathcal{C}$. By (2), (3) and (4) become

(3 R) $$? \otimes_R ??\colon Mod_R \times {}_R\mathcal{C} \to \mathcal{C} ,$$

(4 R) $$\varphi_{M,{}_RB,A}\colon [M \otimes_R {}_RB, A]_{\mathcal{C}} \xrightarrow{\cong} \mathrm{Hom}_R(M, [{}_RB, A]_{\mathcal{C}})$$

with $M \in |Mod_R|$, ${}_RB \in |{}_R\mathcal{C}|$, $A \in |\mathcal{C}|$. From 17.7.2 (a) one gets the isomorphism

(5 R) $${}_RR_R \otimes_R ? \cong 1_{{}_R\mathcal{C}} ,$$

because $H_*\colon R \to Mod_R = Add(R^0, Ab)$ gives exactly ${}_RR_R$ (compare 15.1.8). Applying the forgetful functor to (5 R) yields

(5′ R) $$R_R \otimes_R ? \cong U(?)\colon {}_R\mathcal{C} \to \mathcal{C} ,$$

compare 15.1.4. The above remarks apply, in particular, to $\mathcal{C} = Ab$ and to $\mathcal{C} = Mod_S$, which turns (4 R) into

(4 RS) $$\mathrm{Hom}_S(M_R \otimes_R {}_RS_S, A_S) \cong \mathrm{Hom}_R(M_R, \mathrm{Hom}_S({}_RB_S, A_S)) .$$

Here M_R is the same as what was previously written M. The tensor product defined here is isomorphic to the usual one (see for instance MacLane [21].) This follows from 17.7.2 (c) or also from the fact that the universal properties generally used for its definition are consequences of the adjunction (4 RS). By 17.7.2 (a) and 17.7.3 (c), 17.7.2 (b) allows one to calculate the tensor product.

By 17.7.3 (d), (4 RS) yields isomorphisms if M_R, ${}_RB_S$ or A_S are provided additionally with a module structure for a third ring; as an example, if Mod_R is replaced by ${}_\Gamma Mod_R$, an isomorphism of trifunctors with values in Mod_Γ results, namely

$$(4\,\Gamma RS)\quad \mathrm{Hom}_S\,({}_\Gamma M_R \otimes_R {}_RB_S,\ A_S) \cong \mathrm{Hom}_R\,({}_\Gamma M_R,\ \mathrm{Hom}_S\,({}_RB_S,\ A_S)).$$

17.7.3 (e) produces amongst other similar statements

$$(7\,\Gamma RS)\qquad \begin{aligned}&\mathrm{Hom}_{\Gamma,S}\,({}_\Gamma M_R \otimes_R {}_RB_S,\ {}_\Gamma A_S) \cong\\ &\mathrm{Hom}_{\Gamma,R}\,({}_\Gamma M_R,\ \mathrm{Hom}_S\,({}_RB_S,\ {}_\Gamma A_S)),\end{aligned}$$

where $\mathrm{Hom}_{\Gamma,S}$, $\mathrm{Hom}_{\Gamma,R}$ refer to bimodule homomorphisms, and $? \otimes_R ??: {}_\Gamma Mod_R \times {}_RMod_S \to {}_\Gamma Mod_S$ originates from (3 R) by replacing Mod_R by ${}_\Gamma Mod_R = Add\,(\Gamma, Mod_R)$. 17.7.2 (c) yields further conclusions, such as the existence of a trifunctor isomorphism

$$(9)\qquad (N_R \otimes_R {}_RM_S) \otimes_S {}_SB \cong N_R \otimes_R ({}_RM_S \otimes_S {}_SB)\,.$$

For, by means of (5 R), there is an isomorphism with ${}_RR_R$ instead of N_R and both sides preserve colimits in Mod_R. For $\mathscr{C} = Ab$, 17.7.2 (c) and (e) yield

$$M_R \otimes_R {}_RR_R \cong M_R;\quad M_R \otimes_R {}_RN \cong N_{R^\circ} \otimes_{R^\circ} {}_{R^\circ}M$$

by repeated applications of natural isomorphisms and with the identification ${}_{R^\circ}Mod$ and Mod_R. 16.7.2 (3) thus follows from (4 RS), as does the special case 15.1.8 (12).

17.7.5 For $\mathscr{B} = R$, 17.7.2 (d) states that the functor K: ${}_R\mathscr{C} \to$ $\to Add\,(Mod_R, \mathscr{C})$ described by ${}_RB \mapsto ? \otimes_R {}_RB$, $\alpha \mapsto ? \otimes_R \alpha$ is left adjoint to $[H_*, \mathscr{C}]$: $Add\,(Mod_R, \mathscr{C}) \to {}_R\mathscr{C}$, where $H_*\colon R \to Mod_R$ is the Yoneda embedding, and that K induces an equivalence of ${}_R\mathscr{C}$ with the full subcategory of $Add\,(Mod_R, \mathscr{C})$ whose objects are the functors preserving colimits. For an arbitrary additive functor $S\colon Mod_R \to \mathscr{C}$, one has

$$(10)\qquad [? \otimes_R {}_RB,\ S(?)]_{Add(Mod_R,\,\mathscr{C})} \cong \mathrm{Hom}_R\,({}_RB,\ S({}_RR_R))\,.$$

This orginates from the adjunction isomorphism $[K(TH_*), S] \cong$ $\cong [TH_*, SH_*]$ with $T = ? \otimes_R {}_RR$, for, by (5), $TH_* \cong {}_RB$, $K(TH_*)$ $\cong T$, and $SH_* = S({}_RR_R)$. (10) stands as an isomorphism of contra-co-variant functors.

17.7.6 If R is a commutative ring, then every object of ${}_R\mathscr{C}$ can be regarded as a bimodule object on which the two simple module structures coincide. Thus one obtains from (3 R) as in (4 ΓRS) a tensor product $\otimes_R$: ${}_RMod \times {}_R\mathscr{C} \to {}_R\mathscr{C}$ by restriction to special bimodules, so that one gets

$$(11)\qquad \mathrm{Hom}_R\,({}_RM \otimes_R {}_RB,\ {}_RA) \cong \mathrm{Hom}_R\,({}_RM,\ \mathrm{Hom}_R\,({}_RB,\ {}_RA))$$

with ${}_RM$ in ${}_RMod = Mod_R$, ${}_RA$, ${}_RB$ in ${}_R\mathscr{C}$ and where ${}_RMod$ is the codomain of Hom_R.

For $R = \boldsymbol{Z}$, ${}_{\boldsymbol{Z}}\mathscr{C}$ is canonically isomorphic to $\mathscr{C}$ and ${}_{\boldsymbol{Z}}Mod$ to Ab. Thus one has the tensor product $? \otimes ??: Ab \times \mathscr{C} \to \mathscr{C}$, and (4) takes on the simple form

$$(4\,\boldsymbol{Z}) \qquad [M \otimes B, A]_{\mathscr{C}} \cong [M, [B, A]_{\mathscr{C}}]_{Ab}.$$

17.8 Relatives of the Tensor Product

17.8.1 Let $\mathscr{C}$ be a complete, additive category and $\mathscr{B}$ again a small additive category. We regard $Add\,(\mathscr{B}^0, \mathscr{C}^0)$ as the dual category of $Add\,(\mathscr{B}, \mathscr{C})$ (4.5.6) and we set $F^0 = \mathrm{Op}\,F\,\mathrm{Op}$ for $F\colon \mathscr{B} \to \mathscr{C}$. Then, by 17.7 (4), there is the isomorphism

$$(1') \qquad [M \otimes_{\mathscr{B}^0} F^0, A^0]_{\mathscr{C}^0} \xrightarrow{\cong} [M(?), [F^0(?), A^0]_{\mathscr{C}^0}]_{Add(\mathscr{B}, Ab)},$$

where $M\colon \mathscr{B} \to Ab$, $F\colon \mathscr{B} \to \mathscr{C}$ and $A \in |\mathscr{C}|$. Applying $\mathrm{Op}\colon \mathscr{C}^0 \to \mathscr{C}$ produces a bifunctor from $\otimes_{\mathscr{B}^0}$ which we write as a contra-co-variant functor

$$(2) \qquad \{?, ??\}_{\mathscr{B}} \text{ with } ? \text{ in } Add\,(\mathscr{B}, Ab),\ ?? \text{ in } Add\,(\mathscr{B}, \mathscr{C})$$

and codomain $\mathscr{C}$.

From (1′) one gets in this way

$$(1) \qquad \psi_{A,M,F}\colon [A, \{M, F\}_{\mathscr{B}}]_{\mathscr{C}} \xrightarrow{\cong} [M(?), [A, F(?)]_{\mathscr{C}}]_{Add(\mathscr{B}, Ab)}$$

as an isomorphism of trifunctors where on the right, just as in (1′), one has the additive group of natural transformations $M(?) \to [A, F(?)]$. The functor (2) is called the *symbolic* Hom-*functor* (or *contensor*) over $\mathscr{B}$. $\{?, F\}_{\mathscr{B}}$ takes colimits into limits, $\{M, ?\}_{\mathscr{B}}$ preserves limits by 17.7.2 (e). 17.7 can be reformulated to fit this partially dualized case (Ab is not dualized).

If $\mathscr{B}$ is a ring R, then one has, in particular, the contra-co-variant functor

$$\{?, ??\}_R \text{ with } ? \text{ in } {}_RMod,\ ?? \text{ in } {}_R\mathscr{C} \text{ and codomain } \mathscr{C},$$

and (1) becomes

$$(1\,R) \quad [A, \{{}_RM, {}_RB\}_R]_{\mathscr{C}} \cong \mathrm{Hom}_R\,({}_RM, [A, {}_RB]_{\mathscr{C}}) \text{ with } {}_RM \in |{}_RMod|.$$

17.8.2 For $\mathscr{C} = Ab$, there is the following analogue of $(4\,R\,S)$ in 17.7.4:

$$(3) \qquad \mathrm{Hom}_R\,({}_RM, \mathrm{Hom}_S\,(A_S, {}_RR_S)) \xrightarrow{\cong} \mathrm{Hom}_S\,(A_S, \mathrm{Hom}_R\,({}_RM, {}_RB_S))$$

as an isomorphism of trifunctors. One can prove (3) as follows: Consider both sides to be functors from ${}_RMod$ into the category which is dual to the category of biadditive functors from $Mod_S^0 \times {}_RMod_S$ to Ab. Using 15.1 (9) and (10), one obtains an isomorphism between the restrictions to the full subcategory of ${}_R Mod$ which has ${}_R R$ as its only object. (3) follows then from 17.2.7 using the argument of 17.7.2 (c).

From (3) and (1 R), it follows by the additive version of 4.5.4 that, for $S = \mathbf{Z}$ and $\mathscr{C} = Ab$, $\{?, ??\}_R$ is isomorphic to the Hom-functor of ${}_R Mod$. An analogous isomorphism exists for $\mathscr{C} = Mod_S$, and (1 R), as well as (1), may be regarded as generalizations of (3).

17.8.3 Let $\mathscr{C}$, $\mathscr{D}$, $\mathscr{E}$ be additive categories. For additive functors $U: \mathscr{E} \to \mathscr{D}$, $V: \mathscr{E} \to \mathscr{C}$, $T: \mathscr{D} \to \mathscr{C}$, there is an isomorphism of trifunctors

$$(4) \qquad \mathrm{Nat}\,([U(?), ??]_{\mathscr{D}}, [V(?), E(T, ??)]_{\mathscr{C}}) \xrightarrow{\cong} [V, TU]_{Add(\mathscr{E},\mathscr{C})},$$

where $E: Add\,(\mathscr{D}, \mathscr{C}) \times \mathscr{D} \to \mathscr{C}$ is the evaluation functor and where the transformations on the left are natural transformations of bifunctors. If, in particular, $\mathscr{E}$ is a ring R^0, then one has the isomorphism

$$(5) \qquad [[{}_R X, ??]_{\mathscr{D}}, [{}_R A, E(T, ??)]_{\mathscr{C}}]_{Add(\mathscr{D}, Mod_R)} \xrightarrow{\cong} \mathrm{Hom}_R\,({}_R A, T({}_R X))$$

of trifunctors ${}_R\mathscr{D} \times ({}_R\mathscr{C})^0 \times Add\,(\mathscr{D}, \mathscr{C}) \to Ab$.

Proof. Using 17.6.5, one obtains (4) from the Yoneda isomorphism 4.3.3

$$\mathrm{Nat}\,([X, ??]_{\mathscr{D}}, [A, E(T, ??)]_{\mathscr{C}}) \cong [A, E(T, X)]_{\mathscr{C}}$$

if, in addition, the canonical isomorphism $Add\,(\mathscr{E}, Add\,(\mathscr{D}, \mathscr{C})) \cong$ $\cong Biadd\,(\mathscr{E} \times \mathscr{D}, \mathscr{C})$ is taken into account. For (5), one makes use of $Biadd\,(R^0 \times \mathscr{D}, Ab) \cong Add\,(\mathscr{D}, Mod_R)$ as well.

17.8.4 Proposition. *Let $\mathscr{C}$, $\mathscr{D}$ be additive categories and assume $\mathscr{C}$ to be cocomplete. For additive functors T: $\mathscr{D} \to \mathscr{C}$, ${}_R A \in |{}_R\mathscr{C}|$, ${}_R X \in |{}_R\mathscr{D}|$, there is an isomorphjism*

$$(6) \qquad [[{}_R X, ?]_{\mathscr{D}} \otimes_R {}_R A, T(?)]_{Add(\mathscr{D},\mathscr{C})} \xrightarrow{\cong} \mathrm{Hom}_R\,({}_R A, T({}_R X))$$

of trifunctors ${}_R\mathscr{D} \times ({}_R\mathscr{C})^0 \times Add\,(\mathscr{D}, \mathscr{C}) \to Ab$.

Proof. From the adjunction

$$[M \otimes_R {}_R A, C]_{\mathscr{C}} \xrightarrow{\cong} [M, [{}_R A, C]_{\mathscr{C}}]_{Mod_R}$$

one gets, by 17.6.6,

$$\mathrm{Nat}\,(U \otimes_R {}_R A, T) \xrightarrow{\cong} [U(?), [{}_R A, T(?)]]_{Add(\mathscr{D}, Mod_R)}.$$

(6) follows from this by restricting U to functors $[{}_R X, ?]$, by (5).

17.8.5 With $R = \mathbf{Z}$, i. e., with Ab instead of Mod_R, (6) becomes

$$(7) \qquad [H^X \otimes A, T]_{Add(\mathscr{D}, Ab)} \xrightarrow{\cong} [A, T(X)]_{\mathscr{C}} = [A, E(T, X)]_{\mathscr{C}}.$$

The tensor product $\otimes$ over $\mathbf{Z}$ here is not to be confused with the one in 17.7.2 (5). With $\mathscr{C} = Ab$ and $A = \mathbf{Z}$, (7) again yields the Yoneda lemma. (4) through (7) are generalizations. They can be partially dualized.

17.8.6 Now let $\mathcal{C}$ be complete. If 17.8.4 is applied to $\mathcal{C}^0$ and $\mathcal{D}^0$, then one gets, as in 17.8.1, from (6) and (7)

$$(8) \qquad [T(?), \{[?, {}_RX]_{\mathcal{D}}, {}_RA\}_R]_{Add(\mathcal{D},\mathcal{C})} \cong \mathrm{Hom}_R\left(T_R(X), {}_RA\right),$$

$$(9) \qquad [E(T, X), A]_{\mathcal{C}} = [T(X), A]_{\mathcal{C}} \cong [T, \{H_X, A\}]_{Add(\mathcal{D},\mathcal{C})}$$

with $T: \mathcal{D} \to \mathcal{C}$, $X \in |\mathcal{D}|$, $A \in |\mathcal{C}|$.

For $\mathcal{C} = Ab$ one gets, by 17.8.2,

$$(8') \quad [T(?), \mathrm{Hom}_R([?, {}_RX]_{\mathcal{D}}, {}_RA)]_{Add(\mathcal{D},Ab)} \cong \mathrm{Hom}_R\left(T({}_RX), {}_RA\right),$$

$$(9') \qquad [T(X), A]_{Ab} \cong [T, H_A H_X]_{Add(\mathcal{D},Ab)},$$

where $H_A H_X(?) = [[?, X]_{\mathcal{D}}, A]_{Ab}$.

17.8.7 The non-additive case. The previous statements, starting with 17.7.1, are valid in the non-additive case with *Ens* instead of *Ab* and the corresponding categories of all functors instead of additive ones. The place of a ring is taken by a category with only one object, in particular this may be a group. Where the special case $R = \boldsymbol{Z}$ is considered, a set of one element is to be substituted.

17.8.8 A functor $T: \mathcal{C} \to \mathcal{D}$ need not have a left adjoint. However, one can consider the full subcategory $\mathcal{E}$ of $\mathcal{D}$ for whose objects $[X, T(?)]_{\mathcal{D}}: \mathcal{C} \to Ens$ is representable. One obtains a functor S: $\mathcal{E} \to \mathcal{C}$ such that with the inclusion $J: \mathcal{E} \to \mathcal{D}$

$$[S(?), ??]_{\mathcal{C}} \cong [J(?), T(??)]_{\mathcal{D}}$$

is valid. In a more general context such a situation may exist if J is a given functor from a category $\mathcal{E}$ to a category $\mathcal{D}$. 17.2.9 is formulated to fit this case. Analogously, isomorphisms of the form

$$[S(?, ??), ???]_{\mathcal{C}} \cong [J(?), T(??, ???)]_{\mathcal{D}}$$

can be considered. This and 17.7.2 (a), (b) can be used to define a tensor product $\otimes_R: \mathcal{E} \times {}_R\mathcal{C} \to \mathcal{C}$ for a subcategory $\mathcal{E}$ of ${}_RMod$, even if $\mathcal{C}$ is not cocomplete, as, e. g., for finitely presentable modules if $\mathcal{C}$ is finitely cocomplete. In a similar fashion most of 17.7. and 17.8 can be generalized. For details we refer to Ulmer [67] and [68].

17.9 Problems

17.9.1 Let $\mathcal{Z}$ be a terminal category and $\mathcal{C}$ a cocomplete category. The canonical isomorphism $[\mathcal{Z}, \mathcal{C}] = \mathcal{C}$ yields a functor $T: \mathcal{C} \to Dg(\mathcal{C})$ which is right adjoint to $\mathrm{Colim}: Dg(\mathcal{C}) \to \mathcal{C}$. This generalizes 8.5 (1).

17.9.2 Let $\mathcal{B}$, $\mathcal{C}$, $\mathcal{D}$ be arbitrary categories and $U: \mathcal{B} \to \mathcal{C}$ a functor. If for a given functor $F: \mathcal{B} \to \mathcal{D}$ and for every $A \in |\mathcal{C}|$ a colimit of $F\,Q_A: U/A \to \mathcal{D}$ exists, then the functor $[F, \widetilde{U}(?)]_{[\mathcal{B},\mathcal{D}]} = [F, [U, \mathcal{D}](?)]$ is representable.

Remark. Representing objects of this functor are called *left Kan extensions* of F.

17.9.3 For functors $U: \mathcal{B} \to \mathcal{C}$ and $G: \mathcal{C} \to \mathcal{D}$, the rule $(U, G) \mapsto G U$ extends to a bifunctor $T': [\mathcal{B}, \mathcal{C}] \times [\mathcal{C}, \mathcal{D}] \to [\mathcal{B}, \mathcal{D}]$ by 16.1 (4), (5). If $\mathcal{B}$ is small and $\mathcal{D}$ cocomplete, then 17.1.6 (a) yields an adjoint pair of bifunctors in the sense of 16.7.1.

17.9.4 Carry out the details in the proof of 17.2.3 and show that $\{\theta_{A,B}\}$ and $\{\Omega_{A,B}\}$ are natural transformations between contra-covariant functors (with codomain *ENS*).

17.9.5 Let the category $\mathcal{C}$ have coproducts.

(a) A functor $G: \mathcal{C} \to Ens$ has a left adjoint if and only if G is representable.

(b) A functor $F: Ens \to \mathcal{C}$ has a right adjoint if and only if F preserves coproducts. In that case, F preserves colimits.

17.9.6 Carry out 17.2.10 and check 17.3.4.

17.9.7 Let $P \twoheadrightarrow Q$ be an epimorphism in a cocomplete abelian category $\mathcal{C}$. If P satisfies condition (S) in 17.4.6, then so does Q. If $\mathcal{C}$ is a Grothendieck category and if P_1, P_2 satisfy condition (S), then so does $P_1 \oplus P_2$. If $\mathcal{C}$ is a Grothendieck category with a generating set, then every object satisfying (S) is finitely generated. If the generating set consists of small projective objects, then the converse is also true.

17.9.8 Let P be an object in a cocomplete abelian category. Then the following conditions are equivalent:

(i) If $\{m_i: A_i \rightarrowtail P\}$ is a family of monomorphisms with $\bigcup m_i = 1_P$, then 1_P is the union of a finite number of members of the family.

(ii) P satisfies condition (S) in 17.4.6.

(iii) P satisfies condition (S) restricted to strongly ordered families of monomorphisms. (Hint: Use 14.7.7).

17.9.9 Prove: If an exact category $\mathcal{C}$ has a generator G and if, for every index set I, there exists a coproduct $\coprod_{i \in I} G_i$ with $G_i = G$ for all i, then $\mathcal{C}$ is abelian and cocomplete. Using this, give a variant of 17.4.10.

17.9.10 Check 17.7.3.

17.9.11 Let R be a commutative ring. For $\otimes_R: {}_RMod \times {}_RMod \to {}_RMod$ show that the identity morphism of ${}_RR \otimes_R {}_RR = {}_RR$ extends to an isomorphism: τ: ? $\otimes_R$?? $\xrightarrow{\approx}$?? $\otimes_R$?. τ can be chosen in such a way that $\tau\tau$ is the identity natural transformation for $\otimes_R$.

17.9.12 (*Morita's theorems*). (a) Let $V: Mod_S \to Mod_R$ be an additive functor. If V preserves limits, then there is a bimodule ${}_RP_S$ such that

$V(?) \cong \mathrm{Hom}_S\,({}_RP_S, ?)$. If V preserves colimits, then there is a bimodule ${}_SQ_R$ such that $V(?) \cong ? \otimes_S {}_SQ_R$. If V preserves limits and colimits and is fully faithful, then ${}_SQ_R \otimes_R {}_RP_S \cong {}_SS_S$.

(b) If V is an equivalence, then $? \otimes_R {}_RP_S$ is equivalence-inverse to $?? \otimes_S {}_SQ_R$ and to $\mathrm{Hom}_S({}_RP_S, ??)$. In that case, ${}_RP_S \otimes_S {}_SQ_R \cong$ $\cong {}_RR_R$, ${}_SQ_R \cong \mathrm{Hom}_S({}_RP_S, {}_SS_S)$, ${}_RP_S \cong \mathrm{Hom}_R({}_SQ_R, {}_RR_R)$, ${}_RP_S \otimes_S ?$: ${}_S\mathit{Mod} \to {}_R\mathit{Mod}$ is an equivalence with equivalence-inverses ${}_SQ_R \otimes_R ??$ and $\mathrm{Hom}_R({}_RP_S, ??)$, and ${}_SQ_R \cong \mathrm{Hom}_R({}_RP_S, {}_RR_R)$, ${}_RP_S \cong$ $\mathrm{Hom}_S({}_SQ_R, {}_SS_S)$.

(c) $\mathrm{Hom}_S({}_RP_S, ?)$: $\mathit{Mod}_S \to \mathit{Mod}_R$ is an equivalence if and only if P_S is a small projective generator in Mod_S and if the action of R on P_S is given by an isomorphism $R \cong \mathrm{Hom}_S(P_S, P_S)$, which is an isomorphism of rings. In that case, ${}_RP$ is a small projective generator in ${}_R\mathit{Mod}$ and the action of S on ${}_RP$ is given by an isomorphism $S^0 \cong \mathrm{Hom}_R({}_RP, {}_RP)$.

(d) The center of R is isomorphic to the center of ${}_R\mathit{Mod}$ and to the center of Mod_R (see 3.9.2). If Mod_S and Mod_R are equivalent, then the centers of S and R are isomorphic. In particular, if S and R are commutative rings, then Mod_S is equivalent to Mod_R if and only if S and R are isomorphic.

18. Principles of Universal Algebra

18.1 Algebraic Theories

18.1.1 Preliminary remarks. In 11.1 we did not give an explicit definition of the type $\mathfrak{S}$ of an algebraic structure. It will become evident that this notion of type can be expressed by a special category, so that then objects with an algebraic structure are functors preserving finite products. Here, as in 11, we restrict ourselves to finitary algebraic operations and to structures with only one base object.

18.1.2 Definition. An *algebraic theory* is a category $\mathcal{A}$ with the following properties:

(i) $|\mathcal{A}|$ consists of countably many different objects $A^0, A^1, A^2, A^3, \ldots$.

(ii) For $k \geq 0$, A^k is a product of A^1 with itself k-times with projections $p_1^k, p_2^k, \ldots, p_k^k$ for $k \geq 1$. Here $p_1^1 = 1_{A^1}$ is assumed.

The morphisms $A^n \to A^1$ are called *n-ary operations*. A *theory-morphism* $F\colon \mathcal{A} \to \mathcal{B}$ is a functor satisfying

(1) $F(A^0) = B^0$ and $F(p_j^k) = p_j^k$ for all (k, j) with $1 \leq j \leq k < \infty$.

The *algebraic category* corresponding to $\mathcal{A}$ is the full subcategory $\mathcal{A}^b$ of $[\mathcal{A}, Ens]$ whose objects are the functors which preserve finite products. The objects of $\mathcal{A}^b$ are called *$\mathcal{A}$-algebras*, the morphisms are called *$\mathcal{A}$-homomorphisms*.

If $\mathcal{C}$ is a category with finite products, then one can consider analogously the algebraic category corresponding to $\mathcal{A}$ over $\mathcal{C}$.

18.1.3 In the algebraic theory $\mathcal{A}$, we may consider $[A^m, A^n]$ with $n \geq 1$ as the set of n-tupels $(t^m_{\nu_1}, t^m_{\nu_2}, \ldots, t^m_{\nu_n})$ with $t^m_{\nu_j} \in [A^m, A^1]$ and

(2) $$p^n_j(t^m_{\nu_1}, t^m_{\nu_2}, \ldots, t^m_{\nu_n}) = t^m_{\nu_j} \quad \text{for} \quad n \geq 1 .$$

p^n_0 designates the only morphism from A^n into the terminal object A^0. Here

(3) $$p^n_0\,(t^m_{\nu_1}, t^m_{\nu_2}, \ldots, t^m_{\nu_n}) = p^m_0$$

holds. p^0_0 and $(p^n_1, p^n_2, \ldots, p^n_n)$ are the identity morphisms of A^0 and A^n with $n \geq 1$.

18.1.4 . Definition 18.1.2 is consistent with definition 11.1.9. The commutativity conditions are now the composition rules for morphisms in $\mathcal{A}$. Conversely, an algebraic theory can be constructed from given algebraic operations and diagram conditions. This requires three steps. First, one takes an algebraic theory $\mathcal{N}$ with no operations except the projections of products (18.1.5). Then a free theory is constructed for the given operations (18.1.6) and finally given equations between composed operations are taken into account (18.1.10). The commutativity conditions in 11.1.9 are just such equations, because a morphism into a k-fold product corresponds to a k-tupel of morphisms into the factors. Therefore, algebras in the sense given above are called more precisely *equationally defined algebras.*

18.1.5 For every integer $n \geq 0$, n is the set of integers from 0 to $n-1$, in particular thus $0 = \Phi$, $1 = \{\Phi\}$. (This is the usual recursive set theoretical definition of the finite cardinals.) Let $\boldsymbol{N}$ be the full subcategory of Ens containing the objects n. $\boldsymbol{N}$ has finite coproducts: $n + k$ is a coproduct of n and k, where $in_1\colon n \to n + k$ is the inclusion and $in_2\colon k \to n + k$ is the map $x \mapsto n + x$. In particular, 0 is initial and n is a coproduct of n cofactors 1, where $[1, n]$ consists exactly of the corresponding injections in^n_j with $in^n_j(1) = j - 1$. Let $\mathcal{N}$ be the dual category of $\boldsymbol{N}$. Instead of n^0 we write N^n, and we write $p^n_1, \ldots, p^n_n$ with $p^n_j = (in^n_j)^0$ for the morphisms from N^n to N^1. We regard the morphisms from N^n to N^k for $k \geq 1$ as k-tupels as in 18.1.3.

$\mathcal{N}$ is an *initial theory*. In fact, by (1) there is exactly one theory-morphism $I_{\mathcal{A}}\colon \mathcal{N} \to \mathcal{A}$ for any algebraic theory $\mathcal{A}$.

18.1.6 For every integer $n \geq 0$ let there be a set G_n given in such a way that the sets G_n are pairwise disjoint and disjoint to every set of morphisms of $\mathcal{N}$. We construct a *free algebraic theory* $\mathcal{F}_G$ corresponding to $G = \{G_n\}$ for which the elements of G_n are defining n-ary operations and thus the elements of $\bigcup G_n$ are *defining operations.*

We consider the diagram scheme underlying $\mathcal{N}$, omit, for $k > 1$, all arrows with endpoint N^k, then add G_n as a set of arrows from N^n to N^1, and finally we add, for $k > 1$, all k-tupels of existing arrows $N^n \to N^1$ as arrows $N^n \to N^k$. The arrows that we now have are called formal first stage operations. Formal r-th stage operations are defined recursively for every natural number r. Those from N^k to N^1 are pairs $({}_1t_1^m, {}_{r-1}t_m^n)$, where ${}_1t_1^m$ is a formal first stage operation $N^m \to N^1$, ${}_{r-1}t_m^n$ a formal $(r - 1)$-th stage operation $N^n \to N^m$ and m an arbitrary non-negative integer. For $k > 1$, formal r-th stage operations $N^n \to N^k$ are k-tupels of r-th stage operations $N^n \to N^1$, while formal r-th stage operations $N^n \to N^0$ are pairs $(p_0^m, {}_{r-1}t_m^n)$.

We now define composites $({}_rt_k^m)\,({}_st_m^n)$ of formal r-th and s-th stage operations recursively with respect to r as formal $(r + s)$-th stage operations. First, let $r = 1$. For $k = 0$ or 1, let $({}_1t_k^m)\,({}_st_m^n) = ({}_1t_k^m, {}_st_m^n)$. For $k > 1$, ${}_1t_k^m$ is a k-tupel $(u_1, u_2, \ldots, u_k)$ of formal first stage operations $N^m \to N^1$. Let $({}_1t_k^m)\,({}_st_m^n)$ be the k-tupel $(v_1, v_2, \ldots, v_k)$ with $v_j = (u_j)\,({}_st_m^n)$. For $r > 1$ and $k = 0$ or 1, $({}_rt_k^m) = ({}_1t_k^l, {}_{r-1}t_l^m)$, and $({}_1t_k^l, {}_{r-1}t_l^m)\,({}_st_m^n) = ({}_1k_k^l)\,(({}_{r-1}t_l^m)\,({}_st_m^n))$ is the recursive definition. The definition for $k > 1$ follows from this as above, namely "componentwise" for the k-tupel ${}_1t_k^l$.

For the thus described diagram scheme Σ, we set up the following commutativity conditions:

(a) Every composite $({}_rt_k^m)\,({}_st_m^n)$ is equivalent to the corresponding diagram path of length 2.

(b) p_0^0, p_1^1 and all k-tupels $(p_1^k, p_2^k, \ldots, p_k^k)$ with $k > 1$ are equivalent to the corresponding identity arrows.

(c) Both sides of (2) represent equivalent paths, provided $(t_{v_1}^m, \ldots, t_{v_n}^m)$ is a formal r-th stage operation for any $r \geq 1$.
The same holds for (3).

(d) If $k > 1$ and if a and b are paths from N^n to N^k, then a and b are equivalent provided $(pr_1^k a, pr_1^k b), \ldots, (pr_k^k a, pr_k^k b)$ are equivalent pairs.

Let $\mathcal{F}_G$ be the category $\mathcal{P}(\Sigma/K)$ as in 6.3.1. By (a), every morphism of $\mathcal{F}_G$ is the image of an arrow of Σ. By (b), every formal r-th stage operation is equivalent to a formal $(r + s)$-th stage operation for every natural number s. Thus (c) and (d) imply that N^k is a k-fold product of N^1 in $\mathcal{P}(\Sigma/K)$, which has as projections the images of $p_1^k, \ldots, p_k^k$ for $k \geq 1$.

18.1.7 Proposition. *If $\mathcal{A}$ is an arbitrary algebraic theory and if, for every one of the sets G_n in 18.1.6, a map $\psi_n\colon G_n \to [A^n, A^1]_{\mathcal{A}}$ is given, then these maps can be extended in a unique way to a theory-morphism $\Psi\colon \mathcal{F}_G \to \mathcal{A}$ (more precisely: for $g \in G_n$, $\psi_n(g)$ is the Ψ-image of the morphism of $\mathcal{F}_G$ which is determined by g).*

Taking (1) into account, this follows from the construction of $\mathcal{F}_G$, which justifies calling $\mathcal{F}_G$ "free".

18.1.8 Algebraic categories are the objects and theory-morphisms are the morphisms of the category $\mathfrak{T}$ of algebraic theories. There is a forgetful functor $V\colon \mathfrak{T} \to Ens^\omega$. Here $Ens^\omega = \prod Ens_n$ in *Cat* with $Ens_n = Ens$ for $n = 0, 1, 2, \ldots$ and $pr_n V(\mathcal{A}) = [A^n, A^1]_{\mathcal{A}}$, $pr_n V(F)$ is the map $[A^n, A^1]_{\mathcal{A}} \to [B^n, B^1]_{\mathcal{B}}$ induced by the theory-morphism $F\colon \mathcal{A} \to \mathcal{B}$. Using 18.1.7, one verifies the

Proposition. *The forgetful functor $V\colon \mathfrak{T} \to Ens^\omega$ has a left adjoint X whose value on the object G of Ens^ω is the free algebraic theory $\mathcal{F}_G$.*

18.1.9 Proposition. *The category $\mathfrak{T}$ of algebraic theories has equalizers which can be formed as in cat. $\mathfrak{T}$ has also coequalizers.*

Proof. According to 7.2.4, the first statement follows easily from (1) and (2). Let $F, G\colon \mathcal{A} \to \mathcal{B}$ be theory-morphisms. On the morphism set Mor $\mathcal{B}$ consider the smallest equivalence relation compatible with the composition of morphisms for which

(a) $F(t)$ and $G(t)$ are equivalent for every $t \in$ Mor $\mathcal{A}$.

(b) $u, v\colon B^n \to B^k$ with $k > 1$ are equivalent provided $(p_j^k u, p_j^k v)$ are equivalent for $j = 1, 2, \ldots, k$.

In making the transition to the quotient category as in 6.4.4, the corresponding projection P is a coequalizer of F and G. Obviously, P maps the objects B^k of $\mathcal{B}$ identically, and condition (b) guarantees that B^k remains a k-fold product of B^1, whereby (1) holds accordingly for P.

18.1.10 Let $G = \{G_n\}$ be given as in 18.1.6. For the corresponding free algebraic theory $\mathcal{F}_G$ we assume that a subset R_n of $[N^n, N^1] \times [N^n, N^1]$ for $n = 0, 1, 2, 3, \ldots$ is given; here $[N^n, N^1]$ is a morphism set of $\mathcal{F}_G$. Then there is an epimorphism $P\colon \mathcal{F}_G \to \mathcal{Q}$ in $\mathfrak{T}$ with the following property: If $T\colon \mathcal{F}_G \to \mathcal{A}$ is a theory-morphism such that $T(f) = T(g)$ for every pair $(f, g) \in R = \bigcup R_n$, then T factors through P.

Proof. Let $\mathcal{K}$ be the free algebraic theory for R. Then by 18.1.7 $(f, g) \mapsto f$ and $(f, g) \mapsto g$ determine two theory-morphisms $J_1, J_2\colon \mathcal{K} \to \mathcal{F}_G$. The coequalizer of J_1, and J_2 exists in $\mathfrak{T}$ by 18.1.9. It is the desired epimorphism.

18.1.11 One says that the just constructed algebraic theory $\mathcal{Q}$ is *presented* through the *set G of generating operations* with the *defining set of equations R.*

Every algebraic theory $\mathcal{A}$ (in the sense of 18.1.2) can be obtained in this way. As an example, take $[A^n, A^1]$ as G_n, consider the evident theory-morphism $\Psi_{\mathcal{A}}: \mathcal{F}_G \to \mathcal{A}$ (this is the adjunction transformation $XV \to 1_{\mathfrak{X}}$ belonging to the adjunction in 18.1.8 at $\mathcal{A}$) and take as defining equations all pairs (f, g) of operations such that f and g have the same image under $\Psi_{\mathcal{A}}$. One realizes that every algebraic theory can be presented in infinitely many ways through generating operations and corresponding defining sets of equations.

In presenting $\mathcal{A}$ via a set of generating operations G and a corresponding set of equations R, an element (f, g) of R is usually determined by choosing representatives $\bar{f}, \bar{g}$ of f and g in the diagram scheme Σ of $\mathcal{F}_G$ constructed in 18.1.6, which are then written in terms of the composites introduced there. The obvious interpretation in $\mathcal{A}$ leads to equations $\bar{f} = \bar{g}$ in $\mathcal{A}$, which justifies the misuse of the word equation above.

The commutativity conditions for diagrams in 11.1.9 are such equations. The factors Z which are carried along there are dispensable as shown by 7.3.5, and their only function was to clarify the notation. The arrows are formal first or second stage operations in the sense of 18.1.6. The notation of the type $\mathfrak{S}$ of an algebraic structure used previously now turns out to be a presentation of an algebraic theory.

18.1.12 Among the algebraic categories described by the algebraic theories considered above are categories of modules over a fixed ring and categories of algebras in the usual sense (in particular, Lie, Jordan and associative algebras) over a fixed commutative ring.

Let R be a ring. The effect of an element r of R on objects of ${}_R\mathit{Mod}$ is a unary operation. The module laws $0\,x = 0$, $1\,x = x$, $r_1(r_2\,x) = (r_1\,r_2)\,x$ are equations between unary operations. $0\,x = 0$, for example, expresses the fact that the unary operation consisting of multiplication with the 0 of the ring is the same as the composite of p_0^1 with the nullary operation z, where z is the zero of the additive group of the module. $r_1(r_2\,x) = (r_1\,r_2)\,x$ supplies an equation for every pair (r_1, r_2) of ring elements; more precisely: the multiplication of R is expressed in equations for unary operations. Corresponding statements can be made for the addition and for the ring axioms. The distributive law $(r_1 + r_2)\,x = r_1\,x + r_2\,x$ for modules consists of equations for unary operations, whereas $r(x_1 + x_2) = r\,x_1 + r\,x_2$ involves binary ones. In the case of algebras over a fixed commutative ring R, axioms like $r(x\,y) = (r\,x)\,y = x(r\,y)$ or $r[x, y] = [r\,x, y] = [x, r\,y]$ are also sets of equations between binary operations.

18.1.13 Proposition. *The category $\mathfrak{T}$ of algebraic theories is complete and cocomplete. The free theories $\mathcal{F}_n$, which are generated by one n-ary operation ($n \geq 0$), form a generating set for $\mathfrak{T}$.*

Proof. $\mathfrak{T}$ has a terminal object $\mathcal{P}$. For $\mathcal{P}$, $p_0^n\colon P^n \to P^0$ is an isomorphism, for $n \geq 1$ all projections p_i^n coincide, and for every $n \geq 0$ there is only one n-ary operation. (In 18.1.2 we did not require the projections p_i^n to be different.)

By 18.1.5, $\mathfrak{T}$ has an initial object $\mathcal{N}$, and by 18.1.9, it suffices to show that $\mathfrak{T}$ has products and coproducts with a non-empty index set. Let $\{\mathcal{A}_\nu\}$ be a non-empty family of algebraic theories. The theory $\mathcal{A}$ has as generators all operations of the theories $\mathcal{A}_\nu$, and as defining equations all those which come from all the $\mathcal{A}_\nu$ for the corresponding operations of $\mathcal{A}$ and in addition also those which equate the projections p_j^k of $\mathcal{A}_\nu$ with the p_j^k of $\mathcal{A}$. 18.1.10 shows that theory-morphisms $in_\nu\colon \mathcal{A}_\nu \to \mathcal{A}$ exist. Another application of 18.1.10 shows that, given a family $\{F_\nu\colon \mathcal{A}_\nu \to \mathcal{B}\}$ of theory-morphisms, there is exactly one $F\colon \mathcal{A} \to \mathcal{B}$ with $F\, in_\nu = F_\nu$. Thus $(\mathcal{A}, \{in_\nu\})$ is a coproduct of $\{\mathcal{A}_\nu\}$ in $\mathfrak{T}$.

In the product object $\Pi\, \mathcal{A}_\nu$ in *cat*, consider the full subcategory $\mathcal{A}$, whose objects are families $A^k = \{A_\nu^k | A_\nu^k \in |\mathcal{A}_\nu|\}$. From the projections of $\Pi\, \mathcal{A}_\nu$, one gets projections $pr_\nu\colon \mathcal{A} \to \mathcal{A}_\nu$. $\mathcal{A}$ has the morphisms $p_j^k = \{p_{\nu j}^k | p_{\nu j}^k\colon A_\nu^k \to A_\nu^1\}$. Using them, A^k is a product of k factors A^1 in $\mathcal{A}$ for $k \geq 1$, and A^0 is terminal. $(\mathcal{A}, \{pr_\nu\})$ is a product of the family $\{\mathcal{A}_\nu\}$ in $\mathfrak{T}$. The last statement of the proposition is evident.

18.2 Yoneda Embedding and Free Algebras

18.2.1 Let $S, T\colon \mathcal{A} \to Ens$ be $\mathcal{A}$-algebras and $\alpha\colon S \to T$ an $\mathcal{A}$-homomorphism. To simplify the notation we set

(1) $$S_k = S(A^k), \quad \alpha_k = \alpha_{A^k} \quad \text{for} \quad k = 0, 1, 2, \ldots .$$

For $t_k^n\colon A^n \to A^k$ in $\mathcal{A}$

(2) $$T(t_k^n)\, \alpha_n = \alpha_k\, S(t_k^n)$$

holds, and in particular

(3) $$T(p_j^k)\, \alpha_k = \alpha_1\, S(p_j^k) \quad \text{for} \quad k \geq 1 .$$

As T_k is a (not necessarily canonically chosen) product of k factors T_1, α_k is completely determined by α_1, even for $k = 0$, and for $k \geq 1$ it is the product of k factors α_1. The definition of homomorphism in 18.1.2 thus coincides with the one in 11.3.1.

By 11.5.1 and 11.5.7 the following is true: $\mathcal{A}^b$ is complete and has filtered colimits which commute with finite limits. As a subcategory of $[\mathcal{A}, Ens]$, $\mathcal{A}^b$ is closed with respect to limits and filtered colimits; i. e., these are formed as in $[\mathcal{A}, Ens]$.

There is a forgetful functor $U_{\mathcal{A}}: \mathcal{A}^b \to Ens$ given by $T \mapsto T_1 = T(A^1)$, $\alpha \mapsto \alpha_1$. It assigns to every $\mathcal{A}$-algebra T the *carrier* T_1 and to every $\mathcal{A}$-homomorphism α the *underlying morphism* α_1 of the carriers. By 11.3.3, 11.5.3, 11.5.8, the forgetful functor $U_{\mathcal{A}}$ is faithful and it preserves and reflects isomorphisms, limits and filtered colimits.

18.2.2 Proposition. *Let $J: \mathcal{A}^b \to [\mathcal{A}, Ens]$ be the inclusion. The Yoneda embedding $H^*: \mathcal{A}^0 \to [\mathcal{A}, Ens]$ factors into $H^* = J H^*_\pi$ with*

$$H^*_\pi : \mathcal{A}^0 \to \mathcal{A}^b . \tag{4}$$

H^*_π is a full embedding and it is dense. Furthermore, H^*_π preserves finite coproducts. $H^*_\pi((A^1)^0)$ is a representing object for $U_{\mathcal{A}}$ and it is a generator for $\mathcal{A}^b$.

Proof. The factorization exists, by 10.3.4, and H^*_π is a full embedding which preserves finite coproducts by 10.3.5. Since $\mathcal{A}^b$ is a full subcategory of $[\mathcal{A}, Ens]$, 4.2.4 implies

$$\begin{aligned} [H^*_\pi((A^1)^0), ?] &= [H^*((A^1)^0), J(?)] \\ &\cong E(J(?), A^1) = U_{\mathcal{A}}(?) . \end{aligned} \tag{5}$$

Since $U_{\mathcal{A}}$ is faithful, $H^*_\pi((A^1)^0)$ is a generator (10.5.1). By 17.2.5 and 17.2.4

$$1_{[\mathcal{A}, Ens]} \cong (J H^*_\pi)^\vee = \widetilde{H}^{*0}_\pi J^\vee \tag{6}$$

holds. Since J is fully faithful, $J^\vee J$ is isomorphic to $H_*: \mathcal{A}^b \to [\mathcal{A}^{bo}, Ens]$, by 17.2.5. Thus from (6) and 17.2.4

$$J \cong \widetilde{H}^{*0}_\pi J^\vee J \cong \widetilde{H}^{*0}_\pi H_* = H^{*\vee}_\pi \tag{7}$$

follows. H^*_π is dense, by 17.2.3.

18.2.3 Lemma. *The subcategory $\mathcal{N}^0$ of Ens is finitely cocomplete. The inclusion $I: \mathcal{N}^0 \to Ens$ is dense. For the set M the category I/M formed with respect to I as in 17.1.3 is filtered and the objects (k, a) with injective $a: k \to M$ are the objects of a final full subcategory (9.4.9) of I/M. For $k \in |\mathcal{N}^0|$, $(k, 1_k)$ is terminal in I/k.*

Proof. By 18.1.5, $\mathcal{N}^0$ has finite coproducts. For two morphisms $a, b: n \to k$ there is a coequalizer $c: k \to M$ in *Ens*, where M is a finite set. Thus there is an isomorphism $h: M \to m$ with $m \in |\mathcal{N}^0|$. $h c$ is a coequalizer of a and b in $\mathcal{N}^0$. h may even be chosen in such a way that $h c$ is (weakly) monotone. The remaining statements are easily verified (18.4.4, 17.1.5 (e)). They rest on the fact that every set is a filtered colimit of its finite subsets.

18.2.4 Theorem. *The forgetful functor* $U_{\mathcal{A}}\colon \mathcal{A}^b \to Ens$ *has a left adjoint* $L_{\mathcal{A}}\colon Ens \to \mathcal{A}^b$ *with*

(8) $$L_{\mathcal{A}}\, I = H^*_\pi\, I^0_{\mathcal{A}}\colon \mathcal{N}^0 \to \mathcal{A}^b\,.$$

Here $I\colon \mathcal{N}^0 \to Ens$ *is the inclusion and* $I_{\mathcal{A}}\colon \mathcal{N} \to \mathcal{A}$ *is the (only) theory-morphism.*

Proof. We consider the two contravariant functors $F, G\colon \mathcal{N}^0 \to$ $\to [\mathcal{A}^b, Ens]$ defined by

(9) $$F(?) = [H^*_\pi\, I^0_{\mathcal{A}}(?),\ ??]_{\mathcal{A}^b}\,, \qquad G(?) = [I(?),\ U_{\mathcal{A}}(??)]_{Ens}$$

By 18.1.5 and 18.2.2, $H^*_\pi\, I^0_{\mathcal{A}}$ preserves finite coproducts. The Yoneda embedding $\mathcal{A}^{b0} \to [\mathcal{A}^b, Ens]$ preserves limits, by 10.2.5. Therefore, F takes finite coproducts into finite products. The same holds for G because of the pointwise construction of limits in functor categories, and by 8.7.3. By 18.2.2, $F(1)$ and $G(1)$ are isomorphic. The description of $\mathcal{N}^0$ in 18.1.5 now implies immediately that F and G are isomorphic. By 3.6.3, there is an isomorphism

(10) $$\varrho\colon\ [I(?),\ U_{\mathcal{A}}(??)\,] \overset{\cong}{\Rightarrow} [H^*_\pi\, I^0_{\mathcal{A}}(?),\ ??]$$

of contra-co-variant functors. 17.2.9 can be applied to

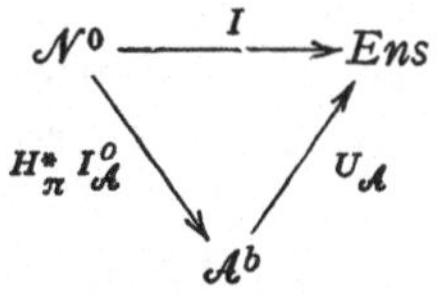

on account of 18.2.3, since $\mathcal{A}^b$ has filtered colimits. Hence $L_{\mathcal{A}}\colon Ens \to \mathcal{A}^b$ exists, with the adjunction isomorphism

(11) $$\varphi\colon\ [L_{\mathcal{A}}(?),\ ??]_{\mathcal{A}^b} \overset{\cong}{\Rightarrow} [?,\ U_{\mathcal{A}}(??)]_{Ens}\,.$$

Since I is an inclusion, one can achieve, according to 16.6.6, that (8) is valid, and in particular, that

(12) $$L_{\mathcal{A}}(k) = [A^k,\ ?]_{\mathcal{A}} \qquad \text{for } k \in |\mathcal{N}^0|$$

is true.

Note. By 16.5.1, for $M \in |Ens|$, $T \in |\mathcal{A}^b|$ and $\alpha\colon L_{\mathcal{A}}(M) \to T$

(13) $$\varphi_{M,T}(\alpha) = \alpha_1\, \eta_M\colon M \to T_1 \quad \text{with} \quad \eta_M\colon M \to U_{\mathcal{A}}\, L_{\mathcal{A}}(M)$$

holds.

18.2.5 Proposition. *If there is an* $\mathcal{A}$*-algebra* S *whose carrier has more than one element, then the adjunction transformation* $\eta\colon 1_{Ens} \to$ $\to U_{\mathcal{A}}\, L_{\mathcal{A}}$ *from (11) is a monomorphism and* $L_{\mathcal{A}}$ *is faithful.*

Proof. Let M be a set. Since one can change over to a suitable product in $\mathcal{A}^b$, and since $U_{\mathcal{A}}$ preserves products, S can be chosen in such a way that there is an injective map $u\colon M \to S_1$. Then by (11) there exists $\alpha: L_{\mathcal{A}}(M) \to S$ with $\varphi_{M,S}(\alpha) = u$. By (13), $u = \alpha_1 \eta_M$. Therefore, η_M is always injective, and $L_{\mathcal{A}}$ is faithful by the dual of 16.5.3.

18.2.6 Up to isomorphisms, there are only two algebraic theories for which the assumptions in 18.2.5 are not satisfied. For a terminal theory the carriers of all the algebras have exactly one element.

If the empty set is the carrier of an $\mathcal{A}$-algebra, then $\mathcal{A}$ has no nullary operations. If the carriers of the remaining $\mathcal{A}$-algebras all consist of one element only, then (12) shows that for $k \geq 1$ all k-ary operations coincide with p_1^k, and p_1^k is then inverse to the diagonal morphism $A^1 \to A^k$.

We call the *algebraic theories* of the two types just described *exceptional.*

18.2.7 Definition. A subset B of the carrier S_1 of the $\mathcal{A}$-algebra S is called a *basis* for S if the following holds:

(*) For every map $u: B \to T_1$ of B into the carrier of an arbitrary $\mathcal{A}$-algebra T there is exactly one $\mathcal{A}$-homomorphism $\alpha: S \to T$ with $\alpha_1 \mid B = u$.

An $\mathcal{A}$*-algebra* S is called *free,* if it has a basis, and it is called free on B, if B is a basis of S.

18.2.8 Proposition. *An $\mathcal{A}$-algebra S is free on $B \subset S_1$ if and only if for a suitable set M there is an isomorphism $\varrho: L_{\mathcal{A}}(M) \to S$ with $\varrho_1 \eta_M(M) = B$. In particular, $L_{\mathcal{A}}(M)$ is free on $\eta_M(M)$.*

Proof. We may assume that $\mathcal{A}$ is not exceptional, since otherwise the statements are trivial. First, let S be free on B and let $j: B \to S_1$ be the inclusion. (*) states that (S, j) is a representation of $[B, U_{\mathcal{A}}(?)]$. By (11), $(L_{\mathcal{A}}(B), \eta_B)$ is also a representation. By 4.4.4, there is an isomorphism $\varrho: L_{\mathcal{A}}(B) \to S$ with $\varrho_1 \eta_B = j$.

Now suppose that there is an isomorphism $\varrho: L_{\mathcal{A}}(M) \to S$ with $\varrho_1 \eta_M(M) = B$. By (11) and (13) there is, for $v = u \varrho_1 \eta_M: M \to T_1$, exactly one $\mathcal{A}$-homomorphism $\beta: L_{\mathcal{A}}(M) \to T$ with

$$(14) \qquad \beta_1 \eta_M = U_{\mathcal{A}}(\beta) \eta_M = v = u \varrho_1 \eta_M .$$

Since η_M is injective (18.2.5), this is equivalent to $\beta_1 \mid \eta_M(M) = u \varrho_1 \mid \mid \eta_M(M)$, and this is equivalent to $\beta_1 \varrho_1^{-1} \mid B = u \mid B = u_1$ since ϱ_1 is bijective. Thus for $\alpha = \beta \varrho^{-1}$, $\alpha_1 \mid B = u$ and α is determined uniquely by this, since by (14) this is the case for $\alpha \varrho$.

18.2.9 Since $L_{\mathcal{A}}$ preserves colimits (16.4.6), and since every set is the coproduct of its one-element subsets, every free $\mathcal{A}$-algebra is a coproduct of cofactors $L_{\mathcal{A}}(\{\emptyset\}) = H_\pi^*((A^1)^0)$, by 18.2.8. (Notice that $\{\emptyset\}$ is the natural number 1). $L_{\mathcal{A}}(\emptyset)$ is initial and free on the empty set. The free algebras which are finitely generated with respect to the generator $L_{\mathcal{A}}(\{\emptyset\})$ are thus precisely those which are isomorphic to an $L_{\mathcal{A}}(k)$ with $k \in |\mathcal{N}^0|$.

The free algebras need not be projective (10.4.7). By definition 18.2.7, they are projective if and only if $U_{\mathcal{A}}$ preserves epimorphisms.

18.2.10 Proposition. *For $k \in |\mathcal{N}^0|$, the carrier of $L_{\mathcal{A}}(k)$ is the set of k-ary operations of $\mathcal{A}$. For $k \geq 1$ the elements p_j^k form a basis. If two k-ary operations of $\mathcal{A}$ yield the same operation for the algebra $L_{\mathcal{A}}(k)$, then they are identical. Thus the equations for operations existing in $\mathcal{A}$ can be recognized by means of the finitely generated free $\mathcal{A}$-algebras. $\mathcal{A}^b$ has a zero object if and only if $\mathcal{A}$ has exactly one nullary operation.*

Proof. The first statement follows immediately from (12). By (5) and 4.2.2 (4), $(L_{\mathcal{A}}(\{\emptyset\}), p_1^1)$ is a representation of $U_{\mathcal{A}}$. The element p_1^1 therefore forms a basis of $L_{\mathcal{A}}(\{\emptyset\})$. Since $L_{\mathcal{A}}$ and H_π^* preserve finite coproducts, one obtains from (8) and (12) that $[p_j^k, ?]: [A^1, ?] \to [A^k, ?]$ are the injections for a representation of $L_{\mathcal{A}}(k)$ as a coproduct. Since $[p_j^k, A^1]\,(p_1^1) = p_j^k$, the second statement follows from the definition of bases and that of coproducts.

Let $t: A^k \to A^1$ be a k-ary operation. 1_{A^k} is an element of $[A_1^k, ?]_k = [A^k, A^k]$. The operation $[A^k, t]$ for $L_{\mathcal{A}}(k)$, which belongs to t, takes it into the element t of the carrier. This implies the third statement.

An $\mathcal{A}$-algebra is terminal in $\mathcal{A}^b$ if and only if its carrier consists of one element (11.2.4). Hence the last statement follows from the fact that $[A^0, A^1]$ is the carrier of the initial algebra $L_{\mathcal{A}}(\emptyset)$.

18.3 Subalgebras and Cocompleteness

18.3.1 We call an *$\mathcal{A}$-algebra T canonical* if, for $k \neq 1$, T_k is a canonically chosen k-fold product of the carrier T_1 with respect to the projections $T(p_j^k)$. In particular, $T_0 = \{\emptyset\}$. The full subcategory of $\mathcal{A}^b$, whose objects are the canonical algebras, we call the *reduced algebraic category* ${}_0\mathcal{A}^b$. For every $\mathcal{A}$-algebra S, there is an isomorphism $\sigma: S \to T$ into a canonical one, even in such a way that σ_1 is the identity map of the carrier. This follows immediately from 18.2.1. One thus has an equivalence $K_{\mathcal{A}}: \mathcal{A}^b \to {}_0\mathcal{A}^b$, for which the following holds:

(1) $${}_0U_{\mathcal{A}} K_{\mathcal{A}} = U_{\mathcal{A}} \quad \text{with} \quad {}_0U_{\mathcal{A}} = U_{\mathcal{A}} \mid {}_0\mathcal{A}^b\,.$$

The transition from $\mathcal{A}^b$ to ${}_0\mathcal{A}^b$ corresponds to the usual way of looking at an $\mathcal{A}$-algebra T as a "set T_1 with a structure", where the products T_k for $k \neq 1$ only play an auxiliary role. However, this transition from $\mathcal{A}^b$ to ${}_0\mathcal{A}^b$ is not needed until 18.5 and 18.6.

For the initial theory $\mathcal{N}$, ${}_0U_{\mathcal{N}}$ is obviously an isomorphism inverse to $K_{\mathcal{N}} L_{\mathcal{N}}$, and $\mathcal{N}^b$ and *Ens* are equivalent by means of $U_{\mathcal{N}}$ and $L_{\mathcal{N}}$.

18.3.2 Proposition. *Let S and T be $\mathcal{A}$-algebras and let $\alpha: S \to T$ be an $\mathcal{A}$-homomorphism. By 10.1.6, there exists the canonical factorization $\alpha = \iota\pi$ in $[\mathcal{A}, Ens]$ with $\pi: S \to R$, $\iota: R \to T$, where π or, resp., ι is* *a surjection or, resp., an inclusion at every place A^k. R is an $\mathcal{A}$-algebra and π and ι are $\mathcal{A}$-homomorphisms. (R, ι) is called the image of α.*

Proof. To simplify matters, we assume that S and T are canonical. The general case can obviously be reduced to this special case. For $k \geq 1$, α_k is the k-fold product of α_1. Since in *Ens* products of epimorphisms are again epimorphisms, one concludes right away that $R(A^k)$ and, resp., π_k, ι_k are products of k factors $R(A^1)$ or, resp., π_1, ι_1. α_0, π_0, ι_0 are identity maps. This proves the statement.

18.3.3 Definition. A *subalgebra* $(S, \sigma)^{\boldsymbol{\cdot}}$ of the $\mathcal{A}$-algebra T is an $\mathcal{A}$-homomorphism $\sigma: S \to T$ for which all σ_k are inclusions in *Ens*. An inclusion between subalgebras of T is a morphism in the category of $\mathcal{A}^b$-morphisms with codomain T.

The set of subalgebras of T is ordered by inclusion. 18.3.2 shows that in $\mathcal{A}^b$ every monomorphism is equivalent to a subalgebra and that every algebraic category is well-powered.

18.3.4 Proposition. *Let M be a subset of the carrier T_1 of the $\mathcal{A}$-algebra T and let $u: M \to T_1$ be the inclusion. Then there is a smallest subalgebra (S, σ) of T whose carrier contains M. It is called the subalgebra of T generated by M. It is the image of the homomorphism $\alpha: L_{\mathcal{A}}(M) \to T$ with $\varphi_{M,T}(\alpha) = u$ as in 18.2.4 (11).*

Proof. By 18.2.4 (13), $M = u(M) = \alpha_1(\eta_M(M))$. Therefore, M is contained in the carrier of the image of α. If (R, ϱ) is a subalgebra of T, whose carrier contains M, then there is a factorization $u = \varrho_1 u'$. Corresponding to u', there is a homomorphism $\alpha': L_{\mathcal{A}}(M) \to R$ with $\varphi_{M,R}(\alpha') = u'$. Now, $\varrho \alpha' = \alpha$, since $\varphi_{M,T}(\varrho \alpha') = \varrho_1 \alpha_1' \eta_M = \varrho_1 u' = u$. Thus α_1 factors through the inclusion ϱ_1, and the image of α is contained in (R, ϱ).

18.3.5 Lemma. *Let $X: \mathcal{A} \to Ens$ by any functor, $T: \mathcal{A} \to Ens$ an $\mathcal{A}$-algebra and $\xi: X \to T$ a natural transformation. Then ξ factors in $[\mathcal{A}, Ens]$ through the subalgebra (S, σ) generated by $M = \xi_1(X(A^1))$.*

Proof. We may assume that T and S are canonical to simplify the argument. For $k \geq 1$, one has $T(p_j^k)\, \xi_k = \xi_1\, X(p_j^k)$. For the product T_k, $T(p_j^k) = pr_j$. Then $M \subset S_1$ implies that there is a factorization $\xi_k = \sigma_k\, \xi_k'$; this is trivially true for $k = 0$. For $t: A^n \to A^k$ in $\mathcal{A}$, $\sigma_k\, \xi_k'\, X(t) = T(t)\, \sigma_n\, \xi_n' = \sigma_k\, S(t)\, \xi_n'$. Since σ_k is a monomorphism, ξ' is a natural transformation.

18.3.6 Theorem. *Let $\mathcal{A}$ be an algebraic theory. The inclusion J: $\mathcal{A}^b \to [\mathcal{A}, Ens]$ has a left adjoint. $\mathcal{A}^b$ is well-powered, complete and cocomplete, and as a subcategory of $[\mathcal{A}, Ens]$ it is closed with respect to limits and filtered colimits. Filtered colimits commute with finite limits.*

Proof. The cocompleteness of $\mathcal{A}^b$ follows by 16.6.1 from the existence of a left adjoint for J. The remaining properties of $\mathcal{A}^b$ are all already known (18.2.1, 18.3.3). The existence of a left adjoint follows from 16.4.5 provided $[X, J(?)]$ is representable for every $X\colon \mathcal{A} \to \mathit{Ens}$, which is what we shall now show.

$\mathcal{A}^b$ is complete, J preserves limits and so does $[X, J(?)]$, by 7.7.3. By 10.3.9, it suffices to show that $[X, J(?)]$ is proper. Since J is the inclusion of a full subcategory, definition 10.3.1 has the following meaning in this case: There is a set $\mathfrak{E}$ of $\mathcal{A}$-algebras such that every natural transformation $\xi\colon X \to T$ into an arbitrary $\mathcal{A}$-algebra T factors through a natural transformation $\alpha'\colon X \to R$ with $R \in \mathfrak{E}$. Then $\{[X, R] \mid R \in \mathfrak{E}\}$ is a dominating set for $[X, J(?)]$. Let C be the carrier of $L_{\mathcal{A}}(X_1)$. Let R be in $\mathfrak{E}$ if and only if R is canonical and R_1 a subset of C. $\mathfrak{E}$ is a set, since for a given carrier M there is only a set of canonical $\mathcal{A}$-algebras. (The products M^k are the objects of a small, full subcategory of Ens, and $\mathcal{A}$ is small). For $\xi_1\colon X_1 \to T_1$ there is, by 18.2.4 (11) and (13), an $\mathcal{A}$-homomorphism $\alpha\colon L_{\mathcal{A}}(X_1) \to T$ with

(2) $$\alpha_1 \eta_{X_1} = \xi_1 .$$

Let (S, σ) be the image of α. By (2), $\xi_1(X_1) \subset S_1$. By 18.3.5, ξ factors through σ. Now, α_1 maps C surjectively onto S_1. Therefore, S is isomorphic to an algebra R of $\mathfrak{E}$, and ξ also factors through R.

Remark. Since it is known that $\mathcal{A}^b$ is cocomplete, a left adjoint of J can also be obtained from 17.3.3 by 18.2.2.

18.3.7 (AB 5)-properties. Since filtered colimits commute with pullbacks in $\mathcal{A}^b$, 7.8.9 implies that filtered colimits of monomorphisms are monomorphisms. Thus 14.6 (1) and (2) follow. The analogue of 14.6.8 is not valid in general as is shown, for instance, by the category of groups.

18.4 Coequalizers and Kernel Pairs

We begin by introducing some auxiliary notions which are of independent interest.

18.4.1 Definition. For $f\colon B \to D$ in the category $\mathcal{C}$ let

(1) $$\begin{array}{ccc} K & \xrightarrow{a} & B \\ {\scriptstyle b}\downarrow & & \downarrow{\scriptstyle f} \\ B & \xrightarrow{f} & D \end{array}$$

be a pullback. The pair (a, b) is called a *kernel pair* of f. A pair of morphisms $a, b\colon K \to B$ is called a kernel pair if it is a kernel pair for a suitable f. Cokernel pairs are defined dually.

Remarks. If (1) is constructed as in 7.8.5, then one obtains for $\mathscr{C} = Ens$ that K is a subset of $B \times B$ consisting of those pairs (x, y) with $x, y \in B$, for which $f(x) = f(y)$. Corresponding statements are valid in the cases of *Top*, *Ab* and other categories whose objects are "sets with a structure". The "structure-compatible equivalence relations" defined by means of elements are thus replaced by kernel pairs in the case of arbitrary categories with pullbacks.

18.4.2 Proposition. *Let (a, b) be a kernel pair of f.*

(a) *a and b are retractions with a unique common coretraction s, that is $a\,s = b\,s = 1_B$. Furthermore, s is an equalizer of a and b.*

(b) *If morphisms $u, v\colon B \to X$ satisfy $u\,a = v\,b$, then $u = v$.*

(c) *f is a monomorphism if and only if $a = b$. In this case, a is an isomorphism.*

Proof. (a) Consider the identity morphism of B into both copies of B in (1). The first statement then follows from the definition of pullbacks. Now let $a\,g = b\,g$ for $g\colon X \to K$. Setting $u = a\,g$, one has $u = a\,s\,u = b\,s\,u$. If one takes $u\colon X \to B$ for both copies of B in (1), then $g = s\,u$ follows, again by the definition of pullbacks. If, in addition, $g = s\,v$, then $u = v$, since s is a monomorphism. Thus s is an equalizer of a and b.

(b) $u\,a = v\,b$ and $a\,s = b\,s = 1_B$ imply $u = v$.

(c) is 7.9.9.

18.4.3 Proposition. *Let (a, b) be a kernel pair of $f\colon B \to D$. Consider (1) and*

(2)
$$\begin{array}{ccc} K & \xrightarrow{\;a\;} & B \\ {\scriptstyle b}\big\downarrow & & \big\downarrow{\scriptstyle c} \\ B & \xrightarrow{\;c\;} & C \end{array} .$$

(a) *c is a coequalizer of a and b if and only if (2) is a pushout. If this is the case, then c is maximal (in the sense of $6.5.4^0$) among the coequalizers with domain B through which f factors.*

(b) *If (a, b) has a coequalizer, then (a, b) is also a kernel pair of its coequalizer.*

(c) *If a coequalizer has a kernel pair, then it is also a coequalizer of its kernel pair.*

Proof. By 18.4.2 (b), the first statement in (a) follows immediately from the definitions. If (2) is a pushout, then there is an $h\colon C \to D$ with $f = h\,c$. Thus (b) follows from (1) and the definition of pullbacks. Now let $f = g\,f'$, and let f' be a coequalizer for $a', b'\colon K' \to B$. It follows from $f\,a' = f\,b'$ and (1) that there is a $t\colon K' \to K$ with $a' = a\,t$, $b' = b\,t$. Hence $c\,a' = c\,b'$. According to the definition of coequalizers, there is a u with $c = u\,f'$. This is the second statement in (a). It implies (c).

18.4.4 In *Ens* every monomorphism is an equalizer, every epimorphism is a coequalizer.

Proof. One verifies easily that every monomorphism is an equalizer of its cokernel pair (compare 7.2.4, 8.8.3). Every epimorphism is a retraction and thus a coequalizer (8.2.2).

18.4.5 Proposition. *An $\mathcal{A}$-homomorphism $f: A \to B$ is a coequalizer in $\mathcal{A}^b$ if and only if $U_{\mathcal{A}}(f)$ is a coequalizer in Ens. A pair $a, b: K \to B$ of $\mathcal{A}$-homomorphisms is a kernel pair in $\mathcal{A}^b$ if and only if $(U_{\mathcal{A}}(a), U_{\mathcal{A}}(b))$ is a kernel pair in Ens.*

Proof. We consider

(3)

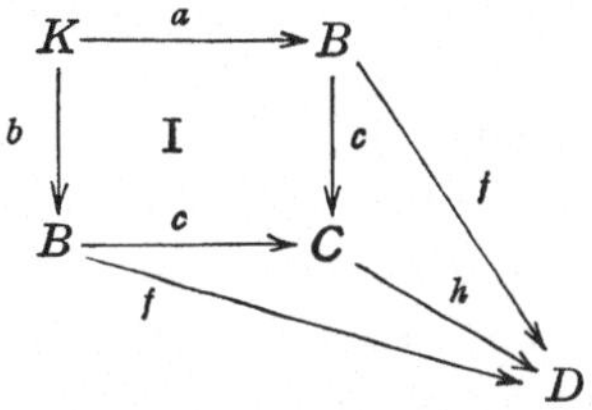

First, let f be a coequalizer in $\mathcal{A}^b$. Construct (3) by forming the pullback (1) and the factorization $f = h\,c$ as in 18.3.2 (i.e., c_1 is an epimorphism, h_1 a monomorphism). The square I is then commutative, because h is a monomorphism. By 18.4.3 (c) and (a), (1) is also a pushout. Therefore, there exists a $g: D \to C$ with $c = g\,f$. $f = h\,g\,f$ and thus $h\,g = 1_D$ follow, since f is an epimorphism. So h is a monomorphic retraction and thus an isomorphism. It follows that $U_{\mathcal{A}}(f) = f_1$ is an epimorphism. By 18.4.4, f_1 is a coequalizer in *Ens*.

Now let f_1 be a coequalizer in *Ens*. Construct (3) by first forming the pullback (1) and then the bicartesian square (2). h exists by construction. By 18.4.3 (a) and what was proved above, c_1 is an epimorphism. By applying $U_{\mathcal{A}}$ one obtains pullbacks from (1) and (2), since $U_{\mathcal{A}}$ preserves limits. By the assumption about f_1 and 18.4.3 (c), one gets a bicartesian square from (1). Therefore, there exists a $g_1: D_1 \to C_1$ with $c_1 = g_1\,f_1$. Furthermore, $f_1 = h_1\,c_1$. Since c_1 and f_1 are epimorphisms, it follows that g_1 is inverse to h_1. And since $U_{\mathcal{A}}$ reflects isomorphisms, h is an isomorphism and thus f is a coequalizer of a and b.

If (a, b) is a kernel pair in $\mathcal{A}^b$, say of f, then (1) is a pullback and (a_1, b_1) is a kernel pair of f_1, since $U_{\mathcal{A}}$ preserves limits. Now, conversely, for (a, b) let (a_1, b_1) be a kernel pair; and let $c_1: B_1 \to C_1$ be a coequalizer of a_1 and b_1. Then

(4)

$$
\begin{array}{ccc}
K_1 & \xrightarrow{a_1} & B_1 \\
{\scriptstyle b_1}\downarrow & & \downarrow{\scriptstyle c_1} \\
B_1 & \xrightarrow[c_1]{} & C_1
\end{array}
$$

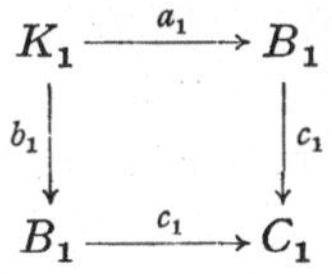

is bicartesian in *Ens*. Since pullbacks commute with products, the k-fold product of (4) is a pullback for $k \geq 1$. Since K_k, B_k, a_k, b_k are (not necessarily canonically chosen) k-fold products of K_1, B_1, a_1, b_1, the pullback

(5)
$$\begin{array}{ccc} K_k & \xrightarrow{a_k} & B_k \\ {\scriptstyle b_k}\downarrow & & \downarrow{\scriptstyle c_k} \\ B_k & \xrightarrow{c_k} & (C_1)^k \end{array}$$

exists in *Ens*, where c_k is a k-fold product of c_1; i.e., for $1 \leq j \leq k$, $pr_j\, c_k = c_1\, B(p_j^k)$ holds. (5) is also valid for $k = 0$, in this case a_0, b_0, c_0 are maps between sets of one element and thus isomorphisms. Since in *Ens* products of epimorphisms are epimorphisms, the c_k - s are epimorphisms and thus coequalizers. By 18.4.3 (c) and (a), for $k \geq 0$, the pullbacks (5) are also pushouts. According to the "pointwise" construction of pushouts in $[\mathcal{A}, Ens]$ one gets a pushout (2) in $[\mathcal{A}, Ens]$ from the diagrams (5). By construction, C lies in $\mathcal{A}^b$, and so does c, because $\mathcal{A}^b$ is a full subcategory of $[\mathcal{A}, Ens]$. So (2) is commutative in $\mathcal{A}^b$ and it is a pullback, since the diagrams (5) are pullbacks. Therefore, (a, b) is a kernel pair of c.

18.4.6 Corollary. *In $\mathcal{A}^b$ the following holds:*

(a) *Composites of coequalizers are coequalizers. (The class of coequalizers is compositive).*

(b) *If*

(6)
$$\begin{array}{ccc} A & \xrightarrow{b} & B \\ {\scriptstyle a}\downarrow & & \downarrow{\scriptstyle d} \\ C & \xrightarrow{c} & D \end{array}$$

is a pullback, and if c is a coequalizers, then b is a coequalizer. (Coequalizers are pullback-closed).

(c) *Products of coequalizers are coequalizers.*

Proof. (a) follows immediately from 18.4.5 and 18.4.4.

(b) From (6) one obtains a pullback in *Ens* by applying $U_{\mathcal{A}}$, and by 18.4.4, c_1 is a coequalizer and thus an epimorphism. By 13.4.4, b_1 is an epimorphism. Thus (b) follows from 18.4.5.

(c) is valid in *Ens* by 18.4.4. By 18.4.5, the statement follows for $\mathcal{A}^b$, since $U_{\mathcal{A}}$ preserves and reflects limits.

Warning. (c) does not state that coequalizers commute with products. This is not even true in *Ens*.

We now consider some interrelations between the statements in 18.3.2, 18.4.5 and 18.4.6.

18.4.7 Proposition. *For every category $\mathscr{C}$ the following statements hold:*

(a) *Coequalizers are pushout-closed.*

(b) *If (6) is commutative, and if b is a coequalizer and c a monomorphism, then there is a uniquely determined $e: B \to C$ such that $a = e\,b$. Furthermore, $d = c\,e$ holds.*

(c) *If a morphism has a factorization into a coequalizer followed by a monomorphism, then the factorization is unique up to an isomorphism.*

(d) *If such a factorization exists for every morphism, then this factorization is natural (in the sense of 12.4.8), and the class of coequalizers is compositive. Furthermore, the analogues of 14.2.5 and 14.2.6 are then valid.*

(e) *Let $\mathscr{C}$ have kernel pairs and coequalizers of kernel pairs. If the class of coequalizers is compositive, then every morphism can be factored into a coequalizer and a monomorphism.*

Proof. (a) Let (6) be a pushout and b a coequalizer of, say, u, v: $X \to A$. Using 1_X and $a\,u$, $a\,v$: $X \to C$, one obtains c as a coequalizer of $a\,u$ and $a\,v$ from the dual of 12.3.5.

(b) Again, let b be a coequalizer of u, v: $X \to A$. Since c is a monomorphism here, $a\,u = a\,v$, and e exists by the definition of coequalizers. Since b is an epimorphism, $d = ce$ follows from $a = eb$.

(c) Let (6) be commutative, and let a, b be coequalizers and c, d monomorphisms. (b) yields $e: B \to C$ and similarly $f: C \to B$. Since b is an epimorphism, $f\,e = 1_B$ follows from $b = f\,a = f\,e\,b$. Similarly, $e\,f = 1_C$.

(d) The first statement follows immediately from (b) (compare with 12.4 (14) and (15)). Now let (6) be commutative, let a, b, d be coequalizers and c a monomorphism; i.e., $c\,a$ is a factorization of $d\,b$. Let d be a coequalizer of, say, u, v: $X \to B$, and let e: $B \to C$ be as in (b). Then $e\,u = e\,v$. By the definition of coequalizers, c is a retraction and, therefore, an isomorphism. Hence, $d\,b$ is a coequalizer, and the class of coequalizers is compositive. The last statement in (d) is easily verified.

(e) Let f: $B \to D$ be any morphism, and let a, b: $K \to B$ be a kernel pair of f and c: $B \to C$ a coequalizer of a, b. Then there is a d: $C \to D$ with $f = d\,c$. Let a', b': $K' \to C$ be a kernel pair of d, and let c' be a coequalizer of a', b'. Now f factors through $c'\,c$. If $c'\,c$ is a coequalizer, then c' is an isomorphism, by 18.4.3 (a). Hence $a' = b'$, and a' is an isomorphism, by 18.4.3 (b). By 7.8.9, d is a monomorphism.

18.4.8 Proposition. *Let the category $\mathscr{C}$ have pullbacks. If the class of coequalizers is pullback-closed, then it is compositive.*

Proof. Let $p: B \to C$ be a coequalizer of, say, $f, g: A \to B$, and let $q: C \to D$ be also a coequalizer. We consider the following diagram

(7)

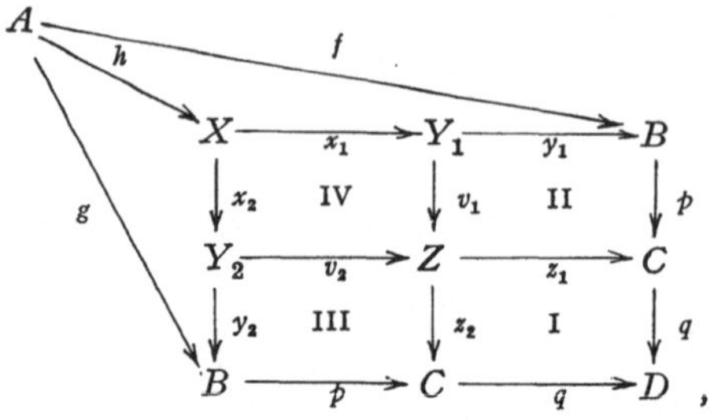

where I, II, III, IV are constructed successively as pullbacks. By 7.8.4, these four squares together form again a pullback. Since $q\,p\,f =$ $= q\,p\,g$, there is an h such that the diagram (7) is commutative. We show that $q\,p$ is a coequalizer of $y_1\,x_1$ and $y_2\,x_2$. Now, let $u: B \to E$ be given with $u\,y_1\,x_1 = u\,y_2\,x_2$. Then $u\,f = u\,g$, and there is a uniquely determined $w: C \to E$ with $w\,p = u$. By assumption, all morphisms in I, II, III, IV are coequalizers. Hence, $v_1\,x_1 = v_2\,x_2$ is an epimorphism and, therefore, $w\,z_1 = w\,z_2$. By 18.4.3, the square I is bicartesian, and there is a $d: D \to E$ with $d\,q = w$. Thus, $d\,q\,p = u$, and d is uniquely determined by u, since $q\,p$ is an epimorphism.

18.4.9 Remarks. Under the assumption of 18.4.7 (d), images of monomorphisms are defined as in 14.4.2 (4) by means of a factorization into a coequalizer and a monomorphism. 14.4.3, 14.4.4 and 14.4.7 are then correspondingly valid. Under the assumptions of 18.4.7 (d) and 18.4.8, there is an analogue of 14.4.6, where $f: A \to B$ is a coequalizer.

In particular, all assumptions in 18.4.7 and 18.4.8 are satisfied by every algebraic category, because of 18.3.2, 18.3.6, 18.4.5 and 18.4.6. 18.3.2, 18.4.3 (c) and 18.4.7 (a) together constitute the statement known as the Homomorphism Theorem (see, e.g., Bourbaki, Algèbre I. 4.4).

Factorizations of morphisms will be discussed more generally in 21.6 (See also 16.8.6).

18.4.10 A pair of morphisms $a, b: K \to B$ uniquely determines a morphism $k: K \to B \sqcap B$ with $pr_1\,k = a$, $pr_2\,k = b$. If a, b is a kernel pair, then k is an equalizer, by 7.8.5. Therefore, for kernel pairs with a fixed codomain, there is a preordering as in the case of monomorphisms. If $a, b: K \to B$ and $a', b': K' \to B$ are kernel pairs with the same codomain B, then $(a, b) \leq (a', b')$ if there is a morphism $h: K \to K'$ with $a'\,h = a$ and $b'\,h = b$. Here h is a monomorphism and even an equalizer.

In categories with kernel pairs and coequalizers an analogue of 12.4.4 holds, on account of 18.4.3 (b), (c). To see this, we consider

(8)

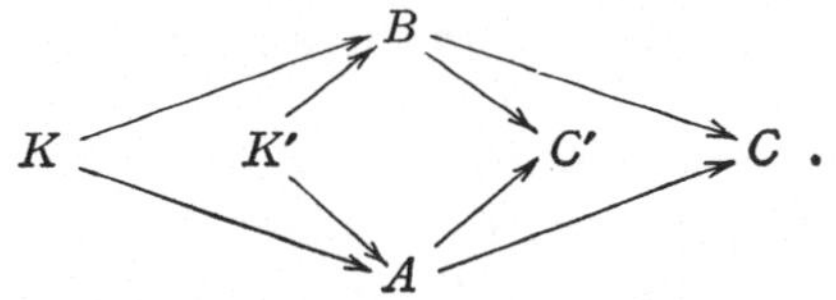

Let the outer and the inner contour be commutative. If the outer contour is a pushout and if $K \to K'$ is a commutative completion, then there exists exactly one commutative completion $C \to C'$. Dually, the converse is true if the inner contour is a pullback. Thus intersections and unions of kernel pairs can be reduced to counions and cointersections of coequalizers, and conversely. Also, an analogue of 14.5.1 holds, namely:

18.4.11 Proposition. *Let the category $\mathscr{C}$ be (finitely) complete, and let $\{(a_j, b_j)\}$ be a (finite) family of kernel pairs with the codomain B, where (a_j, b_j) is a kernel pair of, say, $c_j: B \to C_j$. Then the kernel pair (a, b) of $c: B \to \Pi\, C_j$, where $pr_j\, c = c_j$, is the intersection of the family $\{(a_j, b_j)\}$.*
By 7.8.5, this is in fact a special case of 14.5.1 with $A = B \sqcap B$.

18.4.12 Inverse images of kernel pairs. We consider

(11)
$$
\begin{array}{ccccccc}
B' & \xrightarrow{\ s'\ } & K' & \underset{b'}{\overset{a'}{\rightrightarrows}} & B' & \xrightarrow{\ f'\ } & D' \\
\scriptstyle g'\downarrow & \text{III} & \scriptstyle g''\downarrow & \text{II} & \scriptstyle g'\downarrow & \text{I} & \downarrow\scriptstyle g \\
B & \xrightarrow[s]{} & K & \underset{b}{\overset{a}{\rightrightarrows}} & B & \xrightarrow[f]{} & D
\end{array}
$$

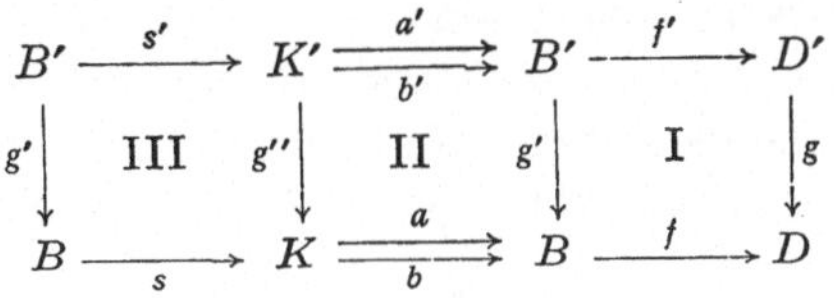

where I is a pullback, (a, b) a kernel pair of f and s the common coretraction for (a, b). (a', b') and s' have the corresponding meaning for f'.

(a) g'' exists with $g'\, a' = a\, g''$ and $g'\, b' = b\, g''$, since $f\, g'\, a' = g\, f'\, a' = f\, g'\, b'$.

(b) I and II together are a pullback.

To show this, we assume that there are $u: X \to K$ and $v: X \to D'$ with $g\, v = f\, a\, u\,(= f\, b\, u)$. Then there exists a unique $x: X \to B'$ with $f'\, x = v$ and $g'\, x = a\, u$, and also $y: X \to B'$ with $f'\, y = v$ and $g'\, y = b\, v$. So then there is a unique $z: X \to K'$ with $a'\, z = x$ and $b'\, z = y$. Further, $a\, g''\, z = g'\, a'\, z = g'\, x = a\, u$ and $b\, g''\, z = b\, u$, which implies $g''\, z = u$.

(c) a, g', a', g'' form a pullback, as do b, g', b', g''. This follows immediately from 7.8.4.

(d) III is a pullback.

This follows from (b) and 7.8.4, since I, II, III together make up I again.

18.5 Algebraic Functors and Left Adjoints

18.5.1 Definition. Let $F\colon \mathcal{A} \to \mathcal{B}$ be a theory-morphism. Since F preserves finite products, the functor $F^b\colon \mathcal{B}^b \to \mathcal{A}^b$ is obtained by restriction from the functor $[F, Ens]\colon [\mathcal{B}, Ens] \to [\mathcal{A}, Ens]$ (compare 16.1.3). Functors of this kind are called *algebraic functors.*

18.5.2 For the $\mathcal{B}$-algebra T, $F^b(T) = T\,F$, and by 18.1.2 (1) one has

(1) $$F^b(T)_k = F^b(T)\,(A^k) = T\,F(A^k) = T(B^k) = T_k\,, \qquad F^k(T)\,(p_j^k) = T(p_j^k)\,.$$

For the $\mathcal{B}$-homomorphism $\alpha\colon S \to T$, $F^b(\alpha) = \alpha * F$. From (1) it follows thus that

(2) $$F^b(\alpha)_k = \alpha_k\,,$$

(3) $$U_{\mathcal{A}}\,F^b = U_{\mathcal{B}}\,.$$

The forgetful functor $U_{\mathcal{A}}$ may be regarded as an algebraic functor in a more general sense, since, for $I_{\mathcal{A}}\colon \mathcal{N} \to \mathcal{A}$, $U_{\mathcal{N}}\,I_{\mathcal{A}}^b = U_{\mathcal{A}}$ is valid by (3), and $U_{\mathcal{N}}$ is an equivalence (18.3.1).

18.5.3 Theorem. *Every algebraic functor $F^b\colon \mathcal{B}^b \to \mathcal{A}^b$ is faithful. It preserves and reflects isomorphisms, limits, monomorphisms and filtered colimits. The analogue of 18.4.5 is valid for F^b (instead of $U_{\mathcal{A}}$). F^b has a left adjoint $F_*\colon \mathcal{A}^b \to \mathcal{B}^b$ which can be chosen so that*

(4) $$\begin{array}{ccc} \mathcal{A}^0 & \xrightarrow{F^0} & \mathcal{B}^0 \\ \downarrow{\scriptstyle H_\pi^*} & & \downarrow{\scriptstyle H_\pi^*} \\ \mathcal{A}^b & \xrightarrow{F_*} & \mathcal{B}^b \end{array}$$

is commutative. In particular, F_* takes the generator for $\mathcal{A}^b$ specified in 18.2.2 into the one for $\mathcal{B}^b$. Furthermore

(5) $$F_*\,L_{\mathcal{A}} \cong L_{\mathcal{B}}\colon Ens \to \mathcal{A}^b$$

holds.

Proof. The first statement follows immediately from (2). The second one is true for $U_{\mathcal{A}}$ and $U_{\mathcal{B}}$ (18.2.1). From this and (3) it also follows for F^b; for instance, a filtered colimit in $\mathcal{B}^b$ is preserved by $U_{\mathcal{B}}$ and it is reflected by $U_{\mathcal{A}}$ in $\mathcal{A}^b$, thus, by (3), it is preserved by F^b. The analogue of 18.4.5 for F^b also follows from (3).

Let $J_{\mathcal{A}}\colon \mathcal{A}^b \to [\mathcal{A}, Ens]$, $J_{\mathcal{B}}\colon \mathcal{B}^b \to [\mathcal{B}, Ens]$ be the inclusions. Let $C_{\mathcal{B}}\colon [\mathcal{B}, Ens] \to \mathcal{B}^b$ be left adjoint to $J_{\mathcal{B}}$ as in 18.3.6. By 16.6.5, we may assume that $C_{\mathcal{B}} J_{\mathcal{B}} = 1_{\mathcal{B}^b}$. By 17.1.8, $\widetilde{F} = [F, Ens]$ has a left adjoint K, for which $K H^* = H^* F^0$. For $S \in |\mathcal{A}^b|$, $T \in |\mathcal{B}^b|$ one then obtains

(6)
$$\begin{aligned}[S, F^b T] &= [J_{\mathcal{A}}(S), J_{\mathcal{A}} F^b(T)] = [J_{\mathcal{A}}(S), \widetilde{F} J_{\mathcal{B}}(T)] \\ &\cong [K J_{\mathcal{A}}(S), J_{\mathcal{B}}(T)] \cong [C_{\mathcal{B}} K J_{\mathcal{A}}(S), T] .\end{aligned}$$

Here the first equation holds because $J_{\mathcal{A}}$ is a full embedding and the second one because F^b is obtained from $[F, Ens]$ by restriction. The two following isomorphisms exist because one has pairs of adjoint functors; and (6) is an isomorphism of contra-co-variant functors. Therefore, $F_* = C_{\mathcal{B}} K J_{\mathcal{A}}$ is left adjoint to F^b. Because of $H^* = J_{\mathcal{A}} H^*_\pi$: $\mathcal{A}^0 \to [\mathcal{A}, Ens]$, the corresponding relation for $\mathcal{B}$ (18.2.2) and $K H^* = = H^* F^0$,

(7)
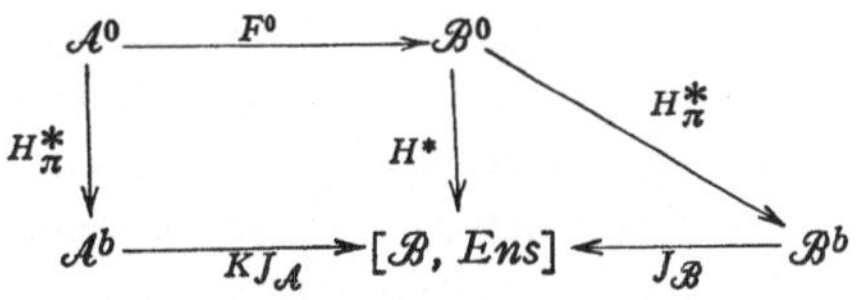

is commutative. (4) is valid, since $C_{\mathcal{B}} J_{\mathcal{B}} = 1_{\mathcal{B}^b}$. From (3) and the definition of $L_{\mathcal{A}}$, $L_{\mathcal{B}}$ (18.2.4) (5) follows by 16.4.4 and

$$[L_{\mathcal{B}}(M), T) \cong [M, U_{\mathcal{B}}(T)] = [M, U_{\mathcal{A}} F^b(T)] \cong [F_* L_{\mathcal{A}}(M), T] .$$

18.5.4 Remarks. F_* can, in fact, be obtained as a restriction of K. For, in the situation of 17.1.6 with $\mathcal{D} = Ens$, one has (using the same notation): If $\mathcal{B}$ has finite products and if these are preserved by U, then V takes functors $\mathcal{B} \to Ens$, which preserve finite products, into such functors from $\mathcal{C}$ to Ens. An analogue for infinite products is valid only under additional assumptions as, e.g., 17.1.5 (e).

Since the transition $K_{\mathcal{A}}$: $\mathcal{A}^b \to {}_0\mathcal{A}^b$ to canonical algebras (18.3.1) is an equivalence with the inclusion as an equivalence-inverse, the preceding material starting with 18.2 yields corresponding results for reduced algebraic categories. From

$${}_0H^*_\pi = K_{\mathcal{A}} H^*_\pi\colon \mathcal{A}^0 \to {}_0\mathcal{A}^b , \qquad {}_0U_{\mathcal{A}} K_{\mathcal{A}} = U_{\mathcal{A}} , \qquad {}_0L_{\mathcal{A}} = K_{\mathcal{A}} L_{\mathcal{A}}$$

one concludes that ${}_0L_{\mathcal{A}}$ is left adjoint to ${}_0U_{\mathcal{A}}$ where, by 18.2.4

(8)
$${}_0L_{\mathcal{A}} I = {}_0H^*_\pi I^0_{\mathcal{A}}\colon \mathcal{N}^0 \to {}_0\mathcal{A}^b .$$

By (1), (2), ${}_0F^b$: ${}_0\mathcal{B}^b \to {}_0\mathcal{A}^b$ is given by restriction, and (3) implies

(9)
$${}_0U_{\mathcal{A}}\, {}_0F^b = {}_0U_{\mathcal{B}} .$$

The left adjoint ${}_0F_*$ of ${}_0F^b$ is formed from F_* by means of $K_{\mathcal{B}}$ and the inclusion ${}_0\mathcal{A}^b \subset \mathcal{A}^b$. (5) is replaced by

(10)
$${}_0F_*\, {}_0L_{\mathcal{A}} \cong {}_0L_{\mathcal{B}}.$$

The following examples are to be understood in the sense of this reduction (compare 18.3.1). Note, that by 18.1.10, 18.1.11 it is sufficient to define $\mathcal{A}$ and $\mathcal{B}$ by generating operations and defining equations and for $F: \mathcal{A} \to \mathcal{B}$ to give the image of the generating operations in $\mathcal{A}$ as operations in $\mathcal{B}$ (which may be composites of generating ones). Compatibility with the equations which are derived from the defining ones has to be guaranteed (compare later 18.6.8).

Examples

18.5.5 Let ${}_0\mathcal{A}^b$ be the category of groups, ${}_0\mathcal{B}^b$ the category of abelian groups, ${}_0F^b$ the obvious forgetful functor. ${}_0F_*$ is the transition to the quotient group by the commutator subgroup.

18.5.6 Let ${}_0\mathcal{A}^b$ be the category of algebras (in the usual sense) over a fixed commutative ring R, let ${}_0\mathcal{B}^b$ be the category of R-algebras with a richer structure; e.g. anticommutative algebras (for every element x, $x^2 = 0$), Lie algebras (anticommutative with the Jacobi identity), associative algebras, commutative associative algebras with a 1, etc. ${}_0F^b$ is always a functor which forgets the additional richness of the structure. ${}_0F_*$ is then the transition to "univeral envelopes" in the richer structure. This remains valid if ${}_0\mathcal{A}^b$ is a category of R-algebras with some additional structure. An example for this is the following: ${}_0\mathcal{A}^b$ is associative R-algebras, ${}_0\mathcal{B}^b$ associative R-algebras with a 1, ${}_0F_*$ is the adjunction of a 1.

The following is a classical example: ${}_0\mathcal{A}^b$ is Lie algebras over R, ${}_0\mathcal{B}^b$ is associative R-algebras with a 1. $F: \mathcal{A} \to \mathcal{B}$ takes the binary operation of $\mathcal{A}$ which corresponds to the Lie product $(x, y) \to [x, y]$ into the operation of $\mathcal{B}$ which corresponds to $(x, y) \mapsto x\,y - y\,x$, the other defining operations are treated in the obvious way. Thus ${}_0F^b$ is the transition from an associative algebra to the associated Lie algebra, ${}_0F_*$ the transition from a Lie algebra to its universal associative envelope.

18.5.7 Let ${}_0\mathcal{A}^b$ be the categories of modules over a fixed commutative ring R, ${}_0\mathcal{B}^b$ a category of R-algebras. ${}_0F^b$ is the obvious forgetful functor, ${}_0F_*$ the transition from a module to the tensor algebra of the corresponding type. Associative tensor algebras with 1 and symmetrical tensor algebras are special cases (if gradation is ignored).

18.5.8 Let R and S be rings, ${}_0\mathcal{A}^b = {}_R Mod$, ${}_0\mathcal{B}^b = {}_S Mod$ and let $\varphi: R \to S$ be a ring homomorphism. Then $(r, x) \to \varphi(r)\,x$ turns every left S-module into a left R-module. ${}_0F^b$ is a change of rings by means of φ; ${}_0F_*$ is usually called an extension of coefficients (by means of φ). ${}_0F_*$ can be represented as a tensor product, ${}_0F_*(M) = {}_S S_R \otimes_R M$, where the right-module structure for S is provided by φ. There are corresponding facts in the case of algebras over commutative rings.

18.5.9 The forgetful functors groups $\to$ monoids, groups $\to H$-sets, abelian groups $\to$ commutative monoids have left adjoints.

The list of examples could be continued.

18.5.10 The transition from a group to its group ring is not left adjoint to an algebraic functor. However, if one choses for ${}_0\mathcal{A}^b$ the category of monoids and for ${}_0\mathcal{B}^b$ the category of rings, then there is the forgetful functor ${}_0F^b$, which forgets addition and the 0. ${}_0F_*$ is then the transition to the monoid-ring. The transition from a group to the group ring is the composite of this ${}_0F_*$ with the forgetful functor groups $\to$ monoids. We refer to [54] for additional information.

18.6 Semantics and Structure

18.6.1 Preliminary remarks. (a) Definition 18.1.2 and the conditions 18.1.2 (1) admit another interpretation. An algebraic theory can be regarded as a functor $I_{\mathcal{A}}: \mathcal{N} \to \mathcal{A}$, which preserves finite products and is bijective on objects. This takes care of the indexing of the projections p_j^k in 18.1.1 (ii). A theory-morphism is then a functor F: $\mathcal{A} \to \mathcal{B}$ for which $I_{\mathcal{B}} = F\, I_{\mathcal{A}}$, which encompasses 18.1.2 (1).

(b) Let $\mathcal{D}$ be a $\mathfrak{V}$-category with finite products, e.g., $\mathcal{D} = [\mathcal{C}, Ens]$. For $X \in |\mathcal{D}|$ there is, up to isomorphisms, exactly one functor P_X: $\mathcal{N} \to \mathcal{D}$ which preserves finite products and for which $P_X(N^1) = X$. P_X need not be injective on objects. If one takes for A^k the pair $(k, P_X(N^k))$ and as morphisms $A^n \to A^k$ triples (n, k, t), where $t: P_X(N^n) \to P_X(N^k)$, with the composition $(k, j, t')\,(n, k, t) = (n, j, t'\,t)$, then a $\mathfrak{V}$-category $\mathfrak{S}(X)$ with finite products and a functor $I_{\mathfrak{S}(X)}: \mathcal{N} \to \mathfrak{S}(X)$ with $I_{\mathfrak{S}(X)}(N^k) = (k, P_X(N^k)) = A^k$ and $I_{\mathfrak{S}(X)}(p_k^j) = (k, 1, P_X(p_j^k))$ are produced. $I_{\mathfrak{S}(X)}$ preserves finite products and is bijective on objects. Thus $\mathfrak{S}(X)$ is an algebraic theory with respect to the universe $\mathfrak{V}$.

$\mathfrak{S}(X)$ can be isomorphic to a $\mathfrak{U}$-category; however, in general there is no particular distinguished isomorphism. This is the reason for the somewhat unwieldy formulations that follow.

(c) If $P_X, P_Y: \mathcal{N} \to \mathcal{D}$ are functors which preserve finite products and such that $P_X(N^1) = X$, $P_Y(N^1) = Y$, and if $\xi: X \to Y$ is an isomorphism in $\mathcal{D}$, then there is an isomorphism $\chi_\xi: \mathfrak{S}(X) \to \mathfrak{S}(Y)$ with

(1) $$I_{\mathfrak{S}(Y)} = \chi_\xi\, I_{\mathfrak{S}(X)}: \mathcal{N} \to \mathfrak{S}(Y)\,.$$

Thus, one obtains isomorphic algebraic theories with respect to $\mathfrak{V}$.

χ_ξ is constructed in the obvious way. For $k \geq 1$, define $\xi^k: P_X(N^k) \to P_Y(N^k)$ by $P_Y(p_j^k)\, \xi^k = \xi\, P_X(p_j^k)$. It is obvious how ξ^0 is to be defined, $\xi^1 = \xi$, and ξ^k is an isomorphism for $k \geq 0$. For $t\colon P_X(N^n) \to \to P_X(N^k)$ set

$$(2) \qquad \chi_\xi(n, k, t) = (n, k, \xi^k\, t\, (\xi^n)^{-1}) .$$

(d) From now on we restrict ourselves to the special case where $\mathcal{D}$ is of the form $[\mathcal{C}, Ens]$, where $\mathcal{C}$ is a $\mathfrak{V}$-category that is isomorphic to a $\mathfrak{U}$-category. For $X\colon \mathcal{C} \to Ens$, let P_X always be fixed such that $P_X(N^k)$ for $k \neq 1$ is the canonically chosen product X^k of k factors X with the projections $P_X(p_j^k)$ for $k > 1$; i.e.,

$$(3) \qquad \begin{aligned} &P_X(N^1) = X, \\ &P_X(N^k)\,(?) = [k, X\,(?)]_{Ens} = X^k\,(?) \quad \text{for} \quad k \neq 1, \\ &P_X(p_j^k) = \varrho\,[i_j^k, X\,(?)] = pr_j^k : X^k \to X \quad \text{for} \quad k > 1, \end{aligned}$$

where ϱ at the object $M \in |Ens|$ is the evident isomorphism $[\{\emptyset\}, M] \to M$. Morphisms $t\colon X^n \to X^k$ are natural transformations of functors $\mathcal{C} \to Ens$. They are elements of $\mathfrak{U}$ only if $\mathcal{C}$ is a small $\mathfrak{U}$-category (3.4.3, 3.6.3).

18.6.2 Definition. The *functor* $X\colon \mathcal{C} \to Ens$ is called *tractable* if $\mathcal{C}$ is a $\mathfrak{V}$-category that is isomorphic to a $\mathfrak{U}$-category, and if $\mathfrak{S}(X)$ is isomorphic to a (necessarily small) $\mathfrak{U}$-category. Here $\mathfrak{S}(X)$ is called the *structure* of X.

18.6.3 Proposition. *Let $\mathcal{C}$ be a $\mathfrak{V}$-category that is isomorphic to a $\mathfrak{U}$-category and let $X\colon \mathcal{C} \to Ens$ be a functor.*

(a) *X is tractable if and only if $[X^n, X^k]_{[\mathcal{C}, Ens]}$ is isomorphic to a $\mathfrak{U}$-set for all non-negative integers n, k. Obviously it suffices to require this for all $[X^n, X]$.*

(b) *If, given X, there is a functor $G\colon \mathcal{N}^0 \to \mathcal{C}$ with an isomorphism*

$$(4) \qquad \varphi\colon [G(?), ??]_{\mathcal{C}} \cong [I(?), X(??)]_{Ens} ,$$

then X is tractable. Here $I\colon \mathcal{N}^0 \to Ens$ is the inclusion.

(c) *If $\mathcal{C}$ is a small $\mathfrak{U}$-category, then X is tractable and $\mathfrak{S}(X)$ is a $\mathfrak{U}$-category.*

Proof. (a) and (c) are evident, by 18.6.1. (b): (4) and (3) imply

$$(5) \qquad H^{G(k)} \cong X^k$$

and thus $[X^k, X] \cong [H^{G(k)}, X] \cong X(G(k))$ by the assumption about $\mathcal{C}$ and the Yoneda lemma.

Remark. If X has a left adjoint $F\colon Ens \to \mathcal{C}$, then (4) is valid for $G = F\,I$. *If*, conversely, $\mathcal{C}$ has coproducts or filtered colimits with respect to $\mathfrak{U}$, then one gets a left adjoint to X from G.

18.6.4 Definition. The *category $\mathcal{K}$ of tractable functors* is the full subcategory of CAT/Ens whose objects are the tractable functors.

Thus morphisms in $\mathcal{K}$ are commutative triangles of functors

(6)

$$\begin{array}{c} \mathcal{C} \xrightarrow{\;\;X\;\;} Ens \\ \scriptstyle f \downarrow \quad\nearrow \scriptstyle Y \\ \mathcal{D} \end{array}$$

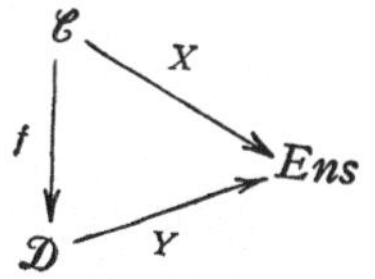

X, Y tractable .

Let $\mathcal{T}$ denote the category whose objects are algebraic theories with respect to $\mathfrak{B}$ which are isomorphic to a theory with respect to $\mathfrak{U}$; morphisms are to be theory-morphisms. 18.1 to 18.5 carry over in an evident fashion; $_0\mathcal{A}^b$ continues to be a subcategory of $[\mathcal{A}, Ens]$.

By 18.2.4, 18.5.4 and 18.6.3, $\mathcal{A} \mapsto {}_0U_{\mathcal{A}}$, $F \mapsto {}_0F^b$ define a contravariant functor "semantics" $\mathfrak{M}: \mathcal{T} \to \mathcal{K}$.

18.6.5 Proposition. (a) *The rule* $X \mapsto \mathfrak{S}(X)$ *can be extended to a contravariant functor "structure"* $\mathfrak{S}: \mathcal{K} \to \mathcal{T}$.
(b) *The tractable functor* $X: \mathcal{C} \to Ens$ *has a factorization*

(7)

$$\begin{array}{c} \mathcal{C} \xrightarrow{\;\;X\;\;} Ens \\ \scriptstyle X' \downarrow \quad\nearrow \\ {}_0\mathfrak{S}(X)^b \xrightarrow{\;{}^U\mathfrak{S}(X)} \end{array}$$

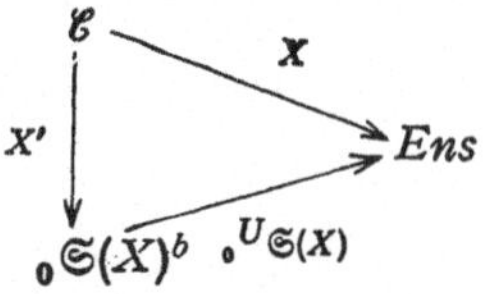

where $X'(C)$ *for* $C \in |\mathcal{C}|$ *comes from "evaluating"* $\mathfrak{S}(X)$ *at* C.
(c) *The rule* $X \mapsto X'$ *is a natural transformation* $\Phi: 1_{\mathcal{K}} \to \mathfrak{M}\,\mathfrak{S}$.

Proof. We may assume that $\mathcal{C}$ is not empty, since otherwise all conditions are trivial.

(a) By (3) and (6), one has

$$P_X(N^k)\,(C) = X^k(C) = \big(X(C)\big)^k = \big(Y\,f(C)\big)^k = Y^k\big(f(C)\big) = P_Y(N^k)\,\big(f(C)\big).$$

Corresponding equations hold for the projections and thus

(8) $P_X = [f, Ens]\,P_Y: \mathcal{N} \to [\mathcal{C}, Ens]$ and $Y^k f = X^k$.

For $u: Y^n \to Y^k$, one sets

(9) $\mathfrak{S}(f)\,(n, k, u) = (n, k, u * f)$.

This gives a theory-morphism $\mathfrak{S}(f): \mathfrak{S}(Y) \to \mathfrak{S}(X)$ by 16.1.1 (4), and (8) and (9) imply that $\mathfrak{S}$ is a contravariant functor.

(b) $X'(C)$ is the canonical $\mathfrak{S}(X)$-algebra described by $(n, k, t) \mapsto t_C$ with carrier $X(C)$. For $g: C \to C'$ in $\mathcal{C}$, $X'(g)$ is the $\mathfrak{S}(X)$-homomorphism whose underlying morphism (18.2.1) is the map $X(g)$. The statement that $X'(g)$ is a homomorphism says exactly that the morphisms of $\mathfrak{S}(X)$ (up to indices) are natural transformations between the powers of X.

(c) It has to be shown that, given (6), the following diagram

$$(10)\qquad \begin{array}{ccc} \mathcal{C} & \xrightarrow{X'} & {}_0\mathfrak{S}(X)^b \\ {\scriptstyle f}\downarrow & & \downarrow{\scriptstyle {}_0\mathfrak{S}(f)^b} \\ \mathcal{D} & \xrightarrow{Y'} & {}_0\mathfrak{S}(Y)^b \end{array}$$

is commutative. A comparison of (6), (8), (9) with 18.5.2 shows that ${}_0\mathfrak{S}(f)^b$ acts on $X'(C)$ in such a way that, amongst the maps between the powers of $X(C) = Y\,f(C)$ which come from $\mathfrak{S}(X)$, only those of the form $u * f$ are taken into account. (The situation is similar to a change of rings for modules). Therefore, (10) is commutative on objects. Thus the statement follows for morphisms, since it holds after composing with the faithful functor ${}_0U_{\mathfrak{S}(Y)}$, by (6), (7) and 18.5.4 (9).

18.6.6 Proposition. *Let $\mathcal{A} \in |\mathcal{T}|$. The rule which for $t^n_k\colon A^n \to A^k$ in $\mathcal{A}$ gives*

$$(11)\qquad t^n_k \mapsto (n, k, \{S(t^n_k) \mid S \in |{}_0\mathcal{A}^b|\})$$

determines an isomorphism $\Psi_{\mathcal{A}}\colon \mathcal{A} \to \mathfrak{S}({}_0U_{\mathcal{A}})$. Here

$$(12)\qquad {}_0(\Psi_{\mathcal{A}}^{-1})^b = ({}_0U_{\mathcal{A}})' \quad \textit{in the sense of (7).}$$

Further, $\Psi = \{\Psi_{\mathcal{A}}\}\colon 1_{\mathcal{T}} \to \mathfrak{S}\,\mathfrak{M}$ is a natural isomorphism.

Proof. Since every $\mathcal{A}$-homomorphism α in ${}_0\mathcal{A}^b$ consists of the powers of $\alpha_1 = {}_0U_{\mathcal{A}}(\alpha)$ and 18.2.1 (2) is valid, (11) certainly determines a theory-morphism $\Psi_{\mathcal{A}}$. As soon as $\Psi_{\mathcal{A}}$ is known to be an isomorphism, (12) follows by means of a comparison of the proof of 18.6.5 (b) with 18.5.2. Also, Ψ is a natural transformation provided, given $F\colon \mathcal{A} \to \mathcal{B}$, the following diagram

$$(13)\qquad \begin{array}{ccc} t^n_k & \overset{\Psi_{\mathcal{A}}}{\mapsto} & (n, k, \{S(t^n_k) \mid S \in |{}_0\mathcal{A}^b|\}) \\ {\scriptstyle F}\downarrow & & \downarrow{\scriptstyle \mathfrak{S}({}_0F^b) = \mathfrak{S}\mathfrak{M}(F)} \\ F(t^n_k) & \overset{\Psi_{\mathcal{B}}}{\mapsto} & (n, k, \{T(F(t^n_k)) \mid T \in |{}_0\mathcal{B}^b|\}) \end{array}$$

is commutative for every $t^n_k \in \operatorname{Mor} \mathcal{A}$. Now, ${}_0F^b\colon {}_0\mathcal{B}^b \to {}_0\mathcal{A}^b$ is defined on objects by the rule $T \mapsto T\,F = {}_0F^b(T)$ so that $\big({}_0F^b(T)\big)\,(t^n_k) = T\big(F(t^n_k)\big)$. $\{S(t^n_k)\}$ is a natural transformation, say $u\colon ({}_0U_{\mathcal{A}})^n \to ({}_0U_{\mathcal{A}})^k$, which at $S \in |{}_0\mathcal{A}^b|$ has the "value" $S(t^n_k)$. And $u * {}_0F^b\colon ({}_0U_{\mathcal{B}})^n \to ({}_0U_{\mathcal{B}})^k$ is that natural transformation which at $T \in |{}_0\mathcal{B}^b|$ has the "value" $\big({}_0F^b(T)\big)\,(t^n_k)$. Therefore, (9) and (11) yield the commutativity of (13).

It remains to be shown that $\Psi_{\mathcal{A}}$ is bijective. Since $\Psi_{\mathcal{A}}$ is known to be a theory-morphism, it suffices to do this for n-ary operations. Now, $\Psi_{\mathcal{A}}$ is injective, since different n-ary operations of $\mathcal{A}$ yield different operations for the free algebra ${}_0L_{\mathcal{A}}(n)$ (18.2.10, 18.3.1). Conversely, if $\mathcal{A}$ is not exceptional, consider first for $n \geq 1$ an n-ary operation τ of $\mathfrak{S}({}_0U_{\mathcal{A}})$ at ${}_0L_{\mathcal{A}}(n)$ singling out what it does to the element x of the n-th power of the carrier which is by $K_{\mathcal{A}}$ from 1_{Ak}; i.e., the one which is taken into the basis element p_j^n of ${}_0L_{\mathcal{A}}(n)$ by pr_j for $j = 1, 2, \ldots, n$ (18.2.10). $\tau(x)$ is a well defined n-ary operation of $\mathcal{A}$, and ${}_0L_{\mathcal{A}}(n)\,(\tau(x)) = \tau_{{}_0L_{\mathcal{A}}(n)}$. If $T \in |{}_0\mathcal{A}^b|$ and $T_1 \neq \Phi$, then there is exactly one $\mathcal{A}$-homomorphism α which takes the basis of ${}_0L_{\mathcal{A}}(n)$ into a given n-tupel of the carrier T_1 of T. Therefore, given an element y of $T(A^n) = (T_1)^n$, there is an $\alpha: {}_0L_{\mathcal{A}}(n) \to T$ with $\alpha_n(x) = y$. It thus follows that τ and $\Psi_{\mathcal{A}}(\tau(x))$ have the same effect on y, for,

$$\begin{aligned}\tau_T(y) &= \tau_T(\alpha_n(x)) = \alpha_1\, \tau_{{}_0L_{\mathcal{A}}(n)}(x)\\ &= \alpha_1({}_0L_{\mathcal{A}}(n)\,(\tau(x))\,(x)) = T(\tau(x))\,(y)\end{aligned}$$

has to hold, by the definition of $\tau(x)$. In the case $n = 0$ the conclusion is simplified, but remains valid. Therefore $\Psi_{\mathcal{A}}$ is also surjective. Finally, if $\mathcal{A}$ is exceptional, then $\Psi_{\mathcal{A}}$ is obviously bijective, since the carriers of all $\mathcal{A}$-algebras consist of at most one element.

Remark. Let $\mathcal{A}$ be an algebraic theory in $\mathfrak{U}$. Since ${}_0H_\pi^*: \mathcal{A}^0 \to {}_0\mathcal{A}^b$ preserves finite coproducts (18.2.2), and $H^*: {}_0\mathcal{A}^{b0} \to [{}_0\mathcal{A}^b, Ens]$ products, $H^*\,{}_0H_\pi^{*0}: \mathcal{A} \to [{}_0\mathcal{A}^b, Ens]$ is a full embedding which preserves finite products. Thus $H^*\,{}_0H_\pi^{*0}(A^1)$ is a functor $X: {}_0\mathcal{A}^b \to Ens$ and $\mathfrak{S}(X) \cong \mathcal{A}$. But, by 18.2.4 (12), $X = [{}_0H_\pi^*((A^1)^0), ??]_{{}_0\mathcal{A}^b} = [{}_0L_{\mathcal{A}}(1), ??]$ and this is a representation of ${}_0U_{\mathcal{A}}$ by 18.2.2. Therefore, $\mathfrak{S}({}_0U_{\mathcal{A}}) \cong \mathfrak{S}(X) \cong \mathcal{A}$. We have described this isomorphism explicitely; one can obtain this from 18.6.1 (c) and 18.6.3.

18.6.7 Theorem. *Op* $\mathfrak{S}: \mathcal{K} \to \mathcal{T}^0$ *is left adjoint to* $\mathfrak{M}$ Op: $\mathcal{T}^0 \to \mathcal{K}$, *and* $\mathfrak{M}$ Op *is fully faithful.*

Proof. As in 16.5.7, we verify the equations 16.5.5 (4), (4°) which take the following form here

$$(\mathfrak{M} * \Psi)\,(\Phi * \mathfrak{M}) = 1_{\mathfrak{M}} \quad \text{in } \mathcal{K}, \tag{14}$$

$$(\mathfrak{S} * \Phi)\,(\Psi * \mathfrak{S}) = 1_{\mathfrak{S}} \quad \text{in } \mathcal{T}, \tag{14^0}$$

since the auxiliary functors Op can be cancelled and since the transition from $\mathcal{T}^0$ to $\mathcal{T}$ reverses the order in the composition. Now, for $\mathcal{A} \in |\mathcal{T}|$

$$(\Phi * \mathfrak{M})_{\mathcal{A}} = \Phi_{{}_0U_{\mathcal{A}}} = ({}_0U_{\mathcal{A}})' = {}_0(\Psi_{\mathcal{A}}^{-1})^b,$$

by 18.6.5 and 18.6.6. Therefore, (14) is valid. Also, in (14^0) one obtains at X,

$$\mathfrak{S}(\Phi_X)\,\Psi_{\mathfrak{S}(X)} = \mathfrak{S}(X')\Psi_{\mathfrak{S}(X)}\,.$$

By (11), $\Psi_{\mathfrak{S}(X)}$ takes the natural transformation $t_k^n: X^n \to X$ into the natural transformation $r_k^n = \{S(t_k^n) \mid S \in |{}_0\mathfrak{S}(X)^b|\}$ which goes from $({}_0U_{\mathfrak{S}(X)})^n$ to $({}_0U_{\mathfrak{S}(X)})^k$. Since $X = {}_0U_{\mathfrak{S}(X)}\,X'$, $\mathfrak{S}(X')$ then produces, by (9), that natural transformation $X^n \to X^k$ which is induced at every place C by r_k^n. By definition of r_k^n, this is precisely t_k^n.

The last statement follows from 16.5.4.

18.6.8 Corollary. (a) *A functor* ${}_0G\colon {}_0\mathcal{B}^b \to {}_0\mathcal{A}^b$ *is algebraic if and only if* ${}_0U_{\mathcal{A}}\,{}_0G = {}_0U_{\mathcal{B}}$; i.e., *if* ${}_0G$ *maps* $\mathcal{B}$*-algebras in such a way that the carriers are pointwise fixed, and if* ${}_0G$ *preserves the maps of the carriers underlying the* $\mathcal{B}$*-homomorphisms.*

(b) *A functor* $G\colon \mathcal{B}^b \to \mathcal{A}^b$ *is isomorphic to an algebraic functor if and only if* $U_{\mathcal{A}}\, G \cong U_{\mathcal{B}}$.

(c) *Different theory-morphisms induce different algebraic functors.*

Proof. Conditions (a) and (c) are equivalent to $\mathfrak{M}$ Op being fully faithful. (b) follows easily from (a), since $U_{\mathcal{A}}$ reflects isomorphisms.

18.6.9 Corollary. *Let* $\mathcal{F}_n$ *be the free algebraic theory which is produced by an n-ary operation. For a tractable* $X\colon \mathcal{C} \to Ens$, *there is a bijection between the n-ary operations of* $\mathfrak{S}(X)$ *and the* $\mathcal{K}$*-morphisms* $X \to {}_0U_{\mathcal{F}_n}$.

Proof. Set $\mathcal{D} = {}_0\mathcal{F}_n^b$ and $Y = {}_0U_{\mathcal{F}_n}$ in (10). Then the statement follows from (12) and the fact that $\mathfrak{M}$ is fully faithful.

18.6.10 Remarks. (a) The effectiveness of 18.6.7 is impaired by the fact that the objects of $\mathcal{K}$ are tractable functors, which are recognized as such only by the criteria in 18.6.3. This condition can not be avoided by permitting algebraic theories to be small $\mathfrak{V}$-categories and by continuing to consider algebraic categories over *Ens*. 18.2.10 shows that then finitely generated free algebras need not exist. 18.6.6, 18.6.7 would thus be void in their present form.

(b) Using the axiom of choice in $\mathfrak{V}$, one obtains, by 16.3.6, that the category $\mathcal{K}$ is equivalent to the full subcategory $\mathfrak{K}$, whose objects are those tractable functors $X\colon \mathcal{C} \to Ens$ for which $\mathcal{C}$ is a $\mathfrak{U}$-category. Analogously, $\mathcal{T}$ is equivalent to the category $\mathfrak{T}$ of algebraic theories in $\mathfrak{U}$. Corresponding to 18.6.7, there is an adjoint situation between $\mathfrak{T}^0$ and $\mathcal{K}$. Going a step further, $\mathfrak{T}$ can be replaced by a skeleton. What is meant by "the" theory of groups, or resp., rings etc. refers to a skeleton of $\mathfrak{T}$.

18.7 The Kronecker Product

18.7.1 Preliminary remark. Let M be an object of a category with finite products and let n, r be natural numbers ≥ 1. The iterated powers $(M^n)^r$ and $(M^r)^n$ are iosmorphic but not identical, not even if the corresponding objects are identical. Let $pr_j\colon\ M^{nr} \to M$, $pr'_h\colon (M^n)^r \to M^n$, $pr''_i\colon M^n \to M$ be the projections. We define the isomorphism $\sigma_{n,r}\colon (M^n)^r \to M^{nr}$ by

$$(1)\qquad pr_{r(i-1)+h}\,\sigma_{n,r} = pr''_i\, pr'_h\,,$$

and the automorphism $\tau_{n,r}\colon M^{nr} \to M^{nr}$ by

$$(2)\qquad pr_{n(h-1)+i}\,\tau_{n,r} = pr_{r(i-1)+h}\,.$$

According to (1) one has $\sigma_{r,n}\colon (M^r)^n \to M^{nr}$ and thus

$\varrho_{n,r} = \sigma_{r,n}^{-1}\, \tau_{n,r}\, \sigma_{n,r}\colon (M^n)^r \to (M^r)^n$, so that

(3) $$q_h'' \, q_i' \, \varrho_{n,r} = pr_i^{\bullet} \, pr_h',$$

where $q_i'\colon (M^r)^n \to M^r$, $q_h''\colon M^r \to M$ again designate projections. $\varrho_{n,r}$ is an interchange of products with products, even when $n = r$. Furthermore, $\varrho_{r,n}\,\varrho_{n,r}$ is always an identity morphism, as (3) shows. By M^1 we shall always mean M with 1_M as projection.

18.7.2 A canonically chosen product in ${}_0\mathcal{B}^b$ means a product in which a canonical choice has been made for the carriers.

Let C be a canonical $\mathcal{B}$-algebra with carrier M, D a canonically chosen product of $n > 0$ factors C in ${}_0\mathcal{B}^b$ and $\alpha\colon D \to C$ a $\mathcal{B}$-homomorphism. For each $u_1^r\colon B^r \to B^1$ in $\mathcal{B}$, there is the following commutative diagram

(4) $$\begin{array}{ccc} (M^n)^r & \xrightarrow{\alpha_r} & M^r \\ {\scriptstyle D(u_1^r)}\downarrow & & \downarrow{\scriptstyle C(u_1^r)} \\ M^n & \xrightarrow{\alpha_1} & M \end{array}$$

For $r \geq 1$, one has $\alpha_r = (\alpha_1)^n = \alpha_1 \sqcap \alpha_1 \sqcap \cdots \sqcap \alpha_1$ (r times), however,

(5) $$D(u_1^r) = [C(u_1^r)]^n \, \varrho_{n,r},$$

which follows by taking α to be the projections $D \to C$. For $u_s^r\colon B^r \to B^s$, $r, s \geq 1$, one has

(6) $$D(u_s^r) = \varrho_{s,n}\,[C(u_s^r)]^n\, \varrho_{n,r}.$$

If D is a product of n factors C in $\mathcal{B}^b$, then (4), (5), (6) are correspondingly valid, since (1), (2), (3) are also valid here; $(\alpha_1)^r$ and $[C(u_1^r)]^n$ are to be defined by means of the projections as morphisms between r- resp. n-fold products.

18.7.3 Let $\mathcal{A}$, $\mathcal{B}$ be algebraic theories. Let ${}_{0\pi}[\mathcal{A}, {}_0\mathcal{B}^b]$ be the full subcategory of $[\mathcal{A}, {}_0\mathcal{B}^b]$ whose objects are those functors $T\colon \mathcal{A} \to {}_0\mathcal{B}^b$ which take finite products of A^1 into canonically chosen products of $T(A^1)$.

For $T \in |{}_{0\pi}[\mathcal{A}, {}_0\mathcal{B}_s^b]|$, let $M = (T_1)_1 \cdot T_n$ is a canonical $\mathcal{B}$-algebra with $(T_n)_r = (M^n)^r$. For $t_k^n\colon A^n \to A^k$ in $\mathcal{A}$, $T(t_k^n)$ is a $\mathcal{B}$-homomorphism; i.e., for $u_s^r\colon B^r \to B^s$ in $\mathcal{B}$,

(7) $$\begin{array}{ccc} (M^n)^r & \xrightarrow{T(t_k^n)_r} & (M^k)^r \\ {\scriptstyle T_n(u_s^r)}\downarrow & & \downarrow{\scriptstyle T_k(u_s^r)} \\ (M^n)^s & \xrightarrow{T(t_k^n)_s} & (M^k)^s \end{array}$$

is commutative. Choosing $D(u_s^r) = T_n(u_s^r)$, $C(u_s^r) = T_1(u_s^r)$, then (6) is valid when $nrs \neq 0$, and $T(t_k^n)_r = T(t_k^n)^r$. Notice that because of (3), for fixed

r, $T(?)_r$ is isomorphic to a canonical $\mathcal{A}$-algebra (even for $r = 0$) whereby $\{T_n(u_s^r) \mid n = 0, 1, 2, \ldots\}$ becomes an $\mathcal{A}$-homomorphism. Furthermore, by means of (1) and (2) one obtains from T a canonical algebra for an algebraic theory $\mathcal{C}$ determined by $\mathcal{A}$ and $\mathcal{B}$. We shall now define this theory $\mathcal{C}$.

18.7.4 Definition. The *Kronecker product* $\mathcal{C} = \mathcal{A} \otimes \mathcal{B}$ of the algebraic theories $\mathcal{A}$ and $\mathcal{B}$ is constructed from the coproduct $\mathcal{A} \sqcup \mathcal{B}$ with injections $in_1 \colon \mathcal{A} \to \mathcal{A} \sqcup \mathcal{B}$, $in_2 \colon \mathcal{B} \to \mathcal{A} \sqcup \mathcal{B}$ in $\mathfrak{T}$ by adjoining the following equations

$$(8) \qquad in_1(t_1^n)\,[in_2(u_1^r)]^n\,\sigma_{r,n}^{-1}\tau_{n,r} = in_2(u_1^r)\,[in_1(t_1^n)]^r\,\sigma_{n,r}^{-1}\colon C^{nr} \to C^1$$

for all operations t_1^n in $\mathcal{A}$ and u_1^r in $\mathcal{B}$. If $n = 0$, then $\sigma_{r,n}$, $\sigma_{n,r}$, $\tau_{n,r}$ and $in_2(u_1^r)^n$ are understood to be the identity morphism of C^0 (which is the only one that exists). The case where $r = 0$ is analogous.

Thus, in $\mathfrak{T}$ there is an epimorphism $p \colon \mathcal{A} \sqcup \mathcal{B} \to \mathcal{A} \otimes \mathcal{B}$, so that one has $h_1 = p\,in_1 \colon \mathcal{A} \to \mathcal{A} \otimes \mathcal{B}$ and $h_2 = p\,in_2 \colon \mathcal{B} \to \mathcal{A} \otimes \mathcal{B}$. More exactly: In $\mathcal{A} \otimes \mathcal{B}$ the equations which are obtained from (8) by replacing in_1, in_2 by h_1, h_2 are valid.

Now we assume that there are also theories $\mathcal{A}'$, $\mathcal{B}'$ with correspondingly defined theory-morphisms $h_1' \colon \mathcal{A}' \to \mathcal{A}' \otimes \mathcal{B}'$, $h_2' \colon \mathcal{B}' \to \mathcal{A}' \otimes \mathcal{B}'$. If $F \colon \mathcal{A} \to \mathcal{A}'$ and $G \colon \mathcal{B} \to \mathcal{B}'$ are theory-morphisms, then there is a uniquely determined theory-morphism

$$(9) \qquad F \otimes G \colon \mathcal{A} \otimes \mathcal{B} \to \mathcal{A}' \otimes \mathcal{B}'$$

with $h_1' F = (F \otimes G)\, h_1$ and $h_2' G = (F \otimes G)\, h_2$.

This follows immediately from (8).

Now, for every pair $(\mathcal{A}, \mathcal{B})$ of algebraic theories choose a coproduct $\mathcal{A} \sqcup \mathcal{B}$ in such a way that it is always the case that $\mathcal{N} \sqcup \mathcal{A} = \mathcal{A}$ with $in_1 = I_{\mathcal{A}}$ and $in_2 = 1_{\mathcal{A}}$, and correspondingly for $\mathcal{A} \sqcup \mathcal{N}$.

18.7.5 Theorem. *The Kronecker product is a bifunctor* $\otimes \colon \mathfrak{T} \times \mathfrak{T} \to \mathfrak{T}$. *Here*

$$(10) \qquad \mathcal{N} \otimes \mathcal{A} = \mathcal{A} = \mathcal{A} \otimes \mathcal{N}\,.$$

There are isomorphisms

$$(11) \qquad \mathcal{A} \otimes \mathcal{B} \cong \mathcal{B} \otimes \mathcal{A}\,,$$

$$(12) \qquad (\mathcal{A} \otimes \mathcal{B}) \otimes \mathcal{C} \cong \mathcal{A} \otimes (\mathcal{B} \otimes \mathcal{C})$$

as isomorphisms of bi- or, resp., trifunctors. Further, there are isomorphisms

$$(13) \qquad {}_0\pi[\mathcal{A}, {}_C\mathcal{B}^b] \cong {}_0(\mathcal{A} \otimes \mathcal{B})^b$$

as isomorphisms of contravariant bifunctors.

Proof. The first statement follows easily from (8), (9), since we fixed $\sqcup$ as a bifunctor above. In (10), (8) does not impose any additional conditions. (11), (12) also follows easily from (8), (9). Note, that for $\mathcal{A} = \mathcal{B}$ in (11) one does not, in general, get the identity theory-morphism for $\mathcal{A} \otimes \mathcal{A}$. By (8), (9), the last statement follows from 18.7.2 and 18.7.3; (8) was precisely motivated by (4), (6), (7).

18.7.6 Remarks. The Kronecker product is a tensor product in the sense of 16.7.3. Since $\mathfrak{S}({}_0U_{\mathcal{A}\otimes\mathcal{B}})$ is isomorphic to $\mathcal{A} \otimes \mathcal{B}$ (18.6.6), 11.6.1 carries over to Kronecker products, a fact that also follows directly from (8). If, in particular, $\mathcal{A}$ is a theory of Hopf objects, $\mathcal{B}$ a theory of groups, then $\mathcal{A} \otimes \mathcal{B}$ is a theory of abelian groups. Notice here, that for every algebraic theory, there are innumerable isomorphic ones in $\mathfrak{T}$, unless $\mathfrak{T}$ is replaced by a skeleton; i.e., isomorphic theories are identified in such a way that different automorphisms remain different.

There is a forgetful functor $V\colon {}_{0\pi}[\mathcal{A}, {}_0\mathcal{B}^b] \to {}_0\mathcal{B}^b$ with $V(T) = T_1$. It is taken into ${}_0h_2^b$ by (13). Note that V has all the properties of an algebraic functor, in particular, it has a left adjoint and ${}_0U_{\mathcal{B}}\, V$ is tractable. By (13), $\mathcal{A} \otimes \mathcal{B} \cong \mathfrak{S}({}_0U_{\mathcal{B}}\, V)$. It would have been convenient to define $\mathcal{A} \otimes \mathcal{B}$ in this way. This would have required a verification of the tractability of ${}_0U_{\mathcal{B}}\, V$ and then a proof of the validity of (13).

${}_\pi[\mathcal{A}, \mathcal{B}^b]$ is equivalent to $(\mathcal{A} \otimes \mathcal{B})^b$. This equivalence follows naturally from (1), (2), (3) and 18.7.3. It is not bijective for object classes. There is no isomorphism. This equivalence and (13) show, however, that algebıaic categories over an algebraic category are essentially such over *Ens* (with respect to other theories). By means of 11.6.1 some negative results are also obtained. For instance, there are only trivial ring objects over *Ab*, and over the category of rings there are only trivial Hopf objects.

18.8 Characterization of Algebraic Categories

The theory $\mathcal{A}$ of an algebraic category $\mathcal{A}^b$ can be reconstructed up to isomorphisms in two ways, namely, as the structure of the forgetful functor $U_{\mathcal{A}}$ (18.6.6), or as the dual of the full subcategory of $\mathcal{A}^b$ whose objects are obtained by restricting the left adjoint $L_{\mathcal{A}}$ of $U_{\mathcal{A}}$ to the subcategory $\mathcal{N}^0$ of *Ens* (18.2.2, 18.2.4). This second aspect leads to a characterization of algebraic categories up to an equivalence, where the relation to $\mathcal{N}^0$ rests on the fact that the algebraic theories considered here have finitary operations only.

18.8.1 Lemma. *Let G: Ens $\to$ Ens be a functor. The following statements are equivalent:*

(a) *G is dominated by $|\mathcal{N}^0|$ (the set of non-negative integers).*

(b) *$Q(GI) \cong G$, where $I\colon \mathcal{N}^0 \to$ Ens is the inclusion and Q left adjoint to $[I, Ens]\colon [Ens, Ens] \to [\mathcal{N}^0, Ens]$ as in 17.1.6.*

Proof. By 17.1.6 (6), $Q(GI)(m) = \operatorname{Colim} G I Q_m$ for $m \in |Ens|$. By 18.2.3, the category I/m belonging to m and I is filtered and $m = \operatorname{Colim} I Q_m$. By 17.1.6 (13), there is a natural transformation $\Psi: Q(GI) \to G$ with Ψ_m: Colim $G I_m \to G(m)$. Here Ψ_m comes from factoring the natural transformation $G * \gamma_m$: $G I Q_m \to G(m)_{I/m}$ of 17.1.3 which, at $f: n \to m$ for $n \in |\mathcal{N}^0|$, is described by

(1) $$(G * \gamma_m)_{(n,f)} = G(f): G(n) \to G(m) .$$

By 10.3.1, (a) means that, given $x \in G(m)$, there is an $f: n \to m$ and a $y \in G(n)$ such that $x = G(f)(y)$. By 9.4.2, this is equivalent to Ψ_m being an epimorphism. If we now show that Ψ_m is a monomorphism in any case, then the proof is complete. This is trivial for $m = \emptyset$, since $I/\emptyset$ has only one morphism, namely $(1\emptyset, 1\emptyset, 1\emptyset)$. Suppose now that $m \neq \emptyset$, and let $[y]$, $[z]$ be elements of Colim $G I Q_m$ with $\Psi_m([y]) = \Psi_m([z]) = x$. By 9.4.6, there are representatives y, z in an object $G I Q_m(n, f) = G(n)$. By 18.2.3, one can assume f to be a monomorphism and $n \neq 0$. Then f is a coretraction in Ens, and hence so is $G(f)$. From (1) and the equalities $G(f)(y) = G(f)(z) = x$, it follows that $y = z$. Hence, Ψ_m is a monomorphism.

18.8.2 Corollary. *If G and F: $Ens \to Ens$ are dominated by $|\mathcal{N}^0|$, then $G \cong F$ if and only if $GI \cong FI$.*

This follows immediately from 18.8.1 (b).

18.8.3 Corollary. *For every algebraic category $\mathcal{A}^b$, $U_{\mathcal{A}} L_{\mathcal{A}}$ is dominated by $|\mathcal{N}^0|$.*

Proof. $U_{\mathcal{A}} L_{\mathcal{A}}$ satisfies condition 18.8.1 (b), since $L_{\mathcal{A}}$ and $U_{\mathcal{A}}$ preserve filtered colimits ($L_{\mathcal{A}}$ as a left adjoint and $U_{\mathcal{A}}$ by 18.2.1).

18.8.4 Lemma. *Let X: $\mathcal{C} \to Ens$ be a functor with a left adjoint K. The following statements are equivalent:*

(a) *$X K$ is dominated by $|\mathcal{N}^0|$.*

(b) *For every morphism u: $K(1) \to K(m)$, there is a factorization $u_2 u_1$: $K_1 \to K(n) \to K(m)$, where $n \in |\mathcal{N}^0|$ and u_2 is of the form $K(f)$.*

Proof. First, let (a) be satisfied. By means of the adjunction $(\varphi, K, X, \mathcal{C}, Ens)$ and 16.5.2, $\varphi_{1,K(m)}(u) = X(u)\,\eta_1 \in [1, X K(m)] \cong X K(m)$. By (a), $X(u)\,\eta_1 = X K(f) \circ y$ for suitable y: $1 \to X K(n)$ and f: $n \to m$ with $n \in |\mathcal{N}^0|$ (compare the proof of 18.8.1). From this and from 16.5.2^0, $u = \varepsilon_{K(m)} \circ K X K(f) \circ K(y) = K(f) \circ \varepsilon_{K(n)} \circ K(y)$ follows. (a) follows from (b) similarly.

18.8.5 Remark. The existence of a left adjoint K for X: $\mathcal{C} \to Ens$ is equivalent to the following: X is representable and a representing object A has arbitrary copowers (i. e., coproducts with identical cofactors) in $\mathcal{C}$. For, if there exists a K, then $K(1)$ is a representing object for X and by 16.4.5, 8.1.3, 10.2.5, the claim follows from the fact that every set is the coproduct of its one-element subsets. Incidentally, this is the non-additive case of 17.8.5.

18.8.6 Let the functor $X\colon \mathscr{C} \to Ens$ have the left adjoint K. Let $\mathscr{C}_\omega$ be the full subcategory of $\mathscr{C}$ with objects $K(n)$ for $n \in |\mathscr{N}^0|$. For the inclusions $I\colon \mathscr{N}^0 \to Ens$, $J\colon \mathscr{C}_\omega \to \mathscr{C}$ and the restriction $K_\omega\colon \mathscr{N}^0 \to \mathscr{C}_\omega$ of K, one has

(2) $$J\,K_\omega = K\,I\colon \mathscr{N}^0 \to \mathscr{C}\,.$$

K_ω and J preserve finite coproducts.

(3) $$\mathrm{Op}\,K_\omega\,\mathrm{Op} = K^0\colon \mathscr{N} \to \mathscr{C}^0$$

makes $\mathscr{C}^0_\omega$ an algebraic theory $\mathscr{B}$, provided K_ω is bijective on objects. If this is not the case, $\mathscr{B}$ can be obtained through proper indexing as in 18.6.1. By 18.6.3 (5), $\mathscr{B}$ is isomorphic to $\mathfrak{S}(X)$. By 8.7.3, the functor $J^\vee\colon \mathscr{C} \to [\mathscr{C}^0_\omega, Ens]$ defined by $A \mapsto [J(?), A]$, $f \mapsto [J(?), f]$ factors through $\mathscr{B}^b$. Thus one gets the socalled *Malcev "embedding"*

(4) $$M\colon \mathscr{C} \to \mathscr{B}^b \quad \text{with } M(A) = [J(?), A]_{\mathscr{C}}\ .$$

M is injective on objects, as is shown by the carriers $[J\,K_\omega(1), A] = [K(1), A]$; however, additional conditions are required to make it faithful.

18.8.7 For the Malcev embedding (3), the following hold:

(a) M is faithful if and only if X is faithful.

(5) $$U_{\mathscr{B}}\,M \cong X\,.$$

(b) $M\,J$ is fully faithful.

(c) If $X\,K$ is dominated by $|\mathscr{N}^0|$, then

(6) $$L_{\mathscr{B}} \cong M\,K \quad \text{and}$$

(7) $$M_{K(m),A}\colon [K(m), A] \to [M\,K(m), M(A)]$$

is an isomorphism for all $m \in |Ens|$ and $A \in |\mathscr{C}|$.

Proof. By the definition of $U_{\mathscr{B}}$, (5) follows from

$$U_{\mathscr{B}}\,M(?) \cong [J\,K_\omega(1), ?]_{\mathscr{C}} \cong [1, X(?)]_{Ens} \cong X(?)\,.$$

Since $U_{\mathscr{B}}$ is faithful, (a) follows from (5).

(b) follows from the fact that $M\,J$ coincides with the Yoneda embedding $H_*\colon \mathscr{C}_\omega \to [\mathscr{C}^0_\omega, Ens]$ up to evident isomorphisms and up to being followed by an inclusion.

(c) Since $[J(?), J\,K_\omega(?\,?)]_{\mathscr{C}} \cong [K^0_\omega(?\,?), ?]_{\mathscr{C}^0_\omega}$, the above, together with (2) and 18.2.4 imply

(8) $$M\,K\,I = M\,J\,K_\omega \cong L_{\mathscr{B}}\,I\,.$$

$L_{\mathscr{B}}$ preserves colimits. Since $m = \mathrm{Colim}\,I\,Q_m$ (compare 18.2.3, 18.8.1), it follows from 17.1.1 and (8) that there is a natural transformation

$\lambda\colon L_{\mathcal{B}} \to M\,K$ and thus one has $U_{\mathcal{B}} * \lambda\colon U_{\mathcal{B}}\,L_{\mathcal{B}} \to U_{\mathcal{B}}\,M\,K \cong X\,K$. This implies (6), by 18.8.2 and 18.8.3, since (8) is obtained from λ by restriction (by construction of λ), and since $U_{\mathcal{B}}$ reflects isomorphisms.

For $m \in |\mathcal{N}^0|$, (7) is an isomorphism, because in this case $M\,K(m) = M\,K\,I(m) = M\,J\,K_\omega(m) \cong H_*\,K_\omega(m)$ and, therefore, $M_{K(m),A}$ can be inverted by the Yoneda isomorphism 4.2.1, as is shown by (4). In particular, this holds for $m = 1$. Since an arbitrary set m is a coproduct of sets 1 by means of the injections $in_x\colon 1 \to m$, and since $M K$ (by (6)) as well as $L_{\mathcal{B}}$ and K preserve coproducts, (7) can be obtained in general from 8.7.3:

$$\begin{array}{ccc} [K(m), A] & \xrightarrow{M_{K(m),A}} & [M\,K(m), M(A)] \\ {\scriptstyle [K(in_x),A]}\downarrow & & \downarrow{\scriptstyle MK(in_x),M(A)} \\ [K(1), A] & \xrightarrow{M_{K(1),A}} & [M\,K(1), M(A)] \end{array}$$

is commutative, since $M_{?,??}\colon [?, ??] \to [M(?), M(??)]$ is a natural transformation of contra-co-variant functors.

18.8.8 Proposition. *A category $\mathcal{C}$ is equivalent to an algebraic category if and only if it has coequalizers and a functor $X\colon \mathcal{C} \to Ens$ with a left adjoint K such that*

(i) *$X\,K$ is dominated by $|\mathcal{N}^0|$.*

(ii) *$f\colon B \to C$ in $\mathcal{C}$ is a coequalizer if and only if $X(f)$ is an epimorphism.*

(iii) *A pair of morphisms, $a, b\colon A \to B$ in $\mathcal{C}$ is a kernel pair provided $(X(a), X(b))$ is a kernel pair.*

Proof. (a) By 18.8.3 and 18.4.5, an algebraic category has the properties named above, with $X = U_{\mathcal{A}}$, $K = L_{\mathcal{A}}$. Thus they hold for every category that is equivalent to $\mathcal{A}^b$. To prove the converse, we make use of 18.8.6 and 18.8.7.

(b) We begin by showing that M is fully faithful. (ii) implies that X reflects epimorphisms. By 16.5.3, X is faithful, and by 18.8.7, M is faithful. By (5), K is left adjoint to $U_{\mathcal{B}}\,M$. For the quasi-inverse adjunction transformations ε'', η'', where $\varepsilon''\colon K\,U_{\mathcal{B}}\,M \to 1_{\mathcal{C}}$, 16.5.5 yields

$$(U_{\mathcal{B}}\,M * \varepsilon'')\,(\eta'' * U_{\mathcal{B}}\,M) = 1_{U_{\mathcal{B}}\,M}.$$

For $A \in |\mathcal{C}|$, $U_{\mathcal{B}}\,M(\varepsilon''_A)$, being a retraction, is epimorphic. Since (ii) is similarly valid for $U_{\mathcal{B}}$ and $U_{\mathcal{B}}M \cong X$, $M(\varepsilon''_A)$ and ε''_A are coequalizers in $\mathcal{B}^b$ or, resp., $\mathcal{C}$. For $\alpha\colon M(A) \to M(B)$ in $\mathcal{B}^b$, we consider

(9) $$\alpha \circ M(\varepsilon''_A)\colon M\,K\,U_{\mathcal{B}}\,M(A) \to M(A) \to M(B)\,.$$

By (7), there is exactly one $f\colon K\,U_{\mathcal{B}}\,M(A) \to B$ in $\mathcal{C}$ with $M(f) = \alpha\,M(\varepsilon''_A)$. Let ε''_A be a coequalizer of $q_1, q_2\colon D \to K\,U_B\,M(A)$. By definition of f, (9) implies $M(f\, q_1) = \alpha\, M(\varepsilon_A''\, q_1) = \alpha\, M(\varepsilon_A''\, q_2) = M(f\, q_2)$. Since M is faithful, $f\, q_1 = f\, q_2$. By the definition of coequalizers, there is exactly one $g\colon A \to B$ with $f = g\,\varepsilon_A''$. (9) implies $\alpha\, M(\varepsilon_A'') = M(g)\, M(\varepsilon_A'')$. Since $M(\varepsilon_A'')$ is an epimorphism, $\alpha = M(g)$. Therefore, M is full.

(c) Now we show that M has a left adjoint G. By (6), $M\, K$ is left adjoint to $U_{\mathscr{B}}$. For the corresponding quasi-inverse adjunction transformations ε, η one has, by 16.5.5, $(U_{\mathscr{B}} * \varepsilon)\,(\eta * U_{\mathscr{B}}) = 1_{U_{\mathscr{B}}}$ with $\varepsilon\colon M\, K\, U_{\mathscr{B}} \to 1_{\mathscr{B}}$. For $S \in |\mathscr{B}^b|$, $U_{\mathscr{B}}(\varepsilon_S)$ is thus a retraction and this makes ε_S a coequalizer, since $U_{\mathscr{B}}$ satisfies (ii). Since $\mathscr{B}^b$ is complete, the pullback

(10)
$$
\begin{array}{ccc}
F & \xrightarrow{\ \pi_2\ } & MKU_{\mathscr{B}} \\
{\scriptstyle \pi_1}\big\downarrow & & \big\downarrow{\scriptstyle \varepsilon} \\
MKU_{\mathscr{B}} & \xrightarrow{\ \varepsilon\ } & 1_{\mathscr{B}^b}
\end{array}
$$

exists in $[\mathscr{B}^b, \mathscr{B}^b]$. It is a bicartesian square, since by 18.4.3, it is bicartesian at every $S \in |\mathscr{B}^b|$. $\varepsilon * F\colon M\, K\, U_{\mathscr{B}}\, F \to F$ is a natural transformation. Since M is fully faithful, there are natural transformations $\alpha_1, \alpha_2\colon K\, U_{\mathscr{B}}\, F \to K\, U_{\mathscr{B}}$ with

(11)
$$
M * \alpha_i = \pi_i(\varepsilon * F) \qquad i = 1, 2 \, .
$$

Since $\mathscr{C}$ has coequalizers, α_1, α_2 have a coequalizer $\beta\colon K\, U_{\mathscr{B}} \to G$ in $[\mathscr{B}^b, \mathscr{C}]$. By (ii), $U_{\mathscr{B}}\, M * \beta \cong X * \beta$ is an epimorphism, and by (ii) for $U_{\mathscr{B}}$, $M * \beta$ is a coequalizer at every place. Since $\varepsilon * F$ is an epimorphism (everywhere), (11) implies $(M * \beta)\,\pi_1 = (M * \beta)\,\pi_2$. (10) being a pushout guarantees the existence of $\eta'\colon 1_{\mathscr{B}^b} \to M\, G$ with

(12)
$$
\eta'\,\varepsilon = M * \beta\colon M\, K\, U_{\mathscr{B}} \to M\, G \, ,
$$

and η' is a coequalizer at every place; this is shown again by applying $U_{\mathscr{B}}$. To prove G to be left adjoint to M with the adjunction transformation η', it suffices, by 16.5.1, to show that for $S \in |\mathscr{B}^b|$, $A \in |\mathscr{C}|$ and $\alpha\colon S \to M(A)$ there is exactly one $f\colon G(S) \to A$ such that $\alpha = M(f)\eta_S'$. First, since M is full, there is an $h\colon K\, U_{\mathscr{B}}(S) \to A$ with $M(h) = \alpha\,\varepsilon_S$.

(13)
$$
\begin{array}{ccc}
MKU_{\mathscr{B}}(S) & \xrightarrow{\ \varepsilon_S\ } & S \\
{\scriptstyle M(\beta_S)}\big\downarrow & \quad\overset{\eta_S'}{\diagdown}\diagup\quad & \big\downarrow{\scriptstyle \alpha} \\
MG(S) & \underset{M(f)}{\xrightarrow{\ \ \ \ }} & M(A)
\end{array}
$$

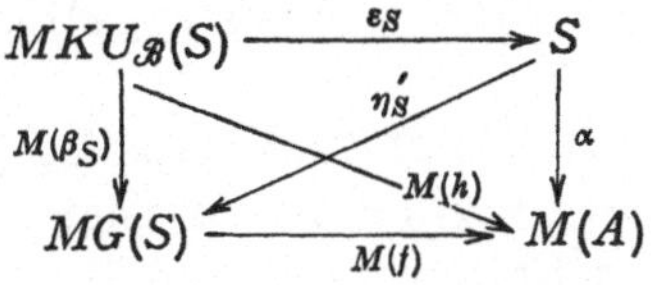

By (10) and (11), $\varepsilon(M * \alpha_1) = \varepsilon(M * \alpha_2)$. Thus
$$
M(h)\, M(\alpha_{1S}) = M(h)\, M(\alpha_{2S}) \, ,
$$

and from this it follows that $h\,\alpha_{1S} = h\,\alpha_{2S}$, since M is faithful. By definition of β, there is an $f\colon G(S) \to A$ with $h = f\,\beta_S$. Therefore, the outer contour of (13) is commutative. Since ε_S is an epimorphism, (12) implies that $\alpha = M(f)\,\eta_S'$. Since η_S' is an epimorphism, this deter-

mines $M(f)$ as well as f uniquely, since M is faithful. So there is, in fact, an adjunction.

(d) By 16.5.4, the adjunction transformation ε', that is quasi-inverse to η', is an isomorphism. The statement then follows, provided η' is shown to be an isomorphism too. By 16.6.1, $\mathscr{C}$ is complete and co-complete, and therefore so is $[\mathscr{B}^b, \mathscr{C}]$. Now, as β is a coequalizer, there exists in $[\mathscr{B}^b, \mathscr{C}]$ the bicartesian square

(14)
$$\begin{array}{ccc} E & \xrightarrow{\gamma_1} & K U_{\mathscr{B}} \\ {\scriptstyle \gamma_2}\downarrow & & \downarrow{\scriptstyle \beta} \\ K U_{\mathscr{B}} & \xrightarrow{\beta} & G \end{array}$$

Applying M to (14), one gets another bicartesian square, since $M * \beta$ is a coequalizer at every place and since M preserves pullbacks (as a right adjoint of G). Thus (10) and (12) imply the existence of a $\mu: F \to M\,E$ with

(15)
$$\pi_i = (M * \gamma_i)\,\mu \qquad i = 1, 2\,.$$

By construction, μ and η' together with the identity transformation of $M\,K\,U_{\mathscr{B}}$ form a natural transformation of bicartesian squares. Thus η' is an isomorphism if and only if μ is an isomorphism. It suffices to prove this for an arbitrary $S \in |\mathscr{B}^b|$. We consider

(16)
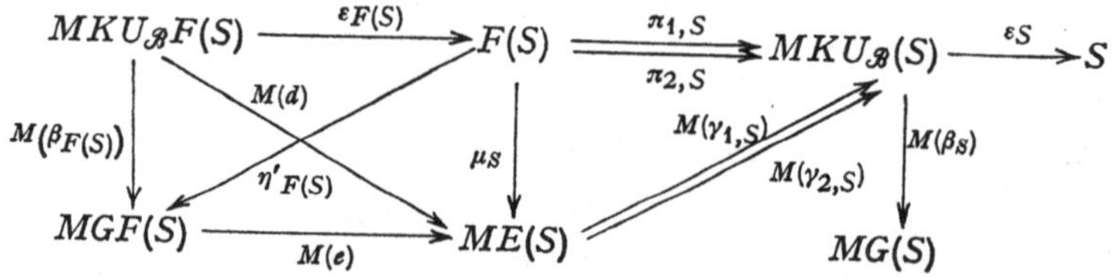

Here, the left half consists of a commutative diagram as in (13) (with a suitable change of notation). By 18.4.10, μ_S is a monomorphism. It follows therefore, that $\eta'_{F(S)}$ is a monomorphic coequalizer and thus an isomorphism (7.2.2). $(M(\gamma_{1,S}\,e),\ M(\gamma_{2,S}\,e))$ is therefore also a kernel pair of ε_S.

Since by assumption $U_{\mathscr{B}}\,M \cong X$ satisfies (iii) and since $U_{\mathscr{B}}$ preserves kernel pairs, $(\gamma_{1,S}\,e, \gamma_{2,S}\,e)$ is a kernel pair in $\mathscr{C}$. It has β_S as a coequalizer, for, by definition of β and by (11), (15), (16), β_S is a coequalizer of $\gamma_{1,S}\,e\,\beta_{F(S)}$ and $\gamma_{2,S}\,e\,\beta_{F(S)}$, and $\beta_{F(S)}$ is an epimorphism. By (14) and 18.4.3 (b), e is an isomorphism and so is μ_S, because $\eta'_{F(S)}$ is an isomorphism. This, finally, concludes the proof.

18.8.9 Generalization. Let $r \geqq 1$ be a regular cardinal (see 9.6.9). If, instead of $\mathscr{N}^0$, one takes the full subcategory $\mathscr{R}^0$ of *Ens*, whose objects are the cardinals $< r$, then according to 18.6.1 algebraic theories of rank r can be defined as functors $I_{\mathcal{A}}: \mathfrak{R} \to \mathcal{A}$ that are bijective on object classes and that preserve those products, where the cardinality of the index set is smaller than r. For such rank r algebraic theories everything from 18.1 on carries over without difficulty. Filtered colimits are replaced by r-filtered colimits. Corresponding to 9.5.1, 9.5.2, one concludes that such colimits commute with the necessary products (index set of cardinality smaller than r) in *Ens* as well as in $\mathcal{A}^b$ (compare 9.6.9 and the remark in 10.1.2).

Such a generalization is of interest for lattices, for example. The theory of lattices is an algebraic theory and, by incorporating certain additional axioms such as the existence of a smallest element 0 and a greatest element 1 and unique complementation (boolean algebra), further algebraic theories are created. By means of the above generalization, lattices which are complete up to r, in particular σ-complete boolean algebras (where r is the first uncountable cardinal), can be treated.

If r and s are regular cardinals with $1 < r < s$, then every rank r theory $\mathcal{A}_r$ can be enlarged to a rank s theory $\mathcal{A}_s$ by means of a construction analogous to the one in 18.1; namely, by using the operations in $\mathcal{A}_r$ as generating ones for $\mathcal{A}_s$ and by preserving the existing equations in $\mathcal{A}_r$. There is then a natural way to show that ${}_0\mathcal{A}_r^b \cong {}_0\mathcal{A}_s^b$ and that $\mathcal{A}_r^b$ and $\mathcal{A}_s^b$ are equivalent. In particular, one obtains in this way a forgetful functor as an algebraic functor from the category of σ-complete boolean algebras into the category of boolean algebras.

A further generalization consists in taking, in the place of $\mathcal{N}^0$, a skeleton of *Ens* (objects are the cardinals in the universe $\mathfrak{U}$) or, instead, *Ens* itself (Linton [56]). In 21, this will turn out to be a special case of a further development of ideas. However, it is not possible to treat complete boolean algebras in this way, since in this case the forgetful functor into *Ens* does not possess a left adjoint.

18.9 Problems

18.9.1 Give a direct description of a free monoid (or, resp., a free ring, a free commutative ring) with a basis of n elements, and interpret its elements as n-ary operations for an arbitrary monoid (or, resp., ring, commutative ring). (Hint: for an arbitrary algebraic theory $\mathcal{A}$, there are n-ary operations that factor through a projection onto a partial product of A^n).

18.9.2 Show that every $\mathcal{A}$-algebra is a filtered colimit of finitely generated subalgebras.

Remark. A subalgebra of a finitely generated algebra need not be finitely generated, as is shown, for instance, by the category of groups or by ${}_R Mod$ for an appropriate ring R. Also, a subalgebra of a free algebra need not be free.

18.9.3 Let R be the ring $[\boldsymbol{Z} \oplus \boldsymbol{Z}, \boldsymbol{Z} \oplus \boldsymbol{Z}]_{Ab}$. By 17.4.10, $\mathrm{Hom}_{\boldsymbol{Z}}({}_R\boldsymbol{Z} \oplus \boldsymbol{Z}, ?) \colon Ab \to Mod_R$ is an equivalence. Show that it does not preserve free objects; in particular, it does not take $\boldsymbol{Z}$ into a free module. Compare this with 18.5 (5).

18.9.4 Verify the assertion in the first sentence of 18.4.9.

18.9.5 Show that the forgetful functor groups $\to$ monoids has a right adjoint which is a coreflector. Hence the functor groups $\to$ rings, which assigns to every group its group ring, has a right adjoint.

18.9.6 Let $J \colon Ens \to [\mathcal{N}, Ens]$ be the functor $? \mapsto [I\,\mathrm{Op}\,(??), ?]_{Ens}$, where $I \colon \mathcal{N}^0 \to Ens$ is the inclusion. (After a change of universe, $J = [I\,\mathrm{Op}, Ens]\,H_*$). Show that, for any algebraic theory $\mathcal{A}$, the following diagram can be completed to a pullback in Cat:

$$\begin{array}{ccc} {}_0\mathcal{A}^b & & [\mathcal{A}, Ens] \\ {\scriptstyle {}_0U_{\mathcal{A}}}\big\downarrow & & \big\downarrow{\scriptstyle [I_{\mathcal{A}}, Ens]} \\ Ens & \xrightarrow{\;J\;} & [\mathcal{N}, Ens] \end{array}$$

($I_{\mathcal{A}} \colon \mathcal{N} \to \mathcal{A}$ is the only theory-morphism).

19. Calculus of Fractions

19.1 Categories of Fractions

19.1.1 Preliminary remarks. If $S \colon \mathcal{C} \to \mathcal{B}$ is a functor, it seems natural to consider the class Σ of those morphisms in $\mathcal{C}$ which are taken into isomorphisms by S. Conversely, if $\Sigma \subset \mathrm{Mor}\,\mathcal{C}$ is a given class of morphisms, one can ask for the functors that take all morphisms of Σ into isomorphisms. In a higher universe $\mathfrak{V}$ there is an initial such functor (19.1.2). Under suitable conditions on Σ reasonable descriptions of it are possible (19.2). More restrictive conditions are closely related to adjoint situations in which one of the two functors is fully faithful and where a change of universe is unnecessary.

19.1.2 Proposition. *Let $\mathcal{C}$ be a category in the universe $\mathfrak{U}$ and Σ a subclass of* Mor $\mathcal{C}$. *In a higher universe $\mathfrak{V}$ there is a small $\mathfrak{V}$-category $\mathcal{C}[\Sigma^{-1}]$ and a functor $P \colon \mathcal{C} \to \mathcal{C}[\Sigma^{-1}]$ with the following properties:*

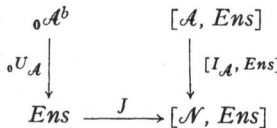

(i) *For every* $s \in \Sigma$, $P(s)$ *is an isomorphism.*

(ii) *If* $F: \mathcal{C} \to \mathcal{D}$ *is a functor such that* $F(s)$ *is an isomorphism for all* $s \in \Sigma$, *then there is exactly one functor* $G: \mathcal{C}[\Sigma^{-1}] \to \mathcal{D}$ *with* $F = G\,P$. *Here* $\mathcal{D}$ *is a category of an arbitrary universe.*

Proof. $\mathcal{C}$ is a small $\mathfrak{V}$-category. Let $\mathcal{C}'$ be the diagram scheme underlying $\mathcal{C}$ (with respect to $\mathfrak{V}$, compare 6.1.2). $\mathcal{C}''$ is obtained from $\mathcal{C}'$ by adding for every arrow s of Σ an arrow s^- with reversed direction; i. e., $o(s^-) = e(s), e(s^-) = o(s)$. For $\mathcal{C}''$ we consider the $\mathfrak{V}$-set K of commutativity conditions which consist of all commutative conditions coming from $\mathcal{C}$ (a trivial extension of $\mathcal{C}''$ as in 6.2.4 is unnecessary) and all pairs $(s^- s, 1_{o(s)})$, $(s\, s^-, 1_{e(s)})$ with $s \in \Sigma$. We set $\mathcal{C}[\Sigma^{-1}] = \mathcal{P}(\mathcal{C}''/K)$ as in 6.3.2. $\mathcal{C}$ and $\mathcal{C}[\Sigma^{-1}]$ have the same objects. If $f \in \operatorname{Mor} \mathcal{C}$, then f is a path of length 1 in $\mathcal{C}''$. The transition from $\mathcal{C}''$ to $\mathcal{C}[\Sigma^{-1}]$ takes it into a morphism $P(f)$. Obviously, this defines P as a functor.

A functor $F: \mathcal{C} \to \mathcal{D}$ can be regarded as a diagram $F': \mathcal{C}' \to \mathcal{D}$. If $F(s)$ is an isomorphism for every $s \in \Sigma$, then F' can be extended uniquely to a diagram $F'': \mathcal{C}'' \to \mathcal{D}$ in such a way that F'' satisfies the commutativity conditions K. The statement thus follows from the construction of P and from 6.3.2.

19.1.3 Definition. The category $\mathcal{C}[\Sigma^{-1}]$ constructed above is called the *category of fractions* of $\mathcal{C}$ with respect to Σ. The functor P: $\mathcal{C} \to \mathcal{C}[\Sigma^{-1}]$ is called the *canonical functor.*

The class Ξ of all morphisms which are taken into isomorphisms by $P: \mathcal{C} \to \mathcal{C}[\Sigma^{-1}]$ is called the *saturation* of Σ. Σ is called *saturated* if $\Sigma = \Xi$.

19.1.4 Proposition. Let $\Sigma \subset \operatorname{Mor} \mathcal{C}$ *and let* $P: \mathcal{C} \to \mathcal{C}[\Sigma^{-1}]$ *be the canonical functor.*

(a) *If* $P': \mathcal{C} \to \mathcal{C}'$ *is a functor satisfying conditions* 19.1.2 (i), (ii), *then the functor* $R: \mathcal{C}[\Sigma^{-1}] \to \mathcal{C}'$, *that is uniquely determined by* $P' = R\,P$, *is an isomorphism.*

(b) *If* Ξ *is the saturation of* Σ *and if* $P': \mathcal{C} \to \mathcal{C}[\Xi^{-1}]$ *is the canonical functor, then there exists exactly one isomorphism* R *with* $P' = R\,P$.

(c) *If* $\Sigma' \subset \operatorname{Mor} \mathcal{C}$ *and if* $P': \mathcal{C} \to \mathcal{C}[\Sigma'^{-1}]$ *is the corresponding canonical functor, then there is an isomorphism* R *with* $P' = R\,P$ *if and only if* Σ *and* Σ' *have the same saturation.*

(d) *For an arbitrary category* $\mathcal{D}$, $[P, \mathcal{D}]: [\mathcal{C}[\Sigma^{-1}], \mathcal{D}] \to [\mathcal{C}, \mathcal{D}]$ *is a full embedding (in a suitable universe). It produces an isomorphism between* $[\mathcal{C}[\Sigma^{-1}], \mathcal{D}]$ *and the full subcategory of* $[\mathcal{C}, \mathcal{D}]$ *whose objects are the functors which take the morphism of* Σ *into isomorphisms.*

Proof. (a) By 19.1.2, one gets functors R and R' with $P' = R\,P$ and $P = R'\,P'$. Here $R'\,R$ and $R\,R'$ are identity functors, again by 19.1.2, and R and R' are uniquely determined.

(b) Every functor $F: \mathscr{C} \to \mathscr{D}$, that takes the morphisms in Σ into isomorphisms, also takes those in Ξ into isomorphisms, by 19.1.2 (ii). In 19.1.2 (ii) one can therefore replace Σ by Ξ, so that the statement follows from (a).

(c) If there is an isomorphism R with $P' = R\,P$, then Σ and Σ' have the same saturation. The converse follows from (b).

(d) P is the identity map on the objects of $\mathscr{C}$. Natural transformations between functors $G, G': \mathscr{C}\,[\Sigma^{-1}] \to \mathscr{D}$ are therefore also natural transformations between $G\,P$ and $G'\,P$. The construction of $\mathscr{C}\,[\Sigma^{-1}]$ by means of the diagram schema $\mathscr{C}''$ shows the converse to be true also. The statement thus follows from 19.1.2. Note that this says that composition with P yields a bijection

$$\mathrm{Nat}\,(G, G') \cong \mathrm{Nat}\,(G\,P,\; G'\,P)\,.$$

19.2 Calculus of Left Fractions

19.2.1 Definition. Let $\mathscr{C}$ be a $\mathfrak{U}$-category and $\Sigma \subset \mathrm{Mor}\,\mathscr{C}$. Σ admits a *calculus of left fractions* if the following hold:

(i) All identity morphisms of $\mathscr{C}$ belong to Σ.

(ii) Every composite of morphisms in Σ (existing in $\mathscr{C}$) belongs to Σ (Σ is closed with regard to composition).

(iii) For every diagram $A' \xleftarrow{s} A \xrightarrow{g} D$ in $\mathscr{C}$ with $s \in \Sigma$ there exists a commutative square

(1)
$$\begin{array}{ccc} A & \xrightarrow{g} & D \\ {\scriptstyle s}\downarrow & & \downarrow{\scriptstyle s'} \\ A' & \xrightarrow{g'} & D' \end{array} \qquad \text{with } s' \in \Sigma\,.$$

(iv) If, given $f, g: A \to B$ in $\mathscr{C}$, there exists an $s \in \Sigma$ with $f\,s = g\,s$, then there exists a $t \in \Sigma$ with $t\,f = t\,g$.

$$\xrightarrow{s} \;\underset{g}{\overset{f}{\rightrightarrows}}\; \overset{t}{\dashrightarrow}$$

Σ admits a *strong calculus of left fractions* if in addition (v) holds:

(v) For every pencil $\{s_j: A \to B_j\}_{j \in J}$, where J is a $\mathfrak{U}$-set and every $s_j \in \Sigma$, there exists a commutative completion $\{f_j: B_j \to C\}$ such that $f_j\,s_j \in \Sigma$ (compare 9.3.4).

For $A \in |\mathscr{C}|$, let A/Σ be the full subcategory of $A/\mathscr{C}$ (compare the dual of 6.5.3) whose objects are the morphisms in Σ with domain A. Σ admits a *terminal calculus of fractions* if (i), (ii), (iii), (iv) are satisfied and if additionally (vi) holds:

(vi) For every $A \in |\mathscr{C}|$, A/Σ has a terminal object.

(vi) obviously implies (v).

A calculus of *right fractions*, a *strong calculus of right fractions* and an *initial calculus of fractions* are all obtained by dualization.

19.2.2 Motivation. For $P\colon \mathscr{C} \to \mathscr{C}[\Sigma^{-1}]$ one certainly gets a morphism $P(A) \to P(B)$ in $\mathscr{C}[\Sigma^{-1}]$ if one of the two following situations is present in $\mathscr{C}$:

$$A \xleftarrow{s} C \xrightarrow{g} B \quad \text{or} \quad A \xrightarrow{g'} D \xleftarrow{s'} B \quad \text{with } s,\, s' \in \Sigma\,.$$

(iii) allows the first situation to be transformed into the second one in such a way that one ends up with the same morphism in $\mathscr{C}[\Sigma^{-1}]$. From (i), (ii), (iii) and the construction of $\mathscr{C}[\Sigma^{-1}]$ one deduces easily:

Every morphism in $\mathscr{C}[\Sigma^{-1}]$ can be represented in the form $P(s)^{-1}P(f)$ with $s \in \Sigma$. We set

(2) $$[s|f] = P(s)^{-1}\, P(f)$$

and call the pair (s, f) a *representative* of $[s|f]$. Here s and f have the same codomain. The domain of f, or, resp., s is the domain, or, resp., the codomain of $[s|f]$; i. e., from $A \xrightarrow{f} C \xleftarrow{s} B$ we get $[s|f]\colon A \to B$.

Now composition of morphisms in $\mathscr{C}[\Sigma^{-1}]$ can be described by representatives. If the codomain of $[s|f]$ is the domain of $[t|g]$, then, by (ii), (1) implies

(3) $$[t|g]\;[s|f] = [s't|g'f]\,,$$

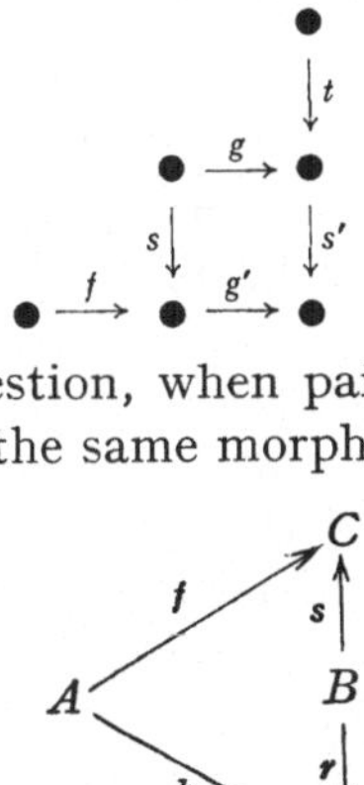

$s,\, s',\, t \in \Sigma\,.$

There remains the question, when pairs (s, f) and (r, h) with $r, s \in \Sigma$ are representatives of the same morphism of $\mathscr{C}[\Sigma^{-1}]$. We consider

(4)

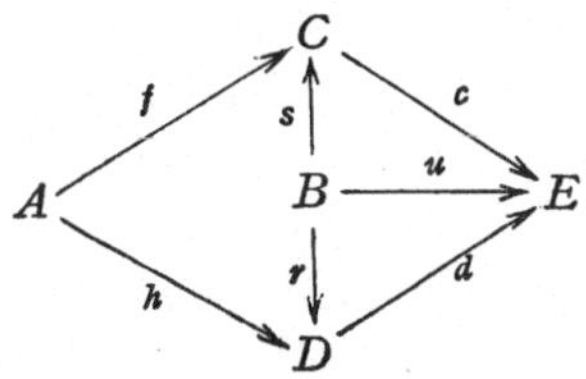

$r,\, s,\, u \in \Sigma\,.$

The left half says that $[s|f]$ and $[r|h]$ have the same domain and codomain. The right half can be obtained in a commutative way with $u \in \Sigma$ by applying (iii) to r and s, thus producing c and d, where, say, $c \in \Sigma$. By (ii), $u = dr = c\,s \in \Sigma$. Now (i), (ii) and (iii) imply in general:

In a diagram (4) with a commutative right half, i. e., with $u = dr = c\,s$.

(5) $$[s|f] = [u|cf]\,; \quad [r|h] = [u|dh]\,.$$

For, in $\mathscr{C}[\Sigma^{-1}]$, $[u|u] = 1_{P(B)}$ and $[u|u]\,[s|f] = [u|cf]$, which is obtained from $1_E\, u = c\, s$ by using (3). The second equation in (5) follows analogously.

By (5), $[s|f] = [r|h]$ will hold if u, c, d in (4) can be chosen in such a way that one gets a commutative diagram with not only $u = dr = cs$, but also $cf = dh$. We shall show (19.2.4) that this condition is also necessary if Σ admits a calculus of left fractions.

19.2.3 Lemma. (a) *If $\Sigma \subset$ Mor $\mathscr{C}$ admits a calculus of left fractions, then A/Σ is filtered for every $A \in |\mathscr{C}|$.*

(b) *Σ admits a strong calculus of left fractions if and only if Σ admits a calculus of left fractions and A/Σ is strongly filtered for every $A \in |\mathscr{C}|$.*

Proof. (a) Conditions 9.3.4 (ii), (iii) for A/Σ follow from 19.2.1 (iv), (ii) or, resp., (iii), (ii). (b) then follows by comparing 9.3.4 (i_s) with 19.2.1 (v).

19.2.4 Assume that $\Sigma \subset$ Mor $\mathscr{C}$ admits a calculus of left fractions. In 19.2.2 we have so far only used 19.1.2 (i) for $\mathscr{C}[\Sigma^{-1}]$ and P, but not 19.1.2 (ii). We construct a new category $\Sigma^{-1}\mathscr{C}$ with a functor P': $\mathscr{C} \to \Sigma^{-1}\mathscr{C}$, so that 19.1.2 (i), (ii) hold. By 19.1.4, there is then an isomorphism

$$R\colon \mathscr{C}[\Sigma^{-1}] \xrightarrow{\cong} \Sigma^{-1}\mathscr{C} \quad \text{with } P' = R\,P\,.$$

The objects of $\Sigma^{-1}\mathscr{C}$ are those of $\mathscr{C}$. For every (ordered) pair (A, B) of objects of $\mathscr{C}$ let $F_{A,B}\colon B/\Sigma \to Ens$ be the functor $H^A\,\Delta^1$ (compare 6.5.2); that is $s \mapsto [A, e(s)]_{\mathscr{C}}$ for $s \in |B/\Sigma|$, where $e(s)$ is the codomain of s. Using the inclusion $Ens \subset ENS$, $F_{A,B}$ has a colimit in ENS. We set $[A, B]_{\Sigma^{-1}\mathscr{C}} = \operatorname{Colim} F_{A,B}$, where this colimit object is constructed as a filtered colimit (19.2.3) according to 9.4.2. An element of $[A, B]_{\Sigma^{-1}\mathscr{C}}$ is an equivalence class of pairs (s, f), where $s \in |B/\Sigma|$ and where f has domain A and the same codomain as s. By 9.4.2, two pairs (s, f) and (r, h) are equivalent if and only if there is a commutative diagram (4) Now let $[s|f]$ be the equivalence class of (s, f). If the codomain of $[s|f]$ is the domain of $[t|g]$, define $[t|g]\,[s|f]$ as in (3) by means of representatives (t, g), (s, f) and using (iii) Using (i), (ii), (iii) and (iv), there is no difficulty in verifying that the composite (as an equivalence class) is well defined; i. e., that for fixed representatives (t, g) and (s, f), it does not depend on the choice of a diagram of the form (1), and also it does not depend on the choice of representatives for $[t|g]$ and $[s|f]$ One verifies then that $\Sigma^{-1}\mathscr{C}$ is a category: that composition of morphisms is associative and that $[1_A|1_A]$ is the identity morphism for A

For $f\colon A \to B$ in $\mathscr{C}$, one sets $P'(f) = [1_B|f]$ and $P'(A) = A$. $P'\colon \mathscr{C} \to \Sigma^{-1}\mathscr{C}$ is thus a functor. For $s\colon A \to B$ in Σ, $P'(s)$ is an isomorphism with the inverse $[s|1_A.]$ If $F: \mathscr{C} \to \mathscr{D}$ has the property that $F(s)$ is an isomorphism for all $s \in \Sigma$, then by means of (4), $[s|f] \mapsto F(s)^{-1} F(f)$ defines a functor $G: \Sigma^{-1} \mathscr{C} \to \mathscr{D}$ with $F = G P'$, where G is uniquely determined, since

$$[s|f] = P'(s)^{-1} P'(f) .$$

19.2.5 Remarks. (a) The morphisms of $\mathscr{C} [\Sigma^{-1}]$ are equivalence classes of paths in the diagram scheme $\mathscr{C}''$ in 19.1.2. If Σ admits a calculus of left fractions, then $\Sigma^{-1} \mathscr{C}$ is obtained from $\mathscr{C} [\Sigma^{-1}]$ by reducing these classes to subclasses, which consist only of paths of the special form $s^- f$ in $\mathscr{C}''$. The isomorphism $R: \mathscr{C} [\Sigma^{-1}] \to \Sigma^{-1} \mathscr{C}$ with $R P = P'$ is simply the restriction on the equivalence classes of paths.

(b) If Σ admits a terminal calculus of fractions, and if a terminal object η_B is chosen in B/Σ, then every morphism $P(A) \to P(B)$ in $\mathscr{C} [\Sigma^{-1}]$ has a special representative (η_B, f') where f' is uniquely determined. For, by the dual of 7.1.8, $[A, e(\eta_B)]$ is also a colimit object of $F_{A,B}$ in 19.2.4. (4) provides a direct proof of this, if one sets $u = \eta_B$ and then considers the case $r = s = \eta_B$.

19.2.6 Lemma. *Let $\Sigma \in$ Mor $\mathscr{C}$ and let $P: \mathscr{C} \to \mathscr{C} [\Sigma^{-1}]$ be the canonical functor.*

(a) *If P is faithful, then Σ consists of bimorphisms.*

(b) *If Σ consists of monomorphisms and if Σ admits a calculus of left fractions, then P is faithful.*

Proof. (a) follows from 13.3.5 (a).

(b) For $f, g: A \to B$ in $\mathscr{C}$, let $P(f) = P(g)$; i.e., $[1_B|f] = [1_B|g]$. By 19.2.2 (4) and 19.2.4, there is a $u \in \Sigma$ with $u f = u g$. $f = g$ thus follows.

Remark. If P is full, then $\mathscr{C} [\Sigma^{-1}]$ is a quotient category of $\mathscr{C}$, compare 6.4.2.

19.2.7 Proposition. *Assume that for the $\mathfrak{U}$-category $\mathscr{C}$, $\Sigma \subset$ Mor $\mathscr{C}$ admits a calculus of left fractions. If A/Σ has a final $\mathfrak{U}$-set for every $A \in |\mathscr{C}|$ (9.2.4), then $\mathscr{C} [\Sigma^{-1}]$ is isomorphic to a $\mathfrak{U}$-category. This applies in particular if Σ admits a terminal calculus of fractions.*

Here the colimits of the functors $F_{A,B}$ of 19.2.4 exist already in *Ens*, see 9.2.3 (compare also 10.1.8).

19.2.8 Theorem. *Let $\mathscr{C}$ be a $\mathfrak{U}$-category, $\Sigma \subset$ Mor $\mathscr{C}$ and let P: $\mathscr{C} \to \mathscr{C} [\Sigma^{-1}]$ be the canonical functor.*

(a) *If Σ admits a calculus of left fractions, then P preserves finite colimits and terminal objects (those that exist). If $\mathscr{C}$ has finite coproducts or, resp., coequalizers, finite colimits, then the corresponding statements hold for $\mathscr{C} [\Sigma^{-1}]$.*

(b) *If Σ admits a strong calculus of left fractions, then P preserves colimits (with respect to $\mathfrak{U}$). If, in addition, $\mathscr{C}$ is cocomplete, then $\mathscr{C}[\Sigma^{-1}]$ is also cocomplete with respect to $\mathfrak{U}$.*

(c) *If Σ admits a terminal calculus of fractions, then P has a fully faithful right adjoint $T: \mathscr{C}[\Sigma^{-1}] \to \mathscr{C}$.*

Proof. (a) is proved in the same way as (b).

(b) First, let B be terminal in $\mathscr{C}$. Then one concludes from (4) with $u = 1_B$ that $[P(A), P(B)]$ has exactly one element and that thus $P(B)$ is terminal in $\mathscr{C}[\Sigma^{-1}]$. If A is initial in $\mathscr{C}$, then (4) implies that $P(A)$ is initial in $\mathscr{C}[\Sigma^{-1}]$.

Let $R: \mathscr{X} \to \mathscr{C}$ be a $\mathfrak{U}$-diagram with colimit (L, λ) in $\mathscr{C}$. By what was just said we may assume that $\mathscr{X}$ is not empty. By 8.7.3, it suffices to show that $([P(L), D], \{[P(\lambda_i), D]\})$ is a limit of $[P\,R(?), D]_{\mathscr{C}[\Sigma^{-1}]}$ for every $D \in |\mathscr{C}|$.

For an arbitrary $B \in |\mathscr{C}|$, $([L, B], H_B\,\lambda)$ is a limit of $H_B\,R = [R(?), B]_{\mathscr{C}}$, by 8.7.3. Further, $[B, D]_{\mathscr{C}[\Sigma^{-1}]}$ is a colimit object for $F_{B,D}: D/\Sigma \to ENS$; in particular, for $B = R(i)$ with $i \in |\mathscr{X}|$ and $B = L$. The first statement in (b) follows thus by interchanging limits with strongly filtered colimits in ENS (9.5.2); with $e(s)$ as codomain of s this situation is described roughly by

$$\begin{aligned}[P(L), D]_{\mathscr{C}[\Sigma^{-1}]} &= \operatorname*{Colim}_{s \in |D/\Sigma|} [L, e(s)]_{\mathscr{C}} = \operatorname*{Colim}_{s \in |D/\Sigma|} [\operatorname*{Colim}_{i \in |\mathscr{X}|} R(i), e(s)]_{\mathscr{C}} = \\ &= \operatorname*{Colim}_{s \in |D/\Sigma|} \operatorname*{Lim}_{i \in |\mathscr{X}|} [R(i), e(s)]_{\mathscr{C}} = \operatorname*{Lim}_{i \in |\mathscr{X}|} \operatorname*{Colim}_{s \in |D/\Sigma|} [R(i), e(s)]_{\mathscr{C}} = \\ &= \operatorname*{Lim}_{i \in |\mathscr{X}|} [P\,R(i), D]_{\mathscr{C}[\Sigma^{-1}]}.\end{aligned}$$

If $\mathscr{C}$ has coproducts with respect to $\mathfrak{U}$, then by what was said above, this is so for $\mathscr{C}[\Sigma^{-1}]$ also, since both categories have the same objects and since those are mapped identically by P.

Now let $\mathscr{C}$ have coequalizers. Let $[s|f]$, $[r|h]: P(A) \to P(B)$ be morphisms in $\mathscr{C}[\Sigma^{-1}]$ with representatives (s, f) or, resp., (r, h). By (5), we may assume that $r = s$. Then let q be a coequalizer of f and h. By what we proved before, $P(q)$ is a coequalizer of $P(f)$ and $P(h)$. Since $P(s) = P(r)$ is an isomorphism, $P(q)\,P(s)$ is a coequalizer of $[s|f] = P(s^{-1})\,P(f)$ and $[s|h] = P(s^{-1})\,P(h)$.

(c) For every $B \in |\mathscr{C}|$, let a terminal object $\eta_B: B \to T(B)$ in B/Σ be chosen. Let $\varepsilon_B = [\eta_B | 1_{T(B)}]$. Since $B = P(B)$, $T(B) = T\,P(B) = P\,T(B)$, and by 16.5.4 (d) and the dual of 16.4.5, it suffices to show that $(T(B), \varepsilon_B)$ is a representation of the contravariant functor $[P(?), B]_{\mathscr{C}[\Sigma^{-1}]}: \mathscr{C} \to Ens$, where, on account of 19.2.7, it is in effect possible to use Ens. By the dual of 16.4.5 one has to show: given any morphism $[s|f]: A \to B$ in $\mathscr{C}[\Sigma^{-1}]$ with codomain B, there is exactly one $h: A \to T(B)$ in $\mathscr{C}$ with $[s|f] = \varepsilon_B\,P(h)$. By 19.2.5 (b) one may assume that $s = \eta_B$. But, since $[\eta_B|f] = [\eta_B|1_{T(B)}]\,[1_{T(B)}|f] = \varepsilon_B\, P(f)$, the statement follows from 19.2.5 (b).

19.3 Factorization of Functors and Saturation

19.3.1 Theorem. *Let $\mathcal{C}$ be a $\mathfrak{U}$-category, $S\colon \mathcal{C} \to \mathcal{B}$ a functor, Σ the class of $\mathcal{C}$-morphisms which are taken into isomorphisms by S, P the canonical functor and $S'\colon \mathcal{C}[\Sigma^{-1}] \to \mathcal{B}$ the uniquely determined functor with $S'P = S$ (as in 19.1.2).*

(a) *If Σ admits a calculus of left fractions, then S' reflects isomorphisms.*
(b) *Assume that Σ admits a calculus of left fractions, that $\mathcal{C}$ has coequalizers and that S preserves them. Then S' preserves coequalizers. Further, S' is faithful and $\mathcal{C}[\Sigma^{-1}]$ is isomorphic to a $\mathfrak{U}$-category, provided $\mathcal{B}$ is also a $\mathfrak{U}$-category.*
(c) *Assume that $\mathcal{C}$ has coequalizers, that S preserves them and is full. Then Σ admits a calculus of left fractions. $\mathcal{C}[\Sigma^{-1}]$ is isomorphic to a $\mathfrak{U}$-category, S' is fully faithful and P is full.*
(d) *If $\mathcal{C}$ is finitely cocomplete and if S preserves finite colimits, then Σ admits a calculus of left fractions and S' preserves finite colimits.*
(e) *If $\mathcal{C}$ is cocomplete and if S preserves colimits, then Σ admits a strong calculus of left fractions and S' preserves colimits with respect to $\mathfrak{U}$.*
(f) *If S has a right adjoint $T\colon \mathcal{B} \to \mathcal{C}$ and if S or T is full, then Σ admits a terminal calculus of fractions. $\mathcal{C}[\Sigma^{-1}]$ is isomorphic to a $\mathfrak{U}$-category and P has a fully faithful right adjoint* (*Compare* 21.8.9).
(g) *If $T\colon \mathcal{B} \to \mathcal{C}$ is right adjoint to S, then T is fully faithful if and only if S' is an equivalence. In this case $P\,T$ is equivalence-inverse to S'.*

Proof. (a) If $S'([s|f])$ is an isomorphism, then $[s|f] = P(s)^{-1}\,P(f)$ implies that $S'P(f) = S(f)$ is also an isomorphism. By the definition of Σ, $P(f)$ is an isomorphism, and therefore, so is $[s|f]$. This conclusion, incidentally, makes no use of 19.2.1 (iv) for Σ.

(b) The first statement follows immediately from the proof of 19.2.8 (b). For $\alpha, \beta\colon A \to B$ in $\mathcal{C}[\Sigma^{-1}]$ let γ be a coequalizer. Then $S'(\gamma)$ is a coequalizer of $S'(\alpha)$ and $S'(\beta)$. If $\alpha \neq \beta$, then γ is not an isomorphism, so, by (a), $S'(\gamma)$ is not an isomorphism, and hence $S'(\alpha) \neq S'(\beta)$. Therefore, S' is faithful. The last statement in (b) follows immediately from this.

(c) Conditions (i), (ii) of 19.2.1 are evident. If, given $f, g\colon A \to B$ in $\mathcal{C}$, there is an $s \in \Sigma$ with $f\,s = g\,s$, then $S(f) = S(g)$. If t is a coequalizer of f and g, then one concludes further that $S(t)$ is an isomorphism. Therefore, 19.2.1 (iv) is valid. Now assume that one has $A' \xleftarrow{s} A \xrightarrow{g} D$ in $\mathcal{C}$ with $s \in \Sigma$. Since S is full, there is a $t \in \Sigma$ with $S(t) = S(s)^{-1}$. Let $s'\colon D \to D'$ be a coequalizer of g and $g\,t\,s$. Since S

preserves coequalizers and since $S(g) = S(g\,t\,s)$, $s' \in \Sigma$. 19.2.1 (iii) then follows with $g' = s'\,g\,t$.

Because for every $s \in |A/\Sigma|$ there is a $t \in \Sigma$ with $S(t) = S(s)^{-1}$ and since $ts \in \Sigma$ also, the endomorphisms in Σ form a final set in A/Σ. Therefore, the second statement in (c) follows from 19.2.7.

S' is full, because $\mathscr{C}$ and $\mathscr{C}\,[\Sigma^{-1}]$ have the same objects and S is full. By (b), S' is fully faithful. This implies that P is full.

(d) is proved analogously to (e).

(e) Conditions 19.2.1 (i), (ii), (iv) are deduced as in (c).

If for $A' \xleftarrow{s} S \xrightarrow{g} D$ in $\mathscr{C}$ with $s \in \Sigma$ the diagram 19.2.1(1) is formed as a pushout, then applying S gives a pushout in $\mathscr{B}$. Therefore, $S(s')$ is an isomorphism; i.e., $s' \in \Sigma$. The following refinement of 19.2.1 (iii) is therefore valid:

Σ is closed with respect to pushouts (compare 18.4.6 (b)).

19.2.1 (v) is obtained analogously, by forming the generalized pushout of the pencil $\{s_j\colon A \to B_j\}$ in $\mathscr{C}$ with $s_j \in \Sigma$. This yields the first statement in (e).

For the second statement, 19.2.8 (b) shows that it is sufficient to prove that S' preserves coequalizers and coproducts with respect to $\mathfrak{U}$. (b) shows the first to be the case, the second follows from the fact that these coproducts are formed in $\mathscr{C}\,[\Sigma^{-1}]$ "as in $\mathscr{C}$" (compare the proof of 19.2.8 (b)).

(f) For the adjunction transformation $\eta\colon 1_{\mathscr{C}} \to TS$, $\eta * T$ and $S * \eta$ are isomorphic, by 16.5.5. For every $A \in |\mathscr{C}|$ one has therefore that $\eta_A \in \Sigma$.

19.2.1 (i), (ii) are again evident. For $f, g\colon A \to B$ in $\mathscr{C}$, let $f\,s = g\,s$ with $s \in \Sigma$. $S(f) = S(g)$ follows and further $\eta_B f = TS(f)\,\eta_A = TS(g)\,\eta_A = \eta_B\,g$. Since $\eta_B \in \Sigma$, 19.2.1 (iv) follows. For $A' \xleftarrow{s} A \xrightarrow{g} D$ in $\mathscr{C}$ with $s \in \Sigma$ one obtains 19.2.1 (iii) with $s' = \eta_D$ and $g' = TS(g)\,T(S(s)^{-1})\circ \circ\eta_{A'}$, since $\eta_{A'}s = TS(s)\,\eta_A$ and $\eta_D\,g = TS(g)\,\eta_A$.

We go on to show that η_A is terminal in A/Σ. For $s\colon A \to D$ in Σ there is a $u\colon D \to TS(A)$ with $u\,s = \eta_A$, namely $T(S(s)^{-1})\,\eta_D$. If one also has $v\,s = \eta_A$, then $S(u) = S(v)$ follows and thus $\eta_{TS(A)}\,u = TS(u)\,\eta_D = \eta_{TS(A)}\,v$, which implies $u = v$. Therefore, η_A is terminal in A/Σ.

The remaining statements in (f) follow from 19.2.7 and 19.2.8 (c).

(g) First, let T be fully faithful. For the adjunction transformations ε, η belonging to S and T, $\varepsilon\colon ST \to 1_{\mathscr{B}}$ is an isomorphism by 16.5.4, and $P * \eta\colon P \to PTS$ is shown to be an isomorphism by the proof of (f). By 19.1.4 (d), $[P, \mathscr{C}\,[\Sigma^{-1}]]$ is fully faithful. Since P and $P\,T\,S = (P\,T\,S')\,P$ are isomorphic functors, $1_{\mathscr{C}[\Sigma^{-1}]}$ is isomorphic to $P\,T\,S'$. Together with $\varepsilon\colon S'\,P\,T \to 1_{\mathscr{B}}$ this implies that S' is equivalence-inverse to $P\,T$.

Now let S' be an equivalence. Then $[S', \mathscr{B}]$ is an equivalence, and by 19.1.4 (d) and $S = S'P$, $[S, \mathscr{B}]\colon [\mathscr{B}, \mathscr{B}] \to [\mathscr{C}, \mathscr{B}]$ is also fully faithful.

For $S * \eta\colon S \to S\,T\,S$ one finds that there is exactly one natural transformation $\eta'\colon 1_{\mathscr{B}} \to S\,T$ (compare 16.1.3) with

$$(1) \qquad \eta' * S = S * \eta\,.$$

We want to show that $S\,T$ is left adjoint to $1_{\mathscr{B}}$ with quasi-inverse adjunction transformations $\varepsilon\colon (S\,T)\,1_{\mathscr{B}} \to 1_{\mathscr{B}}$ and $\eta'\colon 1_{\mathscr{B}} \to 1_{\mathscr{B}}(S\,T)$. If this is known, then ε is an isomorphism by 16.5.4, since $1_{\mathscr{B}}$ is fully faithful, and T is fully faithful, again by 16.5.4. From 16.5.5. (4) and by (1), one obtains by applying S

$$(2) \qquad 1_{S\,T} = (S\,T * \varepsilon)\,(S * \eta * T) = (S\,T * \varepsilon)\,(\eta' * S\,T)\,.$$

From (1) and 16.5.5 (4^0) one gets

$$1_S = (\varepsilon * S)\,(S * \eta) = (\varepsilon * S)\,(\eta' * S) = (\varepsilon\,\eta') * S\,.$$

Since $[S, \mathscr{B}]$ is fully faithful, this implies

$$(3) \qquad 1_{\mathscr{B}} = \varepsilon\,\eta' = (\varepsilon * 1_{\mathscr{B}})\,(1_{\mathscr{B}} * \eta')\,.$$

By (2), (3) and 16.5.5, $S\,T$ is left adjoint to $1_{\mathscr{B}}$.

19.3.2 Corollary. *If $S\colon \mathscr{C} \to \mathscr{B}$ has a right adjoint $T\colon \mathscr{B} \to \mathscr{C}$, then T is fully faithful if and only if $[S, \mathscr{B}]\colon [\mathscr{B}, \mathscr{B}] \to [\mathscr{C}, \mathscr{B}]$ is fully faithful. In this case, $[S, \mathscr{D}]$ is also fully faithful for every category $\mathscr{D}$.*

Compare this with the content of 17.1.6 (c).

Proof. The first statement follows immediately from the preceding proof, the second one from 16.5.11 (b).

19.3.3 Proposition. *Assume that $\Gamma \subset$ Mor $\mathscr{C}$ admits a calculus of left fractions, let $S\colon \mathscr{C} \to \mathscr{C}[\Gamma^{-1}]$ be the canonical functor and Σ the saturation of Γ.*

(a) *A $\mathscr{C}$-morphism u belongs to Σ if and only if there are morphisms v, w, such that $v\,u$ and $w\,v$ exist and belong to Γ.*

(b) *For $u \in \Sigma$, let $u = w\,v$ with an epimorphic v. Then v and w belong to Σ.*

(c) *Σ admits a calculus of left fractions.*

(d) *If Γ admits a strong calculus of left fractions, then so does Σ.*

(e) *If Γ admits a terminal calculus of fractions, then so does Σ.*

Proof. (a) If v, w exist, then one has the following commutative diagram

(4)

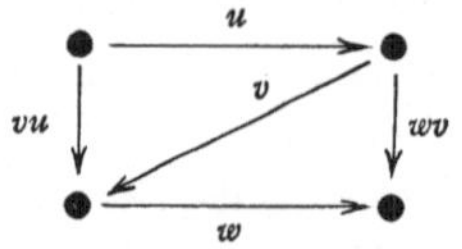

Since $S(v\,u)$ and $S(w\,v)$ are isomorphisms, $S(v)$ is a retraction and a coretraction. It follows that $S(v)$, $S(u)$ and $S(w)$ are isomorphisms.

Conversely, for $u\colon A \to B$, let $S(u)$ be an isomorphism and (s, v) a representative of $S(u)^{-1} = [s \mid v]$. Then $[s \mid v\,u]$ is a representative of $1_{S(A)}$. By 19.2.2 (4), one can assume $s = v\,u$. The existence of w follows analogously from $[1_B \mid u]\,[s \mid v] = 1_{S(B)}$.

(b) By 19.2.8 (a) and the dual of 7.8.9, $S(v)$ is an epimorphism. $S(v)$ is also a coretraction with $S(u)^{-1}\,S(w)$ as a corresponding retraction. Therefore, $v \in \Sigma$ and $w \in \Sigma$.

(c) is proved analogously to (d).

(d) Conditions 19.2.1 (i), (ii) are evident. Let $f\,v = g\,v$ for $v \in \Sigma$. Then $S(f) = S(g)$. By 19.2.2 (4), there is a $u \in \Gamma$ with $u\,f = u\,g$, and $\Gamma \subset \Sigma$. Hence 19.2.1 (iv) is valid.

Suppose that there is $A' \xleftarrow{u} A \xrightarrow{g} D$ with $u \in \Sigma$. By (a), there is a v with $v\,u \in \Gamma$. Thus 19.2.1 (iii) for Σ follows from the assumption for Γ. If there is a pencil $\{u_j\colon A \to B_j\}$ with $u_j \in \Sigma$ for all j, then by (a) there is a v_j with $v_j\,u_j \in \Gamma$ for every j. In this way 19.2.1 (v) for Σ also follows from the assumption for Γ.

(e) For S and the canonical functor $P\colon \mathscr{C} \to \mathscr{C}[\Sigma^{-1}]$ one finds oneself in the situation of 19.3.1, where here S' is an isomorphism, by 19.1.4. The statement thus follows from 19.2.8 (c) and 19.3.1 (f).

19.3.4 Proposition. *Let $\Gamma \subset$ Mor $\mathscr{C}$ and let $S\colon \mathscr{C} \to \mathscr{C}[\Gamma^{-1}]$ be the canonical functor. Then the following are equivalent:*

(a) *The saturation of Γ admits a terminal calculus of fractions.*

(b) *S has a right adjoint T.*

Furthermore, T is fully faithful.

Proof. For S and the saturation Σ of Γ one is in the situation of 19.3.1 with an isomorphic S'. If (a) is valid, then, by 19.2.8 (c), P has a right adjoint T', which implies (b) with $T = T'\,S'^{-1}$. If (b) is valid, then T is fully faithful by 19.3.1 (g), and (a) follows from 19.3.1 (f).

19.3.5 Remarks. (a) Examples of calculi of fractions which need not belong to adjunctions are obtained by a factorization of functors as in 19.3.1. One can, for instance, in the category of pointed topological spaces or in the corresponding homotopy category take for Σ the class of those maps which induce isomorphisms for the homotopy groups. Corresponding statements hold for chain complexes (over Ab or over an arbitrary abelian category) and for maps which induce isomorphisms for the homology. We refer the reader to Gabriel-Zisman [14], Hartshorne [17]. Of other applications, we only mention the investigations of injective objects in abelian categories (Roos [63], Gabriel [43]).

(b) If $\mathscr{C}$ is finitely cocomplete, then $\Sigma \subset \operatorname{Mor} \mathscr{C}$ is a saturated class of morphisms admitting a calculus of left fractions if and only if the following conditions hold:

(i) Σ contains all isomorphisms.

(ii) If $v\,u = w$, and if two of the morphisms u, v, w belong to Σ, then so does the third one. If $v\,u$ and $x\,v$ belong to Σ, then v belongs to Σ.

(iii) Σ is pushout-closed.

(iv) If $f\,s = g\,s$ with $s \in \Sigma$, then the coequalizer of f and g belongs to Σ.

That these conditions are necessary follows from what was said above (proofs of 19.3.3 (a), 19.3.1 (e), (c)). They are sufficient by 19.3.3 (a). This implies ($\mathscr{C}$ is finitely cocomplete) that the intersection of saturated classes of morphisms admitting a calculus of left fractions is again such a class and that, given a class $\Gamma \subset \operatorname{Mor} \mathscr{C}$, there is always a smallest larger one which is saturated and admits a calculus of left fractions.

Corresponding remarks hold for saturated classes admitting a strong calculus of left fractions, provided $\mathscr{C}$ is cocomplete. They also hold for a calculus of left and right fractions, provided $\mathscr{C}$ is finitely complete and finitely cocomplete.

However, there is no corresponding set of conclusions for a terminal calculus of fractions.

(c) Let $(\varphi, S, T, \mathscr{B}, \mathscr{C})$ be an adjoint situation in which S or T is full, let $\eta\colon 1_{\mathscr{C}} \to T\,S$ be the corresponding adjunction transformation and $\Gamma \subset \operatorname{Mor} \mathscr{C}$ be a class of morphisms, which are taken into isomorphisms by S. Also, let $\eta_A \in \Gamma$ for every $A \in |\mathscr{C}|$.

The saturation of Γ is the Σ of 19.3.1, in particular, it admits a terminal calculus of fractions.

To prove this, let $P'\colon \mathscr{C} \to \mathscr{C}[\Gamma^{-1}]$ be the canonical functor. By 19.1.2, the saturation of Γ is contained in Σ. If, conversely, $f\colon A \to B$ belongs to Σ, then $S(f)$ as well as $T\,S(f)$ are isomorphisms. $\eta_B\,f = T\,S(f)\,\eta_A$ implies that $P'(f)$ is also an isomorphism, which concludes the proof.

If, in particular, T is fully faithful, then, by 19.3.1 (g), one obtains an equivalence $S''\colon \mathscr{C}[\Gamma^{-1}] \to \mathscr{B}$ with $S = S''\,P'$. Here $\Gamma = \{\eta_A\}_{A \in |\mathscr{C}|}$ is allowed.

(d) From 19.3.1 (g) and 19.3.4 it follows that for every category $\mathscr{C}$ the adjoint situations $(\varphi, S, T, \mathscr{B}, \mathscr{C})$ with a fully faithful T are those (up to equivalences) which can be obtained through a terminal calculus of fractions with a suitable saturated class of morphisms of $\mathscr{C}$. This applies in particular to the examples in 16.6.8. The dual situation is presented by the examples in 16.6.9, where the corresponding Σ consists of bimorphisms (19.2.6).

Criteria for 19.3.4 (a) are provided by 19.3.3 (a) and later by 19.4.5 (b).

19.3.6 Proposition. *Let the* $\mathfrak{U}$*-category* $\mathscr{C}$ *be finitely complete, well-powered and co-well-powered. Assume that every morphism can be factored into an epi- and a monomorphism. If* $\Sigma \subset \operatorname{Mor} \mathscr{C}$ *admits a calculus of left and of right fractions, then* $\mathscr{C}[\Sigma^{-1}]$ *is isomorphic to a* $\mathfrak{U}$*-category.*

Proof. By 19.3.3 (c), we may assume that Σ is saturated. Let (s, f) be a representative of $\alpha\colon P(A) \to P(B)$ with $s\colon B \to C$ in Σ. One factors s into an epimorphism s' and a monomorphism s'' and forms the pullback for s'' and f, producing s''' and f' with $f s''' = s'' f'$ in this way. By 19.3.3 (b), s' and s'' belong to Σ. By the dual of 19.3.5 (b), s''' belongs to Σ. Also, s''' is a monomorphism (7.8.2). One obtains $\alpha = P(s')^{-1} P(f') P(s''')^{-1}$. Now, one can assume that s' or, resp., s''' belong to a chosen system of representatives for the equivalence classes of epimorphisms with domain B or, resp., monomorphisms with codomain A. This implies that $[P(A), P(B)]$ is isomorphic to a $\mathfrak{U}$-set.

19.4 Interrelation with Subcategories

19.4.1 Definition. A *subcategory* $\mathscr{N}$ of $\mathscr{C}$ is called *strictly full* if every object of $\mathscr{C}$ that is isomorphic to an object of $\mathscr{N}$ belongs to $\mathscr{N}$ ($\mathscr{N}$ is closed with respect to isomorphic objects).

19.4.2 Remarks. (a) Let $\mathscr{X}$ be a full subcategory of $\mathscr{C}$. By 16.3.6, $\mathscr{X}$ is contained in a strictly full subcategory $\mathscr{N}$ of $\mathscr{C}$ in such a way that the inclusion $\mathscr{X} \subset \mathscr{N}$ is an equivalence. The inclusion $\mathscr{X} \subset \mathscr{C}$ has a left adjoint if and only if this is the case for $\mathscr{N} \subset \mathscr{C}$ (16.4.2, 16.5.9).

(b) Let $R\colon \mathscr{C} \to \mathscr{C}$ be a functor with a natural transformation $\eta\colon 1_{\mathscr{C}} \to R$ and let $I\colon \mathscr{X} \to \mathscr{C}$ be the inclusion of a full subcategory such that R factors through I, say $R = I\, S$. S is a left adjoint of I with the adjunction transformation η if and only if the following holds:

If $f\colon A \to Y$ is any $\mathscr{C}$-morphism whose codomain is in $\mathscr{X}$, then there is exactly one $g\colon R(A) \to Y$ with $f = g\, \eta_A$.

(1)

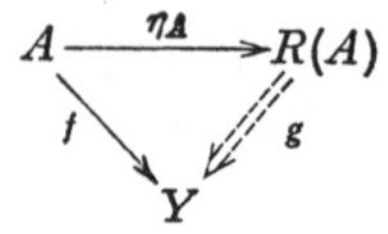

This follows immediately from 16.4.5, the only difference being that here we do not distinguish between $Y \in |\mathcal{X}|$ and $I(Y) \in |\mathcal{C}|$ or between g and $I(g)$.

Let us assume that there is such an adjoint situation. Since I is fully faithful, the adjunction transformation ε, which is quasi-inverse to η, is an isomorphism. By 13.5.5, $\eta * I$ and $S * \eta$ are isomorphisms and $I * \varepsilon * S$ has $\eta * R = R * \eta\colon R \to R\,R$ as its inverse isomorphism. For $Y \in |\mathcal{X}|$, η_Y is an isomorphism; the converse holds, provided $|\mathcal{X}|$ is strictly full. Since

$$(1') \qquad \eta_Y\, g = R(g)\, \eta_{R(A)} = R(g)\, R(\eta_A) = R(f)\,,$$

(1) implies $g = \eta_Y^{-1}\, R(f)$.

(c) We assume that there is an $R\colon \mathcal{C} \to \mathcal{C}$ with a natural transformation $\eta\colon 1_{\mathcal{C}} \to R$ such that $\eta * R\colon R \to R\,R$ is an isomorphism and $\eta * R = R * \eta$. The objects Y, for which η_Y is an isomorphism, are the objects of a strictly full subcategory $\mathcal{X}$ of $\mathcal{C}$. The inclusion $I\colon \mathcal{X} \to \mathcal{C}$ has a left adjoint S with $I\,S = R$ and adjunction transformation $\eta\colon 1_{\mathcal{C}} \to I\,S$.

For, by assumption, the objects $R(A)$ are in $\mathcal{X}$. $\mathcal{X}$ is obviously strictly full, and (1) is valid with $g = \eta_Y^{-1}\, R(f)$, as is shown by (1′) and $R(f)\, \eta_A = \eta_Y\, f$.

Remark. If η is an epimorphism, then $\eta * R = R * \eta$ follows from 16.1.1 (5).

(d) If $\mathcal{X}$ is a full subcategory of $\mathcal{C}$ and if the inclusion $I\colon \mathcal{X} \to \mathcal{C}$ has a left adjoint S, then, by 16.6.5 and 16.6.6, S can be chosen so that $S\,I = 1_{\mathcal{X}}$ and so that $\varepsilon\colon S\,I \to 1_{\mathcal{X}}$ is the identity transformation of $1_{\mathcal{X}}$.

For $R = I\,S$, the following then holds:

$$(2) \qquad R\,R = R\,,$$

$$(3) \qquad R \mid \mathcal{X} = 1_{\mathcal{X}}; \qquad \varepsilon_Y = 1_Y \qquad \text{for } Y \in |\mathcal{X}|\,,$$

$$(4) \qquad \eta_Y = 1_Y \qquad \text{for } Y \in |\mathcal{X}|\,,$$

the latter because of (3), $Y = I(Y)$ and 16.5.5 (4).

(e) Let $\mathcal{N}, \mathcal{M}, \mathcal{C}$ be any categories. If functors $T''\colon \mathcal{N} \to \mathcal{M}$, $T'\colon \mathcal{M} \to \mathcal{C}$ have left adjoints $S''\colon \mathcal{M} \to \mathcal{N}$ and $S'\colon \mathcal{C} \to \mathcal{M}$, then $S = S''\,S'$ is left adjoint to $T = T'\,T''$ (16.4.2). With the evident meaning for η'', η', η and $\varepsilon'', \varepsilon', \varepsilon$ one gets from 16.5.1 and 16.5.1⁰

$$(5) \qquad \eta = (T' * \eta'' * S')\, \eta'\,,$$

$$\varepsilon = \varepsilon''\, (S'' * \varepsilon' * T'')\,.$$

(f) Let $T'\colon \mathcal{M} \to \mathcal{C}$ be fully faithful and $T''\colon \mathcal{N} \to \mathcal{M}$ be given. If $T = T'\,T''$ has a left adjoint S, then $S'' = S\,T'$ is left adjoint to T'', as is shown by the isomorphisms

$$[S\,T'(M),\, N]_{\mathcal{N}} \cong [T'(M),\, T(N)]_{\mathcal{C}} \cong [M,\, T''(N)]_{\mathcal{M}}\,.$$

19.4.3 Definition. Let Σ be a subclass of Mor $\mathscr{C}$. An *object* N of $\mathscr{C}$ is called *left closed* with respect to Σ if $[s, N]_{\mathscr{C}}$ is bijective for every $s \in \Sigma$.

19.4.4 Lemma. *Let $\Sigma \subset$ Mor $\mathscr{C}$ and let $(\varphi, S, T, \mathscr{B}, \mathscr{C})$ be an adjunction. If S takes all the morphisms of Σ into isomorphisms, then $T(X)$ is left closed with respect to Σ for every $X \in |\mathscr{B}|$.*

Proof. For $s \in \Sigma$, $[S(s), X]$ is an isomorphism. φ shows $[s, T(X)]$ to be an isomorphism.

19.4.5 Proposition. *Assume that $\Sigma \subset$ Mor $\mathscr{C}$ admits a calculus of left fractions. Let $P\colon \mathscr{C} \to \mathscr{C}[\Sigma^{-1}]$ be the canonical functor.*

(a) *For $N \in |\mathscr{C}|$, the following are equivalent:*
   (i) *N is left closed with respect to the saturation Ξ of Σ.*
   (ii) *N is left closed with respect to Σ.*
   (iii) *$[s, N]_{\mathscr{C}}$ is surjective for every $s \in \Sigma$ with domain N.*
   (iv) *The objects of N/Σ are coretractions.*
   (v) *$P_{A,N}\colon [A, N]_{\mathscr{C}} \to [P(A), P(N)]_{\mathscr{C}[\Sigma^{-1}]}$ is bijective for every $A \in |\mathscr{C}|$.*

(b) *Σ admits a terminal calculus of fractions if and only if for every $B \in |\mathscr{C}|$ there is an $\eta_B\colon B \to T(B)$ such that $\eta_B \in \Sigma$ and $T(B)$ is left closed with respect to Σ.*

(c) *Assume that Σ admits a terminal calculus of fractions. Let T be right adjoint to P and let $\eta\colon 1_{\mathscr{C}} \to T\,P$ be the corresponding adjunction transformation. $B \in |\mathscr{C}|$ is left closed with respect to Σ if and only if η_B is an isomorphism.*

Further, let $\mathscr{N}$ be the full subcategory of $\mathscr{C}$ whose objects are those that are left closed with respect to Σ and let $I\colon \mathscr{N} \to \mathscr{C}$ be the inclusion. Then there is a factorization $T = I\,T'$ of T, where $T'\colon \mathscr{C}[\Sigma^{-1}] \to \mathscr{N}$ is an equivalence. Furthermore, $T'\,P$ is left adjoint to I and $P\,I$ is equivalence-inverse to T'.

Proof. (a) Trivially, (i) implies (ii) and (ii) implies (iii). If (iii) holds and if $s\colon N \to C$ is in Σ, then there is a $p\colon C \to N$ with $p\,s = 1_N$, and (iv) is true.

Now let (iv) be satisfied. For $f, g\colon A \to N$ in $\mathscr{C}$ let $P(f) = P(g)$, i.e., $[1_N | f] = [1_N | g]$. By 19.2.2 (4), there is a $t\colon N \to C$ in Σ with $t\,f = t\,g$. Since t is a coretraction and thus a monomorphism, $f = g$ follows and $P_{A,N}$ is injective. $P_{A,N}$ is also surjective: For $[s\,|\,f]\colon P(A) \to P(N)$ let p be a retraction corresponding to s. Then $[s\,|\,f] = [p\,s | p\,f] = [1_N\,|\,p\,f] = P(p\,f)$. Thus (v) holds.

Let (v) be satisfied and let $s\colon A \to B$ be in Ξ. Then

$$\begin{array}{ccc} [B, N]_{\mathscr{C}} & \xrightarrow{P_{B,N}} & [P(B), P(N)]_{\mathscr{C}[\Sigma^{-1}]} \\ {\scriptstyle [s,N]}\downarrow & & \downarrow{\scriptstyle [P(s),P(N)]} \\ [A, N]_{\mathscr{C}} & \xrightarrow{P_{A,N}} & [P(A), P(N)]_{\mathscr{C}[\Sigma^{-1}]} \end{array} \tag{6}$$

is commutative and $[s, N]$ is an isomorphism, because $P_{A,N}$, $P_{B,N}$ and $[P(s), P(N)]$ are isomorphisms. Thus we have deduced (i).

(b) First, let the given condition be satisfied. For s in B/Σ, $[s, T(B)]$ is bijective. Therefore, there is exactly one u with $u\ s = \eta_B$, and η_B is terminal in B/Σ.

To prove the converse, let η_B be terminal in B/Σ for every $B \in |\mathscr{C}|$. As in the proof of 19.2.8 (c), $B \mapsto T(B)$ can be extended to a functor T which is right adjoint to P. What remains to be proved follows from 19.4.4.

(c) If η_B is an isomorphism, then B is left closed, since so is $TP(B)$ (19.4.4). Now let B be left closed. By (a) and 19.3.3, we can assume that Σ is saturated. Since $P(\eta_B)$ is an isomorphism, $\eta_B \in \Sigma$. By (iv) in (a), η_B is a coretraction, and by 16.5.4 (b), η_B is an isomorphism.

T has a factorization $T = I\ T'$, by 19.4.4. Since T is fully faithful, so is T'. By 16.3.6 and by what was proved above, T' is an equivalence. Let J be equivalence-inverse to T'. Then $T'\ P$ is left adjoint to $T\ J = I\ T'\ J$ (16.4.2) and therefore also to I. Further, $(P\ I)\ T' = P\ T$ is isomorphic to the identity functor of $\mathscr{C}[\Sigma^{-1}]$, since T is fully faithful. $P\ I$ is then equivalence-inverse to T', because $P\ I \cong P\ I\ T'\ J \cong J$.

19.4.6 Remarks. (a) Let $\mathfrak{N}$ be a class of objects of the category $\mathscr{C}$. Let the class of morphisms Σ consist of all morphisms s such that $[s, N]_{\mathscr{C}}$ is bijective for every $N \in \mathfrak{N}$.

If $\mathscr{C}$ is finitely cocomplete or, resp., cocomplete, then Σ is a saturated class which admits a calculus or, resp., a strong calculus of left fractions. This is verified easily (8.7.4, 19.3.5 (b)).

(b) Let $\Sigma \subset \operatorname{Mor} \mathscr{C}$ and let $\mathfrak{N}$ or, resp., $\mathscr{L}$ be the full subcategory of $\mathscr{C}$ whose objects are those that are left closed with respect to Σ, or, resp., those objects L for which $[s, L]_{\mathscr{C}}$ is a monomorphism for all $s \in \Sigma$. $\mathscr{N}$ and $\mathscr{L}$ are obviously strictly full in $\mathscr{C}$. $\mathscr{N}$ is closed with respect to limits in $\mathscr{C}$ (those which exist); $\mathscr{L}$ also, since limits of monomorphisms are monomorphisms (7.1.9). If further $m\colon A \to L$ is a monomorphism in $\mathscr{C}$ and if $L \in |\mathscr{L}|$, then $A \in |\mathscr{L}|$ ($\mathscr{L}$ is "closed with respect to subobjects"). With some additional conditions on $\mathscr{C}$ one can infer that $\mathscr{L}$ is epireflective in $\mathscr{C}$ (16.6.3 (b)).

(c). (a) and (b) provide a bijection between saturated classes of morphisms of $\mathscr{C}$ which admit a terminal calculus of fractions and strictly full subcategories for which the inclusion has a left adjoint. By 19.4.5 (c) and 19.3.1 (g), this takes care of all adjoint situations $(\varphi, S, T, \mathscr{B}, \mathscr{C})$ with a fully faithful T, at least up to an equivalence.

19.4.7 Proposition. *Assume that in the category $\mathscr{C}$ every morphism can be factored uniquely up to an isomorphism into an epi- and a monomorphism. Let $\mathscr{N}$ be a full subcategory of $\mathscr{C}$, assume that the inclusion I:*

$\mathcal{N} \to \mathcal{C}$ has a left adjoint S, and let η: $1_{\mathcal{C}} \to I\,S$ be the corresponding adjunction transformation.

(a) *The objects M, for which η_M is a monomorphism, are the objects of a strictly full, epireflective subcategory $\mathcal{M}$ of $\mathcal{C}$.*

(b) *The inclusion I'': $\mathcal{N} \to \mathcal{M}$ has a faithful left adjoint S''.*

Proof. (a) $\mathcal{M}$ is obviously strictly full. For $A \in |\mathcal{C}|$, η_A factors into an epimorphism η'_A: $A \twoheadrightarrow R(A)$ and a monomorphism η''_A: $R(A) \to$ $\to I\,S(A)$ in such a way that $\eta'_M = 1_M$ for $M \in |\mathcal{M}|$. Since the factorization of morphisms is natural (12.4.10), given f: $A \to B$, there is exactly one morphism $R(f)$: $R(A) \to R(B)$ with $\eta'_B f = R(f)\,\eta'_A$. Thus R is a functor, and since $R \mid \mathcal{M} = 1_{\mathcal{M}}$, one finds oneself in the situation of 19.4.2 (c), (d).

(b) Obviously, $\mathcal{N} \subset \mathcal{M}$. S'' exists, by 19.2.4 (f), and η'' is the corresponding adjunction transformation (more precisely: $\bar{\eta}$ with $I' * \bar{\eta} = \eta'' = \eta * I'$ for I': $\mathcal{M} \subset \mathcal{C}$). I' reflects monomorphisms. By the dual of 16.5.3, S'' is faithful, which can also be seen directly. For S'', 19.2.6 applies.

19.4.8 Remarks. (a) Obviously, corresponding statements can be made if there is a natural factorization of morphisms in $\mathcal{C}$ into epimorphisms and equalizers or into coequalizers and monomorphisms (or similar situations).

(b) Let Σ be the class of morphisms that are taken into isomorphisms by S. If Σ admits a calculus of right fractions, then under the assumptions of 19.4.7, the category $\mathcal{M}$ coincides with the category $\mathcal{L}$ of 19.4.6 (b), as is easily verified. (Considei the analogues of 19.4.5 (ii) through (v).)

19.5 Additivity and Exactness

19.5.1 Proposition. *Let $\Sigma \subset$ Mor $\mathcal{C}$ admit a calculus of left fractions, let P: $\mathcal{C} \to \mathcal{C}[\Sigma^{-1}]$ be the canonical functor.*

(a) *If $\mathcal{C}$ has a zero object, it is preserved by P.*

(b) *If $\mathcal{C}$ is additive, then there is exactly one additive structure on $\mathcal{C}[\Sigma^{-1}]$ such that P is additive.*

(c) *If Σ is the class of morphisms that is taken into isomorphisms by the additive functor S: $\mathcal{C} \to \mathcal{B}$ (where $\mathcal{C}$* and *$\mathcal{B}$ are additive), then, using* (b), *S': $\mathcal{C}[\Sigma^{-1}] \to \mathcal{B}$ with $S'\,P = S$ is also additive.*

Proof. (a) follows from 19.2.8 (a).

(b) By 19.2.4, the uniqueness of the additive structure follows from the fact that filtered colimits are formed in *AB* the same way as in *ENS* (9.4.7). For every pair (A, B) of objects in $\mathcal{C}$, let $[P(A), P(B)]_{\mathcal{C}[\Sigma^{-1}]}$ have the additive structure provided by 19.2.4 for the filtered colimit in *AB*. It remains to be shown that this addition is distributive on both sides (1.5.1 (4)), which can be done using representatives. For

$[s \mid f], [r \mid h]: P(A) \to P(B)$ in $\mathscr{C}[\Sigma^{-1}]$, we may assume $s = r$, by 19.2.2 (5). Thus left distributivity, $(\alpha + \beta)\,\gamma = \alpha\,\gamma + \beta\,\gamma$ for $\alpha = [s \mid f]$, $\beta = [r \mid h]$ and $\gamma = [t \mid g]$, follows immediately from 19.2.2 (3). Right distributivity follows analogously if one takes into account that for

$$D \xleftarrow{\;t\;} A \underset{h}{\overset{f}{\rightrightarrows}} B$$

with $t \in \Sigma$ there are morphisms f', h', t' with $t' \in \Sigma$ and $f'\,t = t'\,f$, $h'\,t = t'\,h$. This follows from 19.2.1 (ii) and (iii), since B/Σ is filtered. P is additive by construction.

(c) $S'([s \mid f]) = S'(P(s)^{-1}\,P(f)) = S(s)^{-1}\,S(f)$. This and 19.2.2 (5) imply the statement after a simple calculation.

19.5.2 Proposition. *Let $\Sigma \subset \operatorname{Mor} \mathscr{C}$ admit a calculus of left and of right fractions, let $P: \mathscr{C} \to \mathscr{C}[\Sigma^{-1}]$ be the canonical functor.*

(a) *Assume that in $\mathscr{C}$ every morphism can be factored into an epimorphism followed by an equalizer. Then $\alpha \in \operatorname{Mor} \mathscr{C}[\Sigma^{-1}]$ is an epimorphism if and only if there is an isomorphism γ in $\mathscr{C}[\Sigma^{-1}]$ and an epimorphism f' in $\mathscr{C}$ such that $\alpha = \gamma\, P(f')$.*

(b) *If $\mathscr{C}$ is an exact category, then $\mathscr{C}[\Sigma^{-1}]$ is an exact category and P is an exact functor. If $\mathscr{D}$ is another exact category and $G: \mathscr{C}[\Sigma^{-1}] \to \mathscr{D}$ a functor, then G is exact if and only if $G\,P$ is exact.*

(c) *If $\mathscr{C}$ is abelian, then $\mathscr{C}[\Sigma^{-1}]$ is abelian and P is exact and additive.*

Proof. (a) Let (s, f) be a representative of α, α an epimorphism and $f = f''\,f'$, where f' is an epimorphism, f'' an equalizer. One has $P(s)\,\alpha = P(f'')\,P(f')$. Here $P(s)\,\alpha$ is an epimorphism, and thus so is $P(f'')$. Since P preserves equalizers (19.2.8 (a) dual), $P(f'')$ is an epimorphic equalizer and thus an isomorphism. Now $\alpha = P(s)^{-1}\,P(f'')\,P(f')$, $P(s)^{-1}\,P(f'')$ is an isomorphism and $P(f')$ is an epimorphism (19.2.8 (a) and 7.8.9 dual). The converse is evident.

(b) By 19.2.8 (a) and its dual, $\mathscr{C}[\Sigma^{-1}]$ has a zero object and P preserves kernels and cokernels and thus exact sequences. If $\alpha = [s \mid f]$ and if $f = f''\,f'$ is a factorization of f into a cokernel and a kernel, then $P(f')$ and $P(s)^{-1}\,P(f'')$ represent a corresponding factorization for α. Therefore, $\mathscr{C}[\Sigma^{-1}]$ is exact. Let $0 \to X \xrightarrow{\zeta} Y \xrightarrow{\xi} Z \to 0$ be a short exact sequence in $\mathscr{C}[\Sigma^{-1}]$. Then there exists a short exact sequence $0 \to A \xrightarrow{a} Y \xrightarrow{c} C \to 0$ in $\mathscr{C}$ and isomorphisms α, γ in $\mathscr{C}[\Sigma^{-1}]$ such that the following diagram is commutative:

$$(1)\qquad \begin{array}{ccccccccc} 0 & \longrightarrow & P(A) & \xrightarrow{P(a)} & P(Y) & \xrightarrow{P(c)} & P(C) & \longrightarrow & 0 \\ & & \downarrow{\scriptstyle\alpha} & & \| & & \downarrow{\scriptstyle\gamma} & & \\ 0 & \longrightarrow & X & \xrightarrow{\;\xi\;} & Y & \xrightarrow{\;\zeta\;} & Z & \longrightarrow & 0 \end{array}$$

To get this, construct γ and c as in (a) and then $a\colon A \to Y$ as a kernel of c. Since P is exact and γ an isomorphism, there exists an α as indicated, and α is an isomorphism. Now, if $G\,P$ is exact, G is exact, by (1). The converse is trivial.

(c) follows from (b), 19.2.8 (a) and 13.3.2 (d).

19.5.3 Proposition. *Let $\Sigma \subset \operatorname{Mor} \mathcal{C}$ admit a calculus of right fractions and a terminal calculus of fractions. If $\mathcal{C}$ is a Grothendieck category, then $\mathcal{C}[\Sigma^{-1}]$ is also one. If $\mathcal{C}$ has a generator G, then $P(G)$ is a generator of $\mathcal{C}[\Sigma^{-1}]$.*

Proof. By 19.2.8 (c), P has a fully faithful right adjoint T. By 19.5.2, P is exact. T is additive, by 16.5.10. Thus the first statement follows from 16.6.2. The second one follows from the dual of 15.3.4 (a).

19.5.4 Definition. Let $\mathcal{C}$ be an abelian category. A non-empty full *subcategory* $\mathcal{K}$ of $\mathcal{C}$ is called *thick* if the following holds: If

(2) $$0 \to K' \xrightarrow{u} K \xrightarrow{p} K'' \to 0$$

is a short exact sequence in $\mathcal{C}$, then $K \in |\mathcal{K}|$ if and only if K' and K'' belong to $\mathcal{K}$.

Since $\mathcal{K} \neq \emptyset$, all zero objects belong to $\mathcal{K}$, and $\mathcal{K}$ is strictly full in $\mathcal{C}$.

Let the morphism class $\Sigma(\mathcal{K})$ corresponding to $\mathcal{K}$ consist of those morphisms s for which the domain of ker s and the codomain of coker s are in $\mathcal{K}$. We write $\mathcal{C}/\mathcal{K}$ instead of $\mathcal{C}[\Sigma(\mathcal{K})^{-1}]$, which is justified by the following proposition.

19.5.5 Proposition. *Let $\mathcal{K}$ be a thick subcategory of the abelian category $\mathcal{C}$, let $\Sigma(\mathcal{K})$ be the corresponding morphism class and let $P\colon \mathcal{C} \to \mathcal{C}/\mathcal{K}$ be the canonical functor.*

(a) *$\Sigma(\mathcal{K})$ admits a calculus of left and of right fractions. P is exact and additive.*
(b) *$\Sigma(\mathcal{K})$ is saturated. An object K of $\mathcal{C}$ belongs to $\mathcal{K}$ if and only if $P(K)$ is a zero object.*
(c) *If $\mathcal{D}$ is an exact category and $F\colon \mathcal{C} \to \mathcal{D}$ an exact functor, then F can be factored through P if and only if all objects of $\mathcal{K}$ are taken into zero objects by F. Here the factorization $F = G\,P$ is unique.*

Proof. (a) By 19.5.2 (c), the second statement follows from the first one.

(i) $\Sigma(\mathcal{K})$ contains all identity morphisms, because $\mathcal{K} \neq \emptyset$.

(ii) $\Sigma(\mathcal{K})$ is compositive. Let $s\colon A \to B$ and $t\colon B \to C$ be in $\Sigma(\mathcal{K})$.

Then let $k'\colon K' \to A$, $m\colon M \to B$ and $k\colon K \to A$ be kernels of s, t or, resp., $t\,s$. The following diagram is commutative:

(3)

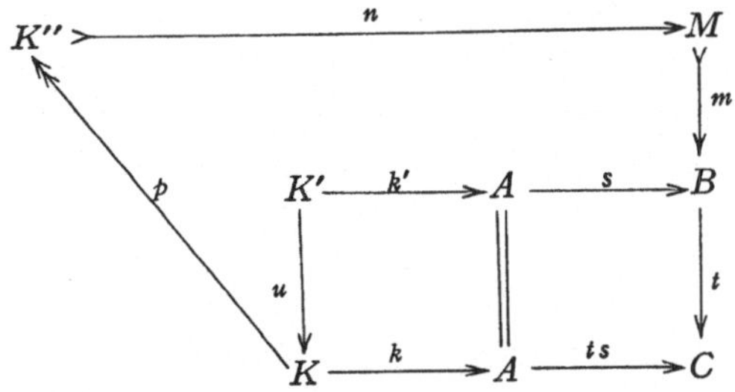

Here, u exists by the definition of kernels, and so does $v\colon K \to M$ with $m\,v = s\,k$. One gets p and n by factoring v into an epi- and a monomorphism.

By 13.1.6 (c), u is a kernel of $s\,k = m\,n\,p$. Since $m\,n$ is a monomorphism, u is a kernel of p. K' and M are assumed to be objects of $\mathcal{K}$. Thus K'' is an object of $\mathcal{K}$ by means of n, and K is an object of $\mathcal{K}$ by means of u and p. This, together with the dual conclusion for cokernels, implies $t\,s \in \Sigma(\mathcal{K})$.

(iii) Let $A' \xleftarrow{s} A \xrightarrow{g} D$ be given with $s \in \Sigma(\mathcal{K})$. If 19.2.1 (1) is formed as a pushout, then according to 13.4.8 one gets isomorphic cokernel objects for s and s' and an epimorphism $h\colon K \twoheadrightarrow L$, where K or, resp., L is a kernel object for s or, resp., s'. Thus $s' \in \Sigma(\mathcal{K})$ follows.

(iv) Given $f\colon A \to B$, we assume that there is an $s \in \Sigma(\mathcal{K})$ with $f\,s = 0$. We consider

(4)
$$\begin{array}{ccccccc} \xrightarrow{s} & A & \xrightarrow{\text{coker}\, s} & A'' & \to & 0 & \\ & \| & & \downarrow u & & & \\ & A & \xrightarrow{\text{coim}\, f} & D & \xrightarrow{\text{im}\, f} & B & \xrightarrow{\text{coker}\, f} C\,. \end{array}$$

Since $f\,s = 0$, $(\text{coim}\, f)\,s = 0$. Therefore, u exists and is an epimorphism. Now $D \in |\mathcal{K}|$ follows from $A'' \in |\mathcal{K}|$. Since $\text{im}\, f = \ker(\text{coker}\, f)$, $\text{coker}\, f \in \Sigma(\mathcal{K})$ follows.

Since $\mathcal{C}$ is additive, (i) through (iv) imply that $\Sigma(\mathcal{K})$ admits a calculus of left fractions. The statement about right fractions is dual.

(b) Let $s\colon A \to B$ be in the saturation of $\Sigma(\mathcal{K})$. By 19.3.3 (a), there is a $t\colon B \to C$ such that $t\,s \in \Sigma(\mathcal{K})$. With the notations in (3) one has $K \in |\mathcal{K}|$ and also $K' \in |\mathcal{K}|$, because both k and u are monomorphisms. Together with the dual conclusion, $s \in \Sigma(\mathcal{K})$ follows. Therefore, $\Sigma(\mathcal{K})$ is saturated.

For $K \in |\mathcal{K}|$, $0\colon K \to K$ is in $\Sigma(\mathcal{K})$. Since P is exact, $P(K)$ is a zero object. The converse follows from the fact that $\Sigma(\mathcal{K})$ is saturated: If $P(K)$ is a zero object, then $0\colon K \to K$ is in $\Sigma(\mathcal{K})$ and thus K is in $\mathcal{K}$.

(c) If F takes all objects of $\mathcal{K}$ into zero objects, then $F(s)$ is an isomorphism for every $s \in \Sigma(\mathcal{K})$, since F is exact. By 19.1.2, there is exactly one functor G with $F = G\,P$. The converse is evident.

19.5.6 Remarks. Let $\mathcal{C}$ be abelian. If $\Gamma \subset \operatorname{Mor}\mathcal{C}$ admits a calculus of left and of right fractions, then by 19.3.3 (c) and its dual, the same is true for the saturation Σ of Γ. For $P\colon \mathcal{C} \to \mathcal{C}[\Sigma^{-1}]$, the objects that are taken into zero objects by P are the objects of a thick subcategory $\mathcal{K}$ of $\mathcal{C}$, as 19.5.2 (c) shows. By 19.5.5 (a), (b), in every abelian category $\mathcal{C}$ there is a bijection between the thick subcategories and those saturated morphism classes which admit a calculus of left and of right fractions.

19.5.7 Proposition. *Let $\mathcal{C}$ be abelian and $\mathcal{K}$ a thick subcategory. Let $I'\colon \mathcal{M} \to \mathcal{C}$ be the inclusion of the strictly full subcategory $\mathcal{M}$ of $\mathcal{C}$ such that $M \in |\mathcal{M}|$ if and only if $[s, M]$ is a monomorphism for all $s \in \Sigma(\mathcal{K})$. Let $\mathcal{N}$ be the strictly full subcategory whose objects are those that are left closed with respect to $\Sigma(\mathcal{K})$.*

(a) *The objects M of $\mathcal{M}$ are characterized by $[K, M] = 0$ for all $K \in |\mathcal{K}|$.*

(b) *$\mathcal{M}$ is closed in $\mathcal{C}$ with respect to limits (as far as they exist), "subobjects" and essential extensions. If*

$$0 \to M' \xrightarrow{m} M \xrightarrow{c} M'' \to 0 \tag{5}$$

is exact and if $M', M'' \in |\mathcal{M}|$, then $M \in |\mathcal{M}|$.

(c) *If (5) is exact, $M \in |\mathcal{M}|$ and $M' \in |\mathcal{N}|$, then $M'' \in |\mathcal{M}|$.*

(d) *If (5) is exact, $M \in |\mathcal{N}|$* and *$M'' \in |\mathcal{M}|$, then $M' \in |\mathcal{N}|$.*

(e) *If $M \in |\mathcal{M}|$ and if $q\colon M \to Q$ is an injective envelope of M in $\mathcal{C}$, then $Q \in |\mathcal{N}|$ and $P(q)\colon P(M) \to P(Q)$ is an injective envelope in $\mathcal{C}/\mathcal{K}$ ($P\colon \mathcal{C} \to \mathcal{C}/\mathcal{K}$ is the canonical functor).*

(f) *If $\mathcal{C}$ is complete and well-powered, then $\mathcal{M}$ is epireflective in $\mathcal{C}$.*

(g) *If, for every $B \in |\mathcal{C}|$, there is a maximal monomorphism $m_B\colon K_B \to B$ with domain in $\mathcal{K}$, then $\mathcal{M}$ is epireflective in $\mathcal{C}$.*

(h) *Let $\mathcal{C}$ be cocomplete and well-powered. Then the condition in (g) is satisfied if and only if $\mathcal{K}$ is closed in $\mathcal{C}$ with respect to coproducts (and thus with respect to colimits).*

Proof. (a) Since H_M takes cokernels into kernels, the statement follows from the definition of $\Sigma(\mathcal{K})$ in 19.5.4 and the fact that, for $K \in |\mathcal{K}|$, $0 \to K$ always belongs to $\Sigma(\mathcal{K})$.

(b) Since H^K preserves limits and is thus left exact, the second statement follows immediately from (a); the first one follows from (a) and 15.2.3 (b).

(c) We consider the diagram

(6)

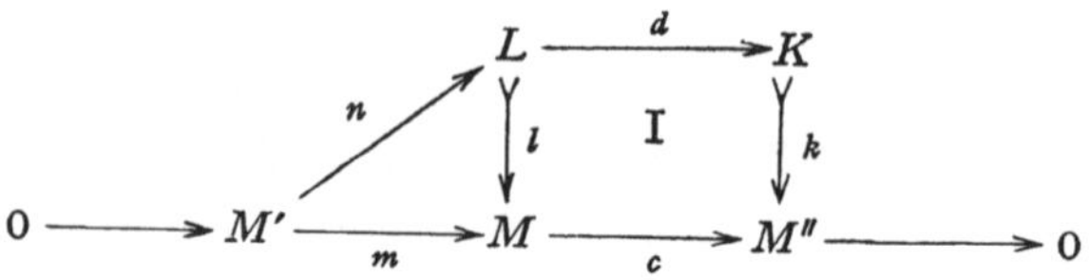

where k is a monomorphism, $K \in |\mathcal{K}|$ and I a pullback. By 7.8.2, l is a monomorphism, and by 13.4.3 (c), d is an epimorphism. n exists and is a kernel of d, by 12.3.4 (d). Now, $n \in \Sigma(\mathcal{K})$. Since $M' \in |\mathcal{N}|$ and by 19.4.5 (iv), n is a coretraction. By the dual of 13.2.4, d is a projection for L as a biproduct of M' and K. With the corresponding injection, $K = 0$ follows, by (a). And again by (a), $M'' \in |\mathcal{M}|$ follows.

(d) We consider

$$\begin{array}{ccccccccc} 0 & \longrightarrow & M' & \xrightarrow{m} & M & \xrightarrow{c} & M'' & \longrightarrow & 0 \\ & & \downarrow{\scriptstyle s} & & \| & & \downarrow{\scriptstyle h} & & \\ & & X & \xrightarrow{u} & M & \xrightarrow{v} & Y & \longrightarrow & 0\,. \end{array} \tag{7}$$

First, let $s \in \Sigma(\mathcal{K})$ be given. According to the definition of $\mathcal{N}$ (19.4.3), and because $M \in |\mathcal{N}|$, u exists. Let v be a cokernel of u. h exists, by the definition of cokernels. Applying P shows that $h \in \Sigma(\mathcal{K})$. h is a monomorphism, since $M'' \in |\mathcal{M}|$, and because $h\,c = v$, h is an isomorphism. Now, by the definition of kernels, u factors uniquely through m, and s is a coretraction. By 19.4.5 (iv), $M' \in |\mathcal{N}|$.

(e) By (b), $Q \in |\mathcal{M}|$. If $s: Q \to A$ is in $\Sigma(\mathcal{K})$, then s is a monomorphism. Since Q is injective, 1_Q shows s to be a coretraction. By 19.4.5 (iv), $Q \in |\mathcal{N}|$.

We show now that $P(q)$ is essential. Since every morphism of $\mathcal{C}/\mathcal{K}$ is of the form $P(s)^{-1}\,P(f)$, it suffices, by 15.2.3, to show the following: If for

$$P(M) \xrightarrow{P(q)} P(Q) \xrightarrow{P(f)} P(N)$$

$P(f\,q)$ is a monomorphism, then $P(f)$ is a monomorphism. Let $v\,u = f\,q$ be a factorization of $f\,q$ into an epimorphic u and a monomorphic v. $P(u)$ is an epimorphism (19.2.8 (a)) and also a monomorphism by the assumption made about $P(f\,q)$. Therefore, u is in $\Sigma(\mathcal{K})$. Since $M \in |\mathcal{M}|$, u is a monomorphism and thus an isomorphism. From this, $f\,q$ is a monomorphism and, by 15.2.3, so is f. This implies that $P(f)$ is a monomorphism.

Finally, we show that $P(Q)$ is injective. Let $\mu: P(A) \rightarrowtail P(B)$ be a monomorphism and suppose that $\alpha: P(A) \to P(Q)$. By the dual of 19.5.2 (a), one can assume that $\mu = P(m)$ with a monomorphic m. By 19.4.5 (v), there is an $f: A \to Q$ with $P(f) = \alpha$. Since Q is injective, there is a $v: B \to Q$ with $v\,m = f$. It thus follows that $P(Q)$ is injective.

(f) follows immediately from (b) and 16.6.3 (c). (Note 12.4.4).

(g) Choosing maximal monomorphisms with domain in $\mathcal{K}$ in such a way that 1_K is chosen for $K \in |\mathcal{K}|$, and choosing cokernels of these monomorphisms such that 1_M is chosen for $M \in |\mathcal{M}|$ gives functors $F: \mathcal{C} \to \mathcal{C}$ and $R: \mathcal{C} \to \mathcal{C}$ and the exact sequence

$$0 \to F \xrightarrow{\varkappa} 1_{\mathcal{C}} \xrightarrow{\eta'} R \to 0\,. \tag{8}$$

Further, $F\,F = F$ and $R\,R = R$ with $F * \varkappa = 1_F$ and $\eta' * R = 1_R$. For $\mathcal{M}$, R, η', (a) shows one to be in the situation of 19.4.2 (c). For $\mathcal{K}$, F, $\varkappa$, it is the dual situation.

(h) If $\mathcal{K}$ is closed with respect to coproducts, then (g) follows from 14.2.5 by forming the coproduct of the domains for a set of representatives of the equivalence classes of monomorphisms with domain in $\mathcal{K}$ and codomain B. The converse follows from the dual of 16.6.7 and the proof of (g).

19.6 Localization in Abelian Categories

19.6.1 Definition. Let $\mathcal{C}$ be an abelian category. A thick *subcategory* $\mathcal{K}$ of $\mathcal{C}$ is called *localizing* if the canonical functor $P\colon \mathcal{C} \to \mathcal{C}/\mathcal{K}$ has a right adjoint T.

19.6.2 Remarks. First, let $\mathcal{K}$ only be thick and let $I\colon \mathcal{N} \to \mathcal{C}$ be the inclusion of the strictly full subcategory whose objects are those which are left closed with respect to $\Sigma(\mathcal{K})$. By 19.4.5 (b), (c), $\mathcal{K}$ is localizing if and only if I has a left adjoint $S\colon \mathcal{C} \to \mathcal{N}$.

Let this be the case. By 19.4.2 (d), we can assume that S and the corresponding adjunction transformation η are chosen such that $\eta_N = 1_N$ for $N \in |\mathcal{N}|$. We then call S a *localizing functor*.

By 19.5.2 and 19.5.4, $\mathcal{N}$ is abelian although cokernels are in general not formed as in $\mathcal{C}$; i.e., I need not be exact. By 11.6.5, the addition in $\mathcal{N}$ as an abelian category is the one coming from $\mathcal{C}$. S is exact and additive. Instead of the canonical functor $P\colon \mathcal{C} \to \mathcal{C}/\mathcal{K}$ we consider from now on always $S\colon \mathcal{C} \to \mathcal{N}$. The corresponding adjunction transformation $\eta\colon 1_{\mathcal{C}} \to I\,S$ can be factored as in 19.4.7. Using the notation employed there, one has the exact sequences

$$(1) \qquad 0 \to K_B \xrightarrow{\ker \eta_B} B \xrightarrow{\eta'_B} I'\,S'(B) \to 0\,,$$
$$0 \to I'\,S'(B) \xrightarrow{\eta''_B} I\,S(B) \xrightarrow{\operatorname{coker} \eta_B} C_B \to 0$$

with $\eta_B = \eta''_B\,\eta'_B$. Since $S * \eta$ is an isomorphism, K_B and C_B are objects of $\mathcal{K}$. If $m\colon A \to B$ is any morphism whose domain is in $\mathcal{K}$, then $S(m) = 0$, by 19.5.5, and $\eta_B\,m = I\,S(m)\,\eta_A = 0$ implies that m factors through $\ker \eta_B$.

Thus for $B \in |\mathcal{C}|$, $\ker \eta_B$ is a maximal monomorphism whith domain in $\mathcal{K}$. Here the categories $\mathcal{M}$ of 19.4.7 and of 19.5.7 are identical, as is shown by a comparison of (1) with 19.5.7. In particular η' in (1) coincides with η' in 19.5.7 (8).

19.6.3 Lemma. *Let $\mathcal{C}$ be abelian and $\mathcal{K}$ a thick subcategory. The two following statements are equivalent:*
(a) *$\mathcal{K}$ is localizing.*

(b) *For every object B of $\mathscr{C}$ the following holds: Among the monomorphisms with codomain B and domain in $\mathscr{K}$ there is a maximal one. If this one is zero, then there is a monomorphism from B to an object that is left closed with resepct to $\Sigma(\mathscr{K})$.*

Proof. By the preceding remarks, (b) follows from (a). Now let (b) be satisfied. By 19.5.7 (g), there exists an epireflector $S'\colon \mathscr{C} \to \mathscr{M}$ with a corresponding adjunction transformation $\eta'\colon 1_{\mathscr{C}} \to I'\,S'$. $\eta'_B\colon B \to I'\,S'(B)$ belongs to $\Sigma(\mathscr{K})$ for all $B \in |\mathscr{C}|$. According to the second condition in (b), there is a monomorphism $i\colon I'\,S'(B) \to N$, so that N is left closed. We consider

(2)
$$\begin{array}{ccccccccc} & & & & N' & \xrightarrow{\;c'\;} & K & & \\ & & & \nearrow^{\eta''_B} & \big\downarrow{\scriptstyle n} & \text{I} & \big\downarrow{\scriptstyle m} & & \\ 0 & \longrightarrow & I'S'(B) & \xrightarrow[\;i\;]{} & N & \xrightarrow[\;c\;]{} & C & \longrightarrow & 0\,. \end{array}$$

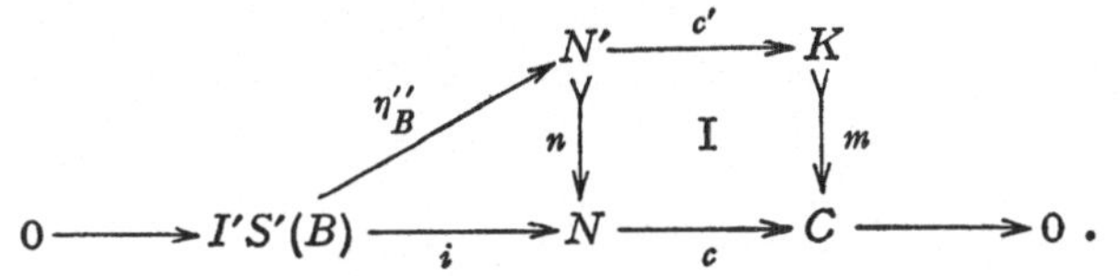

Here c is a cokernel of i, m a maximal monomorphism with domain in $\mathscr{K}$, and I a pullback. It follows, as in 19.5.7 (6), that η''_B exists and is a kernel of c'. $\eta'_C\colon C \to I'\,S'(C)$ is a cokernel of m. By 13.4.3 (c), I is bicartesian, and by the dual of 12.3.4 (d), $\eta'_C\,c$ is a cokernel of n. Since $I'\,S'(C) \in |\mathscr{M}|$ and by 19.5.7 (d), N' is left closed with respect to $\Sigma(\mathscr{K})$. Since $I'\,S'(B) \in |\mathscr{M}|$, $\eta''_B \in \Sigma(\mathscr{K})$, and thus $\eta_B = \eta''_B\,\eta'_B$ also. (a) thus follows from 19.4.5 (b).

19.6.4 Proposition. *Let $\mathscr{C}$ be an abelian category with injective envelopes and $\mathscr{K}$ a thick subcategory.*

(a) *$\mathscr{K}$ is localizing if and only if for every $B \in |\mathscr{C}|$ there is a maximal monomorphism with codomain B and domain in $\mathscr{K}$.*

(b) *If this is the case, then $\mathscr{C}/\mathscr{K}$ also has injective envelopes. The right adjoint of the canonical functor P preserves injective objects.*

(c) *If, in addition, $\mathscr{K}$ is closed with respect to injective envelopes, then P preserves injective objects.*

Proof. (a) follows immediately from 19.6.3 and 19.5.7 (e).

(b) The first statement follows from 19.5.7 (e), for, by (1) and 19.3.3 (b), every object of $\mathscr{C}/\mathscr{K}$ is isomorphic to one of the form $P(M)$ (with $M \in |\mathscr{M}|$). The second statement is valid by 15.3.4 (b).

(c) Let Q be injective, $k\colon K \to Q$ a maximal monomorphism with domain in $\mathscr{K}$ and $h\colon K \to H$ an injective envelope of K. Since h is essential and Q injective, there exists a monomorphism $i\colon H \to Q$ with $k = i\,h$ (15.2.3). By the dual of 10.4.6 and by 12.6.3, Q has a representation as a biproduct $H \oplus J$, where i is an injection. Here J is also injective. Every monomorphism with codomain J and domain in $\mathscr{K}$ is a zero morphism by the choice of K. The injective envelope of J is 1_J, and by 19.5.7 (e), $P(J)$ is injective. If the additional assumption about $\mathscr{K}$ is made, then h is an isomorphism and $P(H) = 0$.

Remark. Without the additional assumption for $\mathcal{K}$ one finds that every injective object Q of $\mathcal{C}$ is isomorphic to one of the form $H \oplus T(J')$, where H is the injective envelope of an object of $\mathcal{K}$ and J' is injective in $\mathcal{C}/\mathcal{K}$. For above, J is left closed with respect to $\Sigma(\mathcal{K})$ and $T\,P(J) \cong$ $\cong J$, by 19.4.5 (c).

19.6.5 Theorem. *Let $\mathcal{C}$ be a Grothendieck category with a generator G and $\mathcal{K}$ a thick subcategory. Then the following are equivalent:*
(a) *$\mathcal{K}$ is localizing.*
(b) *$\mathcal{K}$ is closed with respect to coproducts (and thus colimits).*
If this is the case, then $\mathcal{C}/\mathcal{K}$ is a Grothendieck category with generator $P(G)$.

Proof. By 15.3.7, the conditions in 19.6.4 are satisfied. Furthermore, $\mathcal{C}$ is well-powered (10.6.3). By 19.5.7 (h) and 19.6.4 (a), (a) and (b) are equivalent. (b) follows from (a) more simply by 19.5.5 (b), since $P: \mathcal{C} \to \mathcal{C}/\mathcal{K}$ preserves colimits. The last statement follows from 19.5.3.

Remark. By 10.5.3, the existence of a generator is equivalent to the existence of a generating set.

19.6.6 Lemma. *Let $\mathcal{C}$ be an abelian category. $\mathcal{M}$ is assumed to be a strictly full epireflective subcategory satisfying the following condition:*
(*) *For every $M \in |\mathcal{M}|$ there is a monomorphism $m: M \to Q$ such that $Q \in |\mathcal{M}|$ and Q is injective in $\mathcal{C}$.*

Let $I': \mathcal{M} \to \mathcal{C}$ be the inclusion, $S': \mathcal{C} \to \mathcal{M}$ the epireflector, $R = I'\,S'$ and η' the corresponding epimorphic adjunction transformation. Further, let $\mathcal{K}$ be the full subcategory of $\mathcal{C}$ such that $K \in |\mathcal{K}|$ if and only if $[K, M] =$ $= 0$ for all $M \in |\mathcal{M}|$.
(a) *If $m: M' \rightarrowtail M$ is a monomorphism in $\mathcal{C}$ and $M \in |\mathcal{M}|$, then $M' \in |\mathcal{M}|$. If*

$$0 \to M' \xrightarrow{m} M \xrightarrow{p} M'' \to 0 \tag{3}$$

is exact in $\mathcal{C}$ and if $M', M'' \in |\mathcal{M}|$, then $M \in |\mathcal{M}|$.
(b) *For every $B \in |\mathcal{C}|$,* $\ker \eta'_B$ *is a maximal monomorphism with domain in $\mathcal{K}$.*
(c) *$\mathcal{K}$ is thick, and an application of 19.5.7 (a) to $\mathcal{K}$ produces $\mathcal{M}$ again.*

Proof. (a) For the first statement, η'_M is an isomorphism (16.6.5). Since $\eta'_M\, m = R(m)\, \eta'_{M'}$, $\eta'_{M'}$ is a monomorphism; it is also an epimorphism by assumption. Since $\mathcal{M}$ is strictly full, $M' \in |\mathcal{M}|$.

For the second statement let $m': M' \rightarrowtail Q$ be a monomorphism as in (*). We consider

$$\begin{array}{ccccccccc} 0 & \longrightarrow & M' & \xrightarrow{m} & M & \xrightarrow{p} & M'' & \longrightarrow & 0 \\ & & {\scriptstyle m'}\downarrow & \text{I} & \downarrow{\scriptstyle \bar{m}'} & & \| & & \\ 0 & \longrightarrow & Q & \xrightarrow[\bar{m}]{} & X & \xrightarrow[\bar{p}]{} & M'' & \longrightarrow & 0 \end{array} \tag{4}$$

in $\mathscr{C}$. Here I is constructed as a pushout, p and $\bar{p}$ are cokernels of m and $\bar{m}$ (12.3.4 (d) dual). By the dual of 13.4.3 (c), $\bar{m}$ and $\bar{m}'$ are monomorphisms. Since Q is injective, the lower line in (4) splits (10.4.6 dual). By 13.2.4, $X \overset{\approx}{\Rightarrow} Q \oplus M''$. Thus $X \in |\mathscr{M}|$ follows, because R is additive. By what was proved before, $M \in |\mathscr{M}|$.

(b) Let $k: K \to B$ be a kernel of $\eta'_B: B \to R(B)$. Every morphism $f: A \to B$ with $A \in |\mathscr{K}|$ factors through K, because $\eta'_B f = 0$ by the definition of $\mathscr{K}$. It remains to be shown that $K \in |\mathscr{K}|$. Let $f: K \to M$ be any morphism with $M \in |\mathscr{M}|$. The diagram

$$(5) \qquad \begin{array}{ccccccccc} 0 & \longrightarrow & K & \xrightarrow{k} & B & \xrightarrow{\eta'_B} & R(B) & \longrightarrow & 0 \\ & & {\scriptstyle f}\downarrow & \mathrm{I} & \downarrow{\scriptstyle \bar{f}} & & \| & & \\ 0 & \longrightarrow & M & \xrightarrow[\bar{k}]{} & X & \longrightarrow & R(B) & \longrightarrow & 0 \end{array}$$

is formed the same way as (4), with I as a pushout. By (a), $X \in |\mathscr{M}|$. By 19.4.2 (1), there is a $g: R(B) \to X$ with $g\,\eta'_B = \bar{f}$. $0 = \bar{f}\,k = \bar{k}\,f$ and thus $f = 0$ follows, since $\bar{k}$ is a monomorphism. By the definition of $\mathscr{K}$, $K \in |\mathscr{K}|$.

(c) Let $0 \longrightarrow K' \xrightarrow{k} K \xrightarrow{q} K'' \to 0$ be exact in $\mathscr{C}$. If $K', K'' \in |\mathscr{K}|$, then applying $[?, M]$ with $M \in |\mathscr{M}|$ shows that $K \in |\mathscr{K}|$. Conversely, let $K \in |\mathscr{K}|$. As above, $K'' \in |\mathscr{K}|$ follows. Let $f': K' \to M$ be any morphism with $M \in |\mathscr{M}|$ and $m: M \rightarrowtail Q$ a monomorphism as in (*). Since Q is injective, there is an $f: K \to Q$ with $f\,k = m\,f'$. Since $Q \in |\mathscr{M}|$ and $K \in |\mathscr{K}|$, $f = 0$. m being a monomorphism implies $f' = 0$ and thus $K' \in |\mathscr{K}|$. Therefore, $\mathscr{K}$ is thick. The last statement follows from (b) and the proof of 19.5.7 (g).

Remark. By 19.6.4 (a) and 19.5.7 (b), for an abelian category with injective envelopes there is a bijection between localizing subcategories and those strictly full epireflective subcategories that are closed with respect to injective envelopes.

19.6.7 Example. Let $\mathscr{C} = Ab$ and let $\mathscr{M}$ be the subcategory of torsion-free (additive) groups. The assumptions of 19.6.6 and 19.6.4 are satisfied, because essential extensions of torsion-free groups are obviously torsion-free. Here $\mathscr{K}$ is the subcategory of torsion groups. One shows easily that the objects of $\mathscr{N}$ are divisible (proof of 15.3.2) and are therefore the torsion-free injective groups. This classical example is the model for 19.6.6 and 19.5.7.

We note here that the category of torsion-free additive groups is an example for a complete and cocomplete additive category with a generator which is not abelian. It is not balanced; in it the inclusion $\mathbf{Z} \subset \mathbf{Q}$ is bimorphic, for instance.

The category $\mathscr{C}'$ of finitely generated abelian groups is abelian with a projective generator. It is not complete, not cocomplete, and it has no injective objects. $\mathcal{C}'$ is equivalent to a small category. The subcategory $\mathcal{K}'$ of finitely generated torsion groups is thick, but not localizing in $\mathcal{C}'$. The inclusion $\mathcal{K}' \subset \mathcal{C}'$ has a full right adjoint. Therefore, $\mathcal{K}'$ is equivalent to a quotient category of $\mathcal{C}'$ (remark in 19.2.6). Notice that this is an example of an adjoint situation which is not an equivalence and where one of the two functors is fully faithful and the other one is full.

19.6.8 Theorem. (Gabriel). *Let $\mathcal{B}$ be a small abelian category, $l(\mathcal{B}, Ab)$ the category of left exact functors $\mathcal{B} \to Ab$. The inclusion $l(\mathcal{B}, Ab) \subset$ $\subset Add(\mathcal{B}, Ab)$ has an exact left adjoint. $l(\mathcal{B}, Ab)$ is a Grothendieck category with a generator.*

Remark. The inclusion $l(\mathcal{B}, Ab) \subset Add(\mathcal{B}, Ab)$ exists, by 13.3.2 (d). By 12.2.7, the left exact functors are those preserving finite limits.

Proof. $\mathcal{C} = Add(\mathcal{B}, Ab)$ is a Grothendieck category (14.6.9). It has a generator, which is even in $l(\mathcal{B}, Ab)$ (15.4.1), and thus has injective envelopes (15.3.7). Let $\mathcal{M}$ be the strictly full subcategory of additive monofunctors (15.4.3). $\mathcal{M}$ is closed with respect to products.

If $m: M' \rightarrowtail M$ is a monomorphism in $\mathcal{C}$ with $M \in |\mathcal{M}|$, then $M' \in |\mathcal{M}|$, which follows immediately from 10.1.4. By 10.6.3, 16.6.3 (c) and 15.4.4, $\mathcal{M}$ satisfies the conditions of 19.6.6. The corresponding subcategory $\mathcal{N}$ of $\mathcal{C}$ (as in 19.5.7, 19.6.4) is a Grothendieck category with a generator, by 19.6.5 and 19.6.2. So the statement will follow, provided we show that $\mathcal{N} = l(\mathcal{B}, Ab)$.

So let (3) be an exact sequence in $\mathcal{C}$ with $M \in |\mathcal{M}|$ and M injective in $\mathcal{C}$. For $M' \in |\mathcal{N}|$ or, resp., $M' \in |l(\mathcal{B}, Ab)|$, there is always a monomorphism $M' \rightarrowtail M$ for a suitable M. Further, let

$$0 \to X \xrightarrow{u} Y \xrightarrow{v} Z \to 0$$

be any short exact sequence in $\mathcal{B}$. We consider

$$(6)\qquad \begin{array}{ccccccc} & & 0 & & 0 & & 0 \\ & & \downarrow & & \downarrow & & \downarrow \\ 0 \to & & M'(X) & \to & M'(Y) & \to & M'(Z) \\ & & \downarrow & & \downarrow & & \downarrow \\ 0 \to & & M(X) & \to & M(Y) & \to & M(Z) \to 0 \\ & & \downarrow & & \downarrow & & \\ 0 \to & & M''(X) & \to & M''(Y) & & \\ & & \downarrow & & \downarrow & & \\ & & 0 & & 0 & & \end{array}$$

Here the columns are exact. The middle row is exact, since M is injective (15.4.3). By 13.5.2, the first row is exact if and only if the third row is exact. Now, if $M' \in |\mathcal{N}|$, then by 19.5.7 (c), $M'' \in |\mathcal{M}|$, and by (6), M' is left exact. If, conversely, M' is left exact, $M'' \in |\mathcal{M}|$ follows. By 19.5.7 (e), $M \in |\mathcal{N}|$. By 19.5.7 (d), $M' \in |\mathcal{N}|$, which completes the proof.

19.6.9 Theorem. (Mitchell). *Every small abelian category $\mathcal{B}$ has an exact full embedding in a module category.*

Proof. By 10.3.4, 10.3.5 and 10.3.10, the Yoneda embedding H^*: $\mathcal{B}^0 \to Add(\mathcal{B}, Ab)$ induces a full exact embedding $\mathcal{B}^0 \to l(\mathcal{B}, Ab)$. Let $\mathcal{C}$ be the dual category of $l(\mathcal{B}, Ab)$. One then has a full exact embedding $J: \mathcal{B} \to \mathcal{C}$. By 19.6.8 and 15.3.7, $l(\mathcal{B}, Ab)$ has an injective cogenerator, and thus $\mathcal{C}$ has a projective generator G'. Since $\mathcal{B}$ is small and $\mathcal{C}$ is cocomplete, $G = \coprod G'_e$, where $G'_e = G'$ and $e \in \bigcup [G', J(B)]$, is also a projective generator The union is to be formed over all $B \in |\mathcal{B}|$. Every $J(B)$ is finitely generated (even singly generated) with respect to G. With the exact embedding 17.5.5, $\mathcal{C} \to Mod_R$, one obtains the desired embedding for $\mathcal{B}$.

19.6.10 Remarks. The assumption in 19.6.9 that $\mathcal{B}$ be small can be forced by a change of universe. 15.4.5 turns out to be a corollary of 19.6.9. However, the preliminaries supplied by 15.4 were used here also. 15.4.5 permits the reduction of statements about exactness of diagrams in an abelian category to such statements in Ab, the lemmas in 13.5 are examples of this. 19.6.9 enables one to reduce not only statements regarding exactness in diagrams, but also statements about existence and naturality of additional morphisms to module categories. The connecting homomorphism for the homology of short exact sequences of chain complexes is an example (compare 13.5.9). A change of universe can be avoided by making use of 15.5.7.

However, homological algebra requires, as in the case of spectral sequences, for instance, farther reaching techniques that permit direct proofs of statements involving diagrams, where infinite intersections and unions are also admissible. Exact squares and categories of correspondences are a point of departure for this.

19.7 Characterization of Grothendieck Categories with a Generator

19.7.1 Definition. Let $\mathcal{C}$ be an abelian category and $\mathfrak{N}$ a class of objects of $\mathcal{C}$. .An *object* K of $\mathcal{C}$ is called *negligible* with respect to $\mathfrak{N}$ if the following holds:

(*) If $f: A \to K$ is any morphism with codomain K, then $[\ker f, N]$ is bijective for every $N \in \mathfrak{N}$.

19.7.2 Proposition. *Let $\mathcal{C}$ be an abelian category and $\mathfrak{N}$ a class of objects of $\mathcal{C}$.*

(a) *The objects that are negligible with respect to $\mathfrak{N}$ are the objects of a thick subcategory $\mathcal{K}$ of $\mathcal{C}$.*

(b) *The objects in* $\mathfrak{N}$ *are left closed with respect to* $\Sigma(\mathcal{K})$. *If* $\mathfrak{N}$ *is replaced by the class* $\mathfrak{N}'$ *of all the objects that are left closed with respect to* $\Sigma(\mathcal{K})$, *then the objects of* $\mathcal{K}$ *are also negligible with respect to* $\mathfrak{N}'$.

(c) *If* $\mathcal{C}$ *is a Grothendieck category with a generating set* $\mathfrak{G}$, *then* $K \in |\mathcal{K}|$ *even if* (*) *is required only for every morphism whose domain is a finite coproduct of objects in* $\mathfrak{G}$.

(d) *If* $\mathcal{C}$ *is a Grothendieck category with a generating set* $\mathfrak{G}$ *of small objects, then* $\mathcal{K}$ *is localizing.*

Proof. Here we make continuous use of the fact that $H_N = [?, N]_{\mathcal{C}}$ takes colimits into limits and that an exact sequence $A \to B \to C \to 0$ goes into an exact sequence $0 \to [C, N] \to [B, N] \to [A, N]$.

(a) We prove four auxiliary statements (i) through (iv).

(i) If $K \in |\mathcal{K}|$, then $[K, N] = 0$ for every $N \in \mathfrak{N}$. To verify this, let $f = 1_K$ in (*).

(ii) If $u\colon K' \rightarrowtail K$ is a monomorphism and $K \in |\mathcal{K}|$, then $K' \in |\mathcal{K}|$ too.

This is so, because $f'\colon A \to K'$ and $u f'$ have the same kernels.

(iii) If $p\colon K \to K''$ is an epimorphism and $K \in |\mathcal{K}|$, then $K'' \in |\mathcal{K}|$ also.

For $f\colon A \to K''$ we consider the following diagram:

(1)
$$\begin{array}{ccccccccc} & & & & H & = & H & & \\ & & & & {\scriptstyle h'}\downarrow & & \downarrow{\scriptstyle h} & & \\ & & K' & \xrightarrow{u'} & B & \xrightarrow{p'} & A & \longrightarrow & 0 \\ & & \| & & {\scriptstyle f'}\downarrow & \mathrm{I} & \downarrow{\scriptstyle f} & & \\ 0 & \longrightarrow & K' & \xrightarrow[u]{} & K & \xrightarrow[p]{} & K'' & \longrightarrow & 0\,. \end{array}$$

Here I is a pullback, u, u', h, h' are kernels of p or, resp., p', f, f'. The two equalities are justified by 13.4.8 (e) for a suitable choice of kernels. By (ii), $K' \in |\mathcal{K}|$. By (i), $[p', N]$ is an isomorphism for all $N \in \mathfrak{N}$. Since $[h', N]$ is also an isomorphism, (*) follows for K''.

(iv) If $0 \to K' \xrightarrow{u} K \xrightarrow{p} K'' \to 0$ is exact and if $K', K'' \in |\mathcal{K}|$, then $K \in |\mathcal{K}|$.

For $f\colon A \to K$ we consider

(2)
$$\begin{array}{ccccccccc} & & H & = & H & & & & \\ & & {\scriptstyle h'}\downarrow & & \downarrow{\scriptstyle h} & & & & \\ 0 & \longrightarrow & A' & \xrightarrow{u'} & A & \xrightarrow{p'} & C & \longrightarrow & 0 \\ & & {\scriptstyle f'}\downarrow & \mathrm{I} & \downarrow{\scriptstyle f} & & \downarrow{\scriptstyle m} & & \\ 0 & \longrightarrow & K' & \xrightarrow[u]{} & K & \xrightarrow[p]{} & K'' & \longrightarrow & 0\,. \end{array}$$

Here I is a pullback, h and h' are kernels of f and f', and p' is a cokernel of u'. By 13.4.8 (d), m exists and is a monomorphism. By (ii), $C \in |\mathcal{K}|$.

Therefore, $[u', N]$ is an isomorphism for $N \in \mathfrak{N}$. $[h', N]$ is also an isomorphism. Thus (*) follows for K.

By (ii), (iii), (iv), $\mathcal{K}$ is thick.

(b) Let $0 \to K \xrightarrow{k} A \xrightarrow{f} B \xrightarrow{c} C \to 0$ be exact, K and C in $\mathcal{K}$, and let $f = f'' f'$ be a factorization of f into an epi- and a monomorphism. Since $C \in |\mathcal{K}|$, $[f'', N]$ is an isomorphism for $N \in \mathfrak{N}$. Furthermore, since $[K, N] = 0$, $[f', N]$ is an isomorphism. Thus N is left closed with respect to $\Sigma(\mathcal{K})$. Conversely, let this be the case. For $K \in |\mathcal{K}|$ and $f\colon A \to K$, $\ker f \in \Sigma(\mathcal{K})$, which proves the second statement in (b).

(c) Let $G = \coprod G_i$ be a coproduct of objects of $\mathfrak{G}$, G_J a finite subcoproduct with injection in_J. Let $h'\colon H' \to G$ be a kernel of $g\colon G \to K$ and h'_J a kernel of $g\, in_J$. Now, G is a filtered colimit of its finite subcoproducts G_J and therefore, h' is a filtered colimit of the kernels h'_J. If, for any $N \in \mathfrak{N}$, $[h'_J, N]$ is always an isomorphism, then $[h', N] = \operatorname{Lim}\,[h'_J, N]$ is an isomorphism.

Now take an $f\colon A \to K$. For a suitable coproduct $G = \coprod G_i$ of objects in $\mathfrak{G}$ there is an epimorphism $p\colon G \twoheadrightarrow A$. We consider

(3)
$$\begin{array}{ccccccc}
 & & L & = & L & & \\
 & & {\scriptstyle l'}\downarrow & & \downarrow{\scriptstyle l} & & \\
0 & \longrightarrow & H' & \xrightarrow{h'} & G & \xrightarrow{fp} & K \\
 & & {\scriptstyle p'}\downarrow & \mathrm{I} & \downarrow{\scriptstyle p} & & \| \\
0 & \longrightarrow & H & \xrightarrow[h]{} & A & \xrightarrow[f]{} & K\,.
\end{array}$$

Here h, h', l, l' are kernels of f, fp, p, p'. By 12.3.4 (c), I is a pullback. Therefore, p' is an epimorphism (13.4.3 (c)). If $[h', N]$ is an isomorphism, then $[h, N]$ is an isomorphism. What was said above then implies (c).

(d) It follows from 17.4.5 that a finite coproduct of small objects is small. $\mathcal{K}$ as a thick subcategory is certainly closed with respect to finite coproducts (biproducts). If $K = \coprod K_i$ is a coproduct of objects of $\mathcal{K}$ and G a finite coproduct of objects in $\mathfrak{G}$, then $[\ker f, N]$ is an isomorphism for every morphism $f\colon G \to K$ and every $N \in \mathfrak{N}$, because f factors through the (monomorphic) injection of a finite subcoproduct of K. (c) implies that $K \in |\mathcal{K}|$, and (d) follows from 19.6.5.

19.7.3 Proposition. *Let $\mathcal{C}$ be an abelian category, $\mathcal{K}$ a localizing subcategory, T right adjoint to the canonical functor $P\colon \mathcal{C} \to \mathcal{C}/\mathcal{K}$ and $\mathfrak{N}$ the class of objects $TP(A)$ for $A \in |\mathcal{C}|$. Then the objects of $\mathcal{K}$ are those that are negligible with respect to $\mathfrak{N}$.*

Proof. The objects $TP(A)$ are left closed with respect to $\Sigma(\mathcal{K})$, by 19.4.4. By 19.7.2 (b), the objects of $\mathcal{K}$ are negligible with respect to $\mathfrak{N}$.

Conversely, let K be negligible. Then

$$[P(K), P(K)] \cong [K, TP(K)] = 0 .$$

Thus $P(K)$ is a zero object and therefore, $K \in |\mathcal{K}|$ by 19.5.5 (b).

19.7.4 Definition. Let $\mathcal{C}$ be an arbitrary category, $\Sigma \subset \operatorname{Mor} \mathcal{C}$ and $P\colon \mathcal{C} \to \mathcal{C}[\Sigma^{-1}]$ the canonical functor. A *$\mathcal{C}$-morphism* $f\colon A \to B$ is called *covering* with respect to Σ if $P(f)$ is an epimorphism.

19.7.5 Lemma. (a) *Let $\mathcal{C}$ be an abelian category and $\mathcal{K}$ a thick subcategory. $f\colon A \to B$ is covering with respect to $\Sigma(\mathcal{K})$ if and only if the codomain of* coker f *is in $\mathcal{K}$.*

(b) *Let $\mathcal{C}$ be a Grothendieck category with a generating set $\mathfrak{G}$ of projective objects. $\mathfrak{N}$ and $\mathcal{K}$ are to be given the meaning they have in* 19.7.2. *For $f\colon A \to B$ consider*

$$(4)\qquad \begin{array}{ccccccc} H & \xrightarrow{h'} & A' & \xrightarrow{f'} & G & \xrightarrow{c'} & C' \\ \| & & \downarrow{\scriptstyle b'} & \mathrm{I} & \downarrow{\scriptstyle b} & & \downarrow{\scriptstyle m} \\ H & \xrightarrow{h} & A & \xrightarrow{f} & B & \xrightarrow{c} & C, \end{array}$$

where G is a finite coproduct of objects in $\mathfrak{G}$, b an arbitrary morphism $G \to B$, I *a pullback, and where* (4) *is completed through addition of kernels and cokernels (compare* 13.4.8).

Then f is covering if and only if, for every possible choice of G and b, and for every $N \in \mathfrak{N}$,

$$(5)\qquad 0 \to [G, N] \xrightarrow{[f', N]} [A', N] \xrightarrow{[h', N]} [H, N]$$

is exact (in Ab).

Proof. By 19.5.5 (b), (a) follows immediately from the exactness of P.

(b) Since $[\operatorname{coim} f', N]$ is a kernel of $[h', N]$, (5) is equivalent to $[\operatorname{im} f', N]$ being an isomorphism. $\operatorname{im} f'$ is a kernel of $c\,b = m\,c'$. If f is covering, i.e., $C \in |\mathcal{K}|$, (5) follows. Conversely, since G is projective, every morphism $g\colon G \to C$ is of the form $c\,b$. If (5) is always exact, $C \in |\mathcal{K}|$ follows from 19.7.2 (c).

19.7.6 Theorem (Gabriel-Popescu). *Let $\mathcal{D}$ be a Grothendieck category, $U \in |\mathcal{D}|$ and R the ring $[U, U]$. U is in a canonical way a left module object over R, which gives rise to the functor $T = [_R U, ?]\colon \mathcal{D} \to Mod_R$* (15.1.6). *$T$ has a left adjoint $S\colon Mod_R \to \mathcal{D}$ with $S(?) = ? \otimes_R {}_R U$* (17.7.4). *The following statements are equivalent:*

(a) *U is a generator.*

(b) *T is faithful.*

(c) *T is fully faithful.*

(d) *T is fully faithful and S is exact.*

(e) *The objects annihilated by S are the objects of a localizing subcategory $\mathcal{K}$ of Mod_R, and S is of the form $S = GP$, where $P\colon Mod_R \to \to Mod_R/\mathcal{K}$ is the canonical functor and G an equivalence.*

Remark. Together with 19.6.5 this implies that a category $\mathcal{D}$ is a Grothendieck category with a generator if and only if $\mathcal{D}$ is equivalent to a category which is constructed from a module category by localization. (Cf. 17.4.10 and 17.9.9.)

Proof. (e) implies (d), since P is exact and T differs from the fully faithful right adjoint of P only by an equivalence. Obviously, it remains to be shown that (e) follows from (a). So let U be a generator of $\mathcal{D}$. By lemma 2 in 15.3.7, T is fully faithful. Let $\mathfrak{N}$ be the class of modules of the form $T(X)$ with $X \in |\mathcal{D}|$ and $\mathcal{K}$ the full subcategory of Mod_R whose objects are those that are negligible with respect to $\mathfrak{N}$. By 19.7.2 (d) with $\mathfrak{G} = \{R\}$ and by 17.4.4, $\mathcal{K}$ is localizing. Let $\mathcal{N}$ be the full subcategory of Mod_R whose objects are left closed with respect to $\Sigma(\mathcal{K})$, and let $I\colon \mathcal{N} \to Mod_R$ be the inclusion. By 16.3.8 and 19.7.2 (b), T has a factorization $T = I\,T'$. Let Q be right adjoint to $P\colon Mod_R \to \to Mod_R/\mathcal{K}$. Q has a factorization $Q = I\,Q'$ by 19.4.5 (c), where Q' is an equivalence.

(6)

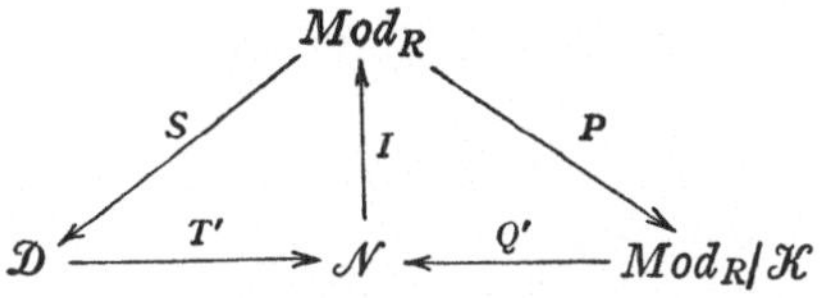

We show that T' is an equivalence. If this is the case, then not only $P' = Q'\,P$ but also $S' = T'\,S$ is left adjoint to I (compare the end of the proof in 19.4.5). It then follows that S' is isomorphic to P' (16.4.4) and that S is isomorphic to a functor obtained from P by following it with an equivalence. By 16.6.6 (with P instead of T), S is of the form $S = G\,P$, where G is also an equivalence. Furthermore, S and P annihilate the same objects of Mod_R. By 19.5.5 (b), these are the objects of $\mathcal{K}$.

It remains to be shown that T' is an equivalence; we do this in five parts (i) through (v) below.

(i) $\mathcal{N}$ is a Grothendieck category with generator R. In Mod_R, $\mathcal{N}$ is closed with respect to limits An $\mathcal{N}$-morphism $f\colon X \to Y$ is an epimorphism in $\mathcal{N}$ if and only if $I(f)$ is covering with respect to $\Sigma(\mathcal{K})$.

Since $R = T(U) \in |\mathcal{N}|$, the first statement follows from the fact that R is a generator in Mod_R and that Q' is an equivalence (19.6.5, 16.2.4). $\mathcal{N}$ is closed with respect to limits, since $\mathcal{N}$ is strictly full in Mod_R and I has a left adjoint. The last statement follows from the fact that $P\,I$ is equivalence-inverse to Q' (19.4.5 (c)).

(ii) $T'\colon \mathcal{D} \to \mathcal{N}$ is exact.

First, T' is left exact, because T preserves limits and I is fully faithful. On account of (i) one still has to show: If $f: A \to B$ in $\mathcal{D}$ is an epimorphism, then $T(f)$ is covering. Now, in Mod_R every finite biproduct with factors R is isomorphic to an object $T(G)$, where G is a finite biproduct of factors U. Every morphism $T(G) \to T(B)$ is of the form $T(b)$, since T is fully faithful. We can therefore form the diagram (4) in $\mathcal{D}$, where f' is also an epimorphism and $C = C' = 0$ (13.4.3). Since T preserves limits,

(4a)
$$\begin{array}{ccccccc} 0 \longrightarrow & T(H) & \xrightarrow{T(h')} & T(A') & \xrightarrow{T(f')} & T(G) \\ & \| & & {\scriptstyle T(b')}\downarrow & \text{I} & \downarrow{\scriptstyle T(b)} \\ 0 \longrightarrow & T(H) & \xrightarrow[T(h)]{} & T(A) & \xrightarrow[T(f)]{} & T(B) \end{array}$$

is commutative with exact rows, and I is a pullback. For an arbitrary $N \in |\mathcal{D}|$, (5) is exact. Since T is fully faithful,

(5a)
$$0 \to [T(G), T(N)] \xrightarrow{[T(f'), T(N)]} [T(A'), T(N)] \xrightarrow{[T(h'), T(N)]} [T(H), T(N)]$$

is exact too. Since R is a small projective generator of Mod_R, 19.7.5 (b) (after changing the notation) yields the fact that $T(f)$ is covering, because in Mod_R every necessary diagram is isomorphic to one of the form (4a).

(iii) If $\{m_i: A_i \rightarrowtail B\}$ is a filtered family of monomorphisms in $\mathcal{D}$ with $\bigcup m_i = 1_B$, then $1_{T'(B)} = \bigcup T'(m_i)$.

Again, let G be a finite biproduct of factors U. The pullbacks

(7i)
$$\begin{array}{ccc} A' & \overset{m_i'}{\rightarrowtail} & G \\ {\scriptstyle b_i'}\downarrow & & \downarrow{\scriptstyle b} \\ A_i & \underset{m_i}{\rightarrowtail} & B \end{array}$$

form a filtered family. Since filtered colimits in $\mathcal{D}$ commute with finite limits, $1_G = \operatorname{Colim} m_i'$. Then, for any $N \in |\mathcal{D}|$,

(8)
$$\operatorname{Lim}[m_i', N] = [\operatorname{Colim} m_i', N] = [1_G, N]$$

holds in Ab. Since T is fully faithful, this implies

(9)
$$[1_{T(G)}, T(N)] = [\operatorname{Colim} T(m_i'), T(N)] .$$

In Mod_R there is a commutative diagram analogous to (4), namely,

(10)
$$\begin{array}{ccccccccc} 0 \longrightarrow & \operatorname{Colim} T(A_i') & \xrightarrow{\operatorname{Colim} T(m_i')} & T(G) & \xrightarrow{c'} & C' & \longrightarrow 0 \\ & \downarrow & \text{I} & \downarrow{\scriptstyle T(b)} & & \downarrow & \\ 0 \longrightarrow & \operatorname{Colim} T(A_i) & \xrightarrow{\operatorname{Colim} T(m_i)} & T(B) & \xrightarrow{c} & C & \longrightarrow 0 . \end{array}$$

Here I is a pullback, because T preserves limits and Mod_R is a Grothendieck category. Furthermore, $\operatorname{Colim} T(m_i)$ and $\operatorname{Colim} T(m_i')$ are monomorphisms. Let c and c' be cokernels in Mod_R. Since I is fully faithful and thus reflects colimits, $I(\text{Colim } T'(m_i)) = \text{Colim } T(m_i)$. By (8) it now follows as in (ii) that Colim $T(m_i)$ is covering. Therefore, Colim $T'(m_i)$ is an isomorphism in $\mathscr{N}$, which implies (iii).

(iv) T' preserves coproducts.

This follows immediately from (iii), since T' preserves finite coproducts (12.2.7, 14.5.4).

(v) T' is an equivalence.

Let $N \in |\mathscr{N}|$. Since R is a generator of $\mathscr{N}$, there is an exact sequence $X \xrightarrow{x} Y \to N \to 0$ in $\mathscr{N}$, where X and Y are suitable coproducts of copies of R. By (iv) and since $R = T(U)$, we may assume that $X = T'(A)$ and $Y = T'(B)$. Then x is of the form $x = T'(f)$, since T' as well as T is fully faithful. By (ii), N is isomorphic to an object $T'(C)$. Thus the statement follows from 16.3.6, which completes the proof.

19.7.7 Remark. In the proof of 19.7.6 and for the auxiliary material for it, 19.6.5 is needed only in the special case $\mathscr{C} = Mod_R$. The general statements in 19.6.5 and 15.3.7 are then corollaries of 19.7.6.

19.8 Problems

19.8.1 Let $\Sigma \subset \text{Mor } \mathscr{C}$. How can the morphisms of $\mathscr{C}\,[\Sigma^{-1}]$ be represented in $\mathscr{C}$? What special form can be chosen, provided 19.2.1 (i) and (ii) are satisfied?

19.8.2 Fill in the details in 19.2.4.

19.8.3 Let $S\colon \mathscr{D} \to \mathscr{C}$ be left adjoint to $T\colon \mathscr{C} \to \mathscr{D}$ with the quasi-inverse adjunction transformations $\varepsilon\colon S\,T \to 1_{\mathscr{C}}$ and $\eta\colon T\,S \to 1_{\mathscr{D}}$. Let $\Sigma = \{\varepsilon_A\}$ and $\Xi = \{\eta_X\}$. Show that $\mathscr{C}[\Sigma^{-1}]$ and $\mathscr{D}[\Xi^{-1}]$ are equivalent.

19.8.4 Carry out 19.4.8 (b).

19.8.5 Let $\mathscr{C}$ be a monoid and $\Sigma \subset \text{Mor } \mathscr{C}$. Investigate $\mathscr{C}\,[\Sigma^{-1}]$ and, in particular, consider the special cases where Σ admits a calculus of left fractions or a calculus of left and of right fractions. What can be said if $\mathscr{C}$ is a commutative monoid? Then consider the analogous additive case, where $\mathscr{C}$ is a ring or, resp., a commutative ring. What is the result for a commutative ring without zero divisors if Σ consists of all elements that are different from zero?

19.8.6 Let $\mathscr{C}$ be an abelian category and $\Sigma \subset \text{Mor } \mathscr{C}$ the class of essential extensions. Then:

(a) Σ admits a calculus of right fractions and is saturated.

(b) Σ admits a calculus of left fractions if and only if Σ consists of isomorphisms only.

(c) Let conditions (a), (b) in 15.5.8 be satisfied for every monomorphism. Show in the given order: In $\mathcal{C}[\Sigma^{-1}]$ every monomorphism is a coretraction, every epimorphism is a retraction, every morphism factors into a retraction and a coretraction, and $\mathcal{C}[\Sigma^{-1}]$ is an abelian category. (Hint: Use 12.6.3). Warning: in general, $P: \mathcal{C} \to \mathcal{C}[\Sigma^{-1}]$ is not exact.

19.8.7 Let $\mathcal{B}$ be a small, additive category and $\mathcal{C}$ a well-powered, complete abelian category. Then the full subcategory of $Add(\mathcal{B}, \mathcal{C})$, whose objects are the monofunctors, is epireflective.

19.8.8 (a) Let $\mathcal{B}$ be an abelian category. For $A \in |\mathcal{B}|$, let $A/\mathcal{M}$ be the full subcategory of $A/\mathcal{B}$ whose objects are the monomorphisms (with domain A). Then $f: A \to B$ gives rise to a functor $f_*: A/\mathcal{M} \to B/\mathcal{M}$ by forming a pushout with f for every $a \in |A/\mathcal{M}|$. If f is a monomorphism, then the functor $f^*: B/\mathcal{M} \to A/\mathcal{M}$ is obtained through composition; i. e., for $b \in |B/\mathcal{M}|$, $f^*(b) = b\,f$. f^* is right adjoint to f_*.

(b) Let $\mathcal{C}$ be a well-powered, cocomplete abelian category and $T: \mathcal{B} \to \mathcal{C}$ an additive functor. Choosing kernels gives a rule $a \mapsto \ker T(a)$, which can be extended to a functor $A/\mathcal{M} \to \mathcal{M}/T(A)$. Then the family $\{\ker T(a)\}_{a \in |A/\mathcal{M}|}$ is filtered. Choosing unions gives

$$\varkappa_{T,A} = \bigcup_{a \in |A/\mathcal{M}|} \ker T(a) : K_T(A) \to T(A)\,.$$

The rule $A \mapsto K_T(A)$ can be extended to a functor $K_T: \mathcal{B} \to \mathcal{C}$ such that $\varkappa_T = \{\varkappa_{T,A}\}$ is a natural transformation $K_T \to T$. [Hint: For $f: A \to B$ and $a \in |A/\mathcal{M}|$, consider $T(f_*(a))$].

(c) Now, let $\mathcal{C}$ be a Grothendieck category. If $f: A \to B$ is a monomorphism, then $K_T(f)$, $T(f)$, $\varkappa_{T,A}$, $\varkappa_{T,B}$ constitute a pullback. Let $\eta'_T: T \to M_T$ be a cokernel of $\varkappa_T$ in $Add(\mathcal{B}, \mathcal{C})$. Then M_T is a monofunctor. Furthermore, the category of monofunctors is epireflective in $Add(\mathcal{B}, \mathcal{C})$ with the adjunction transformation $\eta' = \{\eta'_T\}$.

(d) Now, let $\mathcal{B}$ be small (and $\mathcal{C}$ again a Grothendieck category). If a cokernel, coker a, is assigned to every $a \in |A/\mathcal{M}|$, then one gets a functor $A/\mathcal{M} \to [\mathbf{2}, \mathcal{B}]$, and from $\ker\big(T(\operatorname{coker} a)\big)$ a functor $A/\mathcal{M} \to [\mathbf{2}, \mathcal{C}]$. Factoring $T(a)$ through $\ker\big(T(\operatorname{coker} a)\big)$ yields a functor $A_*: A/\mathcal{M} \to T(A)/\mathcal{C}$. If one forms $R_T(A) = \operatorname{Colim} \Delta^1 A_*$, then this colimit is again a filtered colimit. (Use the addition.) In this way one gets a functor R_T with a natural transformation $\alpha_T: T \to R_T$. Again, R_T is a monofunctor. If T is a monofunctor, then R_T is even left exact. Using R_{R_T}, one finds that the full subcategory of $Add(\mathcal{B}, \mathcal{C})$, whose objects are the left exact functors, can be obtained from *Add* $(\mathcal{B}, \mathcal{C})$ through localizing. (In case of trouble, consult Gabriel [43]).

20. Grothendieck Topologies

20.1 Sieves and Topologies

20.1.1 Conventions. Let $\mathcal{C}$ be a $\mathfrak{U}$-category and $F: \mathcal{C}^0 \to Ens$ a functor. By a *subfunctor* G of F we mean a functor $G: \mathcal{C}^0 \to Ens$ such that $G(X) \subset F(X)$ for all $X \in |\mathcal{C}|$ and such that these inclusions form a natural transformation $i: G \to F$.

We set $\hat{\mathcal{C}} = [\mathcal{C}^0, Ens]$. By means of the Yoneda embedding H_*: $\mathcal{C} \to \hat{\mathcal{C}}$ we regard $\mathcal{C}$ as a full subcategory of $\hat{\mathcal{C}}$ and instead of H_X, H_u we simply write X, u as long as there is no danger of misunderstanding.

20.1.2 Definition. Let $X \in |\mathcal{C}|$. A *sieve* (crible, fr.) R for X is a subfunctor of X (more precisely of H_X).

Note. This identification of an object with the contravariant functor, which is canonically represented by it, is apt to be confusing at first, but it is more than a mere technical convenience. It plays a fundamental role in understanding a number of aspects of category theory. Here what it does is to provide a much more wider notion of "subobject" of an object (together with a canonical choice). Subobjects of X in the usual sense (i. e., in $\mathcal{C}$) clearly determine subfunctors of H_X, but, in general, there are many more. Sieves should be thought of as generalized subobjects of X.

20.1.3 The corresponding morphism class. (a) Let R be a sieve for $X \in |\mathcal{C}|$. For every $Y \in |\mathcal{C}|$, $R(Y)$ is a set of $\mathcal{C}$-morphisms $Y \to X$, since $R(Y) \subset [Y, X]$. If $u: Y \to X$ satisfies $u \in R(Y)$, we say that u belongs to R, and we write $u \in R$. The class of those $\mathcal{C}$-morphisms which belong to R, we call the morphism class corresponding to R. It has the following property:

(S) If $u: Y \to X$ belongs to R and if $v: Z \to Y$ is an arbitrary $\mathcal{C}$-morphism, then $u\,v \in R$.

This follows since $R(v)\,(u) = u\,v$. If, conversely, there is a class R of $\mathcal{C}$-morphisms with codomain X possessing property (S), then one gets a sieve by means of

$$R(Y) = \{u | u\colon\ Y \to X \text{ and } u \in R\}\ ,$$

where $R(v)$ is the map $u \mapsto u\,v$. There is a bijection between the sieves for X and the classes of $\mathcal{C}$-morphisms with codomain X and property (S).

(b) The Yoneda lemma can also be used to describe this correspondence as follows: Let R be a sieve for X. If $u\colon Y \to X$ in $\mathscr{C}$, then $u \in R$ if and only if there is a $\hat{\mathscr{C}}$-morphism $f\colon H_Y \to R$ with

(1) $$i_R f = H_u, \quad i_R\colon R \to H_X \text{ the inclusion}.$$

Here f and u determine each other uniquely. I.e., $u \in R$ if in $\hat{\mathscr{C}}$ u factors through the subobject $R \to X$.

(c) The morphisms $u \in R$ are the objects of a full subcategory $\bar{R}$ of $\mathscr{C}/X$. By (b) and 10.2.1, one has

(2) $$R = \underset{? \in \bar{R}}{\operatorname{Colim}} \left(\Delta^0 H_*(?)\right) \quad \text{in } \hat{\mathscr{C}}.$$

Further, (b) gives a bijection between full subcategories $\bar{R}$ of $\mathscr{C}/X$ with the property (S) for objects and full subcategories of $\mathscr{C}/R$ in $\hat{\mathscr{C}}$ with the pıoperty corresponding to (S).

20.1.4 The sieves for X are ordered by inclusion. X is maximal and the empty sieve $\Phi_{\mathscr{C}}$ is minimal. The correspondence 20.1.3 (a) between sieves and morphism classes in $\mathscr{C}$ is order preserving. This implies that for arbitrary classes of sieves for X intersections and unions always exist, since set theoretical intersections and unions of the corresponding morphism classes in $\mathscr{C}$ always exist.

20.1.5 Inverse images of sieves, change of basis. Let R be a sieve for X with the inclusion $i\colon R \to X$. If $v\colon Y \to X$ is any $\mathscr{C}$-morphism, then $v^{-1}(i)$ can be formed in $\hat{\mathscr{C}}$ in such a way that $v^{-1}(i)$ is the inclusion of a subfunctor $v^{-1}(i)\colon v^{-1}(R) \to Y$ of Y (10.1.6). Therefore, there is the pullback

(3) $$\begin{array}{ccc} v^{-1}(R) & \longrightarrow & R \\ {\scriptstyle v^{-1}(i)}\big\downarrow & & \big\downarrow{\scriptstyle i} \\ Y & \xrightarrow{\;v\;} & X \end{array}$$

in $\hat{\mathscr{C}}$. One says that $v^{-1}(R)$ is obtained from v by a *change of basis* v. We also write $Y \sqcap_X R$ instead of $v^{-1}(R)$, assuming that there is no doubt about the morphisms. Later we shall do the same with other pullbacks (compare 9.5.7).

Since the construction in (3) is pointwise, using 20.1.3 (a), $v^{-1}(R)$ has the following description as a morphism class

(4) $$v^{-1}(R) = \{u \mid u \in \operatorname{Mor} \mathscr{C},\ Y = \operatorname{codomain} u,\ v\,u \in R\}.$$

From this and (S) in 20.1.3, one deduces immediately

(5) $$v^{-1}(R) = Y \quad \text{for } v\colon Y \to X \text{ in } R.$$

20.1.6 Definition. Let $\mathscr{C}$ be a category. A *topology* $\mathfrak{T}$ on $\mathscr{C}$ consists of a class $J(X)$ of sieves for X, for every $X \in |\mathscr{C}|$, such that

(T 1) If $R \in J(X)$ and $v: Y \to X$ is a $\mathcal{C}$-morphism, then $v^{-1}(R) \in J(Y)$ (stability with respect to change of basis).

(T 2) Let $R \in J(X)$ and let R' be another sieve for X. If for every $v: Y \to X$ in R, $v^{-1}(R') \in J(Y)$, then $R' \in J(X)$ (local character).

(T 3) $X \in J(X)$.

The sieves in $J(X)$ are called the *covering sieves* of X or *refinements* of X (with respect to the topology $\mathfrak{T}$). A category $\mathcal{C}$ which is provided with a topology $\mathfrak{T}$ is called a *site*. We denote it by $\mathcal{C}_{\mathfrak{T}}$, or simply by $\mathcal{C}$, as long as $\mathfrak{T}$ remains fixed.

20.1.7 Proposition. *Let $\mathfrak{T}$ be a topology on $\mathcal{C}$, $J(X)$ the class of covering sieves of $X \in |\mathcal{C}|$ with respect to $\mathfrak{T}$.*

(a) *$R \subset R'$ and $R \in J(X)$ imply $R' \in J(X)$.*

(b) *$R, R' \in J(X)$ implies $R \cap R' \in J(X)$.*

I. e., $J(X)$ is a filter in the lattice of sieves for X.

Proof. (a) $R \subset R'$ and (5) imply $v^{-1}(R') = v^{-1}(R)$ for every $v \in R$. Thus the statement follows immediately from (T 2) and (T 3).

(b) First we show

$$(6) \qquad v^{-1}(R') = v^{-1}(R \cap R') \quad \text{foı } v \in R\,.$$

For $f \in v^{-1}(R')$, $vf \in R'$ by (4), and by (S), $vf \in R$; and again by (4), $f \in v^{-1}(R \cap R')$. Conversely, $f \in v^{-1}(R \cap R')$ implies $vf \in R \cap R' \subset R'$ and thus $f \in v^{-1}(R')$.

For R, $R' \in J(X)$, $v^{-1}(R \cap R')$ is a covering sieve for all $v \in R$, by (6) and (T 1). Thus the statement follows from (T 2).

20.1.8 Remarks. (a) By (6), (T 2) is equivalent to the following two conditions:

(T 2i) (T 2) is satisfied if $R' \subset R$.

(T 2ii) $R \in J(X)$ and $R \subset R'$ imply $R' \in J(X)$.

(b) By 20.1.3 (a) and (4), the topology $\mathfrak{T}$ on $\mathcal{C}$ can be described in terms of $\mathcal{C}$ alone, not making any use of $\hat{\mathcal{C}}$. $\hat{\mathcal{C}}$ depends on the choice of the universe $\mathfrak{U}$ (so that $\mathcal{C}$ is a $\mathfrak{U}$-category), whereas this is not so for the sieves as morphism classes, the operations defined on them (20.1.4, 20.1.5) or the possible topologies on $\mathcal{C}$. This follows also from 10.1.8. A proper choice of $\mathfrak{U}$ can always make $\mathcal{C}$ a small $\mathfrak{U}$-category.

(c) In 20.5 we shall elaborate on the relation to topological spaces.

20.1.9 Proposition. *Let $\mathcal{C}$ be a small $\mathfrak{U}$-category and $\tilde{\mathcal{C}}$ a full subcategory of $\hat{\mathcal{C}} = [\mathcal{C}^0, Ens]$. Let the inclusion $I: \tilde{\mathcal{C}} \subset \hat{\mathcal{C}}$ have a left adjoint $A: \hat{\mathcal{C}} \to \tilde{\mathcal{C}}$ which preserves finite limits. If, for $X \in |\mathcal{C}|$, $J(X)$ consists of those sieves $i_R: R \to X$ for which $A(i_R)$ is an isomorphism, then the classes $J(X)$ form a topology on $\mathcal{C}$.*

Proof. (T 3) is evident. (T 1) follows immediately from the fact that A preserves pullbacks. For $i_R\colon R \subset X$, $i_{R'}\colon R' \subset X$, $i\colon R \subset R'$ and $R \in J(X)$, $A(i_R)$ is an isomorphism, and since $i_R = i_{R'}\, i$, $A(i_{R'})$ is a retraction. $A(i_{R'})$ is also a monomorphism, because A preserves monomorphisms. Therefore, (T 2ii) is valid. (T 2i) remains to be shown. So let $R' \subset R \subset X$ and $R \in J(X)$. For an arbitrary $\alpha\colon Y \to R$ in $\hat{\mathscr{C}}$ with $Y \in |\mathscr{C}|$, we consider

(7)
$$\begin{array}{ccccc} Y \sqcap_R R' & \longrightarrow & R' & = & R' \\ \downarrow{\scriptstyle \alpha^{-1}(i)} & & \downarrow{\scriptstyle i} & & \cap \\ Y & \xrightarrow{\alpha} & R & \subset & X. \end{array}$$

Since $|\mathscr{C}|$ is a generating set for $\hat{\mathscr{C}}$ (10.5.2), by 10.5.3, one obtains the following diagram from (7)

(8)
$$\begin{array}{ccc} \coprod\coprod Y_\alpha \sqcap_R R' & \to & R' \\ \downarrow{\scriptstyle \coprod\coprod \alpha^{-1}(i)} & & \downarrow{\scriptstyle i} \\ \coprod\limits_{Y\in|\mathscr{C}|}\ \coprod\limits_{\alpha\in[Y,R]_{\hat{\mathscr{C}}}} Y_\alpha & \twoheadrightarrow & R \end{array}$$

with $Y_\alpha = Y$. i is an equalizer of its cokernel pair (by 18.4.4 and the "pointwise" construction in $\hat{\mathscr{C}}$). Therefore, $A(i)$ is an equalizer. If $A\big(\alpha^{-1}(i)\big)$ is always an isomorphism, then (8) implies that $A(i)$ is an epimorphism too, since A preserves colimits. Therefore, (T 2i) is valid.

20.1.10 We turn now to a proof of the converse of 20.1.9 (20.3.4, 20.3.7). Starting with a topology $\mathfrak{T}$ there will be a construction of the localizing class Σ of the morphisms which are taken into isomorphisms by A. (This is a special case of the non-additive analogue of 19.6.1). Σ is defined in an obvious way (20.2.1). The verification that Σ is, in fact, localizing requires some new considerations.

20.2 Covering Morphisms and Sheaves

Let $\mathscr{C}$ be a $\mathfrak{U}$-category with a topology $\mathfrak{T}$. Recall that in $\hat{\mathscr{C}} = [\mathscr{C}^0, Ens]$ every morphism $c\colon H \to K$ has a canonical factorization into an epimorphism and an inclusion (10.1.6), which we denote by im c.

20.2.1 Definition. The $\hat{\mathscr{C}}$*-morphism* $c\colon H \to K$ is called *covering* (with respect to $\mathfrak{T}$) if the following holds:

For every $X \in |\mathscr{C}|$ and every $\hat{\mathscr{C}}$-morphism $f\colon X \to K$, f^{-1} (im c) is a covering sieve (provided f^{-1} (im c) is chosen as an inclusion).

c is called *bicovering* if c is covering and if the equalizer of the kernel pair of c is covering

Since there is a product in $\hat{\mathscr{C}}/K$ here with a diagonal morphism Δ, we use the following notation

(1)

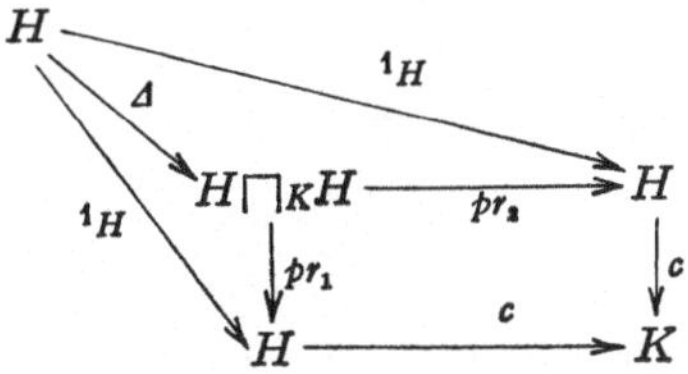

20.2.2 Remarks. (a) By 13.4.4, epimorphisms in $\hat{\mathscr{C}}$ are closed with respect to pullbacks. This, together with 7.8.4 implies

(2) $$f^{-1}(\operatorname{im} c) = \operatorname{im} f^{-1}(c)$$

with the evident definition of $f^{-1}(c)$ as a pullback that we shall make further use of later.

(b) By 7.8.9, every covering monomorphism is bicovering. For $K \in |\mathscr{C}|$ the inclusion of covering sieves are covering monomorphisms, by (T 1); and every covering monomorphism with codomain K is equivalent to the inclusion of a covering sieve. By (T 3), every epimorphism in $\hat{\mathscr{C}}$ is covering, every isomorphism is bicovering.

(c) If m and n are monomorphisms with codomain K, if $m \leq n$ and m is covering, then by 20.2.1 and 20.1.7 (a), n is covering.

20.2.3 Lemma. (a) *For $c\colon H \to K$ in $\hat{\mathscr{C}}$, the following are equivalent:*

(i′) *c is covering,*

(ii′) *for every $f\colon X \to K$ with $X \in |\mathscr{C}|$, there exists a commutative diagram*

(3) $$\begin{array}{ccc} F \twoheadrightarrow R \subset X \\ \downarrow \qquad\qquad \downarrow{\scriptstyle f} \\ H \xrightarrow{\quad c \quad} K, \end{array}$$

where R is a covering sieve of X and $F \twoheadrightarrow R$ is an epimorphism.

(b) *The following are also equivalent:*

(i) *c is bicovering.*

(ii) *c is covering and for*

$G \underset{v}{\overset{u}{\rightrightarrows}} H \xrightarrow{c} K$ with $c\,u = c\,v$

there always exists a covering monomorphism $m\colon F \to G$ with $u\,m = v\,m$. By 20.2.2 (c), the equalizer of u and v is then covering.

Proof. (a) (i′) $\Rightarrow$ (ii′) by definition and (2). (ii′) $\Rightarrow$ (i′): Let $H' \twoheadrightarrow R' \subset X$ be the canonical factorization of $f^{-1}(c)$. From (3) and the definition of $f^{-1}(c)$, one gets the commutative diagram

$$\begin{array}{ccccc} F & \twoheadrightarrow & R & \subset & X \\ \downarrow & & \downarrow & & \| \\ H' & \twoheadrightarrow & R' & \subset & X \\ \downarrow & & \downarrow & & \downarrow{\scriptstyle f} \\ H & \twoheadrightarrow & L & \overset{\mathrm{im}\, c}{\rightarrowtail} & K\,. \end{array}$$

By 20.1.7 (a), (i') thus follows from (ii').

(b) (i) $\Rightarrow$ (ii). Using the notation in (1), there exists a $w: G \to H \sqcap_K H$ with $pr_1\, w = u$, $pr_2\, w = v$. We consider the pullback

$$\begin{array}{ccc} F & \xrightarrow{w'} & H \\ {\scriptstyle m = w^{-1}(\Delta)}\downarrow & & \downarrow{\scriptstyle \Delta} \\ G & \xrightarrow{w} & H \sqcap_K H\,. \end{array}$$

Now, $u\, m = pr_1\, w\, m = pr_1\, \Delta w' = pr_2\, \Delta w' = pr_2\, w\, m = v\, m$. By assumption, Δ is covering. From the definition 20.1.2 it follows immediately that m is covering

(ii) $\Rightarrow$ (i). Consider the special case $G = H \sqcap_K H$, $u = pr_1$, $v = pr_2$ and 20.2.2 (c).

20.2.4 Lemma. *The class of covering morphisms has the following properties:*

(a) *It is closed with respect to pullbacks.*
(b) *It is closed with respect to composition.*
(c) *If $c\, b$ is covering, then c is covering.*

Proof. (a) follows immediately from the definition and (2).

(b) Let $b: G \to H$ and $c: H \to K$ be covering. To show that $c\, b$ is covering, we note that on account of the definition 20.2.1, one is allowed to assume that b is a monomorphism. By (a) and 20.2.1, it then suffices to carry out a proof for $K \in |\mathscr{C}|$.

1st case. Let c be an epimorphism. For $K \in |\mathscr{C}|$, c is a retraction (10.1.7). Let s be a corresponding coretraction. We consider the pullback

$$\begin{array}{ccc} R & \xrightarrow{i} & K \\ {\scriptstyle j}\downarrow & & {\scriptstyle s}\downarrow \uparrow{\scriptstyle c} \\ G & \xrightarrow{b} & H\,. \end{array}$$

Since b is a covering monomorphism, and because $K \in |\mathscr{C}|$, we may assume that $i = s^{-1}(b)$ is the inclusion of a covering sieve R of K. Now one has $c\, b\, j = c\, s\, i = i$. With the canonical factorization $G \twoheadrightarrow R' \subset K$ of $c\, b$, $R \subset R'$ follows. Then, by 20.1.7 (a), $c\, b$ is covering.

2nd case. Let c be an inclusion. For $f\colon Y \to H$ with $Y \in |\mathscr{C}|$ we consider the double pullback

$$\begin{array}{ccccc} R & \longrightarrow & G & = & G \\ {\scriptstyle f^{-1}(b)}\downarrow & & \downarrow{\scriptstyle b} & & \downarrow{\scriptstyle c\,b} \\ Y & \xrightarrow{f} & H & \overset{c}{\rightarrowtail} & K \end{array}$$

Since b is covering, $f^{-1}(b)$ is covering. $c\colon H \to K$ is the inclusion of a covering sieve of $K \in |\mathscr{C}|$. By 20.1.3 (b), $c\,f \in H$. By (T 2), $c\,b$ is covering.

The general case follows by canonically factorizing c.

(c) follows from 20.2.2 (c). since $\operatorname{im} c\,b \leq \operatorname{im} c$.

20.2.5 Lemma. *The class of bicovering morphisms has the following properties:*

(a) *It is closed with respect to pullbacks*
(b) *and with respect to composition.*
(c) *It admits a calculus of right fractions.*
(d) *If, for $b\colon G \to H$ and $c\colon H \to K$, c as well as $c\,b$ are bicovering, then b is bicovering.*

Proof. (a) follows immediately from 18.4.12 and 20.2.4 (a).

(b) Let $b\colon G \to H$ and $c\colon H \to K$ be bicovering. By 20.2.4 (b), $c\,b$ is covering. We show that 20.2.3 (ii) holds. Let $u, v\colon F \to G$ be given with $c\,b\,u = c\,b\,v$. The equalizer $k\colon E \to F$ of $b\,u$ and $b\,v$ is covering, because c is bicovering. The equalizer $k'\colon E' \to E$ of $u\,k$ and $v\,k$ is covering, because b is bicovering. By 20.2.4 (b), $k'\,k$ is covering. Since $u\,k\,k' = v\,k\,k'$, and by 20.2.3 (ii), $c\,b$ is bicovering.

(c) follows from (a), (b) 20.2.2 (b) and 20.2.3 (b).

(d) We consider the following pullback:

$$\begin{array}{ccccc} F & & \xrightarrow{\quad f \quad} & & H \\ {\scriptstyle g}\downarrow & & & & \downarrow{\scriptstyle c} \\ G & \xrightarrow{b} & H & \xrightarrow{c} & K\,. \end{array} \tag{4}$$

By (a), f and g are covering. Let $d\colon D \to F$ be an equalizer of f and $b\,g$. By 20.2.3 (b), d is covering. By 20.2.4 (b), $f\,d = b\,g\,d$ is covering, and 20.2.4 (c) implies that b is covering.

If one has $u, v\colon F \to G$ with $b\,u = b\,v$, then $c\,b\,u = c\,b\,v$, and the equalizer of u and v is covering, since $c\,b$ is bicovering. By 20.2.3 (b), b is then bicovering.

20.2.6 Definition. Let $\mathscr{C}$ be a $\mathfrak{U}$-category with a topology $\mathfrak{T}$. The functors $F\colon \mathscr{C}^0 \to \mathit{Ens}$, i. e., $F \in |\widehat{\mathscr{C}}|$, are called (set-valued) *presheaves*. F is called a (set-valued) *sheaf* or, resp., a *separated presheaf* if, for the inclusion $i_R: R \subset X$ of every covering sieve $R \in J(X)$ for all $X \in |\mathscr{C}|$,

$$(5) \qquad i_R^* = [i_R, F]: [X, F]_{\hat{\mathscr{C}}} \to [R, F]_{\hat{\mathscr{C}}}$$

is bijective or, resp., injective.

20.2.7 Proposition. *Let $\mathscr{C}$ be a small site and $F: \mathscr{C}^0 \to Ens$ a presheaf. Then the two following statements are equivalent:*

(i) *F is a sheaf (separated presheaf).*

(ii) *If $c: H \to K$ is bicovering (covering), then*

$$[c, F]: [K, F]_{\hat{\mathscr{C}}} \to [H, F]_{\hat{\mathscr{C}}}$$

is bijective (injective) (compare 19.4.3, 19.4.6 (b)).

Proof. By definition 20.2.6, (i) follows immediately from (ii) through specializing. (i) $\Rightarrow$ (ii): First, let c be a monomorphism. For $f: X \to K$ with $X \in |\mathscr{C}|$, we consider the pullback

$$(6) \qquad \begin{array}{ccc} R & \overset{\overset{-1}{f}(c)}{\subset} & X \\ \downarrow & & \downarrow{\scriptstyle f} \\ H & \overset{c}{\rightarrowtail} & K . \end{array}$$

By 10.2.1, K is a colimit object of $\Delta^0: \mathscr{C}/K \to \hat{\mathscr{C}}$, and by 10.1.3, colimits in $\hat{\mathscr{C}}$ are universal. By applying $[?, F]$, $[c, F]$ is obtained as a limit of isomorphisms or, resp., monomorphisms, which completes the proof in this special case.

To prove the general case, one factors c canonically into an epimorphism and an inclusion. The latter is bicovering, by 20.2.1 and 20.2.2 (b). Since $[?, F]$ takes epimorphisms into monomorphisms, the general case is now trivial for separated presheaves F. Now let c be bicovering and F a sheaf. By 20.2.5 (d) and what was proved above, we may assume c to be an epimorphism. Using the notation in (1), we consider

$$[K, F] \xrightarrow{[c,f]} [H, F] \underset{[pr_2,F]}{\overset{[pr_1,F]}{\rightrightarrows}} [H \sqcap_K H, F] \xrightarrow{[\Delta, F]} [H, F] .$$

Since c is an epimorphism, c is a coequalizer of pr_1 and pr_2 (18.4.3 (c)) and, therefore, $[c, F]$ is an equalizer of $[pr_1, F]$ and $[pr_2, F]$. Now, Δ is a (bi)-covering monomorphism and $pr_1 \Delta = pr_2 \Delta$. By what was shown above, $[\Delta, F]$ is an isomorphism and thus $[pr_1, F] = [pr_2, F]$. Therefore, $[c, F]$ is an isomorphism.

20.2.8 Remark. The assumption that $\mathscr{C}$ is small, in 20.2.7, can be avoided. $[\mathscr{C}, Ens]$ is a full subcategory of $[\mathscr{C}, ENS]$. Since the inclusion reflects limits and colimits, the transition to colimits with respect to $\mathscr{C}/K$ in (6) yields again c as the colimit of the morphism $f^{-1}(c)$ in $[\mathscr{C}, Ens]$, because there surely is a natural transformation and the colimit in $[\mathscr{C}, ENS]$ exists.

Correspondingly, 20.1.9 is valid if $\mathcal{C}$ is a $\mathfrak{U}$-category. Instead of 20.1.9 (8), a universal colimit has to be considered.

20.3 Sheaves Associated with a Presheaf

20.3.1 The functor $L\colon \hat{\mathcal{C}} \to \hat{\mathcal{C}}$. For the following, let $\mathcal{C}$ be a site that is small with respect to the universe $\mathfrak{U}$. $\hat{\mathcal{C}}$ is then a $\mathfrak{U}$-category. For $X \in |\mathcal{C}|$, let $\mathcal{J}(X)$ be the full subcategory of $\hat{\mathcal{C}}/X$ whose objects are the inclusions of the covering sieves of X. $\mathcal{J}(X)$ is a small $\mathfrak{U}$-category (10.6.4). By 20.1.7, $\mathcal{J}(X)$ is also a set that is directed to the left (14.1.1) and thus cofiltered.

For $v\colon Y \to X$ and $i_R\colon R \subset X$ with $i_R \in |\mathcal{J}(X)|$, the rule $i_R \mapsto v^{-1}(i_R)$ defines the functor $v^*\colon \mathcal{J}(X) \to \mathcal{J}(Y)$, and $i_R \mapsto [i_R^{-1}(v), F]$ for $F \in |\hat{\mathcal{C}}|$ gives the natural transformation

$$[\Delta^0(?), F]_{(? \in \mathcal{J}(X))} \to [\Delta^0(??), F]_{(?? \in \mathcal{J}(Y))} .$$

Thus one has a functor $\mathcal{C}^0 \to Dg\,(Ens)$ (compare 20.1.5 (3) and 17.1.1) and even a bifunctor $\mathcal{C}^0 \times \hat{\mathcal{C}} \to Dg\,(Ens)$ (2.6.8), as is easily verified. A choice of colimits gives the bifunctor $L\colon \mathcal{C}^0 \times \hat{\mathcal{C}} \to Ens$, which at (X, F) is described by

(1) $$LF(X) = \operatorname*{Colim}_{? \in \mathcal{J}(X)} [\Delta^0(?), F]$$

with a filtered colimit. For fixed F, one has the partial functor $LF\colon \mathcal{C}^0 \to Ens$ and thus $L\colon \hat{\mathcal{C}} \to \hat{\mathcal{C}}$ (3.4.4).

For every $i_R \in |\mathcal{J}(X)|$ there is the map

(2) $$\lambda_{F,i_R}\colon [R, F] \to LF(X) .$$

where $\{\lambda_{F,i_R}\}\colon [\Delta^0(?), F] \to LF(X)_{\mathcal{J}(X)}$ is the natural transformation belonging to the colimit (1). If, therefore, $i\colon R \to S$ is a morphism in $\mathcal{J}(X)$, then

(3) $$\lambda_{F,i_R}[i, F] = \lambda_{F,i_S} \quad \text{for} \quad i_R = i_S\, i .$$

For $v\colon X \to Y$, $LF(v)$ is the map of the colimit objects induced by v^*. Therefore (compare 20.1.5 (3)),

(4) $$\begin{array}{ccc} [R, F] & \xrightarrow{\lambda_{F,i_R}} & LF(X) \\ {\scriptstyle [i_R^{-1}(v), F]}\downarrow & & \downarrow{\scriptstyle LF(v)} \\ [v^{-1}(R), F] & \xrightarrow{\lambda_{F,v^{-1}(i_R)}} & LF(Y) \end{array}$$

is commutative. For $i_R = 1_X$, $v^{-1}(1_X) = 1_Y$ and $(1_X)^{-1}(v) = v$. By (4),

(5) $$\lambda_{F,1_X}\colon [X, F] \to LF(X)$$

is therefore a natural transformation of bifunctors. By means of the Yoneda map $[X, F] \to F(X)$, whose inverse we denote by $J_{F,X}$ here,

(5) yields the natural transformation $l\colon 1_{\hat{\mathscr{C}}} \to L$ with

(6) $$l_{F,X} = \lambda_{F,1_X} J_{F,X}\colon F(X) \to LF(X)\,.$$

In the following, F is fixed. To simplify the notation we set

(7) $$j_R = J_{LF,X}\,\lambda_{F,i_R}\colon [R, F] \to [X, LF]\,,$$

thus replacing (3), (4), (6) by

(3′) $$j_R[i, F] = j_S \quad \text{for} \quad i_R = i_S\, i \quad \text{in} \quad \mathscr{J}(X)\,,$$

(4′) $$\begin{array}{ccc} [R, F] & \xrightarrow{\;j_R\;} & [X, LF] \\ {\scriptstyle [i_R^{-1}(v), F]}\downarrow & & \downarrow{\scriptstyle [v, LF]} \\ [v^{-1}(R), F] & \xrightarrow{\;j_{v^{-1}(R)}\;} & [Y, LF]\,, \end{array}$$

(6′) $$j_X\, J_{F,X} = J_{LF,X}\, l_{F,X}\colon F(X) \to [X, LF]\,.$$

Setting $i_S = 1_X$ and taking (6′) into account, (3′) becomes

(3″) $$j_R[i_R, F] = j_X = J_{LF,X}\, l_{F,X}\, J_{F,X}^{-1}\,.$$

20.3.2 Lemma. (a) *For* $i_R\colon R \to X$ in $\mathscr{J}(X)$ *and* $g\colon R \to F$ *in* $\hat{\mathscr{C}}$, *the following diagram is commutative*:

(8) $$\begin{array}{ccc} R & \xrightarrow{\;i_R\;} & X \\ {\scriptstyle g}\downarrow & & \downarrow{\scriptstyle j_R(g)} \\ F & \xrightarrow{\;l_F\;} & LF\,. \end{array}$$

(b) *For every* $u\colon X \to LF$ *there is an* $i_R\colon R \to X$ *in* $\mathscr{J}(X)$ *and a* $g\colon R \to F$ *with* $j_R(g) = u$.

(c) *For* $g, h\colon X \to F$ *with* $X \in |\mathscr{C}|$, *let* $l_F\, g = l_F\, h$. *Then the equalizer of* g *and* h *is an object of* $\mathscr{J}(X)$.

(d) *Let* R *and* R' *be covering sieves of* X. *For* $g\colon R \to F$ *and* $g'\colon R' \to F$, $j_R(g) = j_R(g')$ *if and only if* g *and* g' *coincide on a common refinement (covering* X*) of* R *and* R'. *(More precisely: there is an* R'' *with inclusions* $i_0\colon R'' \to R$, $i_0'\colon R'' \to R'$ *and with* $g\, i_0 = g'\, i_0'$*).*

Proof. (a) It suffices to give a proof at an arbitrary place $Y \in |\mathscr{C}|$. In $\hat{\mathscr{C}}$, we consider

(8′) $$\begin{array}{ccc} [Y, R] & \xrightarrow{\;[Y, i_R]\;} & [Y, X] \\ {\scriptstyle [Y, g]}\downarrow & & \downarrow{\scriptstyle [Y, j_R(g)]} \\ [Y, F] & \xrightarrow{\;[Y, l_F]\;} & [Y, L\,F]\,. \end{array}$$

By (3″), and using the Yoneda map, one has for the lower line here that

(9) $$[Y, l_F] = J_{LF,Y}\, l_{F,Y}\, J_{F,Y}^{-1} = j_Y\,.$$

Now let $v = i_R\,u\colon Y \to X$ for $u \in [Y, R]$. Then by 20.1.5 (5), $v^{-1}(R) = Y$, $v^{-1}(i_R) = 1_Y$ and $i_R^{-1}(v) = u$, so that from (4')

$$(10)\ \jmath_Y(g\,u) = \jmath_Y([u, F]\,(g)) \overset{(4')}{=} [v, L\,F]\,(\jmath_R(g)) = \jmath_R(g)\,v = \jmath_R(g)\,i_R\,u$$

follows. (9) and (10) imply that (8') is commutative. The commutativity of (8) is then implied by the Yoneda map.

(b) follows immediately from (1), (2) and (7).

(c) By (6), $\lambda_{F,1_X}\,J_{F,X}(g) = \lambda_{F,1_X}\,J_{F,X}(h)$. Since (1) is a filtered colimit, 9.4.5 shows that there is an $i_R\colon R \to X$ in $\mathcal{J}(X)$ and an $f \in [R, F]$ with $f = [i_R, F]\,(g) = [i_R, F]\,(h)$, so that $g\,i_R = h\,i_R$. The statement in (c) now follows from 20.3.2 (b) and 20.2.2 (b).

(d) By (7), $\jmath_R(g) = \jmath_{R'}(g')$ is equivalent to $\lambda_{F,i_R}(g) = \lambda_{F,i_{R'}}(g')$. Then, by 9.4.2, this is equivalent to the existence of R'', i_0, i_0' with $[i_0, F]\,(g) = [i_0', F]\,(g')$.

20.3.3 Proposition. *Using the same assumptions and notations as above, the following hold:*

(a) *For $F \in |\widehat{\mathscr{C}}|$, $l_F\colon F \to LF$ is bicovering.*

(b) *$L\colon \widehat{\mathscr{C}} \to \widehat{\mathscr{C}}$ preserves finite limits.*

(c) *LF is a separated presheaf.*

(d) *F is a separated presheaf if and only if l_F is a monomorphism. If this is the case, then LF is a sheaf.*

(e) *F is a sheaf if and only if l_F is an isomorphism.*

Proof. (a) By 20.3.2 (b) and 20.2.3 (a), l_F is covering. For $u, v\colon G \to F$, let $l_F\,u = l_F\,v$ and let $w\colon H \to G$ be an equalizer of u and v. For $f\colon X \to G$ with $X \in |\mathscr{C}|$, let $i_R\colon R \to X$ be an equalizer of $u\,f$ and $v\,f$. By 20.3.2 (c), i_R is covering. We consider

$$\begin{array}{ccccc} R & \xrightarrow{i_R} & X & \overset{uf}{\underset{vf}{\rightrightarrows}} & F \\ \downarrow & \mathrm{I} & \downarrow{\scriptstyle f} & & \| \\ H & \xrightarrow{w} & G & \overset{u}{\underset{v}{\rightrightarrows}} & F, \end{array}$$

where I is a pullback, by 12.3.5. It follows that w is covering (20.2.1). Then by 20.2.3 (b), l_F is bicovering.

(b) $[R, ?]_{\widehat{\mathscr{C}}}$ preserves limits. The colimit (1) is filtered and therefore it commutes with finite limits with respect to the argument F. The statement then follows from the pointwise construction of limits in $\widehat{\mathscr{C}}$.

(c) Let $i_R\colon R \to X$ be the inclusion of a covering sieve. It has to be shown that $[i_R, LF]\colon [X, LF] \to [R, LF]$ is injective. For $u, v\colon X \to LF$, let $u\,i_R = v\,i_R$. By 20.3.2 (b), there exist covering sieves R', R'' of X and morphisms $f\colon R' \to F$, $g\colon R'' \to F$ such that $\jmath_{R'}(f) = u$,

$j_{R''}(g) = v$.

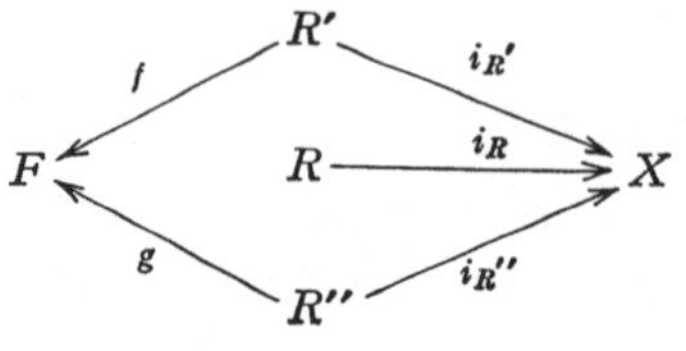

By (3') and 20.1.7 (b), we may assume $R' = R''$ and $R' \subset R$. With $i\colon R' \to R$, 20.3.2 (a) then implies

$$l_F f = j_{R'}(f)\, i_{R'} = j_{R'}(f)\, i_R\, i = u\, i_R\, i = v\, i_R\, i = l_F\, g\,.$$

By (a) and 20.2.3 (b), the equalizer $i'\colon R'' \to R'$ of f and g is covering. By 20.2.5 (b) and 20.2.2 (b), we may assume that $i_{R'}\, i = i_{R''}$ is the inclusion of a covering sieve. By 20.3.2 (d), $j_{R'}(f) = j_{R'}(g)$, so that $u = v$. Therefore, LF is a separated presheaf.

(d) First, let F be a separated presheaf. Then $[i_R, F]\colon [X, F] \to [R, F]$ is a monomorphism for all $i_R \in |\mathcal{J}(X)|$. Taking note of the fact that 1_X is terminal in $\mathcal{J}(X)$, one deduces from (1) and (3) that $\lambda_{F,1_X}$ is a filtered colimit of monomorphisms. Thus $\lambda_{F,1_X}$ is a monomorphism and so by (6), l_F is a monomorphism.

Now let l_F be a monomorphism. We start by showing that LF is a sheaf. By (c), it remains to be shown that $[i_R, LF]$ is surjective, provided $i_R\colon R \subset X$ is covering. For $h\colon R \to LF$, we consider

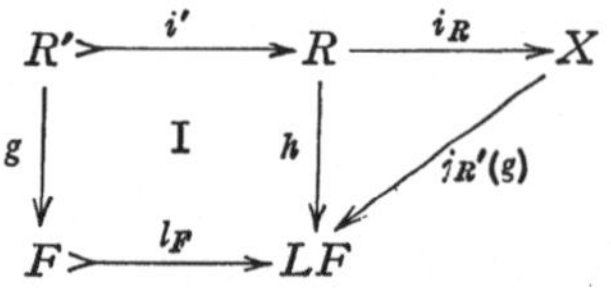

where I is a pullback. By (a), 20.2.5 (a) and by assumption, i' is bicovering. By 20.2.5 (b), $i_R\, i'$ is (isomorphic to) the inclusion of a covering sieve. By 20.3.2 (a), $h\, i' = l_F\, g = j_{R'}(g)\, i_R\, i'$. By (c) and 20.2.7, $[i', L\,F]$ is injective and, therefore, $h = j_{R'}(g)\, i_R$. So LF is a sheaf.

By means of the monomorphism l_F one goes on to conclude that F satisfies the conditions defining separated presheaves.

(e) If l_F is an isomorphism, then (d) guarantees that F is a sheaf. The converse is proved analogously to the beginning of the proof of (d).

20.3.4 Theorem. *Let $\mathcal{C}$ be a small (with respect to $\mathfrak{U}$) site, $\tilde{\mathcal{C}}$ the corresponding category of sheaves. Then:*

(a) *The inclusion $I\colon \tilde{\mathcal{C}} \to \hat{\mathcal{C}}$ has a left adjoint $A\colon \hat{\mathcal{C}} \to \tilde{\mathcal{C}}$ that preserves finite limits. A can be chosen in such a way that $I\,A = L\,L$ and $(l * L)\, l$ is the corresponding adjunction transformation, or such that $A\,I = 1_{\tilde{\mathcal{C}}}$.*

(b) *If $u: F \to G$ is a $\hat{\mathscr{C}}$-morphism, then the following are equivalent:*

(i) *u is bicovering (in the sense of 20.2.1).*

(ii) *For every sheaf H, $[u, H]: [G, H] \to [F, H]$ is bijective.*

(iii) *$A(u)$ is an isomorphism.*

Proof. (a) Let $R = L\,L\colon\ \hat{\mathscr{C}} \to \hat{\mathscr{C}}$ and $\eta = (l * L)\,l\colon 1_{\hat{\mathscr{C}}} \to R$. By 20.3.3 (c), (d), $R(F)$ is a sheaf for every $F \in |\hat{\mathscr{C}}|$, $l * L$ is a monomorphism and $l * R = l * L\,L$ an isomorphism. By 20.3.3 (a) and 20.2.7, $[l_F, L\,L(F)]\colon [L(F), L\,L(F)] \to [F, L\,L(F)]$ is an isomorphism. $(L * l)\,l = (l * L)\,l$ (16.1.1 (5)) implies $[l_F, L\,L(F)]\,(l_{L(F)}) = l_{L(F)}\,l_F = L(l_F)\,l_F = [l_F, L\,L(F)]\,L(l_F)$, and therefore

$$l * L = L * l \tag{12}$$

holds. One goes on to conclude that

$$\eta * R = (l * L\,L\,L)\,(l * L\,L) \overset{(12)}{=} (L\,L * l * L)\,(L\,L * l) = R * \eta\,,$$

and $\eta * R$ is an isomorphism. Further, if η_F is an isomorphism, then, by 20.3.3 (c), (d), $l_{L(F)}$ is a monomorphic retraction and l_F and $l_{L(F)}$ are isomorphisms; therefore, F is a sheaf. Now the existence of A, follows from 19.4.2 (c).

By 20.3.3 (b), A preserves finite limits, since I reflects limits. The remainder of the statement in (a) follows from 19.4.2 (d).

(b) By 20.2.7, (ii) follows from (i), and the adjunction isomorphism $[u, I(?)] \cong [A(u), ?]$ shows (ii) and (iii) to be equivalent. If (iii) is satisfied, then $\eta_G\, u = IA(u)\,\eta_F = IA(u)\,l_{L(F)}\,l_F$ is bicovering, by 20.3.3 (a) and 20.2.5 (b); (i) follows from 20.2.5 (d).

20.3.5 Definition. If $A: \hat{\mathscr{C}} \to \tilde{\mathscr{C}}$ in 20.3.4 is fixed, then $A(F)$ is called *the sheaf associated with* F (with respect to the site $\mathscr{C}$ or, resp., its topology $\mathfrak{T}$).

A $\mathfrak{U}$-category is called a *topos* if it is equivalent to the category of sheaves on a small site.

20.3.6 Remarks. (a) By 19.3.1 (f), (g) and the dual of 19.3.1 (d), 20.3.4 (b) says that the class of bicovering morphisms is a saturated class Σ of morphisms which admits a calculus of right fractions and a terminal calculus of (left) fractions, and that $\tilde{\mathscr{C}}$ is equivalent to $\hat{\mathscr{C}}[\Sigma^{-1}]$. Thus one is dealing with the non-additive case of a localization.

(b) By 20.2.6 and 16.6.3 (c), separated presheaves are the objects of a strictly full, epireflective subcategory $\mathscr{M}$ of $\hat{\mathscr{C}}$ ($\mathscr{C}$ small). But in general, as is shown by 20.3.3 (d), l is not a corresponding adjunction transformation. The situation in 19.4.7 applies to $\hat{\mathscr{C}}$, $\mathscr{M}$, $\tilde{\mathscr{C}}$. In particular, the following holds:

The restriction of $A\colon\ \hat{\mathscr{C}} \to \tilde{\mathscr{C}}$ to the full subcategory of separated presheaves is faithful.

20.3.7 Theorem. *Let $\mathscr{C}$ be a small category, Top the class of topologies on $\mathscr{C}$, and Cad the class of strictly full subcategories of $\hat{\mathscr{C}}$ for which the inclusion has a left adjoint that preserves finite limits. The correspondence φ: Top $\to$ Cad which assigns to every topology the corresponding category of sheaves is a bijection.*

Proof. 20.1.9 supplies a map ψ: *Cad* $\to$ *Top*. From 20.3.4 (b) and 20.2.2 (b), it follows that $\psi\,\varphi$ is the identity map.

Now let $\tilde{\mathscr{C}}$ in *Cad* be given. Let Σ be the class of those $\hat{\mathscr{C}}$-morphisms that are taken into isomorphisms by $A: \hat{\mathscr{C}} \to \tilde{\mathscr{C}}$. By 20.3.6 (a), Σ is a satured class which admits a calculus of right fractions and a terminal calculus of fractions. By 20.3.4 (b) and 19.4.3, the objects of $\tilde{\mathscr{C}}$, regarded as objects of $\hat{\mathscr{C}}$, are exactly those that are left closed with respect to Σ.

Let Σ' be the class of monomorphisms in Σ. By the dual of 19.3.5 (b), Σ' is closed with respect to pullbacks. The proof of 20.2.7 shows that $c \in \Sigma'$ if and only if c is a covering monomorphism with respect to the topology $\psi(\tilde{\mathscr{C}})$ defined in 20.1.9. Since A preserves epimorphisms and finite limits, 20.2.1 (1) shows Σ to be determined completely by Σ'. By 20.2.7, $\tilde{\mathscr{C}}$ and $\varphi\,\psi(\tilde{\mathscr{C}})$ have the same objects. Therefore, $\varphi\,\psi$ is the identity map for *Cad*.

20.3.8 Proposition. *Every category of sheaves $\tilde{\mathscr{C}}$ corresponding to a small site $\mathscr{C}$ has the following properties:*

(a) *$\tilde{\mathscr{C}}$ is complete and cocomplete. Limits are formed as in $\hat{\mathscr{C}}$, i. e. pointwise.*

(b) *Every monomorphism is an equalizer (of its cokernel pair). Monomorphisms are closed with respect to pushouts.*

(c) *Every epimorphism is a coequalizer (of its kernel pair). Epimorphisms are closed with respect to pullbacks.*

(d) *There is a unique natural factorization (up to an isomorphism) of a morphism into an epi- and a monomorphism.*

(e) *$u, v: F \to G$ is a kernel pair if and only if for every $K \in |\tilde{\mathscr{C}}|$, $[K, u]$, $[K, v]$ is a kernel pair.*

(f) *The functor $A\,H_*: \mathscr{C} \to \hat{\mathscr{C}} \to \tilde{\mathscr{C}}$ is dense. In particular, $\{A(X) | X \in |\mathscr{C}|\}$ is a generating set in $\tilde{\mathscr{C}}$.*

(g) *Filtered colimits commute with finite limits.*

(h) *Coproducts commute with pullbacks and with equalizers. Finite products of epimorphisms are epimorphisms.*

(i) *Colimits are universal.*

(j) *Every morphism, whose codomain is an initial object, is an isomorphism.*

(k) *Let* $\coprod F_i$ *be a coproduct with injections* $in_j\colon F_j \to \coprod F_i$. *Then the following squares*

$$(12)\qquad \begin{array}{ccc} F_j & \xrightarrow{1} & F_j \\ {\scriptstyle 1}\downarrow & & \downarrow{\scriptstyle in_j} \\ F_j & \xrightarrow{in_j} & \coprod F_i \end{array} \quad \textit{and} \quad \begin{array}{ccc} 0 & \longrightarrow & F_k \\ \downarrow & & \downarrow{\scriptstyle in_k} \\ F_j & \xrightarrow{in_j} & \coprod F_i \end{array} \quad \textit{for } j \neq k$$

are pullbacks. Here 0 *is an initial object of* $\tilde{\mathscr{C}}$. *("Coproducts are disjoint". Notice that the left hand square is a pullback if and only if* in_j *is a monomorphism.)*

Proof. We may assume, and will do so constantly, that $A\, I = 1_{\tilde{\mathscr{C}}}$.

(a) follows immediately from 16.6.1 and the fact that I reflects limits.

(b) is valid in *Ens* and thus in $\hat{\mathscr{C}}$ (because of the pointwise construction, compare 13.4.4, 18.4.3 and 18.4.4). I preserves monomorphisms and A preserves finite limits and colimits. This implies the statements made for $\tilde{\mathscr{C}}$.

(c) From the first statement in (b) it follows that bimorphisms are isomorphisms (7.2.2). Thus 20.3.4 (b) implies that a $\hat{\mathscr{C}}$-morphism u is covering (in the sense of 20.2.1) if and only if $A(u)$ is an epimorphism (i. e., u is covering in the sense of 19.7.4). The second statement follows thus from 20.2.4 and (a), the first one follows as in part (b) from 19.5.2.

(d) The statement is valid in $\hat{\mathscr{C}}$. The existence of a factorization for morphisms is $\tilde{\mathscr{C}}$ follows by means of a factorization in $\hat{\mathscr{C}}$ and applying A. The uniqueness is guaranteed by 20.2.1 and the proof of (c), the naturalness follows from 12.4.10.

(e) Let p be a cokernel of $I(u)$ and $I(v)$. By 10.4.3, $[H_X, p]$ is a cokernel of $[H_X, I(u)]$ and $[H_X, I(v)]$, for every $X \in |\mathscr{C}|$. If $[K, u]$, $[K, v]$ is a kernel pair for every $K \in |\tilde{\mathscr{C}}|$, then, in particular, $[A\, H_*(X), u]$, $[A\, H_*(X), v]$ is a kernel pair for every $X \in |\mathscr{C}|$. Using the adjunction isomorphism, it follows from 18.4.3 that $[H_*(X), I(u)]$, $[H_*(X), I(v)]$ is a kernel pair of $[H_*(X), p]$. By the Yoneda lemma and the pointwise construction of limits in $\hat{\mathscr{C}}$, $I(u)$, $I(v)$ is a kernel pair of p. Since A preserves finite limits, u, v is a kernel pair of $A(p)$. The converse follows from the fact that $[K, ?]_{\tilde{\mathscr{C}}}$ preserves limits.

(f) The first statement follows from 17.2.6. It implies the second one, by the definition of colimits.

(g), (h) and (i). The statements are valid for $\hat{\mathscr{C}}$. For $\tilde{\mathscr{C}}$ they follow from (a) and the fact that A preserves colimits and finite limits (compare 16.6.2). Note 19.5.2 (a).

(j), (k) are also easily proved.

20.3.9 Remark. There are obvious generalizations of the preceding statements to adjoint situations $(\varphi, S, T, \mathscr{B}, \mathscr{C})$, provided S preserves finite limits, T is fully faithful and $\mathscr{C}$ has corresponding properties.

20.3.10 The additive case. Now let $\mathscr{C}$ be an additive $\mathfrak{U}$-category and $\hat{\mathscr{C}} = Add\,(\mathscr{C}^0, Ab)$. 20.1.1 through 20.1.9, except 20.1.3 (c), carry over smoothly, provided condition (S) in 20.1.3 is completed by the requirement that the morphisms $Y \to X$ belonging to a sieve R form a subgroup of $[Y, X]$. The analogues of 20.1.9, 20.3.4 and 20.3.7 are valid. To show this, it suffices to prove the following.

Proposition. *If $\mathscr{C}$ is a small additive category, then there is a bijection between the class Top of topologies on $\mathscr{C}$ and the class Loc of localizing subcategories of $Add\,(\mathscr{C}^0, Ab) = \hat{\mathscr{C}}$.*

Proof. (a) Let $\mathscr{K}$ be a localizing subcategory of $\hat{\mathscr{C}}$. A sieve R for $X \in |\mathscr{C}|$ with inclusion $i_R\colon R \to X$ is called covering (with respect to $\mathscr{K}$) if the codomain of coker i_R is in $\mathscr{K}$. Since the localizing functor is exact (19.6.1, 19.5.5 (a)), 19.5.5 (b) implies that the covering sieves coincide with those constructed in 20.1.9. $\psi\colon Loc \to Top$ thus is again given by 20.1.9.

(b) Now let a topology $\mathfrak{T}$ be given on $\mathscr{C}$. Let $\mathscr{K}$ be the full subcategory of $\hat{\mathscr{C}}$ whose objects K have the following property:

If $f\colon X \to K$ is any morphism with $X \in |\mathscr{C}|$, then ker f is (isomorphic to) the inclusion of a covering sieve (with respect to $\mathfrak{T}$).

Obviously, $\mathscr{K}$ is strictly full in $\hat{\mathscr{C}}$, by (T 3) it contains all the zero objects, and it is closed with respect to subobjects, because for a monomorphic $m\colon K \to K'$, f and mf have the same kernels. Now we consider

$$(13)\qquad \begin{array}{ccccccccc}
 & & S & \longrightarrow & R & \longrightarrow & R'' & & \\
 & & {\scriptstyle i'}\downarrow & \mathrm{I} & \downarrow{\scriptstyle i} & & \downarrow{\scriptstyle i''} & & \\
 & & Y & \xrightarrow{u} & X & = & X & & \\
 & & {\scriptstyle v}\downarrow & & \downarrow{\scriptstyle f} & & \downarrow{\scriptstyle cf} & & \\
0 & \longrightarrow & K' & \xrightarrow{m} & K & \xrightarrow{c} & K'' & \longrightarrow & 0\,.
\end{array}$$

Here the bottom line is to be exact, with $K', K'' \in |\mathscr{K}|$. Let $f\colon X \to K$ be given with $X \in |\mathscr{C}|$, i, i'' are kernels and inclusions of sieves. By assumption, i'' is covering. Further, let u be in R''; i. e., it is of the form $u = i''g$ (20.1.3). Thus by the definition of kernels, v exists. Let I be a pullback, so that S is a sieve for Y. By 12.3.4 (c), i' is a kernel of v. By (T 2i) and since $K' \in |\mathscr{K}|$, $K \in |\mathscr{K}|$ follows. If, conversely, $K \in |\mathscr{K}|$, $K'' \in |\mathscr{K}|$ follows by (T 2ii), since the objects of $\mathscr{C}$ are projective in $\hat{\mathscr{C}}$. Therefore, $\mathscr{K}$ is thick and thus closed with respect to finite biproducts. Since the objects of $\mathscr{C}$ are small in $\hat{\mathscr{C}}$, $\mathscr{K}$ is closed with respect to coproducts, and by 19.6.5, it is localizing. Thus there is a map $\varphi\colon Top \to Loc$.

(c) Now let $\mathfrak{T}$ be a topology on $\mathcal{C}$ and let $\mathfrak{T}' = \psi\,\varphi(\mathfrak{T})$. (a) and (b) imply immediately that the sieves which are covering with respect to $\mathfrak{T}'$ are also covering with respect to $\mathfrak{T}$. We consider

(14)
$$\begin{array}{ccccccccc}
0 & \longrightarrow & S & \overset{i'}{\rightarrowtail} & Y & \overset{cu}{\rightarrow} & K & \longrightarrow & 0 \\
 & & \downarrow & \mathrm{I} & \downarrow{\scriptstyle u} & & \| & & \\
0 & \longrightarrow & R & \overset{i}{\rightarrowtail} & X & \overset{c}{\twoheadrightarrow} & K & \longrightarrow & 0\,,
\end{array}$$

where R is a covering sieve of X with respect to $\mathfrak{T}$ and u is arbitrary in $\mathcal{C}$. Again by 12.3.4 (c), I is a pullback. Since Y is projective in $\hat{\mathcal{C}}$, $K \in |\mathcal{K}|$ follows from (T 1) as in (b). By (a), R also covers with respect to $\mathfrak{T}'$.

(d) Now let $\mathcal{K}$ again be a localizing subcategory of $\hat{\mathcal{C}}$ and let $\mathcal{L} = \varphi\,\psi(\mathcal{K})$. The conclusion from (14) yields $\mathcal{K} \subset \mathcal{L}$. Now let $L \in |\mathcal{L}|$. We consider all exact sequences $0 \to R_f \xrightarrow{i_f} X_f \xrightarrow{f} L$, where i_f is the inclusion of a covering sieve with respect to $\psi(\mathcal{K})$. We thus get

$$\amalg R_f \xrightarrow{\amalg i_f} \amalg X_f \xrightarrow{\bar{f}} L \text{ with } \bar{f}\,\amalg i_f = 0\,.$$

By localizing with respect to $\mathcal{K}$, $\amalg\, i_f$ goes into an isomorphism, $\ker \bar{f}$ goes into a monomorphic retraction and, therefore, $\bar{f} = \operatorname{coker}(\ker \bar{f})$ into an isomorphism. Now $\mathcal{L} \subset \mathcal{K}$ follows.

20.4 Generation of Topologies

We go back to considering arbitrary categories and *Ens*-valued functors.

20.4.1 The topologies on a category $\mathcal{C}$ considered as classes of sieves are ordered by inclusion. If $\mathfrak{T}_1$, $\mathfrak{T}_2$ are topologies on $\mathcal{C}$, then $\mathfrak{T}_1$ is *greater* than $\mathfrak{T}_2$, and $\mathfrak{T}_2$ *smaller* than $\mathfrak{T}_1$, if $\mathfrak{T}_2$ is contained in $\mathfrak{T}_1$. If $\mathfrak{T}_1$ is greater than $\mathfrak{T}_2$, then all sheaves for $\mathfrak{T}_1$ are also sheaves for $\mathfrak{T}_2$. The corresponding statement can be made for separated presheaves.

There is a greatest topology, in which every sieve is covering This *topology* is called *discrete*. There is also a smallest topology. It is called the *indiscrete*, or *chaotic*, *topology*.

It follows immediately from (T 1), (T 2), (T 3) in 20.1.6 that for every class of topologies on $\mathcal{C}$ the set theoretical intersection is again a topology. Hence, every class of topologies has a greatest lower bound. This implies the existence of a least upper bound (as the intersection of all upper bounds) and even more, namely: given any class $\mathfrak{K}$ of sieves, there is a smallest topology on $\mathcal{C}$ in which the given sieves are all covering. This topology is called the *topology generated by* $\mathfrak{K}$.

20.4.2 Proposition. *Let $\mathscr{C}$ be a $\mathfrak{U}$-category and $F\colon \mathscr{C}^0 \to Ens$ a presheaf. For every $X \in |\mathscr{C}|$, let $J(X)$ be the class of sieves for X such that for the inclusion $i_R\colon R \to X$ the following holds:*

(*) *If $v\colon Y \to X$ is any $\mathscr{C}$-morphism with codomain X, then*

$$[v^{-1}(i_R), F]\colon [Y, F] \to [v^{-1}(R), F]$$

is bijective (or, resp., injective).

These classes $J(X)$ form a topology on $\mathscr{C}$. It is the greatest topology for which F is a sheaf (or, resp., a separated presheaf).

Proof. Conditions (T1) and (T3) are evidently satisfied. (T2i) and (T2ii) have to be verified. The last statement is then evident. So let $i\colon R' \to R$, $i_R\colon R \to X$ be inclusions of sieves such that one of the following two cases applies:

(a) $R \in J(X)$ and $v^{-1}(R') \in J(Y)$ for every $v\colon Y \to X$.

(b) $R' \in J(X)$.

Given $\alpha\colon Y \to R$ in $\hat{\mathscr{C}}$ with $Y \in |\mathscr{C}|$, then in both cases $\alpha^{-1}(i)\colon \alpha^{-1}(R') \to Y$ is in $J(Y)$, by 20.1.3 (b) and because $i_R^{-1}(R') = R'$. Using universality of colimits, 20.1.3 (c) and (*) (compare the proof of 20.2.7 and 20.2.8) yield that $[i, F]\colon [R, F] \to [R', F]$ is bijective (or, resp., injective). Thus it follows in both cases that $[i_R\, i, F]\colon [X, F] \to [R', F]$ is bijective (injective). Since the conditions in (a) and (b) are invariant with respect to an arbitrary change of basis $v\colon Y \to X$ in $\mathscr{C}$ (by 20.1.5 in case (a), by (T1) in case (b)), (*) follows for $i_R\, i\colon R' \to X$.

20.4.3 Corollary. *For every class $\mathfrak{N}$ of presheaves there exists a greatest topology on $\mathscr{C}$ for which all presheaves in $\mathfrak{N}$ are sheaves (or, resp., separated presheaves), namely the intersection of those topologies which correspond (as in 20.4.2) to the individual presheaves in $\mathfrak{N}$.*

20.4.4 Corollary. *For every $X \in |\mathscr{C}|$, let a class $\mathfrak{R}(X)$ of sieves be given such that (T1) is satisfied. A presheaf F is a sheaf (or, resp., separated presheaf) for the topology $\mathfrak{T}$ generated by the classes $\mathfrak{R}(X)$ if and only if for every inclusion $i_R\colon R \to X$ in $\mathfrak{R}(X)$, $[i_R, F]\colon [X, F] \to$ $\to [R, F]$ is a bijection (or, resp., an injection).*

The condition here just states that $\mathfrak{T}$ is smaller than the topology assigned to F by 20.4.2.

20.4.5 Definition. The greatest topology on the category $\mathscr{C}$ for which all $X \in |\mathscr{C}|$ (more precisely, all H_X) are sheaves, is called the *canonical topology* on $\mathscr{C}$.

To be more precise, we distinguish from now on between $X \in |\mathscr{C}|$ and $H_X \in |\hat{\mathscr{C}}|$, between a sieve R as subfunctor of H_X and the corresponding morphism class, whose morphisms are the objects of a full subcategory $\bar{R}$ of $\mathscr{C}/X$. For $v\colon Y \to X$ in $\mathscr{C}$, let $v^{-1}(\bar{R})$ be the subcategory of $\mathscr{C}/Y$ corresponding to $v^{-1}(R)$ (20.1.3, 20.1.5).

20.4.6 Proposition. Let $\mathscr{C}$ be a $\mathfrak{U}$-category and $i_R: R \to H_X$ the inclusion of a sieve in $\hat{\mathscr{C}}$. With respect to the canonical topology on $\mathscr{C}$, R covers H_X if and only if the following holds: For every $\mathscr{C}$-morphism v: $Y \to X$

(1) $$Y = \operatorname*{Colim}_{? \in v^{-1}(\bar{R})} \Delta^0(?) \quad in \quad \mathscr{C}.$$

Proof. By 20.4.3, 20.4.2, R covers H_X if and only if for every v: $Y \to X$ in $\mathscr{C}$ and for every $Z \in |\mathscr{C}|$,

(2) $$[H_v^{-1}(i_R), H_Z]: [H_Y, H_Z] \to [H_v^{-1}(R), H_Z]$$

is an isomorphism. By 20.1.3 (c) and the Yoneda lemma, in ENS

(3) $$[H_v^{-1}(R), H_Z] = \operatorname*{Lim}_{? \in v^{-1}(\bar{R})} [\Delta^0 H_*(?), H_Z]_{\hat{\mathscr{C}}} = \operatorname*{Lim}_{? \in v^{-1}(\bar{R})} [\Delta^0(?), Z]_{\mathscr{C}}$$

is valid. If (2) holds for all $Z \in |\mathscr{C}|$, then the second limit in (3) exists already in Ens, and (1) follows from 8.7.3. If, conversely, (1) is valid, then (2) follows from (3), by considering 20.1.3 (b) and (c). The natural transformation belonging to the colimit (1) is the morphism class corresponding to $v^{-1}(R)$.

20.4.7 Remark. 20.4.6 shows that the canonical topology on $\mathscr{C}$ is independent of the choice of the universe $\mathfrak{U}$ (so that $\mathscr{C}$ is a $\mathfrak{U}$-category).

20.5 Pretopologies

Again we regard $\mathscr{C}$ as a subcategory of $\hat{\mathscr{C}}$.

20.5.1 Let $\{f_\alpha: F_\alpha \to F\}$ be a family of $\hat{\mathscr{C}}$-morphisms with the same codomain, where the indices here and below may be in a $\mathfrak{B}$-set. $\operatorname{im} f_\alpha$ can be regarded as the inclusion of a subfunctor of F. $\bigcup \operatorname{im} f_\alpha$ exists as the inclusion of a subfunctor of F, the *image of the family* $\{f_\alpha\}$. If, in particular, $\{u_\alpha: X_\alpha \to X\}$ is a family of $\mathscr{C}$-morphisms, then one obtains as image in $\hat{\mathscr{C}}$ the inclusion $i_R: R \to X$ of a sieve R. It is called the *sieve* generated by $\{u_\alpha\}$. According to 20.1.3 (a), (b), R is described as a class of $\mathscr{C}$-morphisms by

(1) $\quad \{v | v$ in $\mathscr{C}$, codomain $v = X$, and v factors through some $u_\alpha\}$.

This follows from 20.1.4, if one takes into account that $\operatorname{im} H_*(u_\alpha)$ can be constructed pointwise and that $\operatorname{im} [Y, u_\alpha]$ consists of those $\mathscr{C}$-morphisms $Y \to X$ which factor through u_α.

20.5.2 For every $X \in |\mathscr{C}|$ let a class $Cov(X)$ of families $\{u_\alpha: X_\alpha \to X\}$ be given. The classes $Cov(X)$ form a *pretopology* $\mathfrak{E}$ on $\mathscr{C}$ if the following conditions are satisfied:

(PT 1) If $\{u_\alpha: X_\alpha \to X\} \in Cov(X)$ and $v: Y \to X$ is any $\mathscr{C}$-morphism, then the pullback of v and u_α exists for every α, and

$$\{v^{-1}(u_\alpha): Y \sqcap_X X_\alpha \to Y\} \in Cov(Y) .$$

(PT 2) If $\{u_\alpha: X_\alpha \to X\} \in Cov(X)$ and $\{v_{\alpha\beta_\alpha}: X_{\alpha\beta_\alpha} \to X_\alpha\} \in Cov(X_\alpha)$, then $\{u_\alpha v_{\alpha\beta_\alpha}: X_{\alpha\beta_\alpha} \to X\} \in Cov(X)$.

(PT 3) $\{1_X: X \to X\} \in Cov(X)$.

20.5.3 Proposition. *Let $\mathfrak{C}$ be a pretopology on $\mathscr{C}$.*

(a) *For every $X \in |\mathscr{C}|$, let $J_{\mathfrak{C}}(X)$ be the class of sieves that are generated by the elements of $Cov(X)$, and let $J(X)$ be the class of those sieves which contain a sieve in $J_{\mathfrak{C}}(X)$. The classes $J(X)$ define a topology $\mathfrak{T}$, the one generated by $\mathfrak{C}$. $\mathfrak{T}$ is the smallest topology on $\mathscr{C}$ for which all $J_{\mathfrak{C}}(X)$ consist of covering sieves.*

(b) *$F: \mathscr{C}^0 \to Ens$ is a sheaf (or, resp., a separated presheaf) for $\mathfrak{T}$ if and only if the following holds: If $\{u_\alpha: X_\alpha \to X\}_{\alpha \in A} \in Cov(X)$, then in*

$$F(X) \xrightarrow{f} \prod_{\alpha \in A} F(X_\alpha) \overset{g}{\underset{h}{\rightrightarrows}} \prod_{(\alpha,\beta) \in A \times A} F(X_\alpha \sqcap_X X_\beta) \tag{2}$$

f is an equalizer of g and h (or, resp., f is a monomorphism). Here $pr_\alpha f = F(u_\alpha)$, $pr_{\alpha,\beta}\, g = F(v_{\alpha,\beta})\, pr_\alpha$, $pr_{\alpha,\beta}\, h = F(w_{\alpha,\beta})\, pr_\beta$ and

$$\begin{array}{ccc} X_\alpha \sqcap_X X_\beta & \xrightarrow{w_{\alpha,\beta}} & X_\beta \\ {\scriptstyle v_{\alpha,\beta}}\downarrow & & \downarrow{\scriptstyle u_\beta} \\ X_\alpha & \xrightarrow{u_\alpha} & X \end{array} \tag{3}$$

is a pullback for every $(\alpha, \beta) \in A \times A$. If $\mathscr{C}$ is not small, the products in (2) have to be formed in ENS.

Proof. (a) We verify (T1), (T2), (T3).

(T1): If R is the sieve generated by $\{u_\alpha\}$, then $w \in v^{-1}(R)$ if and only if (20.1.5) $v\, w \in R$; i.e., $v\, w$ factors through some u_α. This is equivalent to w factoring through some $v^{-1}(u_\alpha)$. Thus (T1) follows from (PT1).

(T2): Obviously, it is sufficient to carry out the verification for the case $R' \subset R \in J_{\mathfrak{C}}(X)$, where R' satisfies the conditions in (T2). Let R be generated by $\{u_\alpha\}$. By assumption, $u_\alpha^{-1}(R')$ contains a family in $Cov(X_\alpha)$, let us call it $\{v_{\alpha\beta_\alpha}: X_{\alpha\beta_\alpha} \to X_\alpha\}$. One concludes that the morphisms of the family $\{u_\alpha v_{\alpha\beta_\alpha}: X_{\alpha\beta_\alpha} \to X\}$ belong to R' and thus that (T2) follows from (PT2).

(T3) and the last statement in (a) are evident.

(b) By (PT1), the pullbacks (3) exist and $g f = h f$. Let R be the sieve generated by $\{u_\alpha\}$, $\overline{R}$ the corresponding full subcategory of $\mathscr{C}/X$. $\overline{R}$ and the family $\{u_\alpha\}$ of $\overline{R}$-objects satisfy the assumptions for 9.6.2 (with respect to $\mathfrak{V}$ if $\mathscr{C}$ is not small). By 20.1.3 (c), $[R, F] = \operatorname{Lim}_{? \in \overline{R}} [\Delta^0 \circ H_*(?), F]$. Using $T = [?, F]_{\hat{\mathscr{C}}}$, this limit can be described dually to 9.6.2 (T is here contravariant). With the help of the Yoneda lemma, one deduces from 20.1.3 (b) that $[i_R, F]: [X, F] \to [R, F]$ is an isomorphism (or, resp., a monomorphism) if and only if f is an equalizer of g and h (or, resp., a monomorphism). This shows the necessity of the above condition. Let it be satisfied. $[i_R, F]$ is then an isomorphism (monomorphism) for every $i_R: R \to X$ with $R \in J_{\mathfrak{G}}(X)$ and an arbitrary X. The proof of (a) shows that the classes $J_{\mathfrak{G}}(X)$ satisfy condition (T1), by 20.4.4, F is a sheaf (separated presheaf).

20.5.4 Classical example. Let $\mathscr{C}$ be the ordered set of open sets of a topological space ordered by inclusion. For $X \in |\mathscr{C}|$, let $\{u_\alpha: X_\alpha \to X\} \in Cov(X)$ if and only if $\bigcup X_\alpha = X$. One shows easily that this constitutes a pretopology. By 20.5.3 (a) and 20.4.6, the topology generated by it is the canonical one. 20.5.3 (b) yields the original definition of set-valued sheaves on a topological space (see [16]).

20.5.5 Lemma. *Let the family $\{u_\alpha: G_\alpha \to X\}$ generate the sieve R, let $i_R: R \to H_X = H_*(X)$ be the inclusion and, for every α, let $f_\alpha: H_*(G_\alpha) \to R$ be the $\hat{\mathscr{C}}$-morphism that is uniquely determined by $i_R f_\alpha = H_*(u_\alpha)$. If the indices of the family form a $\mathfrak{U}$-set, then there is a morphism p: $\coprod_\alpha H_*(G_\alpha) \to R$ with $p\, in_\alpha = f_\alpha$, and p is a $\hat{\mathscr{C}}$-epimorphism.*

This follows immediately from 20.5.1 and 14.2.5.

20.6 Characterization of Topos

20.6.1 Definitions. Let $\mathscr{C}$ be a $\mathfrak{U}$-category.

(a) A $\mathfrak{U}$-set $\mathfrak{G}$ of objects in $\mathscr{C}$ is called a *generating set in the sense of Grothendieck* if the following holds:

If $m: A \rightarrowtail B$ is a monomorphism, but not an isomorphism, then, for a suitable $G \in \mathfrak{G}$, there is a morphism $f: G \to B$ that does not factor through m.

(b) Let $\mathfrak{T}$ be a topology on $\mathscr{C}$. A $\mathfrak{U}$-set $\mathfrak{G}$ of objects in $\mathscr{C}$ is called a *topologically generating set* (for $\mathfrak{T}$) if the following holds: For every $X \in |\mathscr{C}|$, the family

$$\{\alpha: G_\alpha \to X \mid \alpha \in \bigcup_{G \in \mathfrak{G}} [G, X]\}$$

generates a covering sieve (with respect to $\mathfrak{T}$).

(c) A site $\mathscr{C}$ is called a $\mathfrak{U}$-*site*, if the underlying category $\mathscr{C}$ is a $\mathfrak{U}$-category and if there is a topologically generating set for its topology.

20.6.2 Remarks. (a) Definition 20.6.1 (a) is equivalent to definition 10.5.1, provided $\mathscr{C}$ is a balanced category that has pushouts and equalizers (compare with 10.7.10). In particular, these definitions are equivalent for a topos, by 20.3.8.

(b) A category with finite intersections and a generating set in the sense of Grothendieck is well-powered. The proof of 10.6.3 carries over with the obvious changes.

(c) If the topology $\mathfrak{T}'$ is greater than the topology $\mathfrak{T}$, then any topologically generating set for $\mathfrak{T}$ is also a topologically generating set for $\mathfrak{T}'$. This follows immediately from the definitions.

(d) If $\mathscr{C}$ is a small category, then $|\mathscr{C}|$ is a topologically generating set for every topology on $\mathscr{C}$.

(e) If $\mathscr{C}$ is a site such that the underlying category $\mathscr{C}$ is a $\mathfrak{U}$-category, then 20.3 is valid for presheaves and sheaves with values in *ENS*. However, we are still interested in presheaves and sheaves with values in *Ens*. To avoid confusions, we then speak of $\mathfrak{U}$-presheaves and $\mathfrak{U}$-sheaves.

20.6.3 Proposition. *Let $\mathscr{C}$ be a $\mathfrak{U}$-site, $\mathfrak{T}$ its topology and $\mathfrak{G}$ a topologically generating set for $\mathfrak{T}$. For every $X \in |\mathscr{C}|$, let $J_{\mathfrak{G}}(X)$ be the class of all those covering sieves that are generated by families $\{u_\alpha: G_\alpha \to X\}$, where each of the domains G_α belongs to $\mathfrak{G}$.*

(a) *If R is a covering sieve for X, then those $\mathscr{C}$-morphisms, which belong to R and whose domain is in $\mathfrak{G}$, form a $\mathfrak{U}$-set. This set generates a covering sieve R', and $R' \subset R$. For every $\mathfrak{U}$-presheaf F, $[R', F]_{[\mathscr{C}^0, Ens]}$ is isomorphic to a $\mathfrak{U}$-set.*

(b) *For every $X \in |\mathscr{C}|$, $J_{\mathfrak{G}}(X)$ is isomorphic to a $\mathfrak{U}$-set. It is initial in the class $J(X)$ of all covering sieves for X. For any $\mathfrak{U}$-presheaf F, the colimit* 20.3.1 (1) *can be chosen in such a way that $LF(X)$ is a $\mathfrak{U}$-set.*

(c) *Theorem* 20.3.4 *is valid for the categories of $\mathfrak{U}$-presheaves and of $\mathfrak{U}$-sheaves and, exept for* 20.3.8 (f), *so are the statements in* 20.3.8 *if limits and colimits are understood to be $\mathfrak{U}$-limits or, resp., $\mathfrak{U}$-colimits. Instead of* 20.3.8 (f) *the following holds:*

(f′) $\{A\, H_*(G) \mid G \in \mathfrak{G}\}$ *is a generating set in $\tilde{\mathscr{C}}$.*

Remark. Since $\tilde{\mathscr{C}}$ is a full subcategory of $[\mathscr{C}^0, Ens]$, $\tilde{\mathscr{C}}$ is not a $\mathfrak{U}$-category.

Proof. (a). The first statement follows immediately from the fact that $\bigcup_{G \in \mathfrak{G}} [G, X]_{\mathscr{C}}$ is a $\mathfrak{U}$-set. Obviously, $R' \subset R$. Let $v: Y \to X$ be a $\mathscr{C}$-morphism in R. By 20.1.5 (4), (5), $v^{-1}(R')$ contains all $\mathscr{C}$-morphisms with codomain Y whose domains are in $\mathfrak{G}$. By the assumption about $\mathfrak{G}$ (20.6.1 (b)) and by (T2), R' is a covering sieve for X. By 20.5.5, there is a $\hat{\mathscr{C}}$-epimorphism $p: \coprod G_\alpha \to R'$. Since $[\coprod G_\alpha, F] \cong \prod F(G_\alpha)$ and since $[p, F]$ is a monomorphism, $[R', F]$ is isomorphic to a $\mathfrak{U}$-set.

(b) The first statement follows from the fact that the sieves in $J_{\mathfrak{G}}(X)$ are generated by subsets of $\bigcup_{G \in \mathfrak{G}} [G, X]_{\mathscr{C}}$. The second statement follows from (a) and 9.4.9, since $J(X)$ is cofiltered. With this, the next statement follows from (a) and 9.2.3.

(c) In the proof of 20.3.4, the assumption that $\mathcal{C}$ be a small category was only used to guarantee the existence of the colimits 20.3.1 (1); for, by 20.2.8, 20.2.7 is valid in any case. Thus, the proofs of 20.3.8, exept for 20.3.8 (f), carry over smoothly. It remains to prove (f'). Now, let $j, k: F \to K$ be two different $\tilde{\mathcal{C}}$-morphism. Then $I(j) \neq I(k)$ and, by the Yoneda lemma, there is an $h: H_X \to I(F)$ in $\hat{\mathcal{C}}$ such that $I(j)\, h \neq \neq I(k)\, h$. The adjunction isomorphism yields $j\, A(h) \neq k\, A(h)$. Now, by 20.6.1 (b) and 20.5.5, there are a covering sieve R with the inclusion $i_R: R \to H_X$ and a $\hat{\mathcal{C}}$-epimorphism $p: \coprod H_*(G_\alpha) \twoheadrightarrow R$. Taking into account that $A(i_R)$ is an isomorphism and that A preserves colimits, it follows that there is a $\tilde{\mathcal{C}}$-epimorphism $q: \coprod A\, H_*(G_\alpha) \to A\, H_*(X)$. Thus, $j\, A(h)\, q \neq k\, A(h)\, q$, and there is an α such that $j\, A(h)\, q\, in_\alpha \neq \neq k\, A(h)\, q\, in_\alpha$; i.e., there is a $\tilde{\mathcal{C}}$-morphism $f: A\, H_*(G_\alpha) \to F$ such that $j\, f \neq k\, f$, as was to be shown.

20.6.4 Theorem (Giraud). *Let $\mathcal{E}$ be a $\mathfrak{U}$-category. The following assertions are equivalent:*

(a) *$\mathcal{E}$ is a topos; i.e., $\mathcal{E}$ is equivalent to the category of sheaves on a small site.*

(b) *$\mathcal{E}$ has the following properties:*

  (i) *$\mathcal{E}$ is finitely complete.*

  (ii) *$\mathcal{E}$ has coproducts. They are disjoint (in the sense of 20.3.8 (k)) and universal (as colimits).*

  (iii) *If $f, g: A \to B$ is a pair of $\mathcal{E}$-morphisms such that $[K, f]$, $[K, g]$ is a kernel pair for every $K \in |\mathcal{E}|$, then (f, g) is a kernel pair and it has a coequalizer which is pullback-closed (see 18.4.6 (b)). ("Equivalence relations are universally effective".)*

  (iv) *$\mathcal{E}$ has a generating set $\mathfrak{G}$ in the sense of Grothendieck.*

(c) *$\mathcal{E}$ has property (iv) and, for the canonical topology on $\mathcal{E}$, the $\mathfrak{U}$-sheaves are precisely the representable functors.*

(d) *There is a small, finitely complete category $\mathcal{C}$ with a topology, which is smaller than the canonical one, such that the corresponding category of $\mathfrak{U}$-sheaves is equivalent to $\mathcal{E}$.*

(e) *There is a small category $\mathcal{C}$ and a fully faithful functor $T: \mathcal{E} \to \to [\mathcal{C}^0, Ens]$ such that T has a left adjoint S which preserves finite limits.*

Proof. (a) $\Leftrightarrow$ (e) by 20.3.7. Obviously, (d) $\Rightarrow$ (a). By 20.3.8 and 20.6.2 (a), (a) $\Rightarrow$ (b). Therefore, it suffices to show that (b) $\Rightarrow$ (c) $\Rightarrow$ (d). We give these proofs as separate lemmas, because they involve several steps. Some elementary details are left to the reader.

20.6.5 Lemma. (b) $\Rightarrow$ (c).

Step 1. Initial objects are strictly initial; i.e., they satisfy 20.3.8 (j).

Let O be an initial object of $\mathscr{E}$. If $f\colon A \to B$ is an arbitrary $\mathscr{E}$-morphism, then the square

(1)
$$\begin{array}{ccc} O & \longrightarrow & O \\ \downarrow & & \downarrow \\ A & \xrightarrow{f} & B \end{array}$$

is a pullback, since the empty coproduct is universal. In particular, if there is a morphism $f\colon A \to O$, then (1) yields an isomorphism $O \to A$.

Step 2. The class of coequalizers is pullback-closed and compositive.

By (i), every coequalizer has a kernel pair. Since $[K, ?]$ preserves limits for every $K \in |\mathscr{E}|$, the class of coequalizers is pullback-closed by (iii) and 18.4.3. By 18.4.8, it is also compositive.

Step 3. Every morphism $f\colon A \to B$ has a factorization $f = h f'$, where f' is a coequalizer of a kernel pair of f, and h is a monomorphism.

The first claim follows immediately from (i) and (iii). Let (a, b) be a kernel pair of h, and let h' be a coequalizer of (a, b). It then follows from step 2 and from 18.4.3 that h', a, b are isomorphisms. By 7.9.9, h is a monomorphism.

Step 4. For any $A \in |\mathscr{E}|$,

(2)
$$p\colon \coprod_{\substack{\alpha \in [G,A] \\ G \in \mathfrak{G}}} G_\alpha \to A \qquad \text{with } p\, in_\alpha = \alpha \text{ is a coequalizer.}$$

This follows immediately from step 3 and condition (iv).

Step 5. Let R be the sieve that is generated by the set $\bigcup\limits_{G \in \mathfrak{G}} [G, A]$. Then $A = \operatorname*{Colim}\limits_{? \in \bar{R}} \Delta^0(?)$.

Applying the universality of coproducts twice, one gets a pullback

(3)
$$\begin{array}{ccc} \coprod\limits_{\alpha,\beta} G_\alpha \sqcap_A G_\beta & \xrightarrow{b} & \coprod\limits_{\beta} G_\beta \\ {\scriptstyle a}\downarrow & & \downarrow{\scriptstyle p} \\ \coprod\limits_{\alpha} G_\alpha & \xrightarrow{p} & A\,, \end{array}$$

where p is defined by (2) and $G_\alpha \sqcap_A G_\beta$ comes from the pullback

(4)
$$\begin{array}{ccc} G_\alpha \sqcap_A G_\beta & \xrightarrow{\gamma_{\alpha,\beta,2}} & G_\beta \\ {\scriptstyle \gamma_{\alpha,\beta,1}}\downarrow & & \downarrow{\scriptstyle \beta} \\ G_\alpha & \xrightarrow{\alpha} & A \end{array} \qquad \text{with} \quad \alpha, \beta \in \bigcup_{G \in \mathfrak{G}} [G, A]\,.$$

By step 4, p is a coequalizer of a and b. Further, the diagonal of the square (4) is a product in $\mathscr{E}/A$ and, by construction of (3),

(5)
$$a\, in_{\alpha,\beta} = in_\alpha\, \gamma_{\alpha,\beta,1}\,, \qquad b\, in_{\alpha,\beta} = in_\beta\, \gamma_{\alpha,\beta,2}\,.$$

With this, the claim follows from 9.6.2.

Step 6. $\mathfrak{G}$ is a topologically generating set for the canonical topology.

Let $f: B \to A$ be an $\mathscr{E}$-morphism and R as in step 5. Further, let R' be the sieve that is generated by the set $\bigcup_{G \in \mathfrak{G}} [G, B]$. By 20.1.5 (4), $R' \subset f^{-1}(R)$. With this, it follows from step 5 and from 20.4.6 that R is a covering sieve for A in the canonical topology. Thus, the claim follows from definition 20.6.1 (b).

Step 7. Let $\tilde{\mathscr{E}}$ be the category of $\mathfrak{U}$-sheaves for the canonical topology on $\mathscr{E}$, and let $I: \tilde{\mathscr{E}} \to \hat{\mathscr{E}} = [\mathscr{E}^0, Ens]$ be the inclusion. By the definition of $\tilde{\mathscr{E}}$, the Yoneda embedding $H_*: \mathscr{E} \to \hat{\mathscr{E}}$ has a factorization $H_* = I\,H'_*$, where H'_* is a full embedding that preserves limits. H'_* also preserves coequalizers of kernel pairs.

Let $a, b: A \to B$ be a kernel pair and $p: B \to P$ a coequalizer of (a, b). Further, let R be the sieve that is generated by the single morphism p. R is a covering sieve for P in the canonical topology. Using step 2, this follows analogously to steps 5 and 6 by means of 9.6.2. Let F be an arbitrary $\mathfrak{U}$-sheaf. By 20.5.3, $F(p)$ is an equalizer of $F(a)$ and $F(b)$, and $[H_p, I(F)]$ is an equalizer of $[H_a, I(F)]$ and $[H_b, I(F)]$, by the Yoneda lemma. Since $\tilde{\mathscr{E}}$ is full in $\hat{\mathscr{E}}$, $H'_*(p)$ is a coequalizer of $H'_*(a)$ and $H'_*(b)$, by 8.7.3.

Step 8. H'_* preserves coproducts.

First, let O be an initial object. By step 1 and 20.4.6, the empty sieve covers O in the canonical topology. For any $\mathfrak{U}$-sheaf F, $F(O)$ is terminal in *Ens*, by 20.5.3. As in step 7, it follows from the Yoneda lemma and from 8.7.3 that $H'_*(O)$ is initial in $\tilde{\mathscr{E}}$. Now, let $A = \coprod A_i$ be a coproduct with a non-empty index set and with injections in_i. Let R be the sieve that is generated by the family $\{in_i\}$. The argument of step 5 shows that $A = \operatorname{Colim}_{?\in \bar{R}} \Delta^0(?)$, where instead of (3), (4) the pullbacks

(6)
$$\begin{array}{ccc} \coprod_{i,j} A_i \sqcap_A A_j & \xrightarrow{\;b\;} & \coprod A_j \\ {\scriptstyle a}\downarrow & & \downarrow{\scriptstyle 1_A} \\ \coprod A_i & \xrightarrow{\;1_A\;} & A \end{array}$$

(7)
$$\begin{array}{ccc} A_i \sqcap_A A_j & \xrightarrow{\;w_{i,j}\;} & A_j \\ {\scriptstyle v_{i,j}}\downarrow & & \downarrow{\scriptstyle in_j} \\ A_i & \xrightarrow{\;in_i\;} & A \end{array}$$

are to be used. Since coproducts are disjoint, $A_i \sqcap_A A_j = O$ for $i \neq j$, and $v_{i,i} = w_{i,i} = 1_{A_i}$ for all i. Let $g: B \to A$ be given. By the universality of coproducts, B is a coproduct with injections $g^{-1}(in_i)$: $g^{-1}(A_i) \to B$. By the above, the argument of step 6 shows that R is a covering sieve for A. Now, let F be any $\mathfrak{U}$-sheaf. 20.5.3 yields an isomorphism $f\colon F(A) \to \Pi\, F(A_i)$, because g and h in 20.5.3 (2) turn out to be isomorphisms. Taking the description of f into account, the claim follows from the Yoneda lemma and from 8.7.3.

Step 9. Every $\mathfrak{U}$-sheaf is representable.

By step 6 and 20.6.3, the inclusion $I\colon \tilde{\mathscr{E}} \to [\mathscr{E}^0, Ens]$ has a left adjoint A, and we can assume that $A\,I = 1_{\tilde{\mathscr{E}}}$. Thus, $H'_* = A\,H_*$. Further, conditions (i) through (iv) are correspondingly valid for $\tilde{\mathscr{E}}$, where $\{H'_*(G) \mid G \in \mathfrak{G}\}$ is a generating set (in either sense by 20.6.2 (a)). Therefore, steps 1 through 5 carry over to $\tilde{\mathscr{E}}$ (as well as steps 6 through 8). If F is any object of $\tilde{\mathscr{E}}$, then $[H'_*(G), F]_{\tilde{\mathscr{E}}}$ is isomorphic to a $\mathfrak{U}$-set by the Yoneda lemma, and so is $\bigcup_{G\in\mathfrak{G}} [H'_*(G), F]_{\tilde{\mathscr{E}}}$. Thus, analogous to (2), there is a coequalizer

$$\pi\colon \coprod_{\mu\in M} H'_*(G_\mu) \to F\,, \tag{8}$$

where M is a $\mathfrak{U}$-set. Now, the coproduct $A = \coprod_{\mu\in M} G_\mu$ exists in $\mathscr{E}$ and, by step 8, one can assume that $\coprod_{\mu\in M} H'_*(G_\mu) = H'_*(A)$. Let $\alpha, \beta\colon G \to H'_*(A)$ be a kernel pair of π. As in (2), there is a coequalizer $\omega\colon H'_*(B) \to G$ for a suitable $B \in |\mathscr{E}|$. Since H'_* preserves limits, α, β determine a morphism $\gamma\colon G \to H'_*(A \sqcap A)$ with $H'_*(pr_1)\,\gamma = \alpha$, $H'_*(pr_2)\,\gamma = \beta$; and γ is a monomorphism by 7.8.5. Since H'_* is a full embedding, there is a $u\colon B \to A \sqcap A$ such that $H'_*(u) = \gamma\,\omega\colon H'_*(B) \to H'_*(A \sqcap A)$. Let $u = w\,v$ be a factorization of u into a coequalizer and a monomorphism, and let D be the domain of w. The factorization exists by step 3. By step 7, $H'_*(v)$ is a coequalizer. Taking into account that $H'_*(w)$ is a monomorphism, it follows from 20.3.8 (d) and 20.6.3 (d) that $H'_*(D)$ is isomorphic to G. So we can assume that $G = H'_*(D)$, $\alpha = H'_*(pr_1\,w)$, $\beta = H'_*(pr_2\,w)$. If E is any object of $\mathscr{E}$, then $[H'_*(E), \alpha]$, $[H'_*(E), \beta]$ is a kernel pair. By the Yoneda lemma, it follows from (iii) that $pr_1\,w$, $pr_2\,w$ is a kernel pair in $\mathscr{E}$ and that it has a coequalizer $p\colon A \to C$. By step 7 and 18.4.3, F is isomorphic to $H'_*(C)$, which was to be shown.

20.6.6 Lemma. (c) $\Rightarrow$ (d).

Let $\tilde{\mathscr{E}}$ be again the category of $\mathfrak{U}$-sheaves for the canonical topology on $\mathscr{E}$. Since $\mathscr{E}$ is equivalent to $\tilde{\mathscr{E}}$, it follows from definition 20.2.6 that $\mathscr{E}$ is complete.

Step 1. Construction of $\mathscr{C}$.

Let $\mathfrak{G}$ be a generating set in the sense of Grothendieck. By 20.6.2 (b), $\mathscr{E}$ is well-powered. For every finite subset of $\mathfrak{G}$ let a product be chosen in such a way that the objects of $\mathfrak{G}$ are the products with only one factor. One now has a set $\mathfrak{G}'$ of objects such that, for any finite subset of $\mathfrak{G}'$, there is a product in $\mathscr{E}$ whose product object belongs to $\mathfrak{G}'$

(compare with 7.7.7). For every object $X \in \mathfrak{G}'$, choose a representative set of monomorphisms with codomain X, including 1_X. Let $\mathcal{C}$ be the full subcategory of $\mathcal{E}$ whose class of objects consists of the domains of the chosen monomorphisms, and let $U: \mathcal{C} \to \mathcal{E}$ be the inclusion. Since $\mathcal{E}$ is well-powered, $|\mathcal{C}|$ is a $\mathfrak{U}$-set, and $\mathfrak{G} \subset |\mathcal{C}|$. Since a product of monomorphisms is a monomorphism, and since "subobjects" of "subobjects" are "subobjects", $\mathcal{C}$ is closed in $\mathcal{E}$ with respect to finite products and "subobjects" up to isomorphisms. By 7.4.2, $\mathcal{C}$ is finitely complete, and U preserves and reflects finite limits. Since $\mathfrak{G} \subset |\mathcal{C}|$, $|\mathcal{C}|$ is again a generating set in the sense of Grothendieck.

Step 2. $|\mathcal{C}|$ is a topologically generating set for the canonical topology on $\mathcal{E}$.

Let $\tilde{\mathcal{E}}_{\mathfrak{V}}$ be the category of sheaves with respect to the universe $\mathfrak{V}$ (where $\mathfrak{U} \in \mathfrak{V}$). $\tilde{\mathcal{E}}_{\mathfrak{V}}$ has all the properties listed in 20.3.8. Clearly, $\tilde{\mathcal{E}}$ is a full subcategory of $\tilde{\mathcal{E}}_{\mathfrak{V}}$, the Yoneda embedding $H_*: \mathcal{E} \to [\mathcal{E}^0, ENS]$ factors through a full embedding $H'_*: \mathcal{E} \to \tilde{\mathcal{E}}_{\mathfrak{V}}$, and H'_* preserves limits. For $A \in |\mathcal{E}|$, let M be the $\mathfrak{U}$-set $M = \bigcup_{X \in |\mathcal{C}|} [X, A]$. For $\mu, \nu \in M$, there are pullbacks

$$\begin{array}{ccc} X_\mu \sqcap_A X_\nu & \longrightarrow & X_\nu \\ \downarrow & & \downarrow{\scriptstyle \nu} \\ X_\mu & \xrightarrow{\mu} & A \end{array} \qquad \begin{array}{ccc} H'_*(X_\mu \sqcap_A X_\nu) & \longrightarrow & H'_*(X_\nu) \\ \downarrow & & \downarrow{\scriptstyle H'_*(\nu)} \\ H'_*(X_\mu) & \xrightarrow{H'_*(\mu)} & H'_*(A) \end{array} \tag{10}$$

and there is an $\tilde{\mathcal{E}}_{\mathfrak{V}}$-morphism

$$\pi: \coprod_{\mu \in M} H'_*(X_\mu) \to H'_*(A) .$$

By 20.3.8, π has a factorization into a coequalizer π' and a monomorphism π''. Since limits and monomorphisms in $\tilde{\mathcal{E}}_{\mathfrak{V}}$ can be constructed as in $[\mathcal{E}^0, ENS]$, since a subfunctor of an *Ens*-valued functor is again *Ens*-valued, and since every $\mathfrak{U}$-sheaf is representable, π', π'' can be chosen in such a way that, for a suitable $B \in |\mathcal{E}|$, $H'_*(B)$ is the domain of π''. Now, $|\mathcal{C}|$ is a generating set in $\mathcal{E}$ and H'_* is a full embedding. From this it follows that π'' is an isomorphism. Therefore, π is a coequalizer. Since coproducts are disjoint in $\tilde{\mathcal{E}}_{\mathfrak{V}}$, there is a bicartesian square

$$\begin{array}{ccc} \bigsqcup H'_*(X_\mu \sqcap_A X_\nu) & \longrightarrow & \coprod H'_*(X_\nu) \\ \downarrow & & \downarrow{\scriptstyle \pi} \\ \coprod H'_*(X_\mu) & \xrightarrow{\pi} & H'_*(A) \end{array}$$

similar to (3). Applying $[?, H'_*(B)]_{\tilde{\mathcal{E}}_{\mathfrak{V}}}$, it follows from the Yoneda lemma, from 20.4.6 and from 9.6.2 that

$$A = \operatorname*{Colim}_{? \in \overline{R}} \Delta^0(?) , \tag{11}$$

where R is the sieve generated by M and where $\overline{R}$ is the corresponding full subcategory of $\mathscr{E}/A$. From this, the claim follows as in 20.6.5, step 6.

Step 3. $U: \mathscr{C} \to \mathscr{E}$ is dense.

For $A \in |\mathscr{E}|$, we consider again the full subcategory $\overline{R}$ of $\mathscr{E}/A$ as in step 2. Let $\overline{R}'$ be the full subcategory of $\overline{R}$ whose objects are the elements of $M = \bigcup_{X \in |\mathscr{C}|} [X, A]$. Now, in (10), $X_\mu \sqcap_A X_\nu$ is a "subobject" of $X_\mu \sqcap X_\nu$. By the construction of $\mathscr{C}$, we can assume that $X_\mu \sqcap_A X_\nu$ is an object of $\mathscr{C}$. The diagonal of the pullback on the left of (10) is a product of μ and ν in $\mathscr{E}/A$, and hence also in $\overline{R}$ and in $\overline{R}'$. With this, it follows easily that $\overline{R}'$ is final in $\overline{R}$ (compare 9.6.2). Thus, the claim follows from 9.2.3 and (11).

Step 4. Conclusion.

By step 2 and 20.6.3, the Yoneda embedding $H_*: \mathscr{E} \to [\mathscr{E}^0, Ens]$ has a left adjoint A that preserves finite limits, and $\mathscr{E}$ is cocomplete. By 17.1.6 and 17.1.6⁰, the functor $U^* = [U^0, Ens]: [\mathscr{E}^0, Ens] \to [\mathscr{C}^0, Ens]$ has a left adjoint $U_!$ which preserves finite limits too. Therefore, $A\, U_!$ is left adjoint to $U^* H_*$, and $A\, U_!$ preserves finite limits. Since $U: \mathscr{C} \to \mathscr{E}$ is dense, $U^* H_*$ is fully faithful, by 17.2.4 and 17.2.3. By 20.3.7, $\mathscr{E}$ is equivalent to the category of sheaves for some topology $\mathfrak{T}$ on $\mathscr{C}$. For every $X \in |\mathscr{C}|$, $[U(?), U(X)]_{\mathscr{E}} = [?, X]_{\mathscr{C}}$ is a sheaf for $\mathfrak{T}$, by construction. Therefore, $\mathfrak{T}$ is smaller than the canonical topology on $\mathscr{C}$, which completes the proofs of 20.6.6 and 20.6.4.

20.7 Problems

20.7.1 Let $\mathscr{C}$ be a $\mathfrak{U}$-category. Determine the category of sheaves for the smallest and for the discrete topology on $\mathscr{C}$.

20.7.2 Let $\mathscr{C}$ be a group G (as a small category). Which Grothendieck topologies exist on $\mathscr{C}$? What follows from this for the sets with a G-operation (non-additive case of 15.1)? Note the special case $G = \{1\}$.

20.7.3 In the additive case (20.3.10), let $\mathscr{C}$ be a ring R (as a small additive category). Describe the additive sieves as subsets of R and as objects of $Add(R^0, Ab) = Mod_R$. How can a topology on $\mathscr{C}$ be described? Consider the special case $R = \mathbf{Z}$. What is the smallest topology on $\mathbf{Z}$ for which a given ideal is a covering sieve? Describe the localizing subcategory of $Ab = Mod_{\mathbf{Z}}$ which corresponds to a topology on $\mathbf{Z}$ and give a survey of all localizations of Ab. Which is the canonical topology on $\mathbf{Z}$?.

In the additive case the topos are exactly the Grothendieck categories with a generating set.

20.7.4 Let $\mathscr{C}$ be a $\mathfrak{U}$-category and let $N \in |\mathscr{C}|$. Then the following statements are equivalent:

(i) N is strictly initial; i.e., N is initial and satisfies 20.3.8 (j).
(ii) In the canonical topology the empty sieve covers N.

20.7.5 Let $\mathscr{D}$ be a category with pullbacks and Σ' a class of $\mathscr{D}$-morphisms.

(a) If
(PT1′) Σ' is closed with respect to pullbacks,
(PT2′) Σ' is closed with respect to composition,
(PT3′) Σ' contains all identity morphisms,
then $Cov(X) = \{\{u\} \mid u \in \Sigma' \text{ and codomain } u = X\}$ (thus every family in $Cov(X)$ consists of exactly one morphism) yields a pretopology on $\mathscr{D}$.

(b) Let Σ' consist of monomorphisms. Then Σ' admits a calculus of right fractions. The topology generated by Σ' remains unchanged if Σ' is replaced by the following class Σ'':
(PT4′) $v \in \Sigma'' \Leftrightarrow v$ is a monomorphism and there is a u with $v\,u \in \Sigma'$.

(c) Let $\mathscr{C}$ be a small category with a topology $\mathfrak{T}$ and $\mathscr{D} = \hat{\mathscr{C}}$. The class of monomorphisms, which are covering with respect to $\mathfrak{T}$ in $\hat{\mathscr{C}}$, satisfies (PT1′), (PT2′), (PT3′), (PT4′) with $\Sigma'' = \Sigma'$, as well as
(PT5′) If a monomorphism m is a colimit object for a colimit of monomorphisms from Σ' (i.e., for a colimit in $[\mathbf{2}, \hat{\mathscr{C}}]$), then $m \in \Sigma'$.

Conversely, every class of monomorphisms in $\hat{\mathscr{C}}$ with these properties is the class of covering monomorphisms for a uniquely determined topology on $\mathscr{C}$.

20.7.6 (a) Find a pretopology $\mathfrak{G}$ on *Ens* such that the topology generated by $\mathfrak{G}$ is the canonical one. Determine also a topologically generating set.

(b) For every $X \in |Ens|$, let $Cov(X)$ consist of the two families $\{1_X\}$ and $\{\emptyset \to X\}$ (which coincide for $X = \emptyset$). This constitutes a pretopology on *Ens*. The topology generated by it is neither smaller nor greater than the canonical one. The $\mathfrak{U}$-sheaves are the constant *Ens*-valued functors, regarded as contravariant functors.

21. Triples

21.1 The Construction of Eilenberg and Moore

21.1.1 Preliminary remarks. Let $(\varphi, F, U, \mathscr{X}, \mathscr{C})$ be an adjoint situation with quasi-inverse adjunction transformations

(1) $$\varepsilon\colon F\,U \to 1_{\mathscr{X}}\,, \qquad \eta\colon 1_{\mathscr{C}} \to U\,F\,.$$

Note that φ: $[F(?),\ ??] \to [?,\ U(??)]$ is uniquely determined by η or by ε (16.5.2). We denote the adjoint situation also by

(2) $$F \overset{\varepsilon}{\underset{\eta}{\dashv}} U(\mathcal{X}, \mathcal{C}) .$$

Setting

(3) $$T = U\,F\,, \quad \eta = 1_{\mathcal{C}} \to T \quad \text{und} \quad \mu = U * \varepsilon * F\colon T\,T \to T\,,$$

by 16.5.12, the rules

(T1) $$\mu\,(\eta * T) = \mu\,(T * \eta) = 1_T\,,$$

(T2) $$\mu\,(\mu * T) = \mu\,(T * \mu)\colon\ T\,T\,T \to T$$

apply. For $X \in |\mathcal{X}|$ and $A = U(X)$, there is the morphism

(4) $$a = (U * \varepsilon)_X = U(\varepsilon_X)\colon\ T(A) \to A\ ,$$

which satisfies the equations

(A1) $$a\,\eta_A = 1_A\,,$$

(A2) $$a\,\mu_A = a\ T(a)\ .$$

The first equation is 16.5.5 (4). To derive the second one, observe that, by 16.1.1 (5),

(5) $$\varepsilon\,(\varepsilon * F\,U) = \varepsilon\,(F\,U * \varepsilon)$$

holds, and applying U to this yields (A2). For $v\colon X \to Y$, $b = U(\varepsilon_Y)$ and $f = U(v)$,

(H) $$\begin{array}{ccc} T(A) & \xrightarrow{T(f)} & T(B) \\ {\scriptstyle a}\downarrow & & \downarrow{\scriptstyle b} \\ A & \xrightarrow{f} & B \end{array}$$

is commutative, since $U * \varepsilon\colon T\,U \to U$ is a natural transformation.

The following concepts are derived from the special case of algebraic categories over *Ens* (18.2) and will find their justification in 21.1.7 and 20.6.10 (a).

21.1.2 Definitions. Let $\mathcal{C}$ be a category. A *tripel* (*monad, triad*) $\boldsymbol{T} = (T, \eta, \mu)$ is an endofunctor $T\colon \mathcal{C} \to \mathcal{C}$ with natural transformations $\eta\colon 1_{\mathcal{C}} \to T$ and $\mu\colon T\,T \to T$ which satisfy (T1) and (T2).

A $\boldsymbol{T}$*-algebra* is a pair (A, a) with $A \in |\mathcal{C}|$ and $a\colon T(A) \to A$ that satisfies (A1) and (A2). A is called the *carrier* or the *underlying object* (in $\mathcal{C}$) of (A, a).

For an arbitrary $A \in |\mathcal{C}|$, $(T(A), \mu_A)$ is always a $\boldsymbol{T}$-algebra, since (T1) and (T2) hold. *Algebras* of this kind are called *free*.

A $\boldsymbol{T}$*-homomorphism* $(A, a) \to (B, b)$ is a 3-tuple (a, b, f) with f: $A \to B$ such that (H) holds. Here f is called the *underlying morphism* (in $\mathcal{C}$).

As long as only one triple is involved, we simply talk of algebras and homomorphisms.

21.1.3 Remarks. (a) By (A2) and (H) there is a canonical homomorphism into the algebra (A, a) given by (μ_A, a, a): $(T(A), \mu_A) \to$ $\to (A, a)$. It will turn out to be a coequalizer of $(\mu_{T(A)}, \mu_A, \mu_A)$ and $(\mu_{T(A)}, \mu_A, T(a))$ (21.4.5).

(b) 21.1.1 states that every adjoint situation yields a triple. We say that it *generates* this triple. Note that the triple is not determined just by the pair of adjoint functors (F, U), but also requires knowing φ, or ε, or η. Different adjoint situations may generate the same triple.

It will be shown that every triple is generated by at least one adjoint situation, in general by many. We consider at once a more general case.

21.1.4 Definition. An *adjoint right triangle* over the category $\mathscr{C}$ consists of two adjoint situations $F' \overset{\varepsilon'}{\underset{\eta'}{-\!\!|}} U'(\mathscr{X}', \mathscr{C})$ and $F \overset{\varepsilon}{\underset{\eta}{-\!\!|}} U(\mathscr{X}, \mathscr{C})$ and a functor $K\colon \mathscr{X}' \to \mathscr{X}$ with $U' = U\,K$. Let the corresponding triples be $\boldsymbol{T}' = (T', \eta', \mu')$ and $\boldsymbol{T} = (T, \eta, \mu)$.

(6) 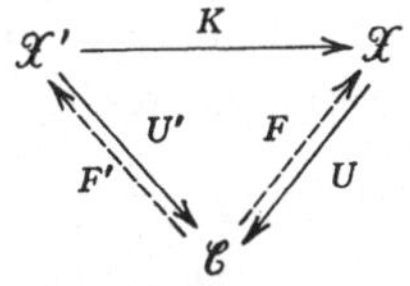

$U' = UK.$

21.1.5 Lemma. *For the adjoint right triangle* (6) *and the natural transformation*

(7) $\theta = (\varepsilon * K\,F')\,(F * \eta')\colon\ F \to K\,F'$ (compare 16.5.2°)

the following rules are valid:

(8) $$\varepsilon * K = (K * \varepsilon')\,(\theta * U')\colon\ F\,U' \to K\,,$$

(9) $$\eta' = (U * \theta)\,\eta\colon\ 1_{\mathscr{C}} \to T'\,,$$

(10) $$(U * \theta)\,\mu = \mu'\,(U * \theta * T')\,(T\,U * \theta)\,.$$

Proof. We make repeated use of 16.1.1 (5).

$$(K * \varepsilon')\,(\theta * U') \overset{(7)}{=\!=} (K * \varepsilon')\,(\varepsilon * K\,F'\,U')\,(F * \eta' * U') =$$
$$= (\varepsilon * K)\,(F\,U\,K * \varepsilon')\,(F * \eta' * U') \overset{(6)}{=\!=} (\varepsilon * K)\,(F\,U' * \varepsilon')\,(F * \eta' * U') =$$
$$= (\varepsilon * K)\,(F * 1_{U'}) = \varepsilon * K\,,$$

where 16.5.5 (4) was used in the last line.

$$(U * \theta)\,\eta \overset{(7)}{=\!=} (U * \varepsilon * K\,F')\,(U\,F * \eta')\,\eta = (U * \varepsilon * K\,F')\,(\eta * U'\,F')\eta'$$
$$\overset{(6)}{=\!=} (U * \varepsilon * K\,F')\,(\eta * U\,K\,F')\,\eta' = (1_U * K\,F')\,\eta' = \eta'\,.$$

Aplpying U to

(11) $$(K * \varepsilon' * F')\,(\theta * U'\,F')\,(F\,U * \theta) \overset{(8)}{=\!=} (\varepsilon * K\,F')\,(F\,U * \theta) = \\ = \theta\,(\varepsilon * F)$$

yields (10).

21.1.6 Definition. Let $\boldsymbol{T}' = (T', \eta', \mu')$ and $\boldsymbol{T} = (T, \eta, \mu)$ be triples over the category $\mathscr{C}$. A *triple-morphism* $\tau\colon \boldsymbol{T} \to \boldsymbol{T}'$ is a natural transformation $\tau\colon T \to T'$ such that the diagrams

(12) $$\begin{array}{ccc} & & T \\ & \overset{\eta}{\nearrow} & \downarrow{\scriptstyle \tau} \\ 1_{\mathscr{C}} & \underset{\eta'}{\searrow} & T' \end{array} \qquad\qquad \begin{array}{ccc} TT & \xrightarrow{\ \mu\ } & T \\ {\scriptstyle (\tau|\tau)}\downarrow & & \downarrow{\scriptstyle \tau} \\ T'T' & \xrightarrow{\ \mu'\ } & T' \end{array}$$

are commutative, where

(13) $$(\tau \mid \tau) = (\tau * T')\,(T * \tau) = (T' * \tau)\,(\tau * T)\,.$$

(9) and (10) say that, given the adjoint right triangle (6), there is a corresponding triple-morphism $U * \theta\colon \boldsymbol{T} \to \boldsymbol{T}'$.

One verifies easily that there is always the identity triple $\mathbf{1} = (1_{\mathscr{C}}, 1_{\mathscr{C}}, 1_{1_{\mathscr{C}}})$ over $\mathscr{C}$ and that for an arbitrary triple $\boldsymbol{T} = (T, \eta, \mu)$ over $\mathscr{C}$, $\eta\colon \mathbf{1} \to \boldsymbol{T}$ is a triple-morphism.

21.1.7 Theorem. *Let $\boldsymbol{T} = (T, \eta, \mu)$ be a triple over the category $\mathscr{C}$.*

(a) *The $\boldsymbol{T}$-algebras and their homomorphisms form a category $\mathscr{C}^{\boldsymbol{T}}$, the Eilenberg-Moore category corresponding to $\boldsymbol{T}$.*

(b) *There is the forgetful functor $U^{\boldsymbol{T}}\colon \mathscr{C}^{\boldsymbol{T}} \to \mathscr{C}$ which assigns to each algebra its carrier and to each homomorphism the underlying $\mathscr{C}$-morphism. $U^{\boldsymbol{T}}$ is faithful.*

(c) *For $f\colon A \to B$ in $\mathscr{C}$, the rule $A \mapsto (T(A), \mu_A)$, $f \mapsto (\mu_A, \mu_B, T(f))$ defines a functor $F^{\boldsymbol{T}}\colon \mathscr{C} \to \mathscr{C}^{\boldsymbol{T}}$.*

(d) *$U^{\boldsymbol{T}}\,F^{\boldsymbol{T}} = T$.*

(14) $$\varepsilon^{\boldsymbol{T}}_{(A,a)} = (\mu_A, a, a)\colon\ (T(A), \mu_A) \to (A, a)$$

defines a natural transformation $\varepsilon^{\boldsymbol{T}}\colon F^{\boldsymbol{T}}\,U^{\boldsymbol{T}} \to 1_{\mathscr{C}^{\boldsymbol{T}}}$, which yields an adjoint situation $F^{\boldsymbol{T}} \underset{\eta}{\overset{\varepsilon^{\boldsymbol{T}}}{-}}\!| \; U^{\boldsymbol{T}}(\mathscr{C}^{\boldsymbol{T}}, \mathscr{C})$ and it generates $\boldsymbol{T}$.

(e) *If the adjoint situation $F' \underset{\eta'}{\overset{\varepsilon'}{-}}\!| \; U'(\mathscr{X}', \mathscr{C})$ generates the triple $\boldsymbol{T}' = (T', \eta', \mu')$ and if there is a triple-morphism $\tau\colon \boldsymbol{T} \to \boldsymbol{T}'$, then there is exactly one functor $K\colon X' \to \mathscr{C}^{\boldsymbol{T}}$ with*

(15) $$U^{\boldsymbol{T}}\,K = U' \quad \text{and} \quad U^{\boldsymbol{T}} * \theta = \tau\,,$$

where $\theta\colon F^{\boldsymbol{T}} \to K\,F'$ is defined by (7).

(f) *If, in particular, $\boldsymbol{T} = \boldsymbol{T}'$ and $\tau = 1_{\boldsymbol{T}}$, then (15) is equivalent to*

(16) $$U^{\boldsymbol{T}}\,K = U' \quad \textit{and} \quad F^{\boldsymbol{T}} = K\,F'\,.$$

Here

(17) $$\theta = 1_F \quad \textit{and} \quad \varepsilon * K = K * \varepsilon'$$

are valid.

Proof. (a) and (b) are evident. (c) follows because μ is a natural transformation.

(d) Since $U^{\boldsymbol{T}}$ is faithful, 21.1.3 (a) and (H) imply that $\varepsilon^{\boldsymbol{T}}$ is a natural transformation and that $U^{\boldsymbol{T}} * \varepsilon^{\boldsymbol{T}} * F^{\boldsymbol{T}} = \mu$. One has further

$$U^{\boldsymbol{T}}(\varepsilon^{\boldsymbol{T}}_{(A,a)})\, \eta_{U^{\boldsymbol{T}}(A,a)} = a\, \eta_A = 1_A = 1_{U^{\boldsymbol{T}}(A,a)}\,,$$
$$U^{\boldsymbol{T}}\big(\varepsilon^{\boldsymbol{T}}_{F^{\boldsymbol{T}}(A)}\, F^{\boldsymbol{T}}(\eta_A)\big) = \mu_A\, T(\eta_A) = 1_{T(A)}\,.$$

Since $U_{\boldsymbol{T}}$ is faithful, this implies (d), by 16.5.7.

(e) We first assume that K exists and that (15) holds. By (15), if $X' \in |\mathfrak{X}'|$, then $K(X')$ must have the carrier $U'(X')$, so $K(X') = (U'(X'), a)$, where by (14), $a = U^{\boldsymbol{T}}(\varepsilon^{\boldsymbol{T}}_{K(X')})$. By (8) and (15), this implies

$$a = U^{\boldsymbol{T}} K(\varepsilon'_{X'})\, U^{\boldsymbol{T}}(\theta_{U'(X')}) = U'(\varepsilon'_{X'})\, \tau_{U'(X')}\,,$$

and thus

(18) $$K(X') = (U'(X'),\, U'(\varepsilon'_{X'})\, \tau_{U'(X')})\,.$$

For $v'\colon X' \to Y'$ in $\mathfrak{X}'$,

(19) $$K(v') = \big(U'(\varepsilon'_{X'})\, \tau_{U'(X')},\, U'(\varepsilon'_{Y'})\, \tau_{U'(Y')},\, U'(v')\big)\,,$$

follows from $U^{\boldsymbol{T}} K = U'$ and (18), which proves the uniqueness of K.

We now show that (18) and (19) define a functor K. (18) is, in fact, a $\boldsymbol{T}$-algebra, for, using 16.1.1 (5),

$$(U' * \varepsilon')\, (\tau * U')\, (\eta * U') \overset{(12)}{=} (U' * \varepsilon')\, (\eta' * U') = 1_{U'}\,,$$

follows and

$$\begin{aligned}(U' * \varepsilon')\, (\tau * U')\, (\mu * U') &\overset{(12)}{=} (U' * \varepsilon')\, (\mu' * U')\, (\tau * T'\, U')\, (T * \tau * U') \\ &\overset{(3)}{=} (U' * \varepsilon')\, (U' * \varepsilon' * F'\, U')\, (\tau * T'\, U')\, (T * \tau * U') \\ &\overset{(5)}{=} (U' * \varepsilon')\, (T'\, U' * \varepsilon')\, (\tau * U'\, F'\, U')\, (T * \tau * U') = \\ &= (U' * \varepsilon')\, (\tau * U')\, (T\, U' * \varepsilon')\, (T * \tau * U')\,,\end{aligned}$$

which at the place X' yields (A1) and (A2) for (18). Furthermore, (19) is a $\boldsymbol{T}$-homomorphism, since

$$\begin{aligned}U'(v')\, U'(\varepsilon'_{X'})\, \tau_{U'(X')} &= U'(\varepsilon'_{Y'})\, U'\, F'\, U'(v')\, \tau_{U'(X')} \\ &= U'(\varepsilon'_{Y'})\, \tau_{U'(Y')}\, T\, U'(v)\,,\end{aligned}$$

and because ε and τ are natural transformations. Thus it follows immediately from (18) and (19) that K is a functor with $U^{\boldsymbol{T}} K = U'$.

Further,

$U * \theta \overset{(7)}{=\!=} (U^{\boldsymbol{T}} * \varepsilon^{\boldsymbol{T}} * KF')(T * \eta') \overset{(14),(18)}{=\!=\!=} (U' * \varepsilon' * F')(\tau * U'F')(T * \eta')$
$\overset{(3),(12)}{=\!=\!=} \mu'(\tau * T')(T * \tau)(T * \eta) \overset{(12)}{=\!=} \tau\,\mu\,(T * \eta) \overset{(T1)}{=\!=} \tau$

so that (15) holds, and (e) is proved.

(f) For $\boldsymbol{T}' = \boldsymbol{T}$ and $\tau = 1_{\boldsymbol{T}}$, it follows from (18) and (3) that

$$KF'(A) = (T(A), \mu_A) = F^{\boldsymbol{T}}(A)\,. \tag{20}$$

For $g: A \to B$ in $\mathscr{C}$, (19) and (20) imply $KF'(g) = (\mu_A, \mu_B, T(g)) = F^{\boldsymbol{T}}(g)$. Thus (16) is valid. Conversely, (16) implies $\theta = (\varepsilon^{\boldsymbol{T}} * F^{\boldsymbol{T}}) \circ (F^{\boldsymbol{T}} * \eta) = 1_{F^{\boldsymbol{T}}}$ on account of (7), $\eta = \eta'$ and 16.5.5 (4°). From this, (15) follows with $\tau = 1_{\boldsymbol{T}}$. By (8), (17) holds.

21.1.8 Definition. For the triple $\boldsymbol{T} = (T, \eta, \mu)$ over $\mathscr{C}$, $\mathscr{C}^{\boldsymbol{T}}$, $U^{\boldsymbol{T}}$, $F^{\boldsymbol{T}}$, $\varepsilon^{\boldsymbol{T}}$ are always to have the meaning given to them in 21.1.7. We call $F^{\boldsymbol{T}} \frac{\varepsilon^{\boldsymbol{T}}}{\eta}| U^{\boldsymbol{T}}(\mathscr{C}^{\boldsymbol{T}}, \mathscr{C})$ the *Eilenberg-Moore situation* corresponding to $\boldsymbol{T}$. The corresponding adjunction isomorphism $[F^{\boldsymbol{T}}(?), ??] \to [?, U^{\boldsymbol{T}}(??)]$ is always denoted by $\varphi^{\boldsymbol{T}}$.

If $F \frac{\varepsilon}{\eta}| U(\mathscr{X}, \mathscr{C})$ generates $\boldsymbol{T}$, then the functor that is determined uniquely by (16) is called the corresponding *comparison functor*. If more than one adjoint situation is being considered, we use the more precise notation $K_{\varepsilon,\eta}$.

21.1.9 Proposition. *Let the two adjoint situations in the adjoint right triangle 21.1.4 (6) be Eilenberg-Moore situations, and let τ be the corresponding triple-morphism. If τ_A is an epimorphism for all $A \in |\mathscr{C}|$, then K is a full embedding.*

Remark. The converse is not true.

Proof. By 21.1.7 (e), K is described by (18) and (19). If $X' = (A, a')$ is a $\boldsymbol{T}'$-algebra, then $K(X') = (A, a'\tau_A)$, by (14) and (18). Since τ_A is an epimorphism, K is injective on objects. For $(a'\tau_A, b'\tau_B, f)$: $(A, a'\tau_A) \to (B, b'\tau_B)$, $f\,a'\tau_A = b'\tau_B\,T(f) = b'\,T'(f)\,\tau_A$. Since τ_A is an epimorphism, $(a', b', f): (A, a') \to (B, b')$ is a homomorphism and, by (19), $K(a', b', f) = (a'\tau_A, b'\tau_B, f)$. Therefore, K is full. By 21.1.7 (b), U' is faithful and thus by (19), so is K.

21.1.10 Theorem. (a) *The adjoint right triangles over the category $\mathscr{C}$, in which all categories are in a fixed universe $\mathfrak{U}$, with the evident composition, are the morphisms of a category $Adj(\mathscr{C})$. The objects of $Adj(\mathscr{C})$ are adjoint situations.*

(b) *Let J_{EM}: $EM(\mathscr{C}) \to Adj(\mathscr{C})$ be the inclusion of the full subcategory $EM(\mathscr{C})$ of $Adj(\mathscr{C})$ whose objects are Eilenberg-Moore situations. There is an adjoint situation.*

$$R \frac{\Psi}{\Phi_{EM}}| J_{EM}(EM(\mathscr{C}), Adj(\mathscr{C})) \qquad \text{with} \qquad \Psi = 1_{RJ_{EM}}, \tag{21}$$

where at the place $F \frac{\varepsilon}{\eta}| \, U(\mathcal{X}, \mathcal{C})$ Φ_{EM} *is the comparison functor* $K_{\varepsilon,\eta}$. *We call* $EM(\mathcal{C})$ *the category of Eilenberg-Moore situations over* $\mathcal{C}$ *and* Φ_{EM} *the corresponding Eilenberg-Moore transformation.*

(c) *The triples over the category* $\mathcal{C}$ *together with the triple-morphisms constitute a category* $Tr(\mathcal{C})$. 21.1.7 (e) *supplies an isomorphism between the dual category of* $Tr(\mathcal{C})$ *and* $EM(\mathcal{C})$.

Proof. (a) is evident. We start by proving (c). Apart from (6), let there be the adjoint right triangle (6') consisting of $F'' \frac{\varepsilon''}{\eta''}| \, U''(\mathcal{X}'', \mathcal{C})$, $F' \frac{\varepsilon'}{\eta'}| \, U'(\mathcal{X}', \mathcal{C})$ and $K' \colon \mathcal{X}'' \to \mathcal{X}'$ with $U'' = U' K'$. With the evident meaning for θ', (7), (8), $U'' = U' K'$ and 16.1.1 (5) then imply

(22) $$(K * \theta')\, \theta \overset{(7)}{=} (K * \varepsilon' * K' F'')\,(K F' * \eta'')\,\theta$$
$$= (K * \varepsilon' * K' F'')\,(\theta * U'' F'')\,(F * \eta'') \overset{(8)}{=} (\varepsilon * K K' F'')\,(F * \eta''),$$

which is transformation (7) corresponding to $K K'$. Now, if all three adjoint situations involved are Eilenberg-Moore situations, then it follows from (22) and 21.1.7 (c) that the composite of triple-morphisms is a triple-morphism and thus that (c) is valid.

(b) From 21.1.7 (f) and (e), one gets for the adjoint right triangle (6)

(23) $$\begin{array}{ccc} \mathcal{X}' & \xrightarrow{K} & \mathcal{X} \\ {\scriptstyle K_{\varepsilon',\eta'}}\downarrow & & \downarrow{\scriptstyle K_{\varepsilon,\eta}} \\ \mathcal{C}^{T'} & \xrightarrow{\overline{K}} & \mathcal{C}^{T} \end{array}$$

where K and $\overline{K}$ correspond to the same triple-morphism $\tau \colon \boldsymbol{T} \to \boldsymbol{T}'$. This determines (23) uniquely, by (22) and by 21.1.7 (e) it is commutative. Thus the rule $K \mapsto \overline{K}$ defines a functor $R \colon Adj(\mathcal{C}) \to EM(\mathcal{C})$ such that $R\, J_{EM} = 1_{EM(\mathcal{C})}$. (23) is a natural transformation $\Phi_{EM} \colon 1_{Adj(\mathcal{C})} \to J_{EM}\, R$. Obviously, $\Phi_{EM} * J_{EM} = 1_{J_{EM}}$ and $R * \Phi_{EM} = 1_R$, so that (b) follows from 16.5.7.

21.1.11 The additive case. For an additive category $\mathcal{C}$, one can consider the categories of additive triples and of additive right triangles over $\mathcal{C}$, in which all categories and functors involved are additive. In this case $\mathcal{C}^{T}$, F^{T}, U^{T} and K in (19) are additive. Thus 21.1.7, 21.1.9 and 21.1.10 are valid correspondingly in the additive case.

21.2 Full Image and Kleisli Categories

21.2.1 Definition. Let $F \colon \mathcal{C} \to \mathcal{X}$ be a functor. The *full image* of F is the category $\mathcal{C}_F$ whose objects are those of $\mathcal{C}$ and whose morphisms $A \to B$ are the 3-tuples $(1_A, 1_B, u)$ with $u \colon F(A) \to F(B)$ and the evident composition. (Compare with 16.2.6, observing 3.3.5). There exist functors

(1) $\operatorname{cl} F \colon \mathcal{C} \to \mathcal{C}_F$, $\quad \operatorname{fim} F \colon \mathcal{C}_F \to \mathcal{X}$ $\quad$ with $F = (\operatorname{fim} F)\,(\operatorname{cl} F)$,

so that the following holds:

(i) cl F is bijective on objects and maps $g: A \to B$ into $(1_A, 1_B, F(g))$,
(ii) fim F is fully faithful.

cl F is usually called the (full) *clone* of F; fim stands for full image.

21.2.2 Proposition. *Let*

$$\begin{array}{ccc} \mathcal{C} & \xrightarrow{F} & \mathcal{X} \\ {\scriptstyle Q}\downarrow & & \downarrow{\scriptstyle R} \\ \mathcal{D} & \xrightarrow{G} & \mathcal{Y} \end{array}$$

be a commutative square in Cat. There is exactly one functor $P: \mathcal{C}_F \to \mathcal{D}_G$ *with*

(2) $$P(\mathrm{cl}\,F) = (\mathrm{cl}\,G)\,Q \quad \textit{and} \quad R(\mathrm{fim}\,F) = (\mathrm{fim}\,G)\,P\,.$$

It is given by

(3) $$P(A) = Q(A) \quad \textit{and} \quad P(1_A, 1_B, u) = (1_{Q(A)}, 1_{Q(B)}, R(u))\,.$$

Proof. (3) defines a functor P: $\mathcal{C}_F \to \mathcal{D}_G$ for which (2) holds, by 21.2.1. Conversely, (3) follows from (2), which proves the uniqueness.

21.2.3 Remarks. (a) For every functor F, the requirements that F' be bijective on objects and F'' fully faithful, determine a factorization $F = F''F'$ uniquely up to an isomorphism. 21.2.1 says that there is such a factorization that is canonical, and 21.2.2 states that this constitutes a natural factorization for the morphisms in *Cat*.

(b) fim F is in general not a monomorphism in *Cat*. However; if two objects A, B of $\mathcal{C}_F$ have the same image under fim F, i.e., if $F(A) = F(B)$, then there is a distinguished isomorphism $(1_A, 1_B, 1_{F(A)})$ between them. If (fim F) $M_1 =$ (fim F) M_2 for $M_1, M_2: \mathcal{B} \to \mathcal{C}_F$, then there is a distinguished isomorphism $\xi: M_1 \to M_2$ for which (fim F) $* \xi$ is the identity natural transformation.

(c) In general cl F is not an epimorphism in *Cat*.

(d) The preceding remarks are valid for *cat* as well as *Cat*. More precisely: If $\mathcal{C}$ is a small category and $\mathcal{X}$ an arbitrary $\mathfrak{U}$-category, then $\mathcal{C}_F$ is also small in $\mathfrak{U}$.

21.2.4 Proposition. *Let* $Cl(\mathcal{C})$ *be the full subcategory of* $\mathcal{C}/Cat$ *whose objects are those functors that are bijective on objects and whose domain is* $\mathcal{C}$. *For the* $\mathcal{C}/Cat$*-morphism*

$$\begin{array}{ccc} & \overset{F}{\nearrow} & \mathcal{X} \\ \mathcal{C} & & \downarrow{\scriptstyle R} \\ & \underset{G}{\searrow} & \mathcal{Y} \end{array}$$

let $R_{F,G}: \mathcal{C}_F \to \mathcal{C}_G$ *be the functor which is determined uniquely by* (2) *with* $Q = 1_{\mathcal{C}}$. *One therefore has*

(4) $$R_{F,G} \circ (\mathrm{cl}\,F) = \mathrm{cl}\,G\,, \qquad R \circ (\mathrm{fim}\,F) = (\mathrm{fim}\,G) \circ R_{F,G}\,.$$

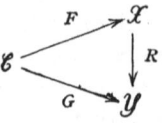

The rules $F \mapsto \mathrm{cl}\, F$, $R \mapsto R_{F,G}$ *define a functor* $L\colon \mathscr{C}/Cat \to Cl(\mathscr{C})$ *which is right adjoint to the inclusion* $I\colon Cl(\mathscr{C}) \to \mathscr{C}/Cat$ *with the adjunction transformation* $\Psi_F = \mathrm{fim}\, F\colon (I\,L)_F \to F$.

Proof. By 21.2.2, L is a functor. fim F is an isomorphism between categories if and only if $F \in |Cl(\mathscr{C})|$. For those F, set $\Phi_F = (\mathrm{fim}\, F)^{-1}$. Thus the statement follows immediately from 16.5.7 or from the dual of 19.4.2 (c).

Remark. By 16.6.6, L can be altered by a natural isomorphism such that LI is the identity functor for $Cl(\mathscr{C})$ (compare 19.4.2 (d)).

21.2.5 If $(\varphi, G, U, \mathscr{X}, \mathscr{C})$ is an adjoint situation, then $(\varphi', \mathrm{cl}\, F, U \circ (\mathrm{fim}\, F))$ is also one with

(5) $\varphi_{A,B} = \varphi_{A,F(B)} \circ (\mathrm{fim})_{A,B}$:

$$[(\mathrm{cl}\, F)(A), B]_{\mathscr{C}_F} \to [F(A), F(B)]_{\mathscr{X}} \to [A, UF(B)]_{\mathscr{C}},$$

since $(\mathrm{cl}\, F)(A) = A$ and $B = (\mathrm{cl}\, F)(B)$ (compare 19.4.2 (f)). It follows easily from (5), 16.5.1 and 21.1.1 (3) that both adjoint situations generate the same triple over $\mathscr{C}$.

21.2.6 Definition. (a) An *adjoint left triangle* over the category $\mathscr{C}$ consists of two adjoint situations $F \frac{\varepsilon}{\eta}| U(\mathscr{X}, \mathscr{C})$ and $F' \frac{\varepsilon'}{\eta'}| U'(\mathscr{X}', \mathscr{C})$ with a functor $N\colon X \to X'$ such that $N\,F = F'$.

(6) 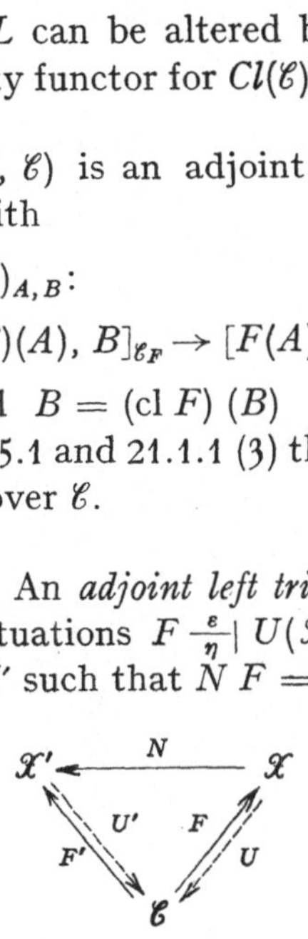

$F' = NF$.

We denote the corresponding triples by $\boldsymbol{T} = (T, \eta, \mu)$ and $\boldsymbol{T}' = (T', \eta', \mu')$.

21.2.7 Lemma. *For the adjoint left triangle* (6) *and the natural transformation*

(7) $$\sigma = (U'\,N * \varepsilon)\,(\eta' * U) : U \to U'\,N$$

the following hold:

(8) $$N * \varepsilon = (\varepsilon' * N)\,(F' * \sigma)\colon F'\,U \to N\,,$$

(9) $$\eta' = (\sigma * F)\,\eta\colon 1_{\mathscr{C}} \to T'\,,$$

(10) $$(\sigma * F)\,\mu = \mu'\,(T' * \sigma * F)\,(\sigma * F\,T)\,,$$

(11) $$\sigma\,(U * \varepsilon) = (U' * \varepsilon' * N)\,(U'\,F' * \sigma)\,(\sigma * F\,U)\,.$$

This is proved by a calculation analogous to the one in 21.1.5. By (9), (10), $\sigma * F$ is a triple-morphism $\boldsymbol{T} \to \boldsymbol{T}'$ corresponding to (6).

21.2.8 Proposition. *Let* $\tau = \sigma * F$ *be the triple-morphism corresponding to* (6). *For* $B \in |\mathscr{C}|$, *the following are equivalent:*

(i) τ_B *is a retraction,*

(ii) $N_{F(A),F(B)}\colon [F(A), F(B)]_{\mathscr{X}} \to [F'(A), F'(B)]_{\mathscr{X}'}$ *is surjective for all* $A \in |\mathscr{C}|$.

Proof. We consider

$$(12) \qquad \chi_{A,B} = \varphi'_{A,F'(B)}\, N_{F(A),F(B)}\, \varphi^{-1}{}_{A,F(B)}\colon$$
$$[A, T(B)] \to [F(A), F(B)] \to [F'(A), F'(B)] \to [A, T'(B)]\,,$$

where φ, φ' are adjunction isomorphisms. By 16.5 (2) and (2°), one gets for $g\colon A \to T(B)$,

$$\chi_{A,B}(g) = U'\, N(\varepsilon_{F(B)}\, F(g))\, \eta'_A \overset{(8)}{=\!=} U'(\varepsilon'_{NF(B)}\, T'(\sigma_{F(B)})\, U'\, N\, F(g)\, \tau_A\, \eta_A =$$
$$= \mu'_B\, T'(\tau_B)\, \tau_{T(B)}\, T(g)\, \eta_A = \tau_B\, \mu_B\, \eta_{T(B)}\, g = \tau_B\, g\,.$$

Thus

$$(13) \qquad \chi_{A,B} = [A, \tau_B]_{\mathscr{C}}\colon [A, T(B)] \to [A, T'(B)]\,,$$

is valid and the statement follows from 5.2.4.

21.2.9 Theorem. *Let* $F' \frac{\varepsilon'}{\eta'}\!| \, U'(\mathscr{X}', \mathscr{C})$ *and* $F \frac{\varepsilon}{\eta}\!| \, U(\mathscr{X}, \mathscr{C})$ *be adjoint situations that generate the triples* $\boldsymbol{T}' = (T', \eta', \mu')$ *or, resp.,* $\boldsymbol{T} = (T, \eta, \mu)$. *Further, let* $F\colon \mathscr{C} \to \mathscr{X}$ *be bijective on objects.*

(a) *For every triple-morphism* $\tau\colon \boldsymbol{T} \to \boldsymbol{T}'$, *there is exactly one functor* $N\colon \mathscr{X} \to \mathscr{X}'$ *with*

$$(14) \qquad N\,F = F' \quad \text{and} \quad \sigma * F = \tau\,,$$

where τ *is defined by* (7).

(b) *If, in particular,* $\boldsymbol{T} = \boldsymbol{T}'$ *and* $\tau = 1_{\boldsymbol{T}}$, *then* (14) *is equivalent to*

$$(15) \qquad N\,F = F' \quad \text{and} \quad U = U'\,N\,.$$

Here

$$(16) \qquad \sigma = 1_U \quad \text{and} \quad N * \varepsilon = \varepsilon' * N$$

hold and N *is fully faithful.*

Proof. (a) We begin by assuming that N exists. By assumption, given $X \in |\mathscr{X}|$, there is exactly one $A \in |\mathscr{C}|$ with $X = F(A)$. By (14),

$$(17) \qquad N(X) = F'(A) \quad \text{and} \quad \sigma_X = \tau_A\,, \qquad X = F(A)$$

has to hold. For $v\colon X \to Y$, $X = F(A)$, $Y = F(B)$,

$$(18) \qquad N(v) = \varepsilon'_{F'(B)}\, F'(\tau_B\, U(v)\, \eta_A)$$

because of

$$N(v) \overset{(17)}{=\!=} N(v)\, 1_{F'(A)} = N(v)\, \varepsilon'_{F'(A)}\, F'(\eta'_A) \overset{(17)}{=\!=} N(v)\, \varepsilon'_{N(X)} F'(\eta'_A) =$$
$$= \varepsilon'_{N(Y)}\, F'\, U'\, N(v)\, F'(\eta'_A) = \varepsilon'_{N(Y)}\, F'\, U'\, N(v)\, F'(\sigma_{F(A)})\, F'(\eta_A)$$

$$\overset{(17)}{=} \varepsilon'_{F'(B)}\, F'\, U'\, N(v)\, F'(\sigma_X)\, F'(\eta_A) = \varepsilon_{F'(B)}\, F'(\sigma_Y)\, F'\, U(v)\, F'(\eta_A)$$

$$= \varepsilon_{F'(B)}\, F'(\tau_B\, U(v)\, \eta_A)\,.$$

We now define N by (17) and (18). Then $N(1_X) = N(1_{F(A)}) = 1_{F'(A)}$. Since $\varepsilon_Y\, F\, U(v) = v\, \varepsilon_X$ and $X = F(A)$, $Y = F(B)$,

$$\mu_B\, T\, U(v) = U(v)\, \mu_A \tag{19}$$

holds. Thus

$$U'\, N(v)\, \tau_A \overset{(18)}{=} \mu'_B\, T'(\tau_B\, U(v)\, \eta_A)\, \tau_A = \mu'_B\, \tau_{T'(B)}\, T(\tau_B\, U(v)\, \eta_A) \tag{20}$$

$$= \tau_B\, \mu_B\, T\, U(v)\, T(\eta_A) \overset{(19)}{=} \tau_B\, U(v)\, \mu_A\, T(\eta_A) \overset{(T1)}{=} \tau_B\, U(v)\,.$$

follows. And for $w\colon Y \to Z$ and $Z = F(C)$, one goes to conclude

$$N(w)\, N(v) \overset{(18)}{=} N(w)\, \varepsilon'_{F'(B)} F'(\tau_B U(v)\, \eta_A) = \varepsilon'_{F'(C)}\, F'\, U'\, N(w)\, F'(\tau_B\, U(v)\, \eta_A)$$

$$\overset{(20)}{=} \varepsilon'_{F'(C)}\, F'(\tau_C\, U(w)\, U(v)\, \eta_A) \overset{(18)}{=} N(w\, v)\,.$$

Thus N is a functor. For $v = F(g)$,

$$\tau_B\, U\, F(g)\, \eta_A = T'(g)\, \tau_A\, \eta_A = T'(g)\, \eta'_A\,.$$

By (18), this implies

$$N\, F(g) = \varepsilon'_{F'(B)}\, F'\, T'(g)\, F'(\eta'_A) = F'(g)\, \varepsilon'_{F'(A)}\, F'(\eta'_A) = F'(g)\,,$$

and thus $N\, F = F'$. Now, (14) follows from this and from

$$\sigma_{F(A)} \overset{(7)}{=} U'\, N(\varepsilon_{F(A)})\, \eta'_{T(A)} \overset{(20)}{=} \tau_A\, U(\varepsilon_{F(A)})\, \eta_{T(A)} = \tau_A\,.$$

(b) By (17), for $\tau = 1_T$, (14) implies $\sigma = 1_U$. Thus (15) follows. (15) and $\eta = \eta'$ imply (16), because of (7) and (8), and thus (15) implies (14) for $\tau = 1_T$. The last statement follows from 21.2.8.

21.2.10 Definitions. (a) Let $F^{\boldsymbol{T}} \frac{\varepsilon^{\boldsymbol{T}}}{\eta}| \, U^{\boldsymbol{T}}(\mathscr{C}^{\boldsymbol{T}}, \mathscr{C})$ be the Eilenberg-Moore situation corresponding to the triple $\boldsymbol{T}$. The *Kleisli category* $\mathscr{C}_{\boldsymbol{T}}$ corresponding to $\boldsymbol{T}$ is the full image of $F^{\boldsymbol{T}}$; i.e., $\mathscr{C}_{\boldsymbol{T}} = \mathscr{C}_{F^{\boldsymbol{T}}}$. We set $F_{\boldsymbol{T}} = \operatorname{cl} F^{\boldsymbol{T}}$, $K_{\boldsymbol{T}} = \operatorname{fim} F^{\boldsymbol{T}}$, $U_{\boldsymbol{T}} = U^{\boldsymbol{T}} K_{\boldsymbol{T}}$. By 21.2.5, there is the adjoint situation $F_{\boldsymbol{T}} \frac{\varepsilon_{\boldsymbol{T}}}{\eta}| \, U_{\boldsymbol{T}}(\mathscr{C}_{\boldsymbol{T}}, \mathscr{C})$ with adjunction isomorphism $\varphi_{\boldsymbol{T}}$ which also generates $\boldsymbol{T}$. We call it the *Kleisli situation* corresponding to $\boldsymbol{T}$. $\mathscr{C}_{\boldsymbol{T}}, F_{\boldsymbol{T}}, U_{\boldsymbol{T}}, \varepsilon_{\boldsymbol{T}}, K_{\boldsymbol{T}}, \varphi_{\boldsymbol{T}}$ are always to have this special meaning.

(b) If $F \frac{\varepsilon}{\eta}| \, U(\mathscr{X}, \mathscr{C})$ also generates $\boldsymbol{T}$, then the functor $N\colon \mathscr{C}_{\boldsymbol{T}} \to \mathscr{X}$ that is uniquely determined by (15) is called the corresponding *comparison functor*. If several adjoint situations are being considered, then we denote it more precisely by $N_{\varepsilon,\eta}$. By 21.1.7,

$$K_{\varepsilon,\eta}\, N_{\varepsilon,\eta} = K_{\boldsymbol{T}} = K_{\varepsilon_{\boldsymbol{T}},\eta} \tag{21}$$

holds.

21.2.11 Theorem. (a) *The adjoint left triangles over the category* $\mathscr{C}$, *in which all categories involved belong to the fixed universe* $\mathfrak{U}$, *are the morphisms of a category* $\overline{Adj}(\mathscr{C})$ *with the evident composition. The objects of* $\overline{Adj}(\mathscr{C})$ *are adjoint situations.*

(b) *Let* J_{Kl}: $Kl(\mathscr{C}) \to \overline{Adj}(\mathscr{C})$ *be the inclusion of the full subcategory of* $\overline{Adj}(\mathscr{C})$ *whose objects are Kleisli situations. There is an adjoint situation*

$$J_{Kl} \frac{\Psi_{Kl}}{\Phi}\Big| \bar{R}(\overline{Adj}(\mathscr{C}), Kl(\mathscr{C})) \quad \textit{with} \quad \Phi = 1_{\bar{R}J_{Kl}}, \tag{22}$$

where Ψ_{Kl} *at the place* $F \frac{\varepsilon}{\eta}| U(\mathscr{X}, \mathscr{C})$ *is the comparison functor* $N_{\varepsilon,\eta}$. *We call* $Kl(\mathscr{C})$ *the category of Kleisli situations over* $\mathscr{C}$ *and* Ψ_{Kl} *the corresponding Kleisli transformation.*

(c) $N_{\varepsilon,\eta}$ *is an isomorphism if and only if* F *is bijective on objects, i.e., if* F *is an object of* $Cl(\mathscr{C})$.

(d) *By* 21.2.9, $Kl(\mathscr{C})$ *is isomorphic to the category* $Tr(\mathscr{C})$ *of triples over* $\mathscr{C}$. *From this and* 21.1.10 (c), *one has an isomorphism* I: $EM(\mathscr{C}) \to (Kl(\mathscr{C}))^0$.

Proof. (a) and (d) are evident.

(b) The proof is analogous to the one for 21.1.10 (b). In particular, analogously to 21.1(22), one obtains for σ: $U \to U' N$ and σ': $U' \to U'' N'$, where all the notations are to have the evident meaning,

$$(\sigma' * N)\, \sigma = (U''\, N'\, N * \varepsilon)\, (\eta'' * U)\,, \tag{23}$$

where the transformation on the right is transformation (7) corresponding to $N'\, N$.

(c) If F is bijective on objects, then by 21.2.9, there is a functor N: $\mathscr{X} \to \mathscr{C}_T$, which also satisfies (15). From the uniqueness statement it follows that N is inverse to $N_{\varepsilon,\eta}$. The converse is trival.|

21.2.12 Remarks. (a) 21.2.9 and 21.2.11 are similarly valid in the additive case, as is easily verified.

(b) By means of φ_T and 16.5.2, $\mathscr{C}_T$ can be replaced by a canonically isomorphic category. The correspondence $v \mapsto U_T(v)\, \eta_A$, for v: $F_T(A) \to F_T(B)$, is a bijection $[F_T(A), F_T(B)] \to [A, T(B)]$. If w: $F_T(B) \to F_T(C)$ and $f = U_T(v)\, \eta_A$, $g = U_T(w)\, \eta_B$, one obtains the composition rule $U_T(w\, v)\, \eta_A = \mu_C\, T(f)\, g$. This is Kleisli's original construction.

(c) 21.2.6 through 21.2.11 are not dual to 21.1 in the sense of 3.6.6 (although duality concepts for bicategories form the foundation). The proofs of 21.1.7 and 21.2.9 are distinctly different, and in contrast to 21.2.11 (c), there is no corresponding simple characterization of Eilenberg-Moore situations up to an isomorphism (see below, 21.5). Dualization in the sense of 3.6.6 creates cotriples from triples, co-adjoint left and right triangles.

21.3 Limits and Colimits in Eilenberg-Moore Categories

21.3.1 Definition. A *triple* (T, η, μ) is called *degenerate* (*idempotent*) if $\mu: T\,T \to T$ is an isomorphism.

21.3.2 Lemma. *Let* $\boldsymbol{T} = (T, \eta, \mu)$ *be a triple over* $\mathscr{C}$ *and* $A \in |\mathscr{C}|$.
(a) *If* $T(\eta_A)$ *or* $\eta_{T(A)}$ *is an isomorphism, then* μ_A *is an isomorphism.*
(b) *If* μ_A *is an isomorphism, then* $\eta_{T(A)} = \mu_A^{-1}$.
(c) *If* $\eta_{T(A)} = T(\eta_A)$, *then* A *is the carrier of a* $\boldsymbol{T}$*-algebra if and only if* η_A *is an isomorphism.*

Proof. (a) and (b) follow immediately from (T1) in 21.1.1.

(c) If $a\,\eta_A = 1_A$, then $\eta_A\,a = T(a)\,\eta_{T(A)} = T(a)\,T(\eta_A) = 1_{T(A)}$, and thus $\eta_A = a^{-1}$. If, conversely, η_A is an isomorphism, then, by (a) and (b), $\eta_A^{-1}\,T(\eta_A^{-1}) = \eta_A^{-1}\,\mu_A$ and, therefore, (A, η_A^{-1}) is an algebra.

21.3.3 Proposition. *If* $\boldsymbol{T} = (T, \eta, \mu)$ *is a degenerate triple over* $\mathscr{C}$, *then* $U^{\boldsymbol{T}}: \mathscr{C}^{\boldsymbol{T}} \to \mathscr{C}$ *induces an isomorphism between* $\mathscr{C}^{\boldsymbol{T}}$ *and the strictly full subcategory* $\mathscr{C}'$ *of* $\mathscr{C}$ *which is determined by* $|\mathscr{C}| = \{A \mid A \in |\mathscr{C}|$ *and* η_A *is isomorphic*$\}$. *Further,* $K_{\boldsymbol{T}}: \mathscr{C}_{\boldsymbol{T}} \to \mathscr{C}^{\boldsymbol{T}}$ *is an equivalence.*

Proof. The first statement follows immediately from 21.3.2; by 16.3.6, the second one follows from the fact that here $(\eta_A^{-1}, \mu_A, \eta_A)$: $(A, \eta_A^{-1}) \to (T(A), \mu_A)$ is an isomorphism.

21.3.4 Remarks. For degenerate triples, by 21.3.3 and 16.6.1, the existence of limits and colimits is inherited by $\mathscr{C}^{\boldsymbol{T}}$ from $\mathscr{C}$; in the general case only the existence of limits is inherited. The corresponding question for colimits gives rise to considerable difficulties and so far it has only been possible to answer it in the affirmative under certain restrictions. We shall return to this later. The following lemma contains a characteristic proof technique, in which monomorphisms or, resp., epimorphisms can later be replaced by limits or, resp., colimits.

21.3.5 Lemma. *Let* (A, a), (B, b), (C, c) *be* $\boldsymbol{T}$*-algebras and* $f: A \to B$, $g: B \to C$ $\mathscr{C}$*-morphisms such that* $(a, c, g\,f)$ *is a homomorphism.*
(a) *If* (b, c, g) *is a homomorphism and* g *a monomorphism in* $\mathscr{C}$, *then* (a, b, f) *is a homomorphism.*
(b) *If* (a, b, f) *is a homomorphism and* $T(f)$ *an epimorphism in* $\mathscr{C}$, *then* (b, c, g) *is a homomorphism.*
(c) *Let* $m: D \rightarrowtail C$ *be a* $\mathscr{C}$*-monomorphism. If there is a* $d: T(D) \to D$ *with* $m\,d = c\,T(m)$, *then* (D, d) *is an algebra and* (d, c, m) *a homomorphism.* (*There is at most one such* d).

The proof is trivial.

21.3.6 Corollary. (a) *If* $\mathscr{C}$ *is well-powered, then* $\mathscr{C}^{\boldsymbol{T}}$ *is well-powered.*
(b) *If* $\mathscr{C}$ *is co-well-powered and if* $U^{\boldsymbol{T}}$ *preserves epimorphisms, then* $\mathscr{C}^{\boldsymbol{T}}$ *is co-well-powered.*

Proof. U^T is faithful and hence it reflects monomorphisms and epimorphisms. U^T preserves monomorphisms. Thus the statements follow from 21.3.5 and the construction of $\mathscr{C}^T$ (compare also 21.3.9 (a) below).

21.3.7 Definitions. Let $U: X \to \mathscr{C}$ be a functor.

(a) A *$\mathscr{C}$-morphism* $f: U(X) \to B$ is *lifted* (*uniquely*) by U if there is a (unique) $\mathscr{X}$-morphism v with domain X such that $U(v) = f$. There are analogous definitions for morphisms $g: A \to U(Y)$ and for natural transformations of diagrams.

(b) U *lifts isomorphisms* (*uniquely*) if, given any isomorphism $f: U(X) \to B$ in $\mathscr{C}$ for an arbitrary $X \in |\mathscr{X}|$, there is a (unique) isomorphism v with domain X such that $U(v) = f$.

U *creates isomorphisms* if U reflects isomorphisms and lifts them uniquely.

(c) Let $D: \Sigma \to \mathscr{X}$ be a diagram in $\mathscr{X}$ and (L, λ) a limit of UD. U *lifts this limit* (*uniquely*) if there is a (unique) limit (Z, ξ) of D such that $U(Z) = L$, $U * \xi = \lambda$.

U *creates limits of* D if the following holds:

(i) $U D$ has a limit,

(ii) U lifts every limit of $U D$ uniquely,

(iii) U reflects limits of D.

U *creates limits* if U creates limits of all diagrams D which satisfy (i).

(d) The dual definitions to (c) are made for colimits.

21.3.8 Remarks. (a) The statement that U lifts isomorphisms (uniquely) is equivalent to the following: for every isomorphism $g: A \to U(Y)$ in $\mathscr{C}$ there is, for an arbitrary $Y \in |\mathscr{X}|$, a (unique) isomorphism w with codomain Y such that $U(w) = g$. Consider g^{-1}.

(b) The fact that U lifts isomorphisms does not exclude the existence of an isomorphism $f: U(X) \to B$ or, resp., $g: A \to U(Y)$ that can be lifted to a non-isomorphic $\mathscr{X}$-morphism, unless U reflects isomorphisms. An analogous remark applies to the lifting of limits and natural transformations.

(c) The statement that U lifts a limit of $U D$ is equivalent to D possessing a limit and U preserving the limits of D. If U lifts isomorphisms (uniquely) and a limit of $U D$, then U lifts all limits of $U D$ (uniquely).

(d) Let Σ be a connected diagram (i. e., the category $\mathscr{P}(\Sigma)$ corresponding to Σ by 6.3.4 is connected). If for all diagrams $D: \Sigma \to \mathscr{X}$ for which $U D$ has a limit, U lifts every limit (uniquely) or, resp., creates limits, then U lifts isomorphisms (uniquely) or, resp., it creates them. This is shown by the constant diagrams of type Σ (9.1.3).

(e) If $U: \mathscr{X} \to \mathscr{C}$ is an equivalence, then U is an isomorphism of categories if and only if U lifts isomorphisms uniquely. This follows from 16.3.6. In this way criteria for isomorphisms between categories can be obtained from criteria for equivalences.

21.3.9 Theorem. *Let* $F^{\boldsymbol{T}} \frac{\varepsilon^{\boldsymbol{T}}}{\eta}\!\dashv U^{\boldsymbol{T}}(\mathscr{C}^{\boldsymbol{T}}, \mathscr{C})$ *be the Eilenberg-Moore situation corresponding to the triple* $\boldsymbol{T} = (T, \eta, \mu)$ *and* $D\colon \Sigma \to \mathscr{C}^{\boldsymbol{T}}$ *a diagram.*

(a) $U^{\boldsymbol{T}}$ *creates limits, in particular isomorphisms.*

(b) *If* $U^{\boldsymbol{T}} D$ *has a colimit which is preserved by* T *and by* $T\,T$, *then* $U^{\boldsymbol{T}}$ *creates colimits of* D.

Proof. (a) The assertion about isomorphisms follows immediately from the construction of $\mathscr{C}^{\boldsymbol{T}}$ and is also a consequence of the following, by 21.3.8 (d).

To simplify the notation we write U, F, ε here instead of $U^{\boldsymbol{T}}$, $F^{\boldsymbol{T}}$, $\varepsilon^{\boldsymbol{T}}$. First, let Σ be non-empty and (L, λ) a limit of $U\,D$. Then $(U * \varepsilon * D)\,(T * \lambda)\colon\; T(L)_\Sigma \to U\,D$ is a natural transformation and there is exactly one $\mathscr{C}$-morphism $l\colon T(L) \to L$ with

(1) $$\lambda\, l_\Sigma = (U * \varepsilon * D)\,(T * \lambda)\,.$$

We show that (L, l) is a $\boldsymbol{T}$-algebra. From (1), 16.1.1 (5) and 16.5.5 (4)

$$\lambda(l\,\eta_L)_\Sigma \overset{(1)}{=} (U * \varepsilon * D)\,(T * \lambda)\,(\eta_L)_\Sigma = (U * \varepsilon * D)\,(\eta * U\,D)\,\lambda = \lambda$$

follows and thus $l\eta_L = 1_L$, by the definition of limits. From (1), 21.1.1 (5) and 16.1.1 (5)

$$\begin{aligned}\lambda(l\,T(l))_\Sigma &\overset{(1)}{=} (U * \varepsilon * D)\,(T * \lambda)\,T(l)_\Sigma \overset{(1)}{=} (U * \varepsilon * D)\,(T\,U * \varepsilon * D)\circ\\ &\circ (T\,T * \lambda) = (U * \varepsilon * D)\,(U * \varepsilon * F\,U\,D)\,(T\,T * \lambda) =\\ &= (U * \varepsilon * D)\,(T * \lambda)\,[U\,(\varepsilon_{F(L)})]_\Sigma \overset{(1)}{=} \lambda\, l_\Sigma(\mu_L)_\Sigma\end{aligned}$$

follows, and again by the definition of limits, one has $l\,T(l) = l\,\mu_L$. Therefore, (L, l) is an algebra.

Now, since $U(\varepsilon_{(A,a)}) = a$ for $(A, a) \in |\mathscr{C}^{\boldsymbol{T}}|$, (1) says that for every $i \in |\Sigma|$, λ_i underlies a uniquely determined homomorphism $(L, l) \to D(i)$. Thus we have the result: U lifts the natural transformation $\lambda\colon L_\Sigma \to UD$ uniquely to a natural transformation $\varrho\colon (L, l)_\Sigma \to D$ with $U * \varrho = \lambda$. We show that $\big((L, l), \varrho\big)$ is a limit of D. So let $\alpha\colon (A, a)_\Sigma \to D$ be a natural transformation. There is exactly one $\mathscr{C}$-morphism $f\colon A \to L$ with

(2) $$\lambda\, f_\Sigma = U * \alpha\,.$$

Since $a = U(\varepsilon_{(A,a)})$,

$$\begin{aligned}\lambda(f\,a)_\Sigma &\overset{(2)}{=} (U * \alpha)\,a_\Sigma = (U * \varepsilon * D)\,(T\,U * \alpha)\\ &\overset{(2)}{=} (U * \varepsilon * D)\,(T * \lambda)\;\,T(f)_\Sigma \overset{(1)}{=} \lambda\, l_\Sigma\, T(f)_\Sigma\end{aligned}$$

follows. By the definition of limits, this implies that (a, l, f) is a homomorphism and that $\big((L, l), \varrho\big)$ is a limit of D. Since the natural transformation λ is lifted uniquely, the statement follows for a non-empty Σ.

If Σ is empty, then every limit object Z of $U\,D$ is terminal in $\mathscr{C}$. One verifies easily that Z is the carrier of a uniquely determined $\boldsymbol{T}$-algebra and that it is terminal in $\mathscr{C}^{\boldsymbol{T}}$.

(b) Now let (L, λ) be a colimit of $U\,D$, $(T(L), T*\lambda)$ a colimit of $T\,U\,D$ and $(T\,T(L), T\,T*\lambda)$ a colimit of $T\,T\,U\,D$.

If Σ is empty, then L and $T(L)$ are initial in $\mathscr{C}$. $T(L)$ is the carrier of an initial algebra. Since η_L is an isomorphism, L is the carrier of a single algebra.

Now let Σ be non-empty. Since $(T(L), T*\lambda)$ is a colimit, there is exactly one $\mathscr{C}$-morphism $l\colon T(L) \to L$ with

(3) $$l_\Sigma\,(T*\lambda) = \lambda(U*\varepsilon*D)\,.$$

Since (L, λ) is a colimit, $l\,\eta_L = 1_L$ follows from

$$(l\,\eta_L)_\Sigma\,\lambda = l_\Sigma\,(T*\lambda)\,(\eta*U\,D) \overset{(3)}{=\!=} \lambda\,(U*\varepsilon*D)\,(\eta*U\,D) = \lambda\,.$$

Since $(T\,T(L), T\,T*\lambda)$ is a colimit, $l\,T(l) = l\,\mu_L$ follows from

$$\begin{aligned}(l\,T(l))_\Sigma\,(T\,T*\lambda) &\overset{(3)}{=\!=} l_\Sigma\,(T*\lambda)\,(T\,U*\varepsilon*D) \overset{(3)}{=\!=}\\ \lambda\,(U*\varepsilon*D)\,(T\,U*\varepsilon*D) &= \lambda(U*\varepsilon*D)\,(U*\varepsilon*F\,U\,D)\\ &\overset{(3)}{=\!=} l_\Sigma\,(T*\lambda)\,(\mu*U\,D) = l_\Sigma\,(\mu_L)_\Sigma\,(T\,T*\lambda)\end{aligned}$$

Thus (L, l) is an algebra and, by (3), U lifts the natural transformation uniquely to a natural transformation $\varrho\colon D \to (L, l)_\Sigma$. If $\alpha\colon D \to (A, a)_\Sigma$ is a natural transformation, then there is exactly one $\mathscr{C}$-morphism $f\colon L \to A$ with

(4) $$f_\Sigma\,\lambda = U*\alpha\,.$$

Since $(T(L), T*\lambda)$ is a colimit, (l, a, f) is a homomorphism, so that the statement follows as in (a).

21.3.10 Theorem. *Let $\mathscr{C}^{\boldsymbol{T}}$ have coequalizers. If $\mathscr{C}$ has colimits of type Σ/K, then so does $\mathscr{C}^{\boldsymbol{T}}$. If $\mathscr{C}$ has (finite) coproducts, then $\mathscr{C}^{\boldsymbol{T}}$ is (finitely) cocomplete.*

Proof. By 8.4.2, the second statement follows from the first one. In the following we shall again use the simplified notation U, F, ε instead of $U^{\boldsymbol{T}}, F^{\boldsymbol{T}}, \varepsilon^{\boldsymbol{T}}$. Let $D\colon \Sigma/K \to \mathscr{C}^{\boldsymbol{T}}$ be a diagram, (L, λ) a colimit of $U\,D$ and (R, ϱ) a colimit of $T\,U\,D$. If Σ is empty, then L is initial in $\mathscr{C}$ and $F(L)$ is initial in $\mathscr{C}^{\boldsymbol{T}}$.

Now let Σ be non-empty. In view of later applications, we divide the proof into several steps.

Step 1. There is exactly one $\mathscr{C}$-morphism $r\colon R \to L$ with

(5) $$r_\Sigma\,\varrho = \lambda\,(U*\varepsilon*D)$$

and exactly one $\mathscr{C}$-morphism $s\colon R \to T(L)$ with

(6) $$s_\Sigma\,\varrho = T*\lambda\,.$$

Here r is an epimorphism. Every natural transformation $\alpha: D \to (A, a)_\Sigma$ determines uniquely an $f: L \to A$ with

(7) $$U * \alpha = f_\Sigma \lambda ,$$

and

(8) $$\begin{array}{ccccc} R & \xrightarrow{s} & T(L) & \xrightarrow{T(f)} & T(A) \\ & \searrow^{r} & & & \downarrow a \\ & & L & \xrightarrow{f} & A \end{array}$$

is commutative.

Proof. (5) and (6) follow from the definition of colimits. If, for $u, v: L \to M$, $u\,r = v\,r$, then

$$u_\Sigma \lambda(U * \varepsilon * D) \overset{(5)}{=\!=} (u\,r)_\Sigma \varrho = (v\,r)_\Sigma \varrho = v_\Sigma \lambda(U * \varepsilon * D) .$$

If one takes into account that $U * \varepsilon * D$ is a retraction (with the coretraction $\eta * U\,D$) and (L, λ) a colimit, then $u = v$ follows. Thus r is an epimorphism. (7) is evident. The commutativity of (8) follows from inserting Σ as an index everywhere and then adding (8) to

(9) $$\begin{array}{ccccc} TUD & \xrightarrow{\varrho} & R_\Sigma & & \\ {\scriptstyle U*\varepsilon*D}\downarrow & & & \searrow^{r_\Sigma} & \\ UD & & \xrightarrow{\lambda} & & L_\Sigma \end{array},$$

because (R, ϱ) is a colimit of $T\,U\,D$.

Step 2. Let (8) be commutative, r a $\mathscr{C}$-epimorphism and (A, a) an algebra. Let $(\mu_L, c, p): (T(L), \mu_L) \to (C, c)$ be a coequalizer of $(\mu_R, \mu_L, T(r))$ and $(\mu_R, \mu_L, \mu_L T(s))$.

(10) $$\begin{array}{cccc} T(R) & \xrightarrow{T(s)} & TT(L) & \\ & \searrow^{T(r)} & \downarrow \mu_L & \\ & & T(L) & \\ & & & \searrow^{p} \quad C. \end{array}$$

Then

(11) $$\begin{array}{ccccccc} R & \xrightarrow{s} & T(L) & \xrightarrow{T(p\eta_L)} & T(C) & \xrightarrow{T(g)} & T(A) \\ & \searrow^{r} & & \searrow^{p} & & & \downarrow a \\ & & L & \xrightarrow[p\eta_L]{} & C & \xrightarrow[g]{} & A \end{array}$$

is commutative, where $(c, a, g): (C, c) \to (A, a)$ is a homomorphism with

(12) $$g\,p\,\eta_L = f .$$

(11) and (12) determine (c, a, g) uniquely.

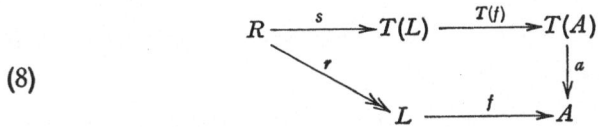

Proof. One has

$$(13)\quad p\,\eta_L\, r = p\, T(r)\,\eta_R = p\,\mu_L\, T(s)\,\eta_R = p\,\mu_L\,\eta_{T(L)}\, s = p\, s\,,$$

$$(14)\quad a\, T(f)\, T(r) \overset{(8)}{=} a\, T(a)\, T\, T(f)\, T(s) = a\,\mu_A\, T\, T(f)\, T(s) = \\ = a\, T(f)\,\mu_L\, T(s)\,.$$

All morphisms in (14) underly homomorphisms. By the definition of p, there is exactly one homomorphism (c, a, g) with

$$(15)\quad a\, T(f) = g\, p\,,$$

and $f\, r \overset{(8)}{=} a\, T(f)\, s = g\, p\, s = g\, p\,\eta_L\, r$ follows. Since r is an epimorphism, (12) follows and the meaning of (13), (14), (15) is that (11) is commutative. Conversely, (15) is implied by (11) and (12). Since (μ_L, c, p) is an epimorphism, (c, a, g) is uniquely determined.

Step 3. $(p\,\eta_L)_\Sigma\,\lambda$ underlies a natural transformation $\gamma\colon D \to (C, c)_\Sigma$ and $((C, c), \gamma)$ is a colimit of D.

Proof. By assumption,

$$(16)\quad p\,\mu_L = c\, T(p)\,,$$

and, using 16.1.1 (5),

$$(p\,\eta_L)_\Sigma\,\lambda(U * \varepsilon * D) \overset{(9)}{=} (p\,\eta_L\, r)_\Sigma\,\varrho \overset{(11)}{=} (p\, s)_\Sigma\,\varrho \overset{(6)}{=} p_\Sigma\,(T * \lambda) = \\ = [p\,\mu_L\, T(\eta_L)]_\Sigma\,(T * \lambda) \overset{(15)}{=} [c\, T(p)\, T(\eta_L)]_\Sigma\,(T * \lambda)$$

follows. This means that $(p\,\eta_L)_\Sigma\,\lambda$ underlies a natural transformation γ. The statement then follows from the uniqueness statements in step 1 and step 2.

21.3.11 Remarks. (a) 21.3.9 (b) and 21.3.10 are due to Linton, as is the following remark.

(b) $(\mu_R, \mu_L, T(r))$ and $(\mu_R, \mu_L, \mu_L\, T(s))$ in 21.3.10 have the common coretraction $(\mu_L, \mu_R, T(t))$, where $t\colon L \to R$ is the uniquely determined $\mathscr{C}$-morphism with $t_\Sigma\,\lambda = \varrho(\eta * U^{\mathbf{T}}\, D)$. It is therefore possible to weaken the assumption in 21.3.10. However, there seems to be little hope of finding criteria for the existence of coequalizers by means of such a weakened condition, because of the following fact:

If a category has finite coproducts and coequalizers for pairs of morphisms with a common coretraction, then it is finitely cocomplete.

This is based on the fact that a pair of morphisms $v, w\colon X \to Y$ has a coequalizer if and only if there is a coequalizer for the following pair of morphisms:

$$X \sqcup Y \sqcup X \sqcup X \underset{q}{\overset{p}{\rightrightarrows}} X \sqcup Y$$

with $p\, in_1 = p\, in_3 = p\, in_4 = j_1$, $p\, in_2 = j_2$, $q\, in_1 = j_1$, $q\, in_2 = j_2$, $q\, in_3 = j_2\, v$, $q\, in_4 = j_2\, w$, where the notation for the corresponding injections is the obvious one. $h\,j_1 = in_1$, $h\,j_2 = in_2$ defines a common coretraction h for p and q.

(c) There still remains the question as to when $\mathscr{C}^{\boldsymbol{T}}$ has coequalizers. By 21.3.9, this is certainly the case if $\mathscr{C}$ has coequalizers and T preserves them.

If, for instance, R is a commutative ring and M an associative R-algebra with a unit $n: R \to M$, then $(? \otimes_R {}_RM_M, \operatorname{Hom}_M({}_RM_M, -))$ is a pair of adjoint functors between Mod_M and Mod_R. Here $\operatorname{Hom}_M({}_RM_M, -)$ is a forgetful functor which preserves colimits. One thus obtains a triple (T, η, μ) over Mod_R in which T preserves colimits.

(d) For degenerate triples there is a further trivial criterion (21.3.4). Below, in 21.3.14, we shall derive a criterion for special triples. In 21.6 criteria under restrive conditions for $\mathscr{C}$ and T are found by means of factorizations of morphisms.

21.3.12 Definition. In the class of ordinals of the universe $\mathfrak{U}$ there is, for every cardinal $\sigma > 0$, a smallest ordinal of its cardinality. We regard σ as this ordinal. A *well-ordered diagram of type* σ is a diagram $D: \Sigma \to \mathscr{C}$, where Σ is a well-ordered set of order type σ. Colimits of such diagrams are called *well-ordered colimits of type* σ.

21.3.13 Remark. If ϱ is an infinite regular cardinal (see 9.6.9) and $T: \mathscr{C} \to \mathscr{D}$ a functor which preserves ϱ-filtered colimits, then, in particular, T preserves well-ordered colimits of every type $\sigma \geq \varrho$.

21.3.14 Theorem. *Let σ be an infinite cardinal of the universe $\mathfrak{U}$, $\mathscr{C}$ a cocomplete $\mathfrak{U}$-category and $\boldsymbol{T} = (T, \eta, \mu)$ a triple over $\mathscr{C}$ such that T preserves well-ordered colimits of type σ. Then the Eilenberg-Moore category $\mathscr{C}^{\boldsymbol{T}}$ is cocomplete.*

Proof. By 21.3.10, it suffices to show that $\mathscr{C}^{\boldsymbol{T}}$ has coequalizers. The proof follows by a transfinite induction whose parts are analogous to the steps in the proof of 21.3.10; however, in step 2 of the proof, p has to be replaced by a coequalizer in $\mathscr{C}$. To give a better general idea of the proof, we omit trivial details and only sketch the main steps.

Step 1. This is step 1 in the proof of 21.3.10, where here D is to be a pair of homomorphisms and λ, ϱ are coequalizers in $\mathscr{C}$.

Step 2. Induction step for a non-limit ordinal. It consists of the second step of the proof of 21.3.10 if p there is chosen to be a coequalizer in $\mathscr{C}$ of $T(r)$ and $\mu_L T(s)$, and thus g is a $\mathscr{C}$-morphism, so that (15) is valid.

Step 3. Induction step for a limit ordinal. Essentially, this is again step 1 of the proof in 21.3.10. $U\,D$ is replaced by the lower horizontal arrows which correspond to λ in (9) and to $p\,\eta_L$ in (11), as well as their composites. $T\,U\,D$ is replaced by the upper horizontal arrows which correspond to ϱ in (9) and to s, $T(p\,\eta_L)$ in (11), together with their composites. $U * \varepsilon * D$ is replaced by the oblique arrows corresponding to r and p in (8), (9), (11). The transition to colimits in $\mathscr{C}$ yields the analogues of (6) through (9), where r is again replaced by an epimorphism as a colimit of epimorphisms.

Step 4. By assumption, one gets for the limit ordinal σ with colimit (C, ω) as the lower horizontal arrows

(17)
$$\begin{array}{ccccc}
\bullet & \xrightarrow{TT*\omega} & TT(C) & & \\
 & & T(c)\big\downarrow\big\downarrow\mu_C & & \\
\bullet & \xrightarrow{T*\omega} & T(C) & \xrightarrow{T(h)} & T(A) \\
\big\downarrow & & \big\downarrow c & & \big\downarrow a \\
\bullet & \xrightarrow{\omega} & C & \xrightarrow{h} & A\,.
\end{array}$$

Here c is obtained as the colimit of the oblique arrows. If the limit ordinals that are smaller than σ are left out, one is left with an index set which is final in the set of all indices (9.4.9), which has no influence on the colimits. The analogues of (10) and (11) imply now that (C, c) is an algebra and (c, a, h) a homomorphism. Now it follows from the uniqueness statements in the single steps that the left half of (17) yields the desired coequalizer in $\mathscr{C}^{\boldsymbol{T}}$.

21.4 Split Forks

21.4.1 Definitions. (a) A *fork* is a diagram

(1) $$A \underset{g}{\overset{f}{\rightrightarrows}} B \xrightarrow{r} C \text{ with } r f = r g\,.$$

It *splits* if there are morphisms

(2) $$A \xleftarrow{j} B \xleftarrow{i} C$$

such that

(3) $$r i = 1_C\,,$$

(4) $$f j = 1_B$$

(5) $$g j = i r\,.$$

Here (2) is called a *splitting* of (1).

(b) Let $U: \mathscr{X} \to \mathscr{C}$ be a functor and $v, w: X \to Y$ a pair of morphisms in $\mathscr{X}$. We say that *the pair is split by* U if $U(v)$, $U(w)$ can be completed to a split fork.

21.4.2 Proposition. (a) *If* (2) *is a splitting of* (1), *then* r *is a coequalizer of* f *and* g.

(b) *If* $r: B \to C$ *is a retraction and* $A \underset{g}{\overset{f}{\rightrightarrows}} B$ *a kernel pair of* r, *then* (1) *is a split fork.*

(c) *Every functor preserves splittings of forks.*

Proof. (a) Let $h\,f = h\,g$. Then

$$h \overset{(4)}{=} h\,f\,j = h\,g\,j \overset{(5)}{=} h\,i\,r\,.$$

By (3), r is a retraction. Therefore, $k = h\,i$ is uniquely determined by $k\,r = h$, which yields statement (a).

(b) Let $i\colon C \to B$ be a coretraction for r. Since $r\,1_B = r\,i\,r$, the definition of kernel pairs says that there is a unique morphism $j\colon B \to B$ with $f\,j = 1_B$ and $g\,j = i\,r$.

(c) is evident.

21.4.3 Proposition. *Let $F \dashv_{\eta}^{\varepsilon} U(\mathcal{X}, \mathcal{C})$ be an adjoint situation and $\boldsymbol{T} = (T, \eta, \mu)$ the corresponding triple.* (a)

(6) $$U\,F\,U\,F\,U \underset{UFU*\varepsilon}{\overset{U*\varepsilon*FU}{\rightrightarrows}} U\,F\,U \xrightarrow{U*\varepsilon} U\,,$$

is a split fork in $[\mathcal{X}, \mathcal{C}]$ with the splitting

(7) $$T\,T\,U \xleftarrow{\eta*TU} T\,U \xleftarrow{\eta*U} U\,.$$

For every $X \in |\mathcal{X}|$, the pair $\varepsilon_{FU(X)}$, $F\,U(\varepsilon_X)\colon F\,U\,F\,U(X) \to F\,U(X)$ is therefore split by U, and $U(\varepsilon_X)$ is a coequalizer of $T\,U(\varepsilon_X)$ and $\mu_{U(X)} = U(\varepsilon_{FU(X)})$.

(b) *Let $v\colon X \to Y$ be an $\mathcal{X}$-morphism such that $U(v)$ is a retraction and i a corresponding coretraction. Then*

$$T\,U(X) \underset{U(\varepsilon_X F(i) FU(v))}{\overset{U(\varepsilon_X)}{\rightrightarrows}} U(X) \xrightarrow{U(v)} U(Y)$$

has the splitting

$$T\,U(X) \xleftarrow[\eta_{U(X)}]{} U(X) \xleftarrow[i]{} U(Y)$$

Proof. (a) (6) is a fork by 21.1.1 (5). (3) and (4) follow from 16.5.5 (4) and, by 16.1.1 (5), (5) is valid. The last statement follows from 21.4.2 (a).

(b) is proved by a trivial computation.

21.4.4 Proposition. *Let $F^{\boldsymbol{T}} \dashv_{\eta}^{\varepsilon\boldsymbol{T}} U^{\boldsymbol{T}}(\mathcal{C}^{\boldsymbol{T}}, \mathcal{C})$ be the Eilenberg-Moore situation corresponding to the triple $\boldsymbol{T} = (T, \eta, \mu)$ over $\mathcal{C}$.*

(a) *$U^{\boldsymbol{T}}$ creates coequalizers for those pairs of morphisms that are split by $U^{\boldsymbol{T}}$.*

(b) *If (a, b, f), (a, b, g) and (b, c, p) are homomorphisms, if p is a coequalizer of f and g, and if $T(p)$ is an epimorphism, then (b, c, p) is a coequalizer of (a, b, f) and (a, b, g).*

(c) *Let (a, b, f), $(a, b, g)\colon (A, a) \to (B, b)$ be a pair of $\mathcal{C}^{\boldsymbol{T}}$-morphisms. If f, g is a kernel pair in $\mathcal{C}$ and if it has a retraction r as a coequalizer, then (a, b, f), (a, b, g) form a kernel pair of their (existing) coequalizer.*

Proof. (a) follows immediately from 21.4.2 and 21.3.9 (b).

(b) If (b, d, q) is a homomorphism with $q\,f = q\,g$, then there is exactly one h with $q = h\,p$. Thus the statement follows from 21.3.5 (b).

(c) By 18.4.3, f, g is a kernel pair of r. By (a) and 21.4.2, (b, c, r) is a coequalizer of (a, b, f) and (a, b, g). The statement now follows from the fact that $U^{\mathbf{T}}$ reflects limits (21.3.9 (a)).

21.4.5 Corollary. *$\varepsilon^{\mathbf{T}}: F^{\mathbf{T}} U^{\mathbf{T}} \to 1_{\mathscr{C}^{\mathbf{T}}}$ is a coequalizer of $\varepsilon^{\mathbf{T}} * F^{\mathbf{T}} U^{\mathbf{T}}$ and $F^{\mathbf{T}} U^{\mathbf{T}} * \varepsilon^{\mathbf{T}}$. For every algebra (A, a), (μ_A, a, a) is a coequalizer of $(\mu_{T(A)}, \mu_A, \mu_A)$ and $(\mu_{T(A)}, \mu_A, T(a))$, and a is a coequalizer of μ_A and $T(a)$.*

Proof. By 21.4.2 (a), the first statement follows immediately from 21.4.3 (a) and 21.4.4 (a). By 21.1.7 (14), the second statement follows similarly. The third statement follows again from 21.4.4 (a).

21.4.6 Proposition. *Let $F \frac{\varepsilon}{\eta}| U(\mathscr{X}, \mathscr{C})$ generate the triple $\mathbf{T} = (T, \eta, \mu)$ and let $K: \mathscr{X} \to \mathscr{C}^{\mathbf{T}}$ be the comparison functor.*

(a) *For every $X \in |\mathscr{X}|$, $K(\varepsilon_X)$ is a coequalizer of $K(\varepsilon_{FU(X)})$ and $K F U(\varepsilon_X) = F^{\mathbf{T}} U(\varepsilon_X)$.*

(b) *For every $A \in |\mathscr{C}|$ and $Y \in |\mathscr{X}|$, $K_{F(A), Y}: [F(A), Y] \to [K F(A), K(Y)]$ is bijective.*

(c) *The following statements are equivalent:*

(i) *K is fully faithful.*

(ii) *If $v, w: X \to Y$ and $p: Y \to Z$ are $\mathscr{X}$-morphisms such that $U(p)$ is a coequalizer of $U(v)$, $U(w)$ and a retraction, then p is a coequalizer of v and w.*

(iii) *For every $X \in |\mathscr{X}|$, ε_X is a coequalizer of $FU(\varepsilon_X)$ and $\varepsilon_{FU(X)}$.*

(iv) *Every ε_X is a coequalizer.*

(v) *If $p: Y \to Z$ is an $\mathscr{X}$-morphism such that $U(p)$ is a retraction, then p is a coequalizer.*

(vi) *U reflects coequalizers of pairs of morphisms in $\mathscr{X}$ which are split by U.*

(vii) *U reflects coequalizers of $\varepsilon_{FU(X)}$ and $F U(\varepsilon_X)$.*

If one of these conditions is satisfied, then U reflects limits and isomorphisms, and U is faithful.

If coequalizers of $\varepsilon_{FU(X)}$ and $F U(\varepsilon_X)$ exist for every $X \in |\mathscr{X}|$ (or, more generally: if $\mathscr{X}$ has coequalizers for every pair of morphisms that is split by U), then the following is also equivalent:

(viii) *U reflects isomorphisms.*

Proof. (a) From 21.1.7 (16) and (17) it follows that

$$\begin{aligned} K * \varepsilon &= \varepsilon^{\mathbf{T}} * K, \\ K * \varepsilon * F U &= \varepsilon^{\mathbf{T}} * F^{\mathbf{T}} U^{\mathbf{T}} K, \\ K F U * \varepsilon &= F^{\mathbf{T}} U^{\mathbf{T}} * \varepsilon^{\mathbf{T}} * K. \end{aligned} \tag{8}$$

Thus the statement follows from 21.4.5 at all places $K(X)$ with $X \in |\mathscr{X}|$.

(b) Since $K F = F^{\mathbf{T}}$ and $U = U^{\mathbf{T}} K$, one has the isomorphism

$$\varphi^{\mathbf{T}}_{A, K(Y)}: [K F(A), K(Y)] \to [A, U(Y)].$$

For $v\colon F(A) \to Y$, one obtains by 16.5.5 (2),

$$(\varphi^{\boldsymbol{T}}_{A,K(Y)} K_{F(A),Y})(v) = \varphi^{\boldsymbol{T}}_{A,K(Y)}(K(v)) = U^{\boldsymbol{T}} K(v)\, \eta_A = \\ = U(v)\, \eta_A = \varphi_{A,Y}(v)\,.$$

Thus $K_{F(A),Y} = (\varphi^{\boldsymbol{T}}_{A,K(Y)})^{-1} \varphi_{A,Y}$ is bijective.

(c) (i) $\Rightarrow$ (ii), by $U = U^{\boldsymbol{T}} K$, 21.4.4 (b) and 8.7.6.

(ii) $\Rightarrow$ (iii), by 21.4.3 (a).

(iii) $\Rightarrow$ (iv) is trivial

(iv) $\Rightarrow$ (i). By 16.5.3, U is faithful. Since $U = U^{\boldsymbol{T}} K$, K is faithful. Now let $\alpha\colon K(X) \to K(Y)$ be a $\mathscr{C}^{\boldsymbol{T}}$-morphism. By (b), there is exactly on $\mathscr{X}$-morphism $v\colon F\,U(X) \to Y$ with $K(v) = \alpha\, K(\varepsilon_X)$.

Let ε_X be a coequalizer of $q_1, q_2\colon Z \to F\,U(X)$. Then

$$K(v\, q_1) = \alpha\, K(\varepsilon_X\, q_1) = \alpha\, K(\varepsilon_X\, q_2) = K(v\, q_2)\,.$$

Since K is faithful, $v\, q_1 = v\, q_2$. Therefore, there is exactly one $w\colon X \to Y$ with $v = w\, \varepsilon_X$, and $K(w\, \varepsilon_X) = \alpha\, K(\varepsilon_X)$. By (a), $K(\varepsilon_X)$ is an epimorphism. Therefore, $K(w) = \alpha$ and K is full.

(ii) $\Rightarrow$ (v), by 21.4.3 (b).

(v) $\Rightarrow$ (iv) is trivial.

(ii) $\Rightarrow$ (vi), by 21.4.2.

(vi) $\Rightarrow$ (vii), by 21.1.1 (5) and 21.4.3 (a).

(vii) $\Rightarrow$ (iii) the same way.

(i) $\Rightarrow$ (viii), since $U = U^{\boldsymbol{T}}K$.

(viii) $\Rightarrow$ (vii), by 21.1.1 (5).

21.4.7 Remark. 21.4.6 (b) shows that the Kleisli situation corresponding to the triple $\boldsymbol{T}$ can be obtained (up to an isomorphism) as a full image from every adjoint situation that generates $\boldsymbol{T}$.

21.5 Characterization of Eilenberg-Moore Situations

21.5.1 Preliminary remarks. (a) If $\boldsymbol{T} = (T, \eta, \mu)$ is a triple over the category $\mathscr{C}$ and $\tau\colon T \to T'$ an isomorphism of endofunctors of $\mathscr{C}$, then

$$\boldsymbol{T}' = \big(T', \tau\, \eta,\, \tau\, \mu\, (T * \tau^{-1})\, (\tau^{-1} * T')\big)$$

is a triple and τ an isomorphism of triples, as is verified easily by using 16.1.1 (5).

(b) If $F\colon \mathscr{C} \to \mathscr{X}$ is left adjoint to $U\colon \mathscr{X} \to \mathscr{C}$, then the triples over $\mathscr{C}$ which are obtained through all possible choices of the adjunction, are all isomorphic to each other, and the isomorphisms are automorphisms of $T = U\,F$. They can be obtained by means of the automorphisms of F or those of U. This follows immediately from 16.4.3.

(c) Let the adjoint situation $F \frac{\varepsilon}{\eta}| U(\mathcal{X}, \mathcal{C})$ generate the triple $\boldsymbol{T} = (T, \eta, \mu)$ over $\mathcal{C}$. If $\overline{\boldsymbol{T}} = (\overline{T}, \overline{\eta}, \overline{\mu})$ is a triple over $\mathcal{X}$ then $\boldsymbol{T}' = (T', \eta', \mu')$ with

$$\begin{aligned} T' &= U \overline{T} F, \\ \eta' &= (U * \overline{\eta} * F)\, \eta, \\ \mu' &= (U * \overline{\mu} * F)\,(U \overline{T} * \varepsilon * \overline{T} F) \end{aligned} \tag{1}$$

is a triple over $\mathcal{C}$, and $\tau = U * \overline{\eta} * F$ is a triple-morphism $\tau\colon \boldsymbol{T} \to \boldsymbol{T}'$.

If $\overline{\tau}\colon \overline{\boldsymbol{T}} \to \overline{\boldsymbol{T}}'$ is a triple-morphism of triples over $\mathcal{X}$, then (1) together with $\overline{\tau} \mapsto U * \overline{\tau} * F$ provides a functor from the category $Tr(\mathcal{X})$ of triples over $\mathcal{X}$ into the category $Tr(\mathcal{C})$ of triples over $\mathcal{C}$; more precisely, one gets a functor $Tr(\mathcal{X}) \to \boldsymbol{T}/Tr(\mathcal{C})$ (compare the dual of 6.5.3).

(d) If $J \frac{\overline{\varepsilon}}{\overline{\eta}}| K(\mathcal{X}', \mathcal{X})$ generates the triple $\overline{\boldsymbol{T}}$ over $\mathcal{X}$, then K (or, resp., J) and (1) provide an adjoint right (or, resp., left) triangle over $\mathcal{C}$ with $U' = U K$ and $F' = J F$. By (1), 21.1.5 and 21.2.7 give natural transformations

$$\begin{aligned} \theta &= \overline{\eta} * F\colon\ F \to K J F = K F'; \\ \sigma &= U * \overline{\eta}\colon\ U \to U K J = U' J', \end{aligned} \tag{2}$$

and thus $U * \theta = \sigma * F = U * \overline{\eta} * F = \tau\colon\ T \to T'$.

21.5.2 Definition. We call the triple $\overline{\boldsymbol{T}}$ an *inverse image* of the triple (1), and (1) the *image* of $\overline{\boldsymbol{T}}$ by $F \frac{\varepsilon}{\eta}| U(\mathcal{X}, \mathcal{C})$, and we say that the functors $Tr(\mathcal{X}) \to Tr(\mathcal{C})$, $Tr(\mathcal{X}) \to \boldsymbol{T}/Tr(\mathcal{C})$ defined in 21.5.1 (c) are *induced* by $F \frac{\varepsilon}{\eta}| U(\mathcal{X}, \mathcal{C})$.

21.5.3 Theorem (Dubuc). *Suppose that there is the adjoint right triangle* 21.1.4 (6):

(3)

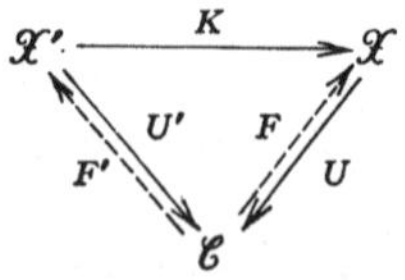

$U' = UK.$

We assume that

(i) *For every* $X \in |\mathcal{X}|$, ε_X *is a coequalizer of* $FU(\varepsilon_X)$ *and* $\varepsilon_{FU(X)}$.

(ii) *For every* $X \in |\mathcal{X}|$, *there exists a coequalizer* $\varkappa_X\colon\ F'U(X) \to J(X)$ *of* $F'U(\varepsilon_X)$ *and* $\varepsilon'_{F'U(X)}\ F'U(\theta_{U(X)})$, *where* θ *is defined by* 21.1.5 (7). *Then there is an adjoint situation*

$$J \frac{\overline{\varepsilon}}{\overline{\eta}}| K(\mathcal{X}', \mathcal{X})\,. \tag{4}$$

Proof. Choosing coequalizers in (ii) determines the functor J: $\mathcal{X} \to \mathcal{X}'$, where $\varkappa = \{\varkappa_X\}\colon F'\,U \to J$ is a coequalizer of $F'\,U * \varepsilon$ and

$(\varepsilon' * F\,U)\,(F'\,U * \theta * U)$, by 7.5.3, 8.5.1. By (i), ε is a coequalizer of $F\,U * \varepsilon$ and $\varepsilon * F\,U$. In the following, we make continuous use of 16.1.1 (5) and $U\,K = U'$. First we consider

(5)

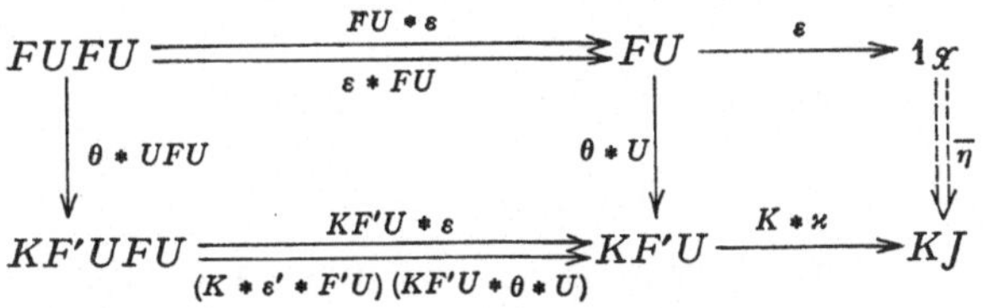
FUFU
FU * ε
ε * FU
FU
ε
1𝔛
θ * UFU
θ * U
η̄
KF'UFU
KF'U * ε
(K * ε' * F'U) (KF'U * θ * U)
KF'U
K * ϰ
KJ

The left half consists of two commutative squares; the commutativity for the lower arrows of the horizontal pairs follows from 21.1.5 (11), for the upper ones from 16.1.1 (5). By (i), (ii), there exists a unique $\bar{\eta}$ such that the right half of (5) is commutative. Now we look at

(6)

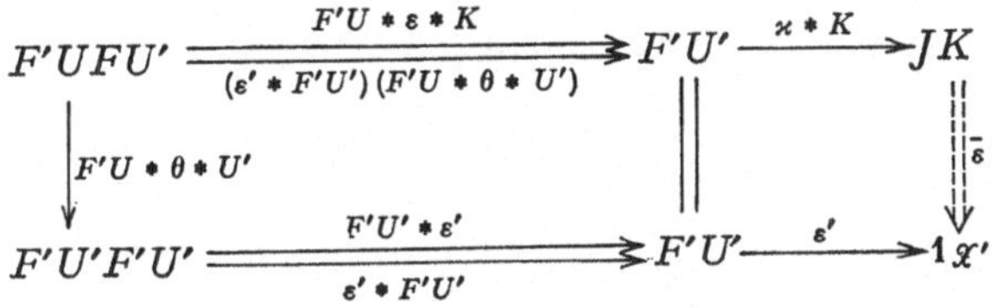

Here commutativity for the upper arrows of the horizontal pairs follows from 21.1.5 (8), for the lower ones there is an identity. Also, the lower part of (6) is a fork, and $\varkappa$ is a coequalizer at all places $K(X')$. Therefore, there exists a unique $\bar{\varepsilon}$, so that the right half of (6) is commutative. (5), (6) and 21.1.5 (8) now imply

$$(K * \bar{\varepsilon})\,(\bar{\eta} * K)\,(\varepsilon * K) = (K * \bar{\varepsilon})\,(K * \varkappa * K)\,(\theta * U\,K) =$$
$$(K * \varepsilon')\,(\theta * U') = \varepsilon * K\,.$$

By (i), $\varepsilon * K$ is an epimorphism, and one has

(7) $$(K * \bar{\varepsilon})\,(\bar{\eta} * K) = 1_K\,.$$

From (5), (6) and 16.1.1 (5), it follows that

$$(\bar{\varepsilon} * J)\,(J * \bar{\eta})\,\varkappa\,(F'U * \varepsilon) = (\bar{\varepsilon} * J)\,(J * \bar{\eta})\,(J * \varepsilon)\,(\varkappa * F\,U) =$$
$$\stackrel{(5)}{=} (\bar{\varepsilon} * J)\,(J\,K * \varkappa)\,(J * \theta * U)\,(\varkappa * F\,U) =$$
$$(\bar{\varepsilon} * J)\,(J\,K * \varkappa)\,(\varkappa * K\,F'\,U)\,(F'\,U * \theta * U) =$$
$$= (\bar{\varepsilon} * J)\,(\varkappa * K\,J)\,(F'\,U\,K * \varkappa)\,(F'\,U * \theta * U) =$$
$$\stackrel{(6)}{=} (\varepsilon' * J)\,(F'\,U' * \varkappa)\,(F'\,U * \theta * U)$$
$$= \varkappa\,(\varepsilon' * F'\,U)\,(F'\,U * \theta * U) = \varkappa(F'\,U * \varepsilon)\,,$$

where the last equality is implied by the definition of $\varkappa$. Now $\varkappa$, being a coequalizer, is an epimorphism, and $F'\,U * \varepsilon = F' * (U * \varepsilon)$

is a retraction. Thus

$$(\bar{\varepsilon} * J)(J * \bar{\eta}) = 1_J. \tag{8}$$

The theorem now follows from (7), (8) and 16.5.7.

21.5.4 Remarks. (a) If there is an adjoint situation (4), then

$$J F \frac{\bar{\varepsilon}(J*\varepsilon*K)}{(U*\bar{\eta}*F)\eta}| U'(\mathscr{X}', \mathscr{C}) \tag{9}$$

is also one. Together with $1_{U'}$, 16.4.3 gives a uniquely determined isomorphism $\varrho: J F \to F'$ for which, by 16.5.2 (2) and (2^0),

$$\begin{aligned} \eta' &= (U' * \varrho)(U * \bar{\eta} * F)\eta, \\ \varepsilon' &= \bar{\varepsilon}(J * \varepsilon * K)(\varrho^{-1} * U'). \end{aligned} \tag{10}$$

is valid. Together with 21.1.5 (7), this implies

$$\theta = (K * \varrho)(\bar{\eta} * F), \tag{11}$$

$$\varrho = (\bar{\varepsilon} * F')(J * \theta) \tag{12}$$

and thus

$$\begin{aligned} (\varrho * U)(J F U * \varepsilon) &= (F' U * \varepsilon')(\varrho * U F U), \\ (\varrho * U)(J * \varepsilon * F U) &= (\varepsilon' * F' U)(F' U * \theta * U)(\varrho * U F U). \end{aligned} \tag{13}$$

(b) If, in 21.5.3, condition (i) is satisfied, then (ii) is necessary for the existence of a left adjoint J of K. Since J preserves colimits, this follows from (13). At the same time, this supplies the idea of the proof (see (5), (6)).

(c) By 21.4.6 (c), condition (i) in 21.5.3 can be replaced by an equivalent one. It is satisfied, in particular, if one has an Eilenberg-Moore situation on the right in (3).

(d) 21.1.5 (9), $F' U * \varepsilon$ and $(\varepsilon' * F' U)(F' U * \theta * U)$ have the common coretraction $F' * \eta * U$. For $X \in |\mathscr{X}|$, the pair $F'U(\varepsilon_X)$, $\varepsilon'_{F'U(X)} F' U(\theta_{U(X)})$ is in general not split by U'.

(e) Corresponding to 21.5.3 there is a pseudodual statement for the adjoint left triangle 21.2.6. A right adjoint for N exists if again ε_X is a coequalizer of $\varepsilon_{FU(X)}$ and $F U(\varepsilon_X)$ for every $X \in |\mathscr{X}|$, and if for every $X' \in |\mathscr{X}'|$ a coequalizer of $\varepsilon_{FU'(X')}$ and $F U'(\varepsilon'_{X'}) F(\sigma_{FU'(X)})$ exists and is preserved by N.

21.5.5 Corollary. *Let $\boldsymbol{T} = (T, \eta, \mu)$, $\boldsymbol{T}' = (T', \eta', \mu')$ be triples over the category $\mathscr{C}$ and $\tau: \boldsymbol{T} \to \boldsymbol{T}'$ a triple morphism. If $\boldsymbol{T}'$ is generated by an adjoint situation $F' \frac{\varepsilon'}{\eta'}| U'(\mathscr{X}', \mathscr{C})$ in which $\mathscr{X}'$ has coequalizers, then $\boldsymbol{T}'$ is isomorphic to the image of a triple $\overline{\boldsymbol{T}}$ on $\mathscr{C}^{\boldsymbol{T}}$.*

Proof. By 21.1.7 (e), there is a unique functor $K: \mathscr{X}' \to \mathscr{C}^{\boldsymbol{T}}$ with $U^{\boldsymbol{T}} K = U'$ and $(U^{\boldsymbol{T}} * \varepsilon^{\boldsymbol{T}} * K F')(T * \eta) = \tau$. By 21.5.3 and 21.5.4

(c), K has a left adjoint J. In this way one gets a triple over $\mathscr{C}^{\boldsymbol{T}}$ whose image is isomorphic to $\boldsymbol{T}'$ (21.5.4 (a)).

21.5.6 A special case. We consider the special case of 21.5.3 given by the condition

$$F = K\,F' \quad \text{and} \quad \eta = \eta'\,. \tag{14}$$

By 21.1.5, both adjoint situations now generate the same triple $\boldsymbol{T}$ and

$$\theta = 1_F \quad \text{and} \quad \varepsilon * K = K * \varepsilon' \tag{15}$$

holds. 21.5.3 (ii) now states that for every $X \in |\mathscr{X}|$ there exists a coequalizer $\varkappa_X\colon\ F'U(X) \to J(X)$ of $F'U(\varepsilon_X)$ and $\varepsilon'_{F'U(X)}$. By (14) and (15), one has

$$\begin{aligned} U'\,F'\,U * \varepsilon &= U\,F\,U * \varepsilon\,, \\ U' * \varepsilon' * F'\,U &= U * \varepsilon * F\,U\,. \end{aligned} \tag{16}$$

Thus $F'\,U * \varepsilon$ and $\varepsilon' * F'\,U$ are split by U' (more precisely, by $[\mathscr{X}, U']$), as are $F'\,U' * \varepsilon'$ and $\varepsilon' * F'\,U$ (21.4.3 (a)). There are the common coretractions $F' * \eta * U$ or, resp., $F' * \eta * U'$.

(a) Let 21.5.3 (i) or an equivalent condition be satisfied; i. e., assume that $K_{\varepsilon,\eta}\colon\ \mathscr{X} \to \mathscr{C}^{\boldsymbol{T}}$ is fully faithful (21.4.6 (c)). Since $K_{\varepsilon,\eta}\,K = K_{\varepsilon',\eta}$ (21.1.10 (23)), K is fully faithful if and only if $K_{\varepsilon',\eta}$ is. The conditions in 21.4.6 (c) with U', ε' instead of U, ε are thus equivalent for the K considered here.

(b) Now let the assumptions of 21.5.3 be satisfied in the special case under consideration. By (11), (12), (15), $\bar{\varepsilon}_{F'(A)}$ and $\bar{\eta}_{F(A)}$ are isomorphisms for all $A \in |\mathscr{C}|$. For $X \in |\mathscr{X}|$, the following are equivalent:

(i) $\bar{\eta}_X$ is an isomorphism.
(ii) K preserves the coequalizer $\varkappa_X$ of $F'\;U(\varepsilon_X)$ and $\varepsilon'_{FU(X)}$.
(iii) $U(\bar{\eta}_X)$ is an isomorphism.
(iv) U' preserves the coequalizer in (ii).
(v) $U'(\varkappa_X)$ and $U(\varepsilon_X)$ are equivalent epimorphisms.

If one of these conditions is satisfied, then, in addition,

(vi) X is isomorphic to an object $K(X')$ for a suitable $X' \in |\mathscr{X}'|$.

If U' takes coequalizers of pairs of morphisms that are split by U' into epimorphisms, then (i) follows from (vi). J is fully faithful if and only if (i) or an equivalent condition is satisfied for all $X \in |\mathscr{X}|$.

Proof. (i) $\Leftrightarrow$ (ii), by (5), (14), (15).

(i) $\Leftrightarrow$ (iii). By 21.4.6 (c), U reflects isomorphisms, and it is trivial that it preserves them.

(iii) $\Leftrightarrow$ (iv). Since $U' = U\,K$, and by 21.4.3 (a), this follows from (5) by applying U.

(iv) $\Leftrightarrow$ (v), the same way.

(i) implies (vi), with $X' = J(X)$. Since $\bar{\eta}$ is a natural transformation, it suffices for the converse to consider the case $X = K(X')$. If U' satisfies the condition above, then $U\,K(\varepsilon_X) = U'(\varkappa_X)$ is an epimorphism and, therefore, $U(\bar{\eta}_X)$ is an epimorphism, by (5). Since U is faithful, $\bar{\eta}_X$ is an epimorphism. For $X = K(X')$, $\bar{\eta}_{K(X')}$ is a coretraction and, therefore, (i) holds.

The last statement follows from 16.5.4.

(c) If U' creates coequalizers of $F'\,U * \varepsilon$ and $\varepsilon' * F'\,U$, then, by (16) and 21.4.3 (a), a coequalizer $\varkappa$ of $F'\,U * \varepsilon$ and $\varepsilon' * F'\,U$ exists with

$$U' * \varkappa = U * \varepsilon\,, \tag{17}$$

and by (15), $U' * \varkappa * K = U' * \varepsilon'$ follows. If U' creates coequalizers of $F'\,U' * \varepsilon'$ and $\varepsilon' * F'\,U'$, then (6) implies that $J\,K = 1_{\mathfrak{X}'}$. If further U creates coequalizers of $F\,U * \varepsilon$ and $\varepsilon * F\,U$, then (17) and (5) imply that $1_{\mathfrak{X}} = K\,J$.

21.5.7 Theorem (Beck). *Let the adjoint situation $F \frac{\varepsilon}{\eta}| \, U(\mathfrak{X}, \mathfrak{C})$ generate the triple $\boldsymbol{T} = (T, \eta, \mu)$ over $\mathfrak{C}$, and let $K\colon\ \mathfrak{X} \to \mathfrak{C}^{\boldsymbol{T}}$ be the comparison functor* (21.1.8). *We consider the pairs of morphisms in $\mathfrak{X}$ which are split by U.*

(a) *K is fully faithful if and only if U reflects coequalizers of these pairs of morphisms.*

(b) *K has a left adjoint J if and only if $F U^{\boldsymbol{T}}(\varepsilon^{\boldsymbol{T}}_{(A,a)})$ and $\varepsilon_{F(A)}$ have a coequalizer for every $(A, a) \in |\mathfrak{C}^{\boldsymbol{T}}|$. This is certainly the case if the pairs of morphisms have coequalizers in $\mathfrak{X}$. If K is fully faithful, then this sufficient condition for the existence of J is also necessary.*

(c) *If there are coequalizers of these pairs of morphisms in $\mathfrak{X}$, and if they are preserved by U, then J is fully faithful.*

(d) *K is an equivalence if and only if there are coequalizers of the pairs of morphisms considered in $\mathfrak{X}$ and if U preserves and reflects these coequalizers.*

(e) *K is an isomorphism of categories if and only if U creates coequalizers of these pairs of morphisms.*

Remark. 21.4.6 (c) supplies variations of this. Note that the assumptions involve only properties of U and $\mathfrak{X}$. Depending on the author, a *functor* $U\colon\ \mathfrak{X} \to \mathfrak{C}$ is called *tripleable* if it has a left adjoint and the conditions in (d) or, resp., (e) are satisfied.

Proof. This is a special case of 21.5.6 with different notation.

(a) is contained in 21.5.6 (a).

(b) The first statement follows from 21.5.3 and 21.5.4 (b), (c), the second one thus follows from (16). Now let K be fully faithful, let J exist, and let $v, w\colon\ X \to Y$ be a pair of morphisms that is split by U.

The pair $K(v)$, $K(w)$ is split by U^T, since $U^T K = U$, and it therefore has a coequalizer (21.4.4). By 16.6.1, v and w have a coequalizer.

(c) is contained in 21.5.6 (b).

(d) A pair of morphisms (u, v) in $\mathcal{X}$ is split by U if and only if $(K(u), K(v))$ is split by U^T. Thus by 21.4.4 (a), 16.5.4 (d) and its dual, the statement follows from (a), (b), (c).

(e) is contained in 21.5.6 (c), because of 21.4.4 (a).

21.5.8 Proposition. (a) *If, in the adjoint right triangle* (3), *both adjoint situations are isomorphic (or, resp., equivalent) to Eilenberg-Moore situations and if K has a left adjoint J, then fixing an adjunction of K and J determines a situation which is isomorphic (or, resp., equivalent) to an Eilenberg-Moore situation over $\mathcal{X}$.*

(b) *Let U in* (3) *be fully faithful and let* (4) *be equivalent to an Eilenberg-Moore situation over $\mathcal{X}$. Then the left adjoint situation in* (3) *is equivalent to an Eilenberg-Moore situation (so is the right one because of* 21.3.3).

Proof. (a) A pair of morphisms in $\mathcal{X}'$ that is split by K is also split by $U' = U\,K$. If U and U' satisfy the condition in 21.5.7 (e), then so does K, as is easily verified. And if U and U' satisfy the condition in 21.5.7 (d), then so does K.

(b) Let $v', w'\colon X' \to Y'$ be a pair of morphisms in $\mathcal{X}'$. If $K(v')$, $K(w')$ have a splitting, it is taken into a splitting of $U'(v')$, $U'(w')$ by U. Conversely, let $r\colon U'(Y') \to C$, $i\colon C \to U'(Y')$, $j\colon U'(Y') \to U'(X')$ be a completion of $U'(v')$, $U'(w')$ to a fork with a splitting. We consider

$$\begin{array}{ccccc} C & \xrightarrow{i} & U'(Y') & \xrightarrow{r} & C \\ {\scriptstyle \eta_C}\downarrow & & \downarrow{\scriptstyle \eta_{U'(Y')}} & & \downarrow{\scriptstyle \eta_C} \\ T(C) & \xrightarrow{T(i)} & TU'(Y) & \xrightarrow{T(r)} & T(C) \end{array}$$

Because U is fully faithful, $\eta_{U'(Y')}$ is an isomorphism, since $U' = U\,K$. It therefore follows that η_C is a coretraction and a retraction and thus an isomorphism. Since $\varepsilon\colon F\,U \to 1_X$ is an isomorphism, v', w' is split by K, as is shown by applying F ($K \cong F\,U\,K = F\,U'$). We thus get the result that the pair v', w' is split by U' if and only if it is split by K. This implies that U' satisfies the condition in 21.5.7 (d), for by assumption, it is valid for K, and according to what was said above, U preserves and reflects splittings.

21.5.9 Remarks. (a) 21.5.8 (b) is a first special case related to the question when the composite of two Eilenberg-Moore situations is again one up to an isomorphism (compare 21.3.8 (e)), for, (4) and the adjoint situation on the right of (3) determine the left one up to an isomorphism. Other special cases are treated later, in 21.6.5, 21.6.6, 21.6.11 (h) and 21.6.12. A counterexample is given in 21.6.10 (d).

(b) The pairs of morphisms in $\mathscr{X}$ considered in 21.5.7 may be restricted to those which are split by U, have a common retraction and whose domain and codomain are image objects of F. This follows immediately from 21.5.6.

(c) By 21.4.2 (b), amongst the pairs of morphisms that are split by U are, in particular, those which are taken into kernel pairs of retractions by U. In 21.5.13 below we furnish a proposition which provides a sufficient condition for 21.5.7 (d). If, here, $\mathscr{C}$ is finitely complete (or already if it has generalized kernel pairs for pairs of morphisms, see 21.5.11 below,) then this condition is also necessary, by 21.3.9 (a). 21.3.8 (e) then provides an isomorphism criterion.

21.5.10 Lemma. *Let*

(18) $$X \underset{w}{\overset{v}{\rightrightarrows}} Y \xrightarrow{r} Z$$

be a fork, so that $r\,v = r\,w$.

(a) *If* (18) *has the splitting*

(19) $$X \xleftarrow{j} Y \xleftarrow{i} Z \quad \textit{with } v\,j = 1_Y,\ w\,j = i\,r,$$

then r *and* $i\,r = w\,j$ *have the same kernel pairs.*

(b) *Let*

(20) $$N \underset{h}{\overset{h'}{\rightrightarrows}} X \xrightarrow{w} Y \quad \textit{be a kernel pair of } w,$$

(21) $$M \underset{k}{\overset{k'}{\rightrightarrows}} Y \xrightarrow{r} Z \quad \textit{a kernel pair of } r.$$

Then there is exactly one morphism $n\colon N \to M$ *with* $v\,h = k\,n$ *and* $v\,h' = k'\,n$.

(c) *If the splitting* (19) *exists, then there is exactly one morphism* $m\colon M \to N$ *with* $j\,k = h\,m$ and $j\,k' = h'\,m$. *Furthermore,* $n\,m = 1_M$.

Proof. (a) follows immediately from the fact that i is a monomorphism. Since $r\,v\,h = r\,w\,h = r\,w\,h' = r\,v\,h'$, (b) follows from the definition of kernel pairs.

(c) As in (b), the first statement follows from $w\,j\,k = i\,r\,k = i\,r\,k' = w\,j\,k'$. This implies the second statement, because $k\,n\,m = v\,h\,m = v\,j\,k = k$ and $k'\,n\,m = k'$.

21.5.11 Definition. Let $\{s_\nu\colon N \to Y\}$ be a non-empty family of morphisms with the same domain N and codomain Y. We call l, l': $L \to N$ a *generalized kernel pair* of this family if

(i) $s_\nu\,l = s_\nu\,l'$ for all ν.

(ii) If u, u': $X \to N$ is a pair of morphisms with $s_\nu\,u = s_\nu\,u'$ for all ν, then there is exactly one morphism v: $X \to L$ with $u = l\,v$ and $u' = l'\,v$.

21.5.12 Remarks. (a) Generalized kernel pairs are limits. If $\Pi\, Y_\nu$ with $Y_\nu = Y$ for all ν exists, then a generalized kernel pair is an ordinary one for $s\colon N \to \Pi\, Y_\nu$ with $s_\nu = pr_\nu\, s$. If, in particular, every s_ν has a kernel pair, then the kernel pair of s is the intersection of these kernel pairs (18.4.11). We only need the special case $\nu = 1, 2$ here.

(b) If for $s, s'\colon N \to Y$ a morphism $r\colon Y \to Z$ with $r\, s = r\, s'$ exists and if r has the kernel pair (21), then l, l' is a generalized kernel pair of s, s' if and only if l, l' is a kernel pair of the uniquely (by the definition of kernel pairs) determined morphism $p\colon N \to M$ with $s = k\, p$ and $s' = k\, p'$.

For, $p\, f = p\, f'$ implies $s\, f = s\, f'$ and $s'\, f = s'\, f'$, and this, conversely, implies $p\, f = p\, f'$, by (21).

(c) l, l' is a kernel pair of s if l, l' is a generalized kernel pair of the pair of morphisms s, s.

21.5.13 Proposition (Duskin). *Let $U\colon \mathfrak{X} \to \mathfrak{C}$ be a functor. Let $\mathfrak{C}$ have kernel pairs for retractions and $\mathfrak{X}$ generalized kernel pairs for pairs of morphisms. Assume that U preserves these generalized kernel pairs. Then the following are equivalent:*

(i) *$\mathfrak{X}$ has coequalizers for all pairs of morphisms that are split by U, and U preserves and reflects these coequalizers.*

(ii) *$\mathfrak{X}$ has coequalizers for all pairs of morphisms which are taken into kernel pairs of retractions by U, and U preserves and reflects these coequalizers.*

Proof. By 21.4.2 (b), (i) implies (ii). Now let (ii) be satisfied.

(a) The assumptions made about $\mathfrak{X}$ and U together with 21.5.12 (c) guarantee that $\mathfrak{X}$ has kernel pairs, and these are preserved by U.

(b) First, let (18) be a fork in $\mathfrak{X}$ which is taken into a split fork by U. By (a), the kernel pairs (20), (21) exist in $\mathfrak{X}$ with the morphism $n\colon N \to M$ of 21.5.10 (b). Since U preserves these kernel pairs (see (a)), $U(n)$ is a retraction, by 21.5.10 (c). Further, a kernel pair $l, l'\colon L \to N$ of n exists in $\mathfrak{X}$.

(22)
$$\begin{array}{ccccc} X & \overset{v}{\underset{w}{\rightrightarrows}} & Y & \overset{r}{\longrightarrow} & Z \\ & & h\uparrow\uparrow h' \quad & k\uparrow\uparrow k' & \\ L & \overset{l}{\underset{l'}{\rightrightarrows}} & N & \overset{n}{\longrightarrow} & M \end{array}$$

By (ii) and 21.4.2 (b), U reflects n as a coequalizer of l, l' and r as a coequalizer of k, k'. Since n is an epimorphism, r is also a coequalizer of $k\, n = v\, h$ and $k'\, n = v\, h'$. If one now has $p\colon Y \to P$ with $p\, v = p w$, then $p\, v\, h = p\, w\, h = p\, w\, h' = p\, v\, h'$, and p factors uniquely through r. Thus r is a coequalizer of v and w, and U reflects this coequalizer (compare 21.4.2 (a)).

(c) Now let $v, w\colon X \to Y$ be a pair of morphism in $\mathfrak{X}$ which is split by U. The kernel pair (20) of w and a generalized kernel pair $l, l'\colon L \to N$ of $v\,h$ and $v\,h'$ exist. The assumption about v, w and 21.4.2 (a) guarantee the existence of a coequalizer $\bar{r}\colon U(Y) \to C$ of $U(v)$, $U(w)$. Since $\bar{r}$ is a retraction, there exists (in $\mathscr{C}$) a kernel pair $f, f'\colon A \to U(Y)$ of $\bar{r}$. By the assumption about U and by 21.5.10 (b), (c) with changed notation, there is exactly one morphism $g\colon U(N) \to A$ with $f\,g = U(v\,h)$ and $f'\,g = U(v\,h')$, and g is a retraction.

(23)
$$\begin{array}{ccccc} & U(X) & \underset{U(w)}{\overset{U(v)}{\rightrightarrows}} & U(Y) & \xrightarrow{\bar{r}} C \\ & {\scriptstyle U(h)}\uparrow\uparrow{\scriptstyle U(h')} & & {\scriptstyle f}\uparrow\uparrow{\scriptstyle f'} & \\ U(L) \underset{U(l')}{\overset{U(l)}{\rightrightarrows}} & U(N) & \xrightarrow{\quad g \quad} & A\,. & \end{array}$$

Again by the assumption about U, and by (20) and 21.5.12 (b), $U(l)$, $U(l')$ is a kernel pair of g. By (ii), there exists a coequalizer $n\colon N \to M$ of l, l', and $U(n)$ and g are isomorphic in $U(N)/\mathscr{C}$. After correcting f and f' with this isomorphism, we can assume $U(n) = g$ and thus, in particular, $A = U(M)$. By the definition 21.5.11 of generalized kernel pairs and the definition of n, there are morphisms k, $k'\colon M \to Y$ with $k\,n = v\,h$ and $k'\,n = v\,h'$ in $\mathfrak{X}$. Since $g = U(n)$ is an epimorphism, $U(k) = f$ and $U(k') = f'$ follow. By 21.4.2 (b) and (ii), there exists a coequalizer r of k, k', and U preserves it. $U(r)$ and $\bar{r}$ are therefore isomorphic in $U(Y)/\mathscr{C}$, and we may assume $\bar{r} = U(r)$. As in (b), one now concludes that r is a coequalizer of v, w and that $U(r)$ is a coequalizer of $U(v)$ and $U(w)$. Together with (b), this implies (i), and the proof is complete.

21.6 Consequences of Factorizations of Morphisms

In view of the intended applications, we restrict ourselves in the following to factorizations of morphisms into epi- and monomorphisms. (The general case, where uniqueness has to be postulated in (F2) below and in (1), can be found in 16.8.6, compare also 21.2.2).

21.6.1 Definitions. In the category $\mathscr{C}$, let $\mathfrak{E}$ be a class of epimorphisms and $\mathfrak{M}$ a class of monomorphisms. We assume that all isomorphisms belong to $\mathfrak{E}$ and to $\mathfrak{M}$ and that $\mathfrak{E}$ and $\mathfrak{M}$ are closed with respect to composition with isomorphisms, which is no loss of generality.

(a) Let $f\colon A \to B$ be a $\mathscr{C}$-morphism. An *$\mathfrak{E}$-$\mathfrak{M}$-factorization* of f is a representation of f in the form $f = f''\,f'$ with $f' \in \mathfrak{E}$, $f'' \in \mathfrak{M}$.

(b) The following possible conditions for $\mathfrak{E}$ and $\mathfrak{M}$ will turn up:

(F1) Every morphism has an $\mathfrak{E}$-$\mathfrak{M}$-factorization.

(F2) The *factorization* is *natural* where it exists; i.e., if $f = f'' f'$ and $g = g'' g'$ are $\mathfrak{E}$-$\mathfrak{M}$-factorizations of f and g and if

$$\begin{array}{ccccc} A & \xrightarrow{f'} & P & \xrightarrow{f''} & B \\ {\scriptstyle a}\downarrow & & & & \downarrow{\scriptstyle b} \\ C & \xrightarrow{g'} & Q & \xrightarrow{g''} & D \end{array}$$

is commutative, then there is a morphism $d\colon P \to Q$ which fills in the diagram commutatively. The uniqueness of d follows from $f' \in \mathfrak{E}$ or from $g'' \in \mathfrak{M}$.

(F2) is obviously equivalent to the following *diagonal condition*: if

(1)
$$\begin{array}{ccc} A & \xrightarrow{e} & P \\ {\scriptstyle h}\downarrow & & \downarrow{\scriptstyle k} \\ Q & \xrightarrow[m]{} & R \end{array}$$

is commutative, $e \in \mathfrak{E}$, $m \in \mathfrak{M}$, then there is a commutative fill in $d\colon P \to Q$.

(F3) $\mathfrak{E}$ is *pointwise small*; i.e.: for every $A \in |\mathcal{C}|$, the equivalence classes of morphisms in $\mathfrak{E}$ with domain A have a set of representatives. Obviously, (F3) holds if $\mathcal{C}$ is co-well-powered, since $\mathfrak{E}$ consists of epimorphisms. In general, however, the converse is not true.

(F4) $\mathfrak{E}$ is closed with respect to composition; i.e.: if $f \in \mathfrak{E}$, $g \in \mathfrak{E}$ and if $g\,f$ exists, then $g\,f \in \mathfrak{E}$.

(F5) If $e\,p \in \mathfrak{E}$, then $e \in \mathfrak{E}$.

(F5′) If $e\,p \in \mathfrak{E}$, and if p is an epimorphism, then $e \in \mathfrak{E}$.

(c) An *epimorphism* p is called *extremal* if the following holds: if there is a factorization $p = m\,q$ with a monomorphic m, then m is an isomorphism.

If $\mathcal{C}$ has equalizers, then the condition given above implies that p is an epimorphism. If $\mathcal{C}$ is balanced, then every epimorphism is extremal.

21.6.2 Proposition. (a) *The class of coequalizers satisfies* (F5′). *Every coequalizer is an extremal epimorphism; in particular, so is every retraction.*

(b) *The class of extremal epimorphisms satisfies* (F5). *It also satisfies* (F4), *provided* $\mathcal{C}$ *has pullbacks.*

(c) (F1) *implies that all extremal epimorphisms belong to* $\mathfrak{E}$ (*since* $\mathfrak{M}$ *consists of monomorphisms*).

(d) (F2) *implies that the intersection of* $\mathfrak{E}$ *and* $\mathfrak{M}$ *consists of all isomorphisms and that factorization is unique up to an isomorphism.*

(e) (F1) *and* (F2) *imply* (F4) *and* (F5) (*since* $\mathfrak{M}$ *consists of monomorphisms*).

(f) *Suppose that* (F1) *and* (F2) *hold.* $\mathfrak{M}$ *consists of all monomorphisms if and only if* $\mathfrak{E}$ *consists of all extremal epimorphisms* (*for,* $\mathfrak{M}$ *consists of monomorphisms,* $\mathfrak{E}$ *of epimorphisms*).

(g) *Let* $\mathcal{C}$ *be complete and well-powered. If* $\mathfrak{M}$ *is the class of all monomorphisms and* $\mathfrak{E}$ *the class of all extremal epimorphisms, then* (F1) *and* (F2) *are valid.* (F1) *and* (F2) *are also valid if* $\mathfrak{E}$ *is the class of all epimorphisms and* $\mathfrak{M}$ *the class of all extremal monomorphisms* (*dual of* 21.6.1 (c)).

(h) *Let* $\mathcal{C}$ *have kernel pairs and coequalizers of kernel pairs, and let* $\mathfrak{E}$ *be the class of all coequalizers and* $\mathfrak{M}$ *the class of all monomorphisms. If* $\mathfrak{E}$ *satisfies* (F4), *then* $\mathfrak{E}$ *is identical with the class of all extremal epimorphisms and* (F1), (F2) *are valid.* (*Compare with* 18.4.7 *and* 18.4.8).

(i) *Let* $\mathcal{C}$ *have products and let* (F1) *and* (F3) *be valid. Then* $\mathcal{C}$ *has coequalizers.*

(j) *Let* $\mathcal{C}$ *be finitely complete. If the class of equalizers satisfies the dual of* (F3), *then the class of coequalizers satisfies* (F3).

Proof. (a) Let c be a coequalizer of f and g, and let $c = q\,p$. If p is an epimorphism, then q is a coequalizer of $p\,f$ and $p\,g$. If q is a monomorphism, then $p\,f = p\,g$ and, therefore, q is a monomorphic retraction and thus an isomorphism.

(b) The first statement follows immediately from the definition. Let $p: A \to B$, $q: B \to C$ be extremal epimorphisms, and let $q\,p = m\,r$ with a monomorphic m. The pullback for q and m gives a monomorphism through which p can be factored. Since p is extremal, one finds that q factors through m. Since q is extremal, m is an isomorphism, and (F4) holds.

(c) and (d) are easily verified.

(e) We start by proving (F5). Let $e = m\,e'$ with $e' \in \mathfrak{E}$, $m \in \mathfrak{M}$. $e\,p \in \mathfrak{E}$ and (F2) imply that m is a monomorphic retraction.

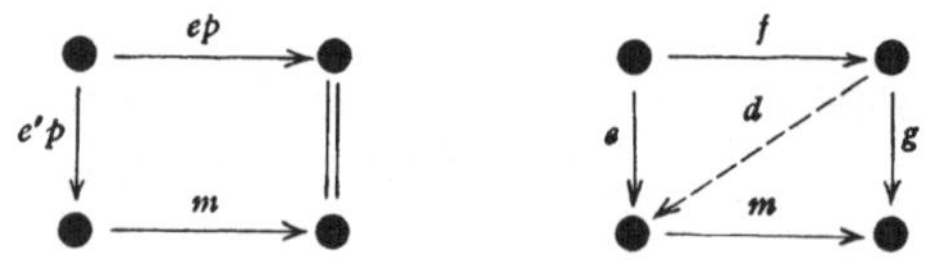

Therefore, (F5) follows.

Now let $f \in \mathfrak{E}$, $g \in \mathfrak{E}$ and $g\,f = m\,e$ with $e \in \mathfrak{E}$, $m \in \mathfrak{M}$. By (F2), $f \in \mathfrak{E}$, $m \in \mathfrak{M}$, there is a d with $d\,f = e$ and $g = m\,d$. (F5) and (d) imply that m is an isomorphism. (F4) thus follows.

(f) follows easily from (c), (d) and definition 21.6.1 (c).

(g) First, let $\mathfrak{E}$ be the class of all extremal epimorphisms and $\mathfrak{M}$ the class of all monomorphisms. For $f: A \to B$, consider the equivalence classes of those monomorphisms with codomain B through which f factors.

Let $\{m_\nu\}$ be a set of representatives of these classes. $\bigcap m_\nu = m: M \to B$ exists. Here m is a monomorphism and $f = m\, e$. If $e = n\, q$ with a monomorphic n, then n is an isomorphism, by the construction of m. By 21.6.1 (c), e is an extremal epimorphism. Hence (F1) is valid. We now show that the diagonal condition of 21.6.1, and thus (F2), holds. Let (1) be commutative, m a monomorphism and e an extremal epimorphism. If we form the pullback of m and k, then we get morphisms $\overline{m}$, $\overline{k}$, l with $e = \overline{m}\, l$, $h = \overline{k}\, l$ and $k\, \overline{m} = m\, \overline{k}$. Here $\overline{m}$ is a monomorphism. Since e is extremal, $\overline{m}$ is an isomorphism. The diagonal $\overline{k}\, \overline{m}^{-1}$ fills in (1) commutatively.

Now let $\mathfrak{E}$ be the class of all epimorphisms and $\mathfrak{M}$ the class of all extremal monomorphisms. In (1) let e be an epimorphism and m an extremal monomorphism. Among those monomorphisms with codomain R through which k and m factor simultaneously there is a smallest one, say n, since $\mathscr{C}$ is well-powered and complete. Now there are morphisms m', k' with $m = n\, m'$, $k = n\, k'$. Since n is a monomorphism, $k'\, e = m'\, h$ follows. We claim that m' is an isomorphism. Since m' is an extremal monomorphism, it suffices to show that m' is an epimorphism. Let morphisms f, g be given such that $f\, m' = g\, m'$. Since e is an epimorphism, $f\, k' = g\, k'$ follows from $k'\, e = m'\, h$. Therefore, m' and k' factor through an equalizer of f and g. By the minimality of n, this equalizer must be an isomorphism. Hence, $f = g$ and m' is an epimorphism. $d = m'^{-1}\, k'$ fills in (1) commutatively. Therefore, (F2) is valid.

If, in particular, h is also an extremal monomorphism, then e must be an isomorphism. Hence the class of extremal monomorphisms is closed with respect to composition.

We now show that the class of extremal monomorphisms with a fixed codomain B is closed with respect to intersections. Let $\{m_\nu: M_\nu \to B\}$ be a family of extremal monomorphisms, $m = \bigcap m_\nu: M \to B$ and $m = n\, e$, where e is an epimorphism. Since (F2) is valid, there are morphisms d_ν with $n = m_\nu\, d_\nu$ for all ν. By the definition of m as a limit, e is an epimorphic coretraction. Therefore, m is an extremal monomorphism.

Now let $f: A \to B$ be given. By what was just proved, there is a smallest extremal monomorphism, say m, through which f factors. Let $f = m\, e$. e is an epimorphism. This follows from the minimality of m and the fact that the class of extremal monomorphisms is closed with respect to composition, since equalizers are extremal monomorphisms, by the dual of (a). Therefore, (F1) is valid, and the proof of (g) is complete.

(h) Follows immediately from (f) and 18.4.7.

(i) Given $f, g\colon A \to B$, we consider all epimorphisms p in $\mathfrak{E}$ whose domain is B and for which $p f = p g$. Let $\{p_\nu\colon B \to P_\nu\}$ be a set of representatives for the equivalence classes of these epimorphisms. There is a morphism $q\colon B \to \prod P_\nu$ with $pr_\nu q = p_\nu$ for all ν. Let $q = m p$ with $p \in \mathfrak{E}$ and $m \in \mathfrak{M}$. We show that p is a coequalizer of f and g.

$p_\nu f = p_\nu g$ for all ν implies $q f = q g$ and thus $p f = p g$, since m is a monomorphism. For $h\colon B \to H$, let $h f = h g$. Let $h = h'' h'$ with $h' \in \mathfrak{E}$ and $h'' \in \mathfrak{M}$. Now, there is a ν and an isomorphism j_ν such that $h' = j_\nu p_\nu$. One thus has $h = h'' j_\nu pr_\nu m p$, so that h is of the form $h = k p$, and k is determined uniquely by h, because p is an epimorphism.

(j) follows immediately from 18.4.3 and 18.4.10.

21.6.3 Theorem. *Let the category $\mathscr{C}$ have a natural $\mathfrak{E}$-$\mathfrak{M}$-factorization of morphisms* ((F1), (F2) *in* 21.6.1). *Let $\boldsymbol{T} = (T, \eta, \mu)$ be a triple over $\mathscr{C}$ and $F^{\boldsymbol{T}} \frac{\varepsilon^{\boldsymbol{T}}}{\eta}| U^{\boldsymbol{T}}(\mathscr{C}^{\boldsymbol{T}}, \mathscr{C})$ be the corresponding Eilenberg-Moore situation. $\mathfrak{E}^{\boldsymbol{T}}$ or, resp., $\mathfrak{M}^{\boldsymbol{T}}$ is to be the class of those homomorphisms which are taken into $\mathfrak{E}$ or, resp., $\mathfrak{M}$ by $U^{\boldsymbol{T}}$. Suppose T maps $\mathfrak{E}$ into itself. Then the following holds:*

(a) *$U^{\boldsymbol{T}}$ creates an $\mathfrak{E}^{\boldsymbol{T}}$-$\mathfrak{M}^{\boldsymbol{T}}$-factorization of the morphisms in $\mathscr{C}^{\boldsymbol{T}}$*; i.e., *every $\mathfrak{E}$-$\mathfrak{M}$-factorization of a $\mathscr{C}$-morphism of the form $U^{\boldsymbol{T}}(\alpha)$ is lifted uniquely to an $\mathfrak{E}^{\boldsymbol{T}}$-$\mathfrak{M}^{\boldsymbol{T}}$-factorization of α; and $U^{\boldsymbol{T}}$ reflects such factorizations. $\mathfrak{E}^{\boldsymbol{T}}$ consists of epimorphisms, $\mathfrak{M}^{\boldsymbol{T}}$ of monomorphisms. The $\mathfrak{E}^{\boldsymbol{T}}$-$\mathfrak{M}^{\boldsymbol{T}}$-factorization satisfies* (F1) *and* (F2)

(b) *$U^{\boldsymbol{T}}$ reflects coequalizers.*

(c) *If $\mathfrak{E}$ consists of all coequalizers, then $U^{\boldsymbol{T}}$ preserves coequalizers of kernel pairs (as far as they exist).*

(d) *If $\mathscr{C}$ has kernel pairs and if $\mathfrak{E}$ consists of all coequalizers, then $\mathfrak{E}^{\boldsymbol{T}}$ consists of all coequalizers. Thus a homomorphism (b, c, p) is a coequalizer if and only if p is a coequalizer.*

(e) *If* (F3) *is valid for $\mathfrak{E}$, i.e., if $\mathfrak{E}$ is pointwise small, then* (F3) *is valid for $\mathfrak{E}^{\boldsymbol{T}}$.*

(f) *If $\mathscr{C}$ has products and* (F3) *is valid for $\mathfrak{E}$, then $\mathscr{C}^{\boldsymbol{T}}$ has coequalizers.*

(g) *Let $F \frac{\varepsilon}{\eta}| U(\mathscr{X}, \mathscr{C})$ be an adjoint situation which generates $\boldsymbol{T}$, $K\colon \mathscr{X} \to \mathscr{C}^{\boldsymbol{T}}$ the corresponding comparison functor* (21.1.8) *and $\mathfrak{E}_{\mathscr{X}}$ the class of those $\mathscr{X}$-morphisms which are taken into morphisms in $\mathfrak{E}$ by U. Then the following holds:*

F maps $\mathfrak{E}$ into $\mathfrak{E}_{\mathscr{X}}$. An $\mathscr{X}$-morphism $v\colon X \to Y$ belongs to $\mathfrak{E}_{\mathscr{X}}$ if and only if $K(v) \in \mathfrak{E}^{\boldsymbol{T}}$. Further, the following are equivalent (see also 21.4.6 (c)):

(i) *K is fully faithful,*

(ii) *K reflects coequalizers,*

(iii) *U reflects coequalizers.*

Proof. (a) Let $(a, b, f)\colon (A, a) \to (B, b)$ be a homomorphism and $p\colon A \to C$, $m\colon C \to B$ with $f = m\, p$ an $\mathfrak{E}$-$\mathfrak{M}$ factorization of f. We consider

$$(2)\qquad \begin{array}{ccccc} T(A) & \overset{T(p)}{\twoheadrightarrow} & T(C) & \overset{T(m)}{\to} & T(B) \\ {\scriptstyle a}\big\downarrow & & {\scriptstyle c}\Downarrow & & \big\downarrow{\scriptstyle b} \\ A & \overset{p}{\to} & C & \overset{m}{\rightarrowtail} & B\,. \end{array}$$

Since $T(p) \in \mathfrak{E}$, $m \in \mathfrak{M}$ and because of the diagonal condition that is equivalent to (F2), there is exactly one $c\colon T(C) \to C$ with

$$(3)\qquad c\, T(p) = p\, a \quad \text{and} \quad m\, c = b\, T(m)\,.$$

By 21.3.5 (c), (C, c) is an algebra, and by (3), (a, c, p) is a homomorphism. Since $m \in \mathfrak{M}$, c is determined uniquely by (3). U^{T} reflects $\mathfrak{E}^{T}$-$\mathfrak{M}^{T}$-factorizations by the definition of $\mathfrak{E}^{T}$ and $\mathfrak{M}^{T}$. Since U^{T} is faithful, $\mathfrak{E}^{T}$ consists of epimorphisms, $\mathfrak{M}^{T}$ of monomorphisms. By what was just proved, the $\mathfrak{E}^{T}$-$\mathfrak{M}^{T}$-factorization satisfies (F1). (F2) follows from 21.3.5 (a), (b).

(b) Let (a, b, f), $(a, b, g)\colon (A, a) \to (B, b)$, $(b, c, p)\colon (B, b) \to (C, c)$ be homomorphisms, and let p be a coequalizer of f and g. Then let $(b, d, q)\colon (B, b) \to (D, d)$ be a homomorphism with $q\, f = q\, g$. There is exactly one $\mathscr{C}$-morphism $h\colon C \to D$ with $q = h\, p$. By 21.6.2 (a), (c), $p \in \mathfrak{E}$ and thus $T(p) \in \mathfrak{E}$. (a) and 21.3.5 (b) imply that (b, c, p) is a coequalizer of (a, b, f) and (a, b, g).

(c) Let (a, b, f), $(a, b, g)\colon (A, a) \to (B, b)$ be a kernel pair with the coequalizer $(b, c, p)\colon (B, b) \to (C, c)$. By 18.4.3, (a, b, f), (a, b, g) is a kernel pair of (b, c, p). Since U^{T} preserves limits, f, g is a kernel pair of p. By (a) and 21.6.2 (a), (c), $(b, c, p) \in \mathfrak{E}^{T}$ and $p \in \mathfrak{E}$, and thus p is a coequalizer in $\mathscr{C}$. p is a coequalizer of its kernel pair f, g, again by 18.4.3.

(d) If $(b, c, p) \in \mathfrak{E}^{T}$, then p is a coequalizer of its kernel pair. Since U^{T} creates limits, (b) implies that (b, c, p) is a coequalizer of its kernel pair. If, conversely, (b, c, p) is a coequalizer, then $(b, c, p) \in \mathfrak{E}^{T}$ by 21.6.2 (a), (c).

(e) follows from (a), because U^{T} creates isomorphisms.

(f) follows from (a), (e), 21.3.9 (a) and 21.6.2 (i).

(g) The first statement follows immediately from the definition of $\mathfrak{E}_{\mathscr{X}}$ and the assumption about T. Since $U = U_{T}\, K$, $v \in \mathfrak{E}_{\mathscr{X}}$, $U(v) \in \mathfrak{E}$ and $K(v) \in \mathfrak{E}^{T}$ are equivalent by the definitions of $\mathfrak{E}_{\mathscr{X}}$ and $\mathfrak{E}^{T}$.

(i) $\Rightarrow$ (ii), by 8.7.6.

(ii) $\Rightarrow$ (iii). Assume that in $\mathscr{X}$ there are $v, w\colon X \to Y$ and $p\colon Y \to Z$ with $p\, v = p\, w$. If $U(p)$ is a coequalizer of $U(v)$ and $U(w)$, then $K(p)$ is a coequalizer of $K(v)$ and $K(w)$, by (b) and since $U = U^{T} K$. (ii) implies that p is a coequalizer of v and w.

(iii) $\Rightarrow$ (i), by 21.4.6 (c).

21.6.4 Theorem. *For every triple $\boldsymbol{T} = (T, \eta, \mu)$ over Ens, T preserves epimorphisms. Every Eilenberg-Moore category $Ens^{\boldsymbol{T}}$ over Ens is complete and cocomplete. It possesses a factorization of morphisms into coequalizers and monomorphisms which satisfies* (F1), (F2), (F3) *and which is created by $U^{\boldsymbol{T}}$. Furthermore, $Ens^{\boldsymbol{T}}$ is well-powered. $U^{\boldsymbol{T}}$ preserves coequalizers of kernel pairs and reflects coequalizers. A pair of morphisms α, β in $Ens^{\boldsymbol{T}}$ is a kernel pair if and only if $U^{\boldsymbol{T}}(\alpha), U^{\boldsymbol{T}}(\beta)$ is a kernel pair.*

$F^{\boldsymbol{T}}$ preserves monomorphisms, and thus so does T.

Proof. Every epimorphism in *Ens* is a retraction. Therefore, epimorphisms are preserved by every functor with domain *Ens*. All the statements above, exept the last one, follow thus from known properties of *Ens* and 21.6.3, taking into account 21.3.9, 21.3.10, 21.3.6 and 21.4.4.

Every monomorphism with non-empty domain in *Ens* is a coretraction and is thus preserved by every functor with domain *Ens*. A map $m: \emptyset \to A$ in *Ens* is always a monomorphism. If $T(\emptyset) = \emptyset$, then $T(m)$ is a monomorphism, and so is $F^{\boldsymbol{T}}(m)$. Now let $T(\emptyset) \neq \emptyset$. Then there is a map $A \to T(\emptyset)$. By means of $\varphi^{\boldsymbol{T}}$, there is a corresponding $Ens^{\boldsymbol{T}}$-morphism $r: F^{\boldsymbol{T}}(A) \to F^{\boldsymbol{T}}(\emptyset)$. Since $F^{\boldsymbol{T}}$ preserves colimits, $F^{\boldsymbol{T}}(\emptyset)$ is initial in $Ens^{\boldsymbol{T}}$. Therefore, r is a retraction with the coretraction $F^{\boldsymbol{T}}(m)$.

21.6.5 Proposition. *Let the category $\mathscr{C}$ have products and an $\mathfrak{E}$-$\mathfrak{M}$-factorization of morphisms which satisfies* (F1), (F2), (F3) *of* 21.6.1. *Let $F^{\boldsymbol{T}} \frac{\varepsilon^{\boldsymbol{T}}}{\eta}| U^{\boldsymbol{T}}(\mathscr{C}^{\boldsymbol{T}}, \mathscr{C})$ be the Eilenberg-Moore situation corresponding to the triple $\boldsymbol{T} = (T, \eta, \mu)$ over $\mathscr{C}$, and assume that T maps $\mathfrak{E}$ into itself.*

(a) *Let $F' \frac{\varepsilon'}{\eta'}| U'(\mathscr{X}', \mathscr{C})$ be the Eilenberg-Moore situation corresponding to the triple $\boldsymbol{T}' = (T', \eta', \mu')$, where T' also maps $\mathfrak{E}$ into itself. Let $\tau: \boldsymbol{T} \to \boldsymbol{T}'$ be a triple-morphism. The corresponding functor K: $\mathscr{X}' \to \mathscr{C}^{\boldsymbol{T}}$ in* 21.1.7 (e) *has a left adjoint J, and $\mathfrak{E}^{\boldsymbol{T}}$ (see* 21.6.3*) is mapped into itself by $K\,J$.*

(b) *Let*

(4) $$J \frac{\bar{\varepsilon}}{\bar{\eta}}| K(\mathscr{X}', \mathscr{C}^{\boldsymbol{T}})$$

be an Eilenberg-Moore situation over $\mathscr{C}^{\boldsymbol{T}}$, where $\mathfrak{E}^{\boldsymbol{T}}$ is mapped into itself by $K\,J$. Then $\mathfrak{E}$ is mapped into itself by $T' = U^{\boldsymbol{T}}\,K\,J\,F^{\boldsymbol{T}}$.

(5) $$J\,F^{\boldsymbol{T}} \frac{\bar{\varepsilon}(J*\varepsilon^{\boldsymbol{T}}*K)}{(K*\eta*J)\bar{\eta}}\Big| U^{\boldsymbol{T}}\,K(\mathscr{X}', \mathscr{C})$$

is isomorphic to an Eilenberg-Moore situation if and only if the following holds:

(i) *K preserves coequalizers of pairs of morphisms that are split by $U^{\boldsymbol{T}}\,K$.*

(c) *If $\mathscr{C}$ is complete, then condition* (i) *in* (b) *can be replaced by*

(ii) *K preserves coequalizers of pairs of morphisms which are taken into kernel pairs of retractions by $U^{\mathbf{T}} K$.*

(d) *If $\mathscr{C}$ is complete, and if $\mathfrak{E}$ is the class of coequalizers, then* (i) *in* (b) *can be replaced by*

(iii) *A pair of morphisms in $\mathscr{X}'$ is a kernel pair if $U^{\mathbf{T}} K$ takes it into a kernel pair of a retraction.*

Remark. Note that all these criteria apply only in the case where $\mathfrak{E}$ is mapped into itself by T and by T'. We shall return to the special case $\mathscr{C} = Ens$ in more detail in 21.6.12.

Proof. By 21.6.3 (f), $\mathscr{X}'$ has coequalizers, and J exists by 21.5.3, 21.5.4 (c). Let $\mathfrak{E}'$ be the class of $\mathscr{X}'$-morphisms which are mapped into $\mathfrak{E}$ by U'. If $\pi: (A, a) \to (B, b)$ is in $\mathfrak{E}^{\mathbf{T}}$, then $U^{\mathbf{T}}(\pi) \in \mathfrak{E}$ and $F' U^{\mathbf{T}}(\pi) \in \mathfrak{E}'$. For $\varkappa: F' U^{\mathbf{T}} \to J$, as in 21.5.3, we have $\varkappa_{(B,b)} F' U^{\mathbf{T}}(\pi) = J(\pi) \varkappa_{(A,a)}$. $J(\pi) \in \mathfrak{E}'$ follows from 21.6.3 and 21.6.2 (a), (c), (e). Thus one has $U^{\mathbf{T}} K J(\pi) = U' J(\pi) \in \mathfrak{E}$ and $K J(\pi) \in \mathfrak{E}^{\mathbf{T}}$.

(b) If $p \in \mathfrak{E}$, then $F^{\mathbf{T}}(p) \in \mathfrak{E}^{\mathbf{T}}$, $K J F^{\mathbf{T}}(p) \in \mathfrak{E}^{\mathbf{T}}$ and $U^{\mathbf{T}} K J F^{\mathbf{T}}(p) \in \mathfrak{E}$. Applying 21.6.3 twice shows that $\mathscr{X}'$ has coequalizers and that $U^{\mathbf{T}} K$ reflects coequalizers. By 21.4.4 (a), $U^{\mathbf{T}}$ preserves coequalizers of pairs of morphisms which are split by $U^{\mathbf{T}}$. Since $U^{\mathbf{T}}$ reflects coequalizers, 21.5.7 (d) implies that (5) is equivalent to an Eilenberg-Moore situation if and only if the given condition for K is satisfied. Since $U^{\mathbf{T}}$ and K lift isomorphisms uniquely, the statement follows from 21.3.8 (e).

(c) By 21.3.9, $\mathscr{C}^{\mathbf{T}}$ and $\mathscr{X}'$ are complete. The statement thus follows from 21.5.7, 21.5.13 and the proof of (b).

(d) By 21.6.3 (d), $\mathfrak{E}^{\mathbf{T}}$ consists of all coequalizers, and by 21.6.3 (c), $U^{\mathbf{T}}$, K and $U^{\mathbf{T}} K$ preserve coequalizers of kernel pairs. (ii) thus follows from (iii). If (ii) is satisfied, then 21.4.4 (c) can be applied to $U^{\mathbf{T}} K$, so that (iii) is valid.

21.6.6 Proposition. *Let the assumptions in* 21.6.5 *be satisfied, and let $\mathscr{C}$ be complete.*

(a) *Let $\tau: \mathbf{T} \to \mathbf{T}'$ with $\mathbf{T}' = (T', \eta', \mu')$ be a triple-morphism such that $\tau_A \in \mathfrak{E}$ for all $A \in |\mathscr{C}|$. Then $\mathfrak{E}$ is mapped into itself by T' too. By* 21.6.5 (a), *one has the adjoint situation* (4). *Here $\bar{\eta}_{(A,a)} \in \mathfrak{E}^{\mathbf{T}}$ for every $(A, a) \in |\mathscr{C}^{\mathbf{T}}|$. By means of K, $\mathscr{X}'$ is isomorphic to the strictly full subcategory $\mathscr{X}$ of $\mathscr{C}^{\mathbf{T}}$ whose objects X are characterized by $\bar{\eta}_X$ being an isomorphism. Among others, $\mathscr{X}$ has the following properties:*

(S) *If $m: (A, a) \to X$ is a morphism in $\mathfrak{M}^{\mathbf{T}}$ with $X \in |\mathscr{X}|$, then $(A, a) \in |\mathscr{X}|$.*

(P) *$\mathscr{X}$ is closed with respect to products in $\mathscr{C}^{\mathbf{T}}$.*

(R) *If $\pi: Y \to Z$ is a $\mathscr{C}^{\mathbf{T}}$-morphism such that $U^{\mathbf{T}}(\pi)$ is a retraction, and if $Y \in |\mathscr{X}|$, then $Z \in |\mathscr{X}|$.*

(b) *Let $K: \mathcal{X} \to \mathcal{C}^{\mathbf{T}}$ be the inclusion of a strictly full subcategory which satisfies* (S), (P) *and* (R). *Then $U^{\mathbf{T}} K$ is the right adjoint in an adjoint situation which is isomorphic to an Eilenberg-Moore situation. K induces a triple-morphism τ such that $\tau_A \in \mathfrak{E}$ for all $A \in |\mathcal{C}|$.*

Proof. The first statement follows immediately from 21.6.2 (e). By 21.5.4 (11), $U^{\mathbf{T}}(\bar{\eta}_{F^{\mathbf{T}}(A)}) \cong \tau_A \in \mathfrak{E}$ and thus $\bar{\eta}_{F^{\mathbf{T}}(A)} \in \mathfrak{E}^{\mathbf{T}}$ for all $A \in |\mathcal{C}|$. By 21.6.5 (a) and 21.6.2 (c), $K J(\varepsilon^{\mathbf{T}}_{(A,a)}) \in \mathfrak{E}^{\mathbf{T}}$ for all $(A, a) \in |\mathcal{C}^{\mathbf{T}}|$. Since

$$KJ(\varepsilon^{\mathbf{T}}_{(A,a)})\, \bar{\eta}_{F^{\mathbf{T}}(A)} = \bar{\eta}_{(A,a)}\, \varepsilon^{\mathbf{T}}_{(A,a)}\,, \qquad \bar{\eta}_{(A,a)} \in \mathfrak{E}^{\mathbf{T}}\,,$$

by 21.6.3 and 21.6.2 (e). By 21.1.9, K is a full embedding and, therefore, $\bar{\mu} = K * \bar{\varepsilon} * J$ is an isomorphism. By 21.5.8 (a) and 21.3.3, K induces an isomorphism between $\mathcal{X}'$ and $\mathcal{X}$.

(S). By 21.6.3, (S) follows from the dual of (F5) according to 21.6.2 (d), (e): If $\bar{\eta}_X$ is an isomorphism, then $KJ(m)\, \bar{\eta}_{(A,a)} = \bar{\eta}_X\, m \in \mathfrak{M}^{\mathbf{T}}$ and thus $\bar{\eta}_{(A,a)} \in \mathfrak{M}^{\mathbf{T}}$. Since $\bar{\eta}_{(A,a)} \in \mathfrak{E}^{\mathbf{T}}$, $\bar{\eta}_{(A,a)}$ is an isomorphism.

(P) By 16.6.1, (P) follows from the fact that $\mathcal{X}$ is a strictly full, reflective subcategory of $\mathcal{C}^{\mathbf{T}}$. (By 21.3.9 (a), $\mathcal{C}^{\mathbf{T}}$ is complete).

(R) Let $\alpha, \beta: (A, a) \to Y$ be a kernel pair of π. By 21.4.4, π is a coequalizer of α, β. For $Y \in |\mathcal{X}|$, $\bar{\eta}_Y$ is an isomorphism, and

$$\bar{\eta}_Y^{-1}\, KJ(\alpha)\, \bar{\eta}_{(A,a)} = \bar{\eta}_Y^{-1}\, \bar{\eta}_Y\, \alpha = \alpha\,, \qquad \bar{\eta}_Y^{-1}\, KJ(\beta)\, \bar{\eta}_{(A,a)} = \beta\,.$$

Since $\bar{\eta}_{(A,a)}$ is an epimorphism, $\pi\, \bar{\eta}_Y^{-1}\, KJ(\alpha) = \pi\, \bar{\eta}_Y^{-1}\, KJ(\beta)$. The definition of kernel pairs implies further that $\bar{\eta}_{(A,a)}$ is a coretraction. Thus $\bar{\eta}_{(A,a)}$ is an isomorphism, and from this $(A, a) \in |\mathcal{X}|$. Since $U^{\mathbf{T}} K$ is isomorphic to the right adjoint of an Eilenberg-Moore situation, π is in $\mathcal{X}$, by 21.4.4 (a). Therefore, $Z \in |\mathcal{X}|$.

(b) Since $\mathcal{C}^{\mathbf{T}}$ has products, one concludes from (S), (P) and 21.6.3 (e), as in the proof of 16.6.3 (b), that there is an adjoint situation (4) with $\mathcal{X}$ instead of $\mathcal{X}'$,where $\bar{\eta}_{(A,a)} \in \mathfrak{E}^{\mathbf{T}}$ for all $(A, a) \in |\mathcal{C}^{\mathbf{T}}|$, and $\bar{\varepsilon}$ is an isomorphism. $J\, F^{\mathbf{T}}$ is left adjoint to $U^{\mathbf{T}} K$. Let $v, w: X \to Y$ be a pair of morphisms in $\mathcal{X}$ which is split by $U^{\mathbf{T}} K$, and let $r: U^{\mathbf{T}} K(Y) \to C$ be a coequalizer of $U^{\mathbf{T}} K(v)$ and $U^{\mathbf{T}} K(w)$. By 21.4.4 (a), there is exactly one $\mathcal{C}^{\mathbf{T}}$-morphism $\pi: K(Y) \to Z$ with $\pi\, K(v) = \pi\, K(w)$. (R) guarantees that $\pi \in \mathcal{X}$. π is a coequalizer of $K(v)$, $K(w)$ in $\mathcal{C}^{\mathbf{T}}$ and also of v, w in $\mathcal{X}$, because K is fully faithful. Therefore, $U^{\mathbf{T}} K$ creates coequalizers of pairs of morphisms that are split by $U^{\mathbf{T}} K$. Thus 21.5.7 implies the first statement in (b). The second one follows from the fact that, by 21.5.4 (11), τ is isomorphic to $U^{\mathbf{T}} * \bar{\eta} * F^{\mathbf{T}}$.

21.6.7 Specializations and remarks. (a) Since every fully faithful functor can be factorized into an equivalence and the inclusion of a strictly full subcategory, 21.6.6 (b) has obvious variations. 21.6.6 is due to Manes [62].

(b) If $\boldsymbol{T}$ in 21.6.6 is the identity triple, then $\tau = \eta'$, (4) and (5) coincide and are adjoint situations generating $\boldsymbol{T}'$. In this special case (S) implies (R), since coretractions belong to $\mathfrak{M}^{\boldsymbol{T}}$. This yields a generalization of 16.6.3 (c) which takes 21.3.3 into account.

(c) If $\mathscr{C} = Ens$, then 21.6.4 applies, $\mathfrak{M}^{\boldsymbol{T}}$ consists of all monomorphisms and condition (R) is simply:
(R′) If $\pi\colon Y \to Z$ is a coequalizer in $Ens^{\boldsymbol{T}}$ and $Y \in |\mathscr{X}|$, then $Z \in |\mathscr{X}|$.

(d) Let the assumptions in 21.6.6 be satisfied again. If $\mathfrak{K}$ is any class of objects in $\mathscr{C}^{\boldsymbol{T}}$, then there is a smallest, strictly full subcategory $\mathscr{X}$ of $\mathscr{C}^{\boldsymbol{T}}$ which contains all the objects of $\mathfrak{K}$ and which satisfies (S), (P), (R).

If, in particular, $\mathscr{X}'$ is a full subcategory of $\mathscr{C}^{\boldsymbol{T}}$ which satisfies (S) and (P), then there is an adjoint situation

$$J' \,\frac{\bar{\varepsilon}'}{\bar{\eta}'}\Big|\, K'(\mathscr{X}', \mathscr{C}^{\boldsymbol{T}}), \tag{4′}$$

where K' is the inclusion, $J'\,K' = 1_{\mathscr{X}'}$, and $\bar{\varepsilon}'$ is the identity natural transformation. Furthermore, $\bar{\eta}'_{(A,a)} \in \mathfrak{E}^{\boldsymbol{T}}$ for all $(A, a) \in |\mathscr{C}^{\boldsymbol{T}}|$. Let $\boldsymbol{T}'$ be the image by $U^{\boldsymbol{T}}$ of the triple over $\mathscr{C}^{\boldsymbol{T}}$ generated by (4′), and let $\tau\colon \boldsymbol{T} \to \boldsymbol{T}'$ be the triple morphism induced by (4′). $\tau_A \in \mathfrak{E}$ for all $A \in |\mathscr{C}|$, and up to an isomorphism the statements in 21.6.6 (b) apply to the Eilenberg-Moore situation of $\boldsymbol{T}'$. (4′) thus factors into two adjoint situations, where the right adjoint functors are inclusions.

In this way one obtains, in fact, the smallest, strictly full subcategory of $\mathscr{C}^{\boldsymbol{T}}$ which contains $\mathscr{X}'$ and satisfies (S), (P), (R). To verify this, it is sufficient to consider the case where (4′) induces the identity triple-morphism. The statement then follows from 21.4.3 (a), since by 21.1.7 (f), $\mathscr{X}'$ contains all the free algebras here.

(e) When considering 21.6.6, one should note:

If $\mathscr{X}$ is a strictly full subcategory of an Eilenberg-Moore category $\mathscr{C}^{\boldsymbol{T}}$ with the inclusion $K\colon \mathscr{X} \to \mathscr{C}^{\boldsymbol{T}}$, then it is possible that $\mathscr{X}$ is isomorphic or equivalent to an Eilenberg-Moore category over $\mathscr{C}$, but that $U^{\boldsymbol{T}}\,K$ does not belong to any such adjoint situation. For instance, let $\mathscr{X}$ be the category of groups, $\mathscr{Y}$ the category of rings and $K\colon \mathscr{X} \to \mathscr{Y}$ the functor which assigns to every group its group ring (over $\mathbf{Z}$). $\mathscr{X}$ and $\mathscr{Y}$ are isomorphic to Eilenberg-Moore categories over Ens (see below 21.6.10 (a)), K is fully faithful and has a right adjoint. K is not an algebraic functor, and 21.6.6 does not apply.

(f) With regard to the special role played by coequalizers in 21.3.10, 21.5.3, 21.6.3, 21.6.4 and 21.6.5, $\mathfrak{E}$-$\mathfrak{M}$-factorizations of morphisms, in which $\mathfrak{E}$ is the class of all coequalizers, are of interest. It is possible in this connection to obtain some statements that apply not only to Eilenberg-Moore situations. (see 21.6.11 below).

21.6.8 Lemma. *Let* $F \,\frac{\varepsilon}{\eta}\Big|\, U(\mathscr{X}, \mathscr{C})$ *generate the triple* $\boldsymbol{T} = (T, \eta, \mu)$ *over* $\mathscr{C}$, *and assume that it satisfies the following condition:*

(A) *If $p: X \to Y$ is a coequalizer in $\mathfrak{X}$, then $U(p)$ is a coequalizer.*
Then the class of coequalizers is mapped into itself by T and
(A') *U preserves coequalizers of kernel pairs.*
If $\mathfrak{X}$ has kernel pairs, then, conversely, (A') *implies* (A).

Proof. The first statement follows immediately from the fact that F preserves coequalizers. Let v, w be a kernel pair in $\mathfrak{X}$ and $p: Y \to Z$ a coequalizer of v and w. v, w is a kernel pair of p, by 18.4.3. Since U preserves limits, $U(v)$, $U(w)$ is a kernel pair of $U(p)$. By (A) and 18.4.3, $U(p)$ is a coequalizer of $U(v)$, $U(w)$. If $p: Y \to Z$ is a coequalizer in $\mathfrak{X}$ and v, w a kernel pair of p, then (A) is similarly implied by (A').

21.6.9 Theorem. *Let the category $\mathscr{C}$ be complete and have an $\mathfrak{E}$-$\mathfrak{M}$-factorization of morphisms in which $\mathfrak{E}$ is to be the class of all coequalizers and* (F1), (F2), (F3) *are satisfied. Let $F \frac{\varepsilon}{\eta}| U(\mathfrak{X}, \mathscr{C})$ generate the triple $\boldsymbol{T} = (T, \eta, \mu)$, and let $K_{\varepsilon,\eta}: \mathfrak{X} \to \mathscr{C}^{\boldsymbol{T}}$ be the corresponding comparison functor. Suppose T maps $\mathfrak{E}$ into itself.*

(a) *$K_{\varepsilon,\eta}$ is fully faithful if and only if the following condition is satisfied:*
(B) *If p is an $\mathfrak{X}$-morphism such that $U(p)$ is a retraction, then p is a coequalizer.*
(b) *If $K_{\varepsilon,\eta}$ is fully faithful, then U is faithful and U reflects isomorphisms, limits and coequalizers.*
(c) *If $K_{\varepsilon,\eta}$ is fully faithful and if $\mathfrak{X}$ has kernel pairs, then*
(B') *if p is an $\mathfrak{X}$-morphism such that $U(p)$ is a coequalizer, then p is a coequalizer.*
(d) *$K_{\varepsilon,\eta}$ has a left adjoint, provided the following condition is satisfied:*
(C) *$\mathfrak{X}$ has coequalizers for pairs of morphisms that are split by U.*
(e) *If $K_{\varepsilon,\eta}$ is fully faithful, and if $K_{\varepsilon,\eta}$ has a left adjoint, then $\mathfrak{X}$ is complete and $\mathfrak{X}$ has coequalizers. In particular,* (B') *and* (C) *are satisfied.*
(f) *If* (B) *and* (C) *are satisfied, then the following are equivalent:*
(A) *in* 21.6.8,
(A') *in* 21.6.8,
(A'') *$\mathfrak{X}$ has a factorization of morphisms into coequalizers and monomorphisms which is preserved by U.*
If (A), (B), (C) *are satisfied, then the factorization* (A'') *satisfies* (F1), (F2), (F3) *of* 21.6.1 *and U reflects it.*
(g) *$K_{\varepsilon,\eta}$ is an equivalence if and only if* (A), (B), (C) *are satisfied and if in addition the following holds:*
(D) *If, v w is a pair of morphisms in $\mathfrak{X}$ such that $U(v)$, $U(w)$ is the kernel pair of a retraction, then v, w is a kernel pair.*
(h) *$K_{\varepsilon,\eta}$ is an isomorphism of categories if and only if* (A), (B), (C), (D) *are satisfied and if also*
(E) *U lifts isomorphisms uniquely.*

Proof. (a) follows from 21.4.6 (c).
(b) follows from 21.4.6 (c) and 21.6.3 (g).

(c) Let v, w be a kernel pair of p in $\mathscr{X}$ and $U(p)$ a coequalizer. Since U preserves limits, $U(v)$, $U(w)$ is a kernel pair of $U(p)$. By 18.4.3, $U(p)$ is a coequalizer of $U(v)$, $U(w)$. (b) implies (B').

(d) and (e) follow immediately from 21.5.7, 21.3.9, 21.6.3 (f), 16.6.1 and from (c).

(f) By the above and by 21.6.8, (A) and (A') are equivalent. Now let (A) be valid. Let v be an $\mathscr{X}$-morphism. By the above, there is a kernel pair p, q of v and a coequalizer v' of p, q. There is exactly one $\mathscr{X}$-morphism v'' with $v = v'' v'$. By 18.4.3, p, q is also a kernel pair of v'. Since U preserves limits ,$U(p)$, $U(q)$ is a kernel pair of $U(v)$ and of $U(v')$. By (A) and 18.4.3, $U(v')$ is a coequalizer of $U(p)$ and $U(q)$, and $U(v'')$ is a monomorphism. U reflects monomorphisms, because of (b), and $v = v'' v'$ is the desired factorization of v. (A'') thus holds. The converse is evident.

By (b) and (c), U reflects the factorization of morphisms (A''). 21.6.3 (a) thus implies that it is preserved and reflected by $K_{\varepsilon,\eta}$. Since $K_{\varepsilon,\eta}$ is fully faithful, (F1), (F2), (F3) follow from 21.6.3.

(g) If (A) through (D) are satisfied, then $K_{\varepsilon,\eta}$ is an equivalence, by (b), (f), 21.5.13 and 21.5.7 (d). The converse follows from 21.4.4 and 21.6.3 (d).

(h) By 21.3.9, $U^{\mathbf{T}}$ lifts isomorphisms uniquely. Therefore, (E) is valid if and only if $K_{\varepsilon,\eta}$ lifts isomorphisms uniquely, and the statement follows from 21.3.8 (e).

21.6.10 Examples and remarks. (a) Let $\mathscr{A}$ be an algebraic theory (with finitary operations) in the sense of 18.1.2. By 21.6.9, one then has an isomorphism with a corresponding Eilenberg-Moore situation for the reduced algebraic category ${}_0\mathscr{A}^b$ (18.3.1) with the forgetful functor ${}_0U_{\mathscr{A}}$ and the left adjoint ${}_0F_{\mathscr{A}}$. This follows from 21.6.4, 18.3.6, 18.4.5 and the fact that ${}_0U_{\mathscr{A}}$ lifts isomorphisms uniquely.

This example has led one to regard Eilenberg-Moore situations (or equivalent adjoint situations) as algebraic situations. It motivates the terminology in 21.1.2. This way of looking at things is, in fact, correct, provided $\mathscr{C} = Ens$ (see later in 21.7.7). If $\mathscr{C} \neq Ens$, it can lead to errors; in general, the forgetful functor in 11.3.2, 11.3.3 is not tripleable.

(b) Let $\mathscr{X}$ be the category of compact (Hausdorff) spaces with continuous maps. The forgetful functor $U\colon \mathscr{X} \to Ens$ has a left adjoint which assigns to every set the Stone-Čech-compactification of the corresponding discrete topological space. Taking 21.6.9 (f) into account, one verifies without trouble that $\mathscr{X}$ and U satisfy conditions (A) through (E).

(c) Let $\mathscr{X} = Top$ be the category of topological spaces and $U: \mathscr{X} \to Ens$ the forgetful functor. U has a left adjoint F which provides every set with the discrete topology. $UF = 1_{Ens}$. The corresponding triple is the identity triple. (A) and (C) are valid, but not (B), (D) or (E). For every subcategory of Top which contains a bijective, continuous, but not homeomorphic map, the forgetful functor violates (B).

For every subcategory of Top which contains the full subcategory of discrete spaces, one gets an adjoint situation over Ens which generates the identity triple. This example shows two things; namely, that there is little hope of surveying all adjoint situations corresponding to a given triple and, also, that Eilenberg-Moore situations are not the only interesting adjoint situations.

(d) Let $\mathscr{X}$ be the category of torsion-free abelian groups. Starting with the forgetful functor $V: Ab \to Ens$ with left adjoint "free abelian group", the restriction $U = V|\mathscr{X}$ produces an adjoint situation generating the same triple. $K: \mathscr{X} \to Ab$ is the inclusion. U satisfies (A), (B), (C), (E), but not (D). Incidentally, this shows that the composite of two Eilenberg-Moore situations need not be isomorphic to an Eilenberg-Moore situation.

(e) As is shown by the proof of 21.6.8, not all assumptions were needed for the partial statements in 21.6.9.

(f) In the following proposition we do not state explicitly all the facts which can be deduced according to the following scheme:

If a property is preserved by U' (or, resp., reflected) and reflected (or, resp., preserved) by U, then it is preserved (or, resp., reflected) by K. If it is preserved by U and K (or, resp., reflected), then by U' also.

21.6.11 Proposition. *Let the category $\mathscr{C}$ satisfy the assumptions in* 21.6.9. *Let*

(6) 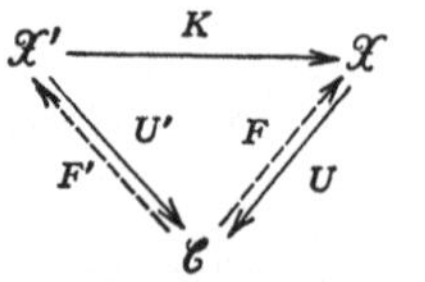

$U' = UK$

be an adjoint right triangle and let $\mathbf{T}' = (T', \eta', \mu')$ *and* $\mathbf{T} = (T, \eta, \mu$ *be the corresponding triples. We consider the conditions* (A), (B), (B'), (C *of* 21.6.8, 21.6.9 *for U and, correspondingly, for U' and K.*

(a) *If U' satisfies* (B), *then U' and K are faithful, and ε' is a coequalizer of $\varepsilon' * F'U'$ and $F'U' * \varepsilon'$.*

(b) *If U satisfies* (A), (B'), *then U' satisfies* (A) *or resp.,* (B') *if and only if K satisfies* (A) *or, resp.,* (B').

(c) *If U and U' satisfy* (A), (B'), *then K preserves and reflects isomorphisms, limits, coequalizers of kernel pairs and factorizations of $\mathscr{X}$-morphisms into coequalizers and monomorphisms.*

(d) *Let T' map* $\mathfrak{E}$ *into itself. If U' satisfies* (B), (C) *and if U satisfies* (B), *then there is an adjoint situation*

(7) $$J \frac{\bar{\varepsilon}}{\bar{\eta}}| K(\mathcal{X}', \mathcal{X}) .$$

(e) *Suppose that there is the adjoint situation* (7) *and that U and* $U' = U K$ *satisfy* (B). *If* $U(\bar{\eta}_{F(A)})$ *is an epimorphism for every* $A \in |\mathcal{C}|$, *then K is fully faithful.*

(f) *Suppose that T maps* $\mathfrak{E}$ *into itself and that U satisfies* (B), (C). *If there is the adjoint situation* (7) *with a fully faithful K, then U' and K satisfy* (B'), (C).

(g) *Let U satisfy* (A), (B), (C). *If U' satisfies* (A), (B), (C), *then* $\mathcal{X}$ *and* $\mathcal{X}'$ *are complete,* $\mathcal{X}$ *and* $\mathcal{X}'$ *have coequalizers,* (7) *exists, and K satisfies* (A), (B'), (C). *If, conversely,* (7) *exists and if K satisfies* (A), (B), (C), *then U' satisfies* (A), (B), (C).

(h) *Let T map* $\mathfrak{E}$ *into itself. Assume that there is the adjoint situation* (7) *and that the adjoint situation on the right in* (6) *is equivalent to an Eilenberg-Moore situation. The left one is equivalent to an Eilenberg-Moore situation, in which* $\mathfrak{E}$ *is mapped into itself by T', if and only if K satisfies* (A), (B), (C) *and also one of the following two conditions (which are equivalent here):*

(D') *K preserves coequalizers of pairs of morphisms which are taken into kernel pairs of retractions by* $U K = U'$.

(D'')*A pair of morphisms in* $\mathcal{X}'$ *is a kernel pair if it taken into a kernel pair of a retraction by* $U K = U'$.

Remark. If K is fully faithful, then (f) makes conditions (B), (C) in (g) superfluous.

Proof. (a) follows from 21.4.6 (c) and $U' = U K$.

(b) follows as in the scheme of 21.6.10 (f).

(c) 21.6.8 and 21.6.9 (a) can be applied to both adjoint situations in (6). The statements then follow as in the scheme of 21.6.10 (f).

(d) 21.6.9 (a), (d), (e) apply to the adjoint situation on the left in (6), so that the statement follows from (a) for U and from 21.5.3.

(e) By (a) for U and by 21.5.4 (b), there exists the coequalizer of 21.5.3, and $U * \theta$ from there is isomorphic to $U * \bar{\eta} * F$, by 21.5.4 (11). It follows thus that $F' U * \theta * U'$ is an epimorphism at every place $X' \in |\mathcal{X}'|$. By (a) and 21.5.3 (6), $\bar{\varepsilon}$ is an isomorphism. 16.5.4 shows that K is fully faithful.

(f) From 21.6.9 (a), (d), (e) and 16.6.1 it follows that $\mathcal{X}$ and $\mathcal{X}'$ are complete and have coequalizers. In particular, (C) is valid for U' and for K. By means of kernel pairs, (B') follows from the fact that K preserves limits and reflects colimits and that U satisfies (B') (21.6.9 (c)).

(g) By 21.6.8 and 21.6.9, the assumptions in 21.6.9 are satisfied by $\mathcal{X}$ instead of $\mathcal{C}$, and (B) can be replaced by (B'). The first statement follows from (b), (d), by 21.6.8. The second statement follows similarly by applying 21.6.9 to (7).

(h) U satisfies (A) through (D) of 21.6.8, 21.6.9. If K satisfies (A), (B), (C), then K reflects limits and coequalizers, by 21.6.9 for (7). If (D′) is valid, then (D″) follows from 21.4.4 and 18.4.3. If (D″) is valid, then by 21.6.9 (g), (7) is equivalent to an Eilenberg-Moore situation. The statement thus follows from 21.6.5, 21.5.8 (a) and from (g).

21.6.12 Theorem. *Let $F \frac{\varepsilon}{\eta}| U(\mathscr{X}, Ens)$ and (7) be equivalent (or, resp., isomorphic) to Eilenberg-Moore situations over Ens or, resp., $\mathscr{X}$. The following statements are equivalent:*

(a) *The composite of the two situations is equivalent (or, resp., isomorphic) to an Eilenberg-Moore situation over Ens.*

(b) *K preserves coequalizers of pairs of morphisms which are taken into kernel pairs by K.*

(c) *The following holds:*

(D‴) *A pair of morphisms v', w' in $\mathscr{X}'$ is a kernel pair if $K(v')$, $K(w')$ is a kernel pair,*

and also one of the following conditions (which are equivalent here):

(A) *If v' is a coequalizer in $\mathscr{X}'$, then $K(v')$ is a coequalizer.*

(A′) *K preserves coequalizers of kernel pairs.*

(A‴) *$K J$ maps the class of coequalizers into itself.*

Proof. By 21.6.4, $\mathscr{X}$ is complete and cocomplete, and U satisfies (A) through (D) in 21.6.8, 21.6.9. Also, $\mathscr{X}$ satisfies the assumptions in 21.6.9. By the assumption about (7), $\mathscr{X}'$ is complete (21.3.9), and by 21.6.8 and 21.6.3 (c), (A), (A′), (A‴) are equivalent for (7).

(a) ⇒ (c). By 21.6.4 and 21.6.5 (a), (A‴) is valid, and (D‴) follows from 21.4.4 (c) for the composite.

(c) ⇒ (b) is obvious.

(b) ⇒ (a). Since K preserves kernel pairs, (A′) and thus (A‴) is valid. Therefore, 21.6.5 (c) can be applied.

21.7 Eilenberg-Moore Categories as Functor Categories

21.7.1 Lemma. *Let the functor $G: \mathscr{B} \to \mathscr{A}$ be bijective on objects. $\mathscr{E}$ is another category.*

(a) *$[G, \mathscr{E}]: [\mathscr{A}, \mathscr{E}] \to [\mathscr{B}, \mathscr{E}]$ creates isomorphisms and is faithful.*

(b) *Let Σ be a diagram scheme with commutativity conditions K. If $\mathscr{E}$ has limits (colimits) of type Σ/K, then $[G, \mathscr{E}]$ creates limits (colimits) of this type.*

The proof of (a) is simple (compare 16.6.6 and 17.1.6 (d)). (b) follows immediately from the pointwise construction of limits and colimits in functor categories.

21.7.2 Proposition. *Let*

(1)
$$\begin{array}{ccc} \mathscr{X} & \xrightarrow{J} & \mathscr{Y} \\ {\scriptstyle U}\downarrow & & \downarrow{\scriptstyle V} \\ \mathscr{C} & \xrightarrow{H} & \mathscr{B} \end{array}$$

be a pullback in Cat or in CAT.

(a) *If H has one of the properties: faithful, full, injective on objects, surjective on objects, then J has the same property.*

(b) *If isomorphisms are lifted (uniquely) or, resp., if they are created by V, then the same is true for U.*

(c) *Let $D: \Sigma \to \mathscr{X}$ be a diagram for which $U\,D$ has a limit (L, λ) which is preserved by H. If $(H(L), H * \lambda)$ is lifted (uniquely) as a limit of $J\,D$, then U lifts the limit (L, λ) (uniquely). If V creates limits of $J\,D$, then U creates limits of D. Corresponding statements hold for colimits.*

Proof. Up to an isomorphism, (1) can be described as in *Ens*: the objects or, resp., the morphisms of $\mathscr{X}$ are pairs (A, Y) with $H(A) = V(Y)$ or, resp., (f, u) with $H(f) = V(u)$. Here one gets U and J by restricting the projections $\mathscr{C} \times \mathscr{Y} \to \mathscr{C}$, $\mathscr{C} \times \mathscr{Y} \to \mathscr{Y}$. The statements then follow easily.

21.7.3 Theorem (Linton). *Let $\boldsymbol{T} = (T, \eta, \mu)$ be a triple over the $\mathfrak{U}$-category $\mathscr{C}$. In CAT there is the pullback* (see 21.2.9)

(2)
$$\begin{array}{ccc} \mathscr{C}^{\boldsymbol{T}} & \xrightarrow{J^{\boldsymbol{T}}} & [\mathscr{C}^0_{\boldsymbol{T}}, Ens] \\ {\scriptstyle U^{\boldsymbol{T}}}\downarrow & & \downarrow{\scriptstyle \tilde{F}^0_{\boldsymbol{T}}} \\ \mathscr{C} & \xrightarrow{H_*} & [\mathscr{C}^0, Ens]\,. \end{array}$$

Here H_* is the Yoneda embedding, $\tilde{F}^0_{\boldsymbol{T}} = [\mathrm{Op}\, F_{\boldsymbol{T}}\, \mathrm{Op}, Ens]$, and one gets $J^{\boldsymbol{T}}$ by the isomorphism

(3) $\qquad \xi: K^\vee_{\boldsymbol{T}} \to J^{\boldsymbol{T}}$, where $K^\vee_{\boldsymbol{T}}(?) = [K_{\boldsymbol{T}}(-), ?]_{\mathscr{C}^{\boldsymbol{T}}}$,

which is determined as follows: If $K^\vee_{\boldsymbol{T}}$ and $J^{\boldsymbol{T}}$ are regarded as contra-covariant functors $\bar{K}$, $\bar{J}$ in the canonical way (3.6.3), *then ξ goes into the isomorphism*

($\bar{3}$) $\qquad \bar{\xi}: \bar{K}(-, ?) = [K_{\boldsymbol{T}}(-), ?]_{\mathscr{C}^{\boldsymbol{T}}} \xrightarrow{\cong} \bar{J}(-, ?)$

which is produced from the adjunction isomorphism

$$\varphi^{\boldsymbol{T}}: [F^{\boldsymbol{T}}(??), ?]_{\mathscr{C}^{\boldsymbol{T}}} \xrightarrow{\cong} [??, U^{\boldsymbol{T}}(?)]_{\mathscr{C}}$$

by the bijection $A \mapsto F_{\boldsymbol{T}}(A)$. (Note that $F^{\boldsymbol{T}} = K_{\boldsymbol{T}}\, F_{\boldsymbol{T}}$). One thus has

(4) $$\bar{\xi}_{F_{\boldsymbol{T}}(A),(B,b)} = \varphi^{\boldsymbol{T}}_{A,(B,b)}\,.$$

Proof. We start with the remark that it follows from the construction of $\mathscr{C}^{\boldsymbol{T}}$ and $\mathscr{C}_{\boldsymbol{T}}$ that these categories are again $\mathfrak{U}$-categories and that, therefore, $[\mathscr{C}^0_T, Ens]$ is, in fact, a small $\mathfrak{V}$-category. (4) determines $\bar{\xi}_{A,(B,b)}$ uniquely, and thus $\bar{J}$ and J^T are determined uniquely by $(\bar{3})$ and (3). Now we form the pullback

$$(5)\qquad \begin{array}{ccc} \mathscr{X} & \xrightarrow{J} & [\mathscr{C}^0_T, Ens] \\ {\scriptstyle U}\downarrow & & \downarrow{\scriptstyle \tilde{F}^0_T} \\ \mathscr{C} & \xrightarrow{H_*} & [\mathscr{C}^0, Ens] \end{array}$$

in CAT. By 21.7.2 (a), J is a full embedding. It has to be shown that (2) is isomorphic to (5). This is done in three steps.

Step 1. U has a left adjoint F with $U\,F = T$ and the adjunction transformation $\eta\colon 1_{\mathscr{C}} \to T$.

Let $H'_*\colon \mathscr{C}_T \to [\mathscr{C}^0_T, Ens]$ be the Yoneda embedding of $\mathscr{C}_T$. Then

$$(6)\qquad \tilde{F}^0_T\, H'_*\, F_T(?) = [F_T(-), F_T(?)]_{\mathscr{C}_T}$$

and

$$(7)\qquad H_*\; T(?) = [-, T(?)]_{\mathscr{C}}\,.$$

The functors (6), (7) $\mathscr{C} \to [\mathscr{C}^0, Ens]$ are isomorphic. More precisely: The adjunction isomorphism φ_T yields an isomorphism for every $B \in |\mathscr{C}|$:

$$(8)\qquad (\varphi_T)_{-,F_T(B)}\colon \tilde{F}^0_T\, H'_* F_T(B) \cong\!\!\Rightarrow H_*\; T(B)\,.$$

By 21.7.1 (a), there is exactly one isomorphism in $[\mathscr{C}^0_T, Ens]$

$$(9)\qquad \varrho_B\colon H'_*\, F_T(B) \cong\!\!\Rightarrow S(B)$$

with

$$(10)\qquad (\tilde{F}^0_T * \varrho_B)_A = (\varphi_T)_{A,F_T(B)}\colon [F_T(A), F_T(B)]_{\mathscr{C}_T} \to [A, T(B)]_{\mathscr{C}}\,.$$

By (9), $B \mapsto S(B)$ can be extended uniquely to a functor $S\colon \mathscr{C} \to [\mathscr{C}^0_T, Ens]$ (compare 16.6.6) in such a way that

$$\varrho = \{\varrho_B\} : H'_*\, F_T \to S\,.$$

is an isomorphism of functors. By construction, $\tilde{F}^0_T\, S = H_*\; T$.

In (5) there exists thus a (unique) functor $F\colon \mathscr{C} \to \mathscr{X}$ with $U\,F = T$ and $J\,F = S$. Since J is fully faithful, the Yoneda isomorphism 4.2.4 and the evaluation functor E (3.7) provide isomorphisms

$$[F(-), ?]_{\mathscr{X}} \xrightarrow{J_{F(-),?}} [J\,F(-), J(?)]_{[\mathscr{C}^0_T, Ens]} =$$
$$= [S(-), J(?)] \xrightarrow{[\varrho, J]} [H'_*\, F_T(-), J(?)] = E(J(?), F_T(-))$$
$$= (\tilde{F}^0_T\, J(?))\,(-) = (H_*\; U(?))\,(-) = [-, U(?)]_{\mathscr{C}}\,.$$

At the place $(A, F(A))$ one first gets

$$1_{F(A)} \mapsto 1_{JF(A)} = 1_{S(A)} \mapsto \varrho_A \in [H'_*\, F_T(A), S(A)]\,.$$

By means of the Yoneda isomorphism, ϱ_A goes into

$$\varrho_A(1_{F_T(A)}) \in S(A)\ (F_T(A)) = (\tilde{F}_T^0\, S(A))\ (A) = [A, T(A)]\,.$$

By (10), $\varrho_A(1_{F_T(A)}) = (\varphi_T)_{A,F_T(A)}(1_{F_T(A)}) = \eta_A$. The statement of step 1 is thus proved.

Step 2. There is a functor $M: \mathscr{C}^T \to \mathscr{X}$ with $J^T = J\,M$, $U^T = U\,M$ and $M\,F^T = F$.

In the same manner as S is obtained from $H'_*\,F_T$ by lifting φ_T, J^T is obtained from $\check{K}_T$ by lifting φ^T, and $\tilde{F}_T^0\,J^T = H_*\,U^T$. (5) thus yields a uniquely determined functor M with $J^T = J\,M$ and $U^T = U\,M$. It remains to be shown that $M\,F^T = F$.

Since J is a full embedding, it suffices to show that $J^T\,F^T = S$. If we regard $J^T\,F^T$ and S as contia-co-variant functors $\overline{S}'$, $\overline{S}$, then $F^T = K_T\,F_T$, (3) and (4) imply that there is the isomorphism

(11) $\bar{\xi}'$: $[K^T(?\,?), F^T(?)]_{\mathscr{C}^T} \cong\!\!> \overline{S}'(?\,?, ?)$ with $\bar{\xi}'_{F_T(A),B} = \varphi^T{}_{A,F^T(B)}$.

By 21.2.5, one has

$$(\varphi_T)_{A,\,F_T(B)} = \varphi^T_{A,\,F^T(B)}(K_T)_{F_T(A),F_T(B)}\,.$$

By (10), one thus gets $\overline{S}$ from $[?\,?, F_T(?)]$ by composing the isomorphism

$$(K_T)_{?\,?,?}\colon\ [?\,?, F_T(?)]_{\mathscr{C}_T} \cong\!\!> [K_T(?\,?), K_T\,F_T(?)]_{\mathscr{C}^T}$$

with (11); this implies $\overline{S}' = \overline{S}$, so that $J^T\,F^T = S$ and thus $M\,F^T = F$.

Step 3. M is an isomorphism of categories.

$M\,F^T = F$ and 21.1.5 (with $\eta = \eta'$) imply that the adjoint situation in step 1 generates the triple $\boldsymbol{T}$ and $U * \theta = 1_T$. Since H_* preserves split forks, 21.7.1 and 21.7.2 imply that U creates coequalizers for those pairs of $\mathscr{X}$-morphisms that are split by U. By 21.4.4, the corresponding statement holds for U^T. By 21.5.7 (e), M is an isomorphism. We remark that the inverse of M is the comparison functor $\mathscr{X} \to \mathscr{C}^T$.

Since M is an isomorphism, and since $J\,M = J^T$ and $U\,M = U^T$, one gets (2) from (5) as a pullback, which is what had to be shown.

21.7.4 Remarks. (a) 21.7.3 implies 21.3.9 (a) again and a weakened version of 21.3.9 (b).

(b) By means of J^T, $\mathscr{C}^T$ is isomorphic to the full subcategory H^T of $[\mathscr{C}_T^0, Ens]$ whose objects aie those functors G for which G Op F_T is a contravariant Hom-functor of $\mathscr{C}$. Let $\mathscr{D}^T$ be the full subcategory of $[\mathscr{C}_T^0, Ens]$ whose objects are the functors G for which G Op F_T is representable. The inclusion $\mathscr{H}^T \to \mathscr{D}^T$ is an equivalence, and one obtains thus an adjoint situation generating the triple $\boldsymbol{T}$ that is equivalent to the Eilenberg-Moore situation. The corresponding right adjoint lifts limits, but no longer uniquely.

(c) $[T^0, Ens]$, $[\eta^0, Ens]$, $[\mu^0, Ens]$ (compare 16.1.1) form a cotriple $\tilde{\boldsymbol{T}}^0$ over $[\mathscr{C}^0, Ens]$. It is generated by

$$\tilde{F}^0_{\boldsymbol{T}} \;\frac{\tilde{\eta}^0}{\tilde{\varepsilon}^0}\Big|\; \tilde{U}^0_{\boldsymbol{T}}(\tilde{\mathscr{C}}^0, \tilde{\mathscr{C}}^0_{\boldsymbol{T}}),$$

where $\tilde{F}^0_{\boldsymbol{T}}$ stands for $[\mathrm{Op}\, F_{\boldsymbol{T}}\, \mathrm{Op}, Ens]$, and so on. This adjoint situation is isomorphic to the Eilenberg-Moore situation of the cotriple $\tilde{\boldsymbol{T}}^0$ (21.1.7 dual). This follows from 21.7.1 and the dual of 21.5.7 (e).

(d) For $(A, a) \in |\mathscr{C}^{\boldsymbol{T}}|$, $J^{\boldsymbol{T}}(A, a)$ can be regarded as a contravariant functor $\mathscr{C}_{\boldsymbol{T}} \to Ens$. Since $F_{\boldsymbol{T}}$ is bijective on objects, and since $K_{\boldsymbol{T}}\, F_{\boldsymbol{T}} = F^{\boldsymbol{T}}$ and $U^{\boldsymbol{T}}(A, a) = A$, the construction of $J^{\boldsymbol{T}}$ yields

$$J^{\boldsymbol{T}}(A, a)\,\big(F_{\boldsymbol{T}}(B)\big) = [B, A]_{\mathscr{C}}\,. \tag{12}$$

For $v\colon F_{\boldsymbol{T}}(B) \to F_{\boldsymbol{T}}(C)$ in $\mathscr{C}_{\boldsymbol{T}}$,

$$J^{\boldsymbol{T}}(A, a)\,(v) = \varphi^{\boldsymbol{T}}_{B,(A,a)}\,[K_{\boldsymbol{T}}(v), (A, a)]\,(\varphi^{\boldsymbol{T}}_{C,(A,a)})^{-1};$$

this is due to (3), (4). By 16.5.2°, $g \in [C, A]$ is here first taken into $\varepsilon^{\boldsymbol{T}}_{(A,a)}\, F^{\boldsymbol{T}}(g)$, then into $\varepsilon^{\boldsymbol{T}}_{(A,a)}\, F^{\boldsymbol{T}}(g)\, K_{\boldsymbol{T}}(v)$ and finally into $a\, T(g)\, U_{\boldsymbol{T}}(v)\, \eta_B$, because $U^{\boldsymbol{T}}(\varepsilon_{(A,a)}) = a$ and $U^{\boldsymbol{T}}\, K_{\boldsymbol{T}} = U_{\boldsymbol{T}}$. One thus has

$$J^{\boldsymbol{T}}(A, a)\,(v)\colon\; [C, A]_{\mathscr{C}} \to [B, A]_{\mathscr{C}}\,, \qquad g \mapsto a\, T(g)\, U_{\boldsymbol{T}}(v)\, \eta_B\,. \tag{13}$$

Correspondingly, since $a\,\eta_A = 1_A$, what one gets for a $\mathscr{C}^{\boldsymbol{T}}$-morphism $\alpha\colon (A, a) \to (B, b)$ is

$$J^{\boldsymbol{T}}(\alpha)_{F_{\boldsymbol{T}}(C)}\colon\; [C, A]_{\mathscr{C}} \to [C, B]_{\mathscr{C}}\,, \qquad g \mapsto U^{\boldsymbol{T}}(\alpha)\, g\,. \tag{14}$$

Actually, this follows also from the fact that (2) is commutative and $F_{\boldsymbol{T}}$ is bijective on objects.

(e) The functor $S\colon \mathscr{C} \to [\mathscr{C}^0_{\boldsymbol{T}}, Ens]$ in step 1 in the proof of 21.7.3 and $T\colon \mathscr{C} \to \mathscr{C}$ induce for the pullback (2) exactly the functor $F^{\boldsymbol{T}}\colon \mathscr{C} \to \mathscr{C}^{\boldsymbol{T}}$.

(f) Let $\tau\colon \boldsymbol{T} \to \boldsymbol{T}'$ be a triple-morphism, and let $K\colon \mathscr{C}^{\boldsymbol{T}'} \to \mathscr{C}^{\boldsymbol{T}}$ or, resp., $N\colon \mathscr{C}_{\boldsymbol{T}} \to \mathscr{C}_{\boldsymbol{T}'}$ be the corresponding functor as in 21.1.7, or, resp., 21.2.9. If one forms the pullbacks (2) for $\boldsymbol{T}$ and for $\boldsymbol{T}'$, then $F_{\boldsymbol{T}'} = N\, F_{\boldsymbol{T}}$ and the definition of pullbacks imply that $[N^0, Ens]$ induces a functor $L\colon \mathscr{C}^{\boldsymbol{T}'} \to \mathscr{C}^{\boldsymbol{T}}$ with $U^{\boldsymbol{T}}\, L = U^{\boldsymbol{T}'}$, $J^{\boldsymbol{T}}\, L = [N^0, Ens]\, J^{\boldsymbol{T}'}$. We claim that $K = L$.

By 21.1.7 (18) and by (12), (13), if $(A, a') \in |\mathscr{C}^{\boldsymbol{T}'}|$, then

$$J^{\boldsymbol{T}}\, K(A, a')\,\big(F_{\boldsymbol{T}}(B)\big) = J^{\boldsymbol{T}}(A, a'\,\tau_A)\,\big(F_{\boldsymbol{T}}(B)\big) = [B, A]_{\mathscr{C}}\,,$$

$$\big(J^{\boldsymbol{T}}\, K(A, a')\,(v)\big)\,(g) = a'\,\tau_A\, T(g)\, U_{\boldsymbol{T}}(v)\, \eta_B = a'\, T'(g)\, \tau_C\, U_{\boldsymbol{T}'}(v)\, \eta_B\,.$$

By the definition of $[N^0, Ens]$ and by (12), one has

$$[N^0, Ens]\, J^{\boldsymbol{T}'}(A, a')\,\big(F_{\boldsymbol{T}'}(B)\big) = J^{\boldsymbol{T}'}(A, a')\,\big(N\, F_{\boldsymbol{T}'}(B)\big) = [B, A]_{\mathscr{C}}\,,$$

$$[N^0, Ens]\, J^{\boldsymbol{T}'}(A, a')\,(v) = J^{\boldsymbol{T}'}(A, a')\,\big(N(v)\big)\,.$$

By (13) and 21.2.9 (20), this is the map which takes $g \in [C, A]_{\mathscr{C}}$ into

$$a'\, T'(g)\, U_{\mathbf{T}'}\, N(v)\, \eta'_B = a'\, T'(g)\, U_{\mathbf{T}'}\, N(v)\, \tau_B\, \eta_B = a'\, T'(g)\, \tau_C\, U_{\mathbf{T}}(v)\, \eta_B\,,$$

so that

$$J^{\mathbf{T}}\, K(A, a') = [N^0, Ens]\, J^{\mathbf{T}'}(A, a') = J^{\mathbf{T}}\, L(A, a')$$

is valid. Since $J^{\mathbf{T}}$ is an embedding, K and L coincide on the objects of $\mathscr{C}^{\mathbf{T}'}$. The same is true for morphisms. This follows from $U^{\mathbf{T}}\, L = U^{\mathbf{T}'} = U^{\mathbf{T}}\, K$, since $U^{\mathbf{T}}$ is faithful. One could have used (14) instead.

(g) Up to an isomorphism, $\mathscr{C}_{\mathbf{T}}$ can be refound from (2) as a full image as follows: By means of $\varphi^{\mathbf{T}}$, $H_*\, U^{\mathbf{T}}$ in (2) is isomorphic to $(F^{\mathbf{T}})^{\vee} = [\mathrm{Op}\, F^{\mathbf{T}}\, \mathrm{Op}, Ens]\, H''_*$, where H''_* is the Yoneda embedding of $\mathscr{C}^{\mathbf{T}}$. A canonical isomorphism (3.4.5 (6)) will then produce from $H_*\, U^{\mathbf{T}}$ and $(F^{\mathbf{T}})^{\vee}$ functors V and $H^*\, \mathrm{Op}\, F^{\mathbf{T}}\, \mathrm{Op}\colon \mathscr{C}^0 \to [\mathscr{C}^{\mathbf{T}}, Ens]$, where H^* is the Yoneda embedding of $(\mathscr{C}^{\mathbf{T}})^0$. If one takes into account that H^* is a full embedding, then one gets for the full images (21.2.1) isomorphisms P, Q such that

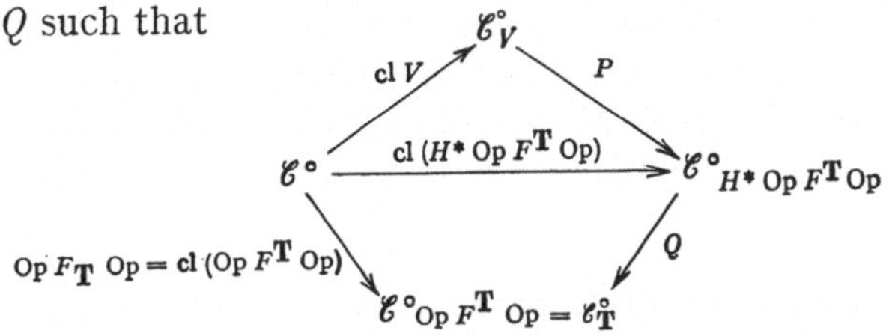

is commutative. Here P is derived from $\varphi^{\mathbf{T}}$ and Q from $H^*\colon (\mathscr{C}^{\mathbf{T}})^0 \to [\mathscr{C}^{\mathbf{T}}, Ens]$.

Note that for the construction of cl V only $U^{\mathbf{T}}$ and $H_*\colon \mathscr{C} \to [\mathscr{C}^0, Ens]$ were used. If $\mathscr{X}$ is any small $\mathfrak{V}$-category and $U\colon \mathscr{X} \to \mathscr{C}$ a functor, then from U and H_* a functor $\mathfrak{S}(U)\colon \mathscr{C}^0 \to \mathscr{Y}^0$, which is bijective on objects, can be constructed the same way. One thus gets a contravariant functor "structure" $\mathfrak{S}\colon CAT/\mathscr{C} \to Cl(\mathscr{C}^0)$, where here $Cl(\mathscr{C}^0)$ is a full subcategory of $\mathscr{C}^0/CAT$ (compare 21.2.4). Using 16.8.5 through 16.8.8 with $[?, Ens]_{CAT}$, one can show that there exists an adjoint contravariant functor "semantics". This adjunction induces an equivalence between $EM(\mathscr{C})$ and the dual category of $Kl(\mathscr{C})$. (Compare 21.2.11 (d)).

21.7.5 Proposition. *Let the* $\mathfrak{U}$*-category* $\mathscr{X}$ *have coproducts.*

(a) *The functor* $U\colon \mathscr{X} \to Ens$ *has a left adjoint if and only if* U *is representable.*

(b) *The functor* $F\colon Ens \to \mathscr{X}$ *has a right adjoint if and only if* F *preserves coproducts.*

Proof. (a) For the one point set $1 = \{\emptyset\}$, $[1, U(?)]_{Ens}$ is isomorphic to U. If U is represented by the object Y, set $F(1) = Y$. Since every set is the coproduct of its one-element subsets, $[M, U(?)]_{Ens}$ is represented by $\coprod Y_m$ with $Y_m = Y$ for all $m \in M$. U thus has a left adjoint

F. If, conversely, F exists, then $[F(1), ?]_{\mathfrak{X}} \cong [1, U(?)]_{Ens} \cong U(?)$ is a representation of U.

(b) Let $[F(1), ?]_{\mathfrak{X}} = U(?)$. According to (a), U has a left adjoint F'. Since F' preserves coproducts, F and F' are isomorphic if and only if F preserves coproducts.

21.7.6 Proposition. *Let $\boldsymbol{T} = (T, \eta, \mu)$ be a triple over Ens. The Eilenberg-Moore category $Ens^{\boldsymbol{T}}$ is equivalent to the full subcategory of $[(Ens_{\boldsymbol{T}})^0, Ens]$ whose objects are the functors which preserve products.*

Proof. By 21.7.4 (b), it suffices to show that a functor $G: (Ens_{\boldsymbol{T}})^0 \to \to Ens$ preserves products if and only if $G\,F^0_{\boldsymbol{T}}: Ens^0 \to Ens$ is representable. If G preserves products, then so does $G\,F^0_{\boldsymbol{T}}$, because $F^0_{\boldsymbol{T}}$ is a right adjoint. Then, by the dual of 21.7.5 (b), $G\,F^0_{\boldsymbol{T}}$ has a left adjoint, and by 21.7.5 (a), it is representable. If, conversely, $G\,F^0_{\boldsymbol{T}}$ is representable, then $G\,F^0_{\boldsymbol{T}}$ preserves products. Since $F^0_{\boldsymbol{T}}$ is bijective on objects and since it preserves products, given a product $Y = \prod Y_i$ in $(Ens_{\boldsymbol{T}})^0$ with projections pr_i, there are objects M_i in Ens^0 with $F^0_{\boldsymbol{T}}(M_i) = Y_i$ and $F^0_{\boldsymbol{T}}(\prod M_i) = Y$. Let $\prod M_i$ have the projections q_i. There is an automorphism v of Y such that $v\,F^0_{\boldsymbol{T}}(q_i) = pr_i$ for all i. It thus follows that G preserves products.

21.7.7 Remarks. (a) Note that in 21.7.6 every functor $F^0: Ens^0 \to \mathcal{Y}$ which preserves products defines a cotriple over Ens^0 and thus a triple over Ens up to an isomorphism, by 21.7.5 dual ($\mathcal{Y}$ is a $\mathfrak{U}$-category). If, in addition, F^0 is bijective on objects, then $F = \mathrm{Op}\,F^0\,\mathrm{Op}$ is the left adjoint of a Kleisli situation up to an isomorphism in CAT. Therefore, Eilenberg-Moore situations over Ens can be regarded as reduced algebraic categories of theories which have operations of arbitrary "arities" (compare 21.6.10 (a)). 21.7.6 says in particular that algebraic categories for such theories are equivalent to $\mathfrak{U}$-categories.

(b) We now return to algebraic categories in the sense of Chapter 18, whose theories are generated by finitary operations. There is an evident generalization of the following to theories generated by operations whose "arity" is less than a regular cardinal (18.8.9). We shall not go into any possible generalizations of partial statements.

21.7.8 Proposition (Thode). *Let $P: \mathcal{A} \to \mathcal{B}$ be a theory-morphism for algebraic theories in the sense of 18.1.2, $P^b: \mathcal{B}^b \to \mathcal{A}^b$ the corresponding algebraic functor between the algebraic categories and P_* left adjoint to P^b. Then the following statements are equivalent:*

a) *If $\mathcal{A}$ is presented by generating operations and defining equations (18.1.11), then one can get a presentation of $\mathcal{B}$ by adding more defining equations.*

(b) *$P: \mathcal{A} \to \mathcal{B}$ is the projection of $\mathcal{A}$ onto a quotient category in the sense of 6.4.1.*

(c) *The restriction of P_* to the full subcategory of $\mathcal{A}^b$, whose objects are the finitely generated free algebras, is full.*

(d) *The restriction of P_* to the full subcategory of $\mathcal{A}^b$, whose objects are the free algebras, is full.*

(e) *The triple-morphism $\tau: U_{\mathcal{A}} L_{\mathcal{A}} \to U_{\mathcal{B}} L_{\mathcal{B}}$ induced by P^b is a retraction at every place $M \in |Ens|$ (i.e., it is "pointwise" a retraction).*

(f) *τ is an epimorphism.*

Proof. (a) $\Leftrightarrow$ (b) follows immediately from 18.1.11 and 6.4.1.

(b) $\Leftrightarrow$ (c) follows from 18.5.3 (4) and 18.2.9.

(c) $\Leftrightarrow$ (d). (c) follows trivially from (d). Now let (c) be satisfied. Let T' and T be free $\mathcal{A}$-algebras. One has to show that

$$(P_*)_{T',T}: [T', T]_{\mathcal{A}^b} \to [P_*(T'), P_*(T)]_{\mathcal{B}^b}$$

is surjective. Let $G = H_\pi^*((A^1)^0) = L_A(\{\emptyset\})$ be the distinguished generator of $\mathcal{A}^b$ (18.2.9). T' is a coproduct with cofactors G, P_* preserves colimits, and contravariant Hom-functors take colimits into limits. Since in *Ens* products of epimorphisms are epimorphisms, it suffices to show that

$$(P_*)_{G,T}: [G, T]_{\mathcal{A}^b} \to [P_*(G), P_*(T)]_{\mathcal{B}^b} \tag{16}$$

is surjective. T is a filtered colimit of finitely generated free algebras. P_* maps free algebras into free ones and finitely generated algebras into finitely generated ones (18.3.5 (5)). Also, $[G, ?]$, $[P_*(G), ?]$ represent the forgetful functors $U_{\mathcal{A}}$, $U_{\mathcal{B}}$, and these preserve filtered colimits. If one takes note of the fact that (16) is the restriction of a natural transformation $[G, ?] \to [P_*(G), P_*(?)]$, then (d) follows from (c).

(d) $\Leftrightarrow$ (e). The triple corresponding to $\mathcal{B}^b$ goes into an isomorphic one if $L_{\mathcal{B}}$ is replaced by $P_* L_{\mathcal{A}}$. By 21.5.1 (d), P^b and P_* then induce the same triple-morphism τ. The statement thus follows from 21.2.8.

(e) $\Leftrightarrow$ (f) by 10.1.4, since in *Ens* every epimorphism is a retraction.

21.7.9 Birkhoff's Theorem. *Let $\mathcal{A}^b$ be the algebraic category for the theory $\mathcal{A}$ and let $\mathcal{X}$ be a strictly full subcategory of $\mathcal{A}^b$. The following statements are equivalent:*

(a) *$\mathcal{X}$ is closed in $\mathcal{A}^b$ with respect to products (including the empty one), subalgebras and homomorphic images.*

(b) *$\mathcal{X}$ is a variety; i.e., $|\mathcal{X}|$ consists of all the $\mathcal{A}$-algebras whose operations satisfy some set of equations.*

(c) *There exists a theory-morphism $P: \mathcal{A} \to \mathcal{B}$, which is a projection of $\mathcal{A}$ onto a quotient category, such that $P^b: \mathcal{B}^b \to \mathcal{A}^b$ induces an equivalence between $\mathcal{B}^b$ and $\mathcal{X}$.*

Proof. The equivalence of (b) and (c) follows immediately from 21.7.8. The equivalence of (a) and (c) is seen to follow from 21.7.8, 21.6.6 and 21.6.7 (c) if one notices that, up to an equivalence, there is an Eilenberg-Moore situation over *Ens* for $\mathcal{A}b$ (21.6.10 (a), 21.7.7 (a)) and that here the factorizations of morphisms derived from 21.6.3 and from 18.3.2 (compare 18.4.5) coincide.

21.7.10 Theorem. *Let* $\tau\colon \boldsymbol{T} \to \mathrm{R}$ *be a triple-morphism for triples over Ens. The following statements are equivalent:*

(a) τ *is an epimorphism.*

(b) *The functor* $N\colon Ens_{\boldsymbol{T}} \to Ens_{\boldsymbol{R}}$ (21.2.9 (a)) *induced by* τ *is full.*

(c) *The functor* $K\colon Ens^{\boldsymbol{R}} \to Ens^{\boldsymbol{T}}$ (21.1.7 (e)) *induced by* τ *induces an equivalence between* $Ens^{\boldsymbol{R}}$ *and a strictly full subcategory* $\mathcal{X}$ *of* $Ens^{\boldsymbol{T}}$ *which is closed with respect to products and "subobjects" and which also satisfies condition* (R′) *in* 21.6.7 (c).

After the previous discussions the proof follows naturally and can be left to the reader. 21.6.6 also contains the converse of (c).

21.8 Problems

21.8.1 Carry out 21.2.7.

21.8.2 In the adjoint left triangle 21.2 (6) let U be fully faithful. Then FU' is right adjoint to N. (Hint: Set $\bar{\eta} = (F * \sigma)\,\varepsilon^{-1}\colon 1_{\mathcal{X}} \to FU'N$ and $\bar{\varepsilon} = \varepsilon'\colon NFU' \to 1_{\mathcal{X}'}$.)

Remark. The pseudo-dual statement for 21.1 (6) with a fully faithful U is a special case of 19.4.2 (f).

21.8.3 Fill in the details in the proof of 21.3.14.

21.8.4 Prove 21.5.4 (e).

21.8.5 Let $\mathscr{C}$ be a well-powered, complete category in which every epimorphism is a retraction (compare 19.8.6). Carry over 21.6.4, 21.6.7 (c), 21.6.12 and 21.7.10.

21.8.6 Let the category $\mathscr{C}$ have an $\mathfrak{E}$-$\mathfrak{M}$-factorization of morphisms which satisfies (F 1) and (F 2) in 21.6.1. Let $\boldsymbol{T} = (T, \eta, \mu)$ be a triple over $\mathscr{C}$ such that $\mathfrak{E}$ is mapped into itself by T. $\tau\colon \boldsymbol{T} \to \boldsymbol{T}'$ is to be a triple-morphism into any triple $\boldsymbol{T}' = (T', \eta', \mu')$ over $\mathscr{C}$. By choosing an $\mathfrak{E}$-$\mathfrak{M}$-factorization $\tau_A = \tau''_A\,\tau'_A$ with $\tau'_A \in \mathfrak{E}$, $\tau''_A \in \mathfrak{M}$ for all $A \in |\mathscr{C}|$, one gets a triple $\boldsymbol{S} = (S, \xi, \nu)$ over $\mathscr{C}$ with triple-morphisms $\tau'\colon \boldsymbol{T} \to \boldsymbol{S}$ and $\tau''\colon \boldsymbol{S} \to \boldsymbol{T}'$. S also maps $\mathfrak{E}$ into itself. Consider also the special case where $\boldsymbol{T}$ is the identity triple, and show that then $\boldsymbol{S}$ is idempotent.

21.8.7 Let $\mathscr{B}$ be a reflective subcategory of $\mathscr{C}$. Show that the inclusion $I: \mathscr{B} \to \mathscr{C}$ preserves extremal monomorphisms. Does I reflect them?

21.8.8 Let

$$(1) \qquad F \frac{\varepsilon}{\eta}\!\mid U(\mathscr{X}, \mathscr{C})$$

be an adjoint situation which generates the triple $\boldsymbol{T} = (T, \varepsilon, \mu)$ over $\mathscr{C}$.

(a) If $f: A \to B$ is a $\mathscr{C}$-morphism, then $F(f)$ is an isomorphism if and only if $T(f)$ is an isomorphism. In that case, η_A factors through f.

(b) If every η_A is an extremal monomorphism, then T reflects isomorphisms.

(c) Let $\mathscr{B}$ be a strictly full reflective subcategory of $\mathscr{C}$, such that $T(A) \in |\mathscr{B}|$ for all $A \in |\mathscr{C}|$. $\mathscr{B}$ determines an idempotent triple $\boldsymbol{S} = (S, \bar{\eta}, 1_S)$ over $\mathscr{C}$, where $S|\mathscr{B} = 1_{\mathscr{B}}$. $\tau = S * \eta: S \to T$ is a triple-morphism. $\mathscr{B}$ contains all objects $X \in |\mathscr{C}|$ for which η_X is a coretraction; in particular, $\mathscr{B}$ contains all carriers of $\boldsymbol{T}$-algebras and U is of the form $U = I\,U'$, where $I: \mathscr{B} \to \mathscr{C}$ is the inclusion. There is an adjoint situation.

$$(2) \qquad U' \frac{\varepsilon}{\eta'}\!\mid FI(\mathscr{X}, \mathscr{B}),$$

where $I * \eta' = \eta * I$ and η, ε have the same meaning as before. (1) and (2) generate the same cotriple over $\mathscr{X}$, and I induces the identity cotriple-morphism.

(d) Let $\mathscr{B}$ have equalizers and a natural $\mathfrak{E}$-$\mathfrak{M}$-factorization of morphisms, where $\mathfrak{M}$ is a class of monomorphisms that contains all equalizers. Factorization of $\tau: S \to T$ at all places $A \in |\mathscr{C}|$ yields a reflector from $\mathscr{C}$ to a strictly full subcategory $\mathscr{B}'$ of $\mathscr{B}$ which also contains all objects $T(A)$.

(e) Let $\mathscr{C}$ have equalizers. Choosing an equalizer of $T(\eta_A)$ and $\eta_{T(A)}$ for every $A \in |\mathscr{C}|$ yields a functor $R: \mathscr{C} \to \mathscr{C}$ and natural transformations $\eta': 1_{\mathscr{C}} \to R$, $\tau: R \to T$ such that $\eta = \tau\,\eta'$. There is a uniquely determined $\nu: R\,R \to R$ such that $\boldsymbol{R} = (R, \eta', \nu)$ is a triple over $\mathscr{C}$ and $\tau: R \to T$ a triple-morphism.

21.8.9 Prove the following

Theorem. *Let $\mathscr{C}$ be a complete, well-powered category, and let* (1) *generate the triple $\boldsymbol{T} = (T, \eta, \mu)$ over $\mathscr{C}$.*

(a) *Among the strictly full, reflective subcategories of $\mathscr{C}$ that contain the carriers of all free $\boldsymbol{T}$-algebras there is a smallest one, say $\mathscr{C}'$. Therefore, there is a best approximation of $\boldsymbol{T}$ by an idempotent triple.*

(b) *Let $\boldsymbol{R} = (R, \bar{\eta}, 1_R)$ be the idempotent triple which is determined by $\mathscr{C}'$. For $f: A \to B$ in $\mathscr{C}$, $T(f)$ is an isomorphism if and only $R(f)$ is an isomorphism. The triple-morphism $R * \eta: R \to T$ is induced by a* *uniquely determined functor* $N: \mathscr{C}_{\mathbf{R}} \to \mathscr{X}$, *where* $\mathscr{C}_{\mathbf{R}}$ *is the Kleisli category for* $\mathbf{R}$. N *is faithful, reflects isomorphisms, and is left adjoint to* $F_{\mathbf{R}} U'$. $\mathscr{C}_{\mathbf{R}}$ *is isomorphic to the category of fractions* $\mathscr{C}[\Sigma^{-1}]$, *where* Σ *is the class of those* $\mathscr{C}$*-morphisms which are taken into isomorphisms by* T.

Hints for a proof: Define $\mathscr{C}'$ as an intersection of subcategories. Let $\mathfrak{K}$ be the class of all those $\mathscr{C}$-morphisms which are extremal monomorphisms in every strictly full, reflective subcategory of $\mathscr{C}$ containing all $T(A)$. Using 21.6.2 (g) and 21.8.8 (d), show that $\eta_X \in \mathfrak{K}$ for $X \in |\mathscr{C}'|$. For $A \in |\mathscr{C}|$, consider all morphisms in $\mathfrak{K}$ through which η_A factors. Use pullbacks for the definition of R.

Hints for another proof (Fakir): Use 21.8.8 (e) and a transfinite induction, which becomes stationary at every place $A \in |\mathscr{C}|$. Which hypotheses are really needed for this proof?

Remark. There is a third proof, which is known as Tierney's tower (see [2], IV).

21.8.10 Prove that the statements in 21.8.9 also hold if $\mathscr{C}$ is a co-complete category with a natural $\mathfrak{E}$-$\mathfrak{M}$-factorization of morphisms, where $\mathfrak{E}$ is a class of epimorphisms and $\mathfrak{M}$ a class of monomorphisms that satisfies the dual of (F3) in 21.6.1.

Hint: Use 21.8.8 (d) for $\mathscr{B} = \mathscr{C}$ and construct a terminal calculus of fractions for $\mathscr{B}'$.

21.8.11 *cat* has a natural $\mathfrak{E}$-$\mathfrak{M}$-factorization of morphisms (i. e., of functors), where $\mathfrak{E}$ is the class of coequalizers, but $\mathfrak{M}$ does not consist solely of monomorphisms. The class of coequalizers is not compositive (i. e., (F4) is not valid). (Hint: 6.6.5., 8.9.5).

21.8.12 The forgetful functor U: *cat* $\to$ *Ens* is the right adjoint of an adjunction (φ, F, U, *cat*, *Ens*) (compare 16.8.10). This adjunction is not equivalent to an Eilenberg-Moore situation. Hence triples do not yield a theory of generalized algebraic structures in the sense 11.7.8.

Bibliography

A. Conference Reports

1. Proceedings of the Conference on Categorical Algebra, La Jolla 1965. Berlin/Heidelberg/New York: Springer 1966.
2. Reports of the Midwest Category Seminar I, II, III, IV, V. Lecture Notes in Math. **47**, **61**, **106**, **137**, **195**. Berlin/Heidelberg/New York: Springer 1967, 1968, 1969, 1970, 1971.
3. Seminar on Triples and Categorical Homology Theory. Lecture Notes in Math. **80**. Berlin/Heidelberg/New York: Springer 1969.
4. Category Theory, Homology Theory and their Applications I, II, III, Lecture Notes in Math. **86**, **92**, **99**. Berlin/Heidelberg/New York: Springer 1969.

B. Books and Lecture Notes

5. Artin, M., Grothendieck, A.: Cohomologie étale des schémas, Seminaire de Géometrie algébrique **4**, 1963/64. IHES, Paris.
6. Brinkmann, H. B., Puppe, D.: Kategorien und Funktoren. Lecture Notes in Math. **18**. Berlin/Heidelberg/New York: Springer 1966.
7. —: Abelsche und exakte Kategorien, Korrespondenzen. Lecture Notes in Math. **96**. Berlin/Heidelberg/New York: Springer 1969.
8. Bucur, I., Deleanu, A.: Categories and Functors. London/New York/Sydney/Toronto: Wiley 1968.
9. Cartan, H., Eilenberg, S.: Homological Algebra. Princeton, N.J.: Princeton Univ. Press 1956.
10. Dold, A.: Halbexakte Homotopiefunktoren. Lecture Notes in Math. **12**. Berlin/Heidelberg/New York: Springer 1966.
11. Dubuc, E.: Kan Extension in Enriched Category Theory. Lecture Notes in Math. **145**. Berlin/Heidelberg/New York 1970.
12. Ehresman, Ch.: Catégories et structures. Paris: Dunod 1965.
13. Freyd, P.: Abelian Categories. Evanston/London: Harper & Row 1964.
14. Gabriel, P., Ulmer, F.: Lokal präsentierbare Kategorien. Lecture Notes in Math. **221**. Berlin/Heidelberg/New York 1971.
15. Gabriel, P., Zisman, M.: Calculus of Fractions and Homotopy Theory. Berlin/Heidelberg/New York: Springer 1967.
16. Godement, R.: Théorie des faisceaux. Paris: Hermann 1958.
17. Hartshorne, R.: Residues and Duality. Lecture Notes in Math. **20**. Berlin/Heidelberg/New York: Springer 1966.
18. Hasse, M., Michler, L.: Theorie der Kategorien. Berlin: VEB Verlag der Wissenschaften 1966.
19. Herrlich, H.: Topologische Reflexionen und Coreflexionen. Lecture Notes in Math. **78**. Berlin/Heidelberg/New York: Springer 1968.
20. Lambek, J.: Completion of Categories. Lecture Notes in Math. **24**. Berlin/Heidelberg/New York: Springer 1966.
21. MacLane, S.: Homology, 2. Aufl. Berlin/Heidelberg/New York: Springer 1967.
22. Mitchell, B.: Theory of Categories. New York/London: Academic Press 1965.

23. Tierney, M.: Categorical Constructions in Stable Homotopy Theory Lecture Notes in Math. **87**. Berlin/Heidelberg/New York: Springer 1969.
24. —: Elementary Topos (to appear).

C. Papers

25. Appelgate, H., Tierney, M.: Categories with Models. In [3].
26. Barr, M.: Coequalizers and Free Triples. Math. Z. **116**, 307—332 (1970).
27. Bénabou, J.: Catégories avec multiplication. C. R. Acad. Sci. Paris **256**, 1887—1890 (1963).
28. Beck, J.: Triples, Algebras and Cohomology. Diss., Columbia University, 1967.
29. Buchsbaum, D. A.: Exact categories and duality. Trans. Am. Math. Soc. **80**, 1—34 (1955).
30. Day, B. J., Kelly, G. M.: Enriched Functor Categories. In [2], III.
31. Dubuc, E.: Adjoint Triangles. In [2], II.
32. Duske, J.: Analogie zwischen *k*-Räumen und bornologischen Räumen. Diss. Kiel 1967.
33. Duskin, J.: Variations on Beck's Tripleability Criterion. In [2], III.
34. Eckmann, B., Hilton, P. J.: Group-like structures in general categories I, II, III. Math. Ann. **145**, 227—255 (1961); **151**, 150—186 (1963); **150**, 165—187 (1963).
35. —, —: Commuting limits with colimits. J. of Alg. **11**, 116—144 (1969).
36. Eilenberg, S., Kelly, G. M.: Closed categories. In [1].
37. Eilenberg, S., MacLane, S.: Group extensions and homology. Ann. Math. **43**, 757—831 (1942).
38. —, —: General theory of natural equivalences. Trans. Am. Math. Soc. **58**, 231—294 (1945).
39. Eilenberg, S., Moore, J.: Adjoint functors and triples. Ill. J. Math. **9**, 381—398 (1965).
40. Ertel, H.-G.: Struktur-Semantik-Funktoren. Diplomarbeit, Kiel 1970.
41. Fisher, J. L.: The tensor product of functors, satellites, and derived functors. J. of Alg. **8**, 277—294 (1968).
42. Gabriel, P., Popescu, N.: Caractérisation des catégories abéliennes avec générateurs et limites inductives exactes. C. R. Acad. Sc. Paris **258**, 4188—4190 (1964).
43. Gabriel, P.: Des catégories abéliennes. Bull. Soc. Math. France **90**, 323—448 (1962).
44. Grothendieck, A.: Sur quelques points d'algèbre homologique. Tôhoku Math. J. 2, **9**, 119—221 (1957).
45. Herrlich, H.: Algebraic categories; an axiomatic approach. Mimeographed Notes, University of Florida 1969.
46. Hilton, P. J.: Correspondences and exact squares. In [1].
47. Isbell, J.: Subobjects, adequacy, completenes and categories of algebras. Rozprawy Mat. **36**, 1—32 (1964).
48. Kan, D. M.: Adjoint functors. Trans. Am. math. Soc. **87**, 194—329 (1958).
49. Kelly, G. M.: Monomorphisms, Epimorphisms and Pullbacks. J. of Australian Math. Soc. Nr. **9** (1969).
50. —: Adjunction for Enriched Categories. In [2], III.
51. Kleisli, H.: Every standard construction is induced by a pair of adjoint functors. Proc. Amer. Math. Soc. **16** (1965).

52. Lawvere, F. W.: The category of categories as a foundation for mathematics. In [1].
53. —: Functorial semantics of algebraic theories. Proc. Nat. Ac. Sci. **50**, 869—872 (1963).
54. —: Some algebraic problems in the context of functorial semantics of algebraic theories. In [2], II.
55. Linton, F. E. J.: Autonomous categories and duality of functors. J. of Alg. **2**, 315—341 (1965).
56. —: Some aspects of equational categories. In [1].
57. —: An outline of functorial semantics. In [3].
58. —: Applied functorial semantics. In [3].
59. —: Coequalizers in categories of algebras. In [3].
60. MacLane, S.: Natural associativity and commutativity. Rice Univ. Studies **49**, 28—46 (1963).
61. —: Categorical algebra. Bull. Am. Math. Soc. **71**, 40—106 (1965).
62. Manes, E.: A triple theoretic construction of compact algebras. In [3].
63. Roos, J.-E.: Locally distributive spectral categories and strongly regular rings. In [2], I.
64. Schumacher, D.: Tripelerzeugung und Tripelbarkeit. Diss., Freiburg i. B. 1968.
65. Thode, Th.: Bruchrechnung in Kategorien. Diplomarbeit, Kiel 1969.
66. —: Kategorielle Form des Satzes von G. Birkhoff über die Charakterisierung von Varietäten. Diss., Universität Düsseldorf 1970.
67. Ulmer, F.: Properties of dense and relative adjoint functors. J. of Alg. **8**, 77—95 (1968).
68. —: Representable functors with values in arbitrary categories. J. of Alg. **8**, 96—129 (1968).
69. Verdier, J. L.: Exposés I, II, III in [5].
70. Volger, H.: Kategorien von Algebren über algebraischen Theorien. Diplomarbeit, Freiburg i. B. 1967.
71. Yoneda, N.: On the homology theory of modules. J. Fac. Sci. Univ. Tokyo Sect. I, **7**, 193—227 (1954).

Index